Nests, Eggs, and Incubation

Nests, Eggs, and Incubation

New Ideas About Avian Reproduction

EDITED BY

D. C. Deeming

&

S. J. Reynolds

OXFORD
UNIVERSITY PRESS

Nests, Eggs, and Incubation. Edited by: D.C. Deeming and S.J. Reynolds,
Oxford University Press (2015). © Oxford University Press.
DOI 10.1093/acprof:oso/9780198718666.001.0001

OXFORD

UNIVERSITY PRESS

Great Clarendon Street, Oxford, OX2 6DP,
United Kingdom

Oxford University Press is a department of the University of Oxford.
It furthers the University's objective of excellence in research, scholarship,
and education by publishing worldwide. Oxford is a registered trade mark of
Oxford University Press in the UK and in certain other countries

First published 2015
First published in paperback 2016

Published in the United States of America by Oxford University Press
198 Madison Avenue, New York, NY 10016, United States of America

British Library Cataloguing in Publication Data

Data available

Library of Congress Cataloging in Publication Data

Data available

ISBN 978–0–19–871866–6 (Hbk.)

ISBN 978–0–19–879168–3 (Pbk.)

Dedication

Amos Ar in 2007.
(Painting by Amnon David Ar [Amos' youngest son]. See Plate 1.)

Amos Ar in 2012.
(Photograph by D.C. Deeming. See Plate 2.)

This book is dedicated to Professor Amos Ar (born in Palestine [now Israel] in 1937) of the George S. Wise Faculty of Life Sciences, Tel Aviv University, Israel for his major contribution to our understanding of avian eggs, embryos, and nests. Amos studied Zoology at Tel Aviv University during the 1960s and after being awarded his PhD in 1970 he secured a postdoctoral position in the Physiology Department, School of Medicine, State University of New York in Buffalo, New York, USA. There, he started working on egg physiology with Hermann Rahn and Charles Paganelli. Returning to Tel Aviv, he started his lecturing career in the Department of Zoology in 1972 reaching a full Professorship in 1987 and an Emeritus position in 2004. Over a career spanning more than 40 years Amos has been an author on over 150 papers and produced key publications on gas exchange across avian eggshells and many other aspects of embryo physiology and nest biology. In 1977 Amos received The Payton Award (with Hermann Rahn and Charles Paganelli) from the Cooper Ornithological Society in Los Angeles, California, USA for 'the most important contribution to the knowledge of the avian egg in the past quarter century'. In 1981 he received The Coues Award (again with Charles Paganelli and Hermann Rahn) from the American Ornithologists' Union in Santa Barbara, California, USA for 'path breaking insights into the field of avian embryonic respiration'. Amos has also worked on invertebrate physiology and the physiology of the mammalian middle ear. He continues in his important research and he continues to publish during his retirement. I have been privileged to consider Amos as a friend and colleague for a long part of my own career.

Charles Deeming (Lincoln)
December 2014

Foreword

It has been said that we know more about birds than about virtually any other group of animals. It may be surprising then to learn that there are aspects of bird biology where there is still a great deal left to discover. One of those that is ripe for discovery is covered by this volume: nests, eggs, and incubation. Our lack of knowledge on these subjects may seem paradoxical when we consider the almost unimaginable volume of research conducted by the poultry industry on two of these topics: eggs and incubation. But that paradox is easily explained when you realize that most poultry research has focussed on the domestic fowl and a handful of other domestic species, has been motivated by maximizing production for commercial purposes, and has focussed largely on mechanisms rather than on the evolutionary significance of particular phenomena. The risk of the poultry industry's particular focus is analogous to the medical approach to reproduction: by concentrating almost entirely on humans, they create a narrow—and limited—view of the world.

This is not to take anything away from poultry researchers (or human reproductive biologists): their work provides a solid set of foundations from which other biologists can launch their research. As this book makes clear, the comparisons between different bird species, made with the benefit of both mechanistic and evolutionary perspectives, reveals the incredible range of inter-related adaptations with respect to nests, eggs, and incubation. Indeed, one might hope that results obtained from investigating the eggs and incubation of non-domesticated bird species in their natural environment might inspire poultry researchers.

It isn't simply the comparison between different bird species that is revealing. Rather, it is the integration of different aspects of birds' life cycles—the study of nests, eggs, and incubation together—that generates real insights into the biology of reproduction.

In startling contrast to what the poultry industry has achieved, ornithology has barely begun to scratch the surface of these topics, despite Charles Deeming's almost single-handed efforts over the last 20 years. The reason for the ornithologists lagging behind is both intriguing and sad: for the last 40 years or more children have been actively discouraged from examining the nests and eggs of wild birds for fear of disturbing them. The intellectual cost is that we end up with a generation of biologists either indifferent to, or uninterested in, such topics, and with no sense of the wonder of nature that might have made them into biologists.

Deeming is a *rara avis*, a poultry biologist turned ornithologist, and this places him in the almost unique position of being able to view this field in its entirety, and with Jim Reynolds to bring together a set of authors capable of providing a series of up to date, informative and extremely valuable reviews.

It amazes me that so few ornithologists recognize that the single thing that makes birds distinct and utterly fascinating is that parent birds—of almost all species—incubate their eggs directly by their own body heat. The consequences of this behaviour are far reaching. As we now know, birds are the direct descendants of dinosaurs, and dinosaur eggs appear to have been deposited and incubated in the ground or in decomposing vegetation. We do not know when the switch from indirect to direct incubation occurred, but that switch was accompanied by a massive change in selection pressures that brought about dramatic changes in bird biology, including the construction of nests, the internal design and external colouration of eggshells, and the internal composition of eggs in terms of yolk and albumen. The diversity of breeding adaptations in birds is both extraordinary and inspiring. Through a collection of scholarly overviews this book provides an entrée into that world. It will inform those already working in this exciting area and hopefully encourage others who aren't now working in this field to do so. This is an area of biology with enormous potential.

Tim Birkhead (Sheffield)
December 2014

Preface

Early in 2013 Charles Deeming (DCD) was asked by a colleague, Carl Soulsbury, if he was ever going to produce a second edition of *Avian Incubation*. The immediate response was 'probably not' largely because much of the information in the book was still valid, it would be time-consuming to update those chapters that needed it, and it was unlikely that the original contributing authors would want to be involved. Shortly after this brief discussion DCD attended a production meeting for a television programme about hatching of eggs and he was struck by the enthusiasm of the people involved directed towards the processes of development and incubation in birds. It was on the homeward train journey following this meeting that DCD came up with the concept of a sequel to *Avian Incubation* (now available to access for free at: www.oup.co.uk/companion/avianincubation2002 [username: avianincubation password: BH9WLxbh]) that brought the field up to date and complemented the original work. Having sketched a few titles for chapters, DCD then sought some feedback from friends and colleagues about the quality of the idea and questioned whether they would like to contribute to a new book about avian incubation? One such colleague was Jim Reynolds (SJR) who was very enthusiastic and made the rash suggestion that perhaps he could be involved in some way—certainly to produce a chapter or two at least or even to be an editor. Perhaps it was a moment of weakness on SJR's part, but having co-written papers in the past and sharing similar research interests, DCD took him up on his offer.

Thereafter, together we planned the layout of the book and approached prospective authors. Once we were close to a finished plan, we approached Ian Sherman, commissioning editor at Oxford University Press (OUP), the publisher of *Avian Incubation*, to sound him out on the idea. Receiving a generally positive response, we pressed on with our plans and presented a preliminary detailed plan for the book in which more than half of the chapters had initial commitments from contributing authors. Fortunately, following review by OUP editors and expert referees, the publishers decided to commit to the project and we spent 2014 commissioning chapter drafts and liaising with authors to produce a completed set of chapters by the start of 2015. We are very pleased with the finished volume.

We are grateful to all of the authors who have contributed to this new book, without whom it would be much less of a reference work of value and would have taken much longer to prepare. Very many thanks to the following reviewers who assisted (sometimes at very short notice and within tight timeframes) with our assessment of the content of individual chapters: Glenn Baggott, Helga Gwinner, Daniel Hanley, Mike Hansell, Helen James, Doug Mock, Ruedi Nager, Richard Sibley, Tim Sparks, Patrick Walsh, and Simone Webber. We thank Ian Sherman and Lucy Nash at OUP for their support throughout this project. Many thanks go to Oliver Smart for the cover image.

DCD must thank Carl Soulsbury for planting the seed of this idea, which has proved to be very interesting and rewarding. DCD is always grateful for the patience of his family, Roslyn, Katherine, and Emily for persevering with his working at home. If, however, you want to pity someone then spare a thought for Millie the dog who during 2014 perhaps didn't have quite as many walks as she would have liked! Most of all DCD would like to thank SJR for his enthusiastic help on this project.

SJR must thank DCD for his initial approach about contributing to the book and for making it sound so exciting that he could not resist asking additional questions! SJR thanks his family, Laine, Tom, and Drew for their endless patience while countless hours were invested in writing, reviewing, and editing chapters. Without them, this never would have happened.

Charles Deeming (Lincoln) and Jim Reynolds
(Birmingham)
December 2014

Contents

Contributors   xiii

1 Incubating new ideas about avian reproduction   1
S.J. Reynolds & D.C. Deeming

2 The fossil record and evolution of avian egg nesting and incubation   8
D.C. Deeming

3 Nest construction behaviour   16
S.D. Healy, K.V. Morgan, & I.E. Bailey

4 Functional properties of nests   29
D.C. Deeming & M.C. Mainwaring

5 The influence of predation on the location and design of nests   50
M.C. Mainwaring, S.J. Reynolds, & K. Weidinger

6 Nest construction and incubation in a changing climate   65
M.C. Mainwaring

7 Microbiology of nests and eggs   75
A. West, P. Cassey, & C.M. Thomas

8 Control of invertebrate occupants of nests   82
I. López-Rull & C. Macías Garcia

9 Egg allometry: influences of phylogeny and the altricial–precocial continuum   97
G.F. Birchard & D.C. Deeming

10 Egg quality, embryonic development, and post-hatching phenotype: an integrated perspective   113
T.D. Williams & T.G.G. Groothuis

11 Egg signalling: the use of visual, auditory, and chemical stimuli   127
K. Brulez, T.W. Pike, & S.J. Reynolds

12 Improvements in our understanding of behaviour during incubation   **142**
V. Marasco & K.A. Spencer

13 The energetic costs of incubation   **152**
A. Nord & J.B. Williams

**14 Influence of incubation temperature on offspring phenotype
and fitness in birds**   **171**
G.R. Hepp, S.E. DuRant, & W.A. Hopkins

15 Advances in techniques to study incubation   **179**
J.A. Smith, C.B. Cooper, & S.J. Reynolds

16 Applications of incubation science to aviculture and conservation   **196**
D.C. Deeming & N.S. Jarrett

17 The role of citizen science in studies of avian reproduction   **208**
C.B. Cooper, R.L. Bailey, & D.I. Leech

18 Perspectives on avian nests and eggs   **221**
D.C. Deeming & S.J. Reynolds

References   227
Index   287

Contributors

Ida E. Bailey Schools of Biology and Psychology, University of St Andrews, St Mary's College, South Street, St Andrews, Fife KY16 9JP, UK

Robyn L. Bailey Cornell Lab of Ornithology, 159 Sapsucker Woods Road, Ithaca, New York 14850, USA

Geoffrey F. Birchard Department of Biology, George Mason University, Fairfax, Virginia 22030, USA

Kaat Brulez School of Biosciences, College of Life & Environmental Sciences, University of Birmingham, Edgbaston, Birmingham B15 2TT, UK Present address: Game & Wildlife Conservation Trust, Burgate Manor, Fordingbridge, Hampshire SP6 1EF, UK

Phill Cassey School of Earth & Environmental Sciences, G41a Mawson Laboratories, The University of Adelaide, North Terrace, South Australia 5005, Australia

Caren B. Cooper North Carolina Museum of Natural Sciences, 11 West Jones Street, Raleigh, North Carolina 27601, USA

D. Charles Deeming School of Life Sciences, University of Lincoln, Joseph Banks Laboratories, Lincoln LN6 7DL, UK

Sarah E. DuRant Department of Zoology, Oklahoma State University, Stillwater, Oklahoma 74078, USA

Ton G.G. Groothuis Behavioural Biology, Institute of Behavioural Neuroscience, University of Groningen, Nijenborgh 7, 9747 AG Groningen, The Netherlands

Susan D. Healy Schools of Biology and Psychology, University of St Andrews, St Mary's College, South Street, St Andrews, Fife KY16 9JP, UK

Gary R. Hepp School of Forestry and Wildlife Sciences, Auburn University, Auburn, Alabama 36849, USA

William A. Hopkins Department of Fish and Wildlife Conservation, Virginia Tech University, Blacksburg, Virginia 24061, USA

Nigel S. Jarrett Wildfowl & Wetlands Trust, Slimbridge, Gloucestershire GL2 7BT, UK

Dave I. Leech The British Trust for Ornithology, The Nunnery, Thetford, Norfolk IP24 2PU, UK

Isabel López-Rull Centro Tlaxcala de Biología de la Conducta, Universidad Autónoma de Tlaxcala, Tlaxcala, México

Constantino Macías Garcia Departamento de Ecología Evolutiva, Instituto de Ecología, Universidad Nacional Autónoma de México, México DF 4510, México

Mark C. Mainwaring Lancaster Environment Centre, Lancaster University, Lancaster LA1 4YQ, UK

Valeria Marasco Institute of Biodiversity, Animal Health and Comparative Medicine, University of Glasgow, Glasgow G12 8QQ, UK

Kate V. Morgan Schools of Biology and Psychology, University of St Andrews, St Mary's College, South Street, St Andrews, Fife KY16 9JP, UK Present address: School of Animal Biology, The University of Western Australia (M092), 35 Stirling Highway, Crawley WA 6009, Australia

Andreas Nord Department of Biology, Lund University, Sölvegatan 37, SE-223 62 Lund, Sweden and Department of Arctic and Marine Biology, Arctic Animal Physiology, Arktisk biologibygget, University of Tromsø, 9037 Tromsø, Norway

Tom W. Pike School of Life Sciences, University of Lincoln, Joseph Banks Laboratories, Lincoln LN6 7DL, UK

S. James Reynolds School of Biosciences, College of Life & Environmental Sciences, University of Birmingham, Edgbaston, Birmingham B15 2TT, UK

Jennifer A. Smith School of Natural Resources, Hardin Hall, 3310 Holdrege Street, University of Nebraska-Lincoln, Lincoln, Nebraska 68583–0961, USA Present address: Department of Biological Sciences, Virginia Polytechnic Institute and State University, 4103 Derring Hall, 1405 Perry Street, Blacksburg, VA 24061, USA

Karen A. Spencer School of Psychology and
Neuroscience, University of St Andrews, South
Street, St Andrews, Fife KY16 9JP, UK

Christopher M. Thomas School of Biosciences, College
of Life & Environmental Sciences, University of
Birmingham, Edgbaston, Birmingham B15 2TT, UK

Karel Weidinger Department of Zoology and
Laboratory of Ornithology, Palacky University, 17
listopadu 50, CZ-771 46 Olomouc, Czech Republic

Allan West School of Biosciences, College of Life &
Environmental Sciences, University of Birmingham,
Edgbaston, Birmingham B15 2TT, UK

Joseph B. Williams Department of Evolution, Ecology
& Organismal Biology, Ohio State University, 318
West 12th Avenue, Columbus, Ohio 43210, USA

Tony D. Williams Department of Biology, Simon Fraser
University, Burnaby, British Columbia, Canada
V5A 1S6

CHAPTER 1

Incubating new ideas about avian reproduction

S.J. Reynolds and D.C. Deeming

1.1 Scientific study of eggs and nests

One of our earliest engagements with the natural world, and more specifically with 'wild animals', is as children when many of us have discovered a nest full of eggs (and often a vociferous bird defending it against all intruders). It is small wonder that we are fascinated by nests that in some cases can be marvels of architecture and construction, while in others they encapsulate functional simplicity. Contained within nests are eggs that have long captivated our attention with their marked variation across species in number, size, shape, and colour. For some the mesmerism of nests and eggs endures into adulthood as evidenced by the success of the Nest Record Scheme (NRS) of the British Trust for Ornithology (BTO). Every spring and summer some 500 volunteer nest recorders (amateur and professional ornithologists alike) visit approximately 30,000 nests of many different species (Greenwood 2012; Chapter 17). In addition there are some museum collections (detailed for Europe by Roselaar 2003) that hold exceptionally large collections of nests and eggs. For example, the Natural History Museum in Tring (NHM Tring), UK contains approximately 4,000 nests and some 400,000 sets of eggs (Russell et al. 2013).

Despite deep-rooted interest from field ornithologists, nests and eggs have received relatively little attention as the main focus of study and it is only comparatively recently that they have come to the fore (Figure 1.1; Collias and Collias 1984; Hansell 2000; Deeming 2002a; Ferguson-Lees et al. 2011). We suspect that the dramatic increase from the 1990s to the present day in the number of these publications does not reflect a sea change in interest in nests and eggs *per se*. Rather, it results from a wholesale increase in the number of technologies now available to the field biologist, in the number of outlets, i.e. journals, for such

published work, in our abilities to collate information about such publications, and in interest in general ornithology 'across the board'. The latter point was acknowledged by Birkhead et al. (2014) who stated, 'In 2011 there were as many papers on birds published as there had been during the entire period between Darwin's *Origin* [Darwin 1859] and 1955.' Although we are not writing a history of ornithology as part of this introductory chapter, as an important historical reference point it is worth noting that according to Haffer (2007) so-called 'New Avian Biology', i.e. the transformation of the discipline of ornithology from natural history into a modern [and mainstream] biological discipline, was attributable to one man. He was Erwin Stresemann (1889–1972) who was the Curator of Ornithology at the Museum of Natural History in Berlin, Germany. Haffer (2007) argued that Stresemann was responsible for transforming ornithology into the modern biological discipline we recognize today, because he sought to unite systematic and field ornithologists by pursuing answers to functional questions related to the morphology, physiology, and behaviour of birds.

Despite this 'new dawn' for modern ornithology, as described by Haffer (2007), we contend that the study of nests and eggs has not received the degree of support from professional ornithology that it warrants. Based upon our contention, it is tempting to predict that the number of published studies about the nest and egg stages of avian life history has lagged behind those addressing other traits. However, when we examined the published literature since 1990, i.e. the threshold decade for escalation of published outputs about nesting biology (Figure 1.1), a key word search in *Web of Science* using life-history traits (*sensu* Bennett and Owens 2002) revealed no apparent 'lag' in the relative numbers of published papers addressing reproductive events of birds at the nest *versus* those outside of it (Figure 1.2). It

Reynolds, S.J. and Deeming, D.C., *Incubating new ideas about avian reproduction*. In: *Nests, Eggs, and Incubation*. Edited by: D.C. Deeming and S.J. Reynolds, Oxford University Press (2015). © Oxford University Press. DOI 10.1093/acprof:oso/9780198718666.003.0001

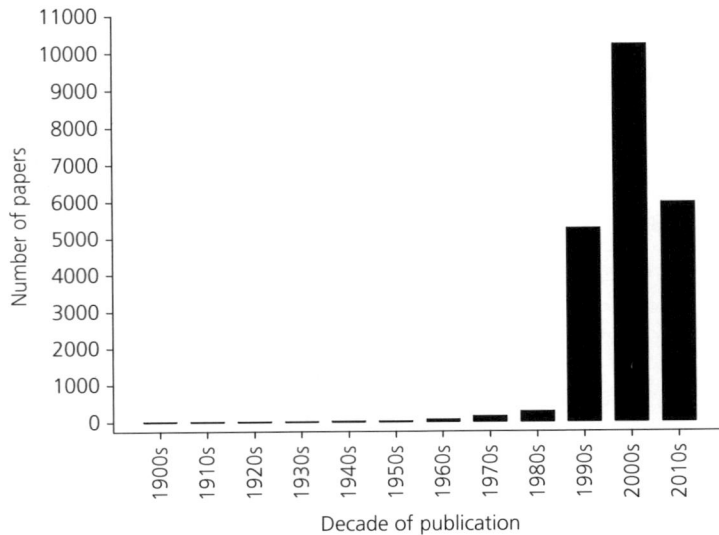

Figure 1.1 The results of a search of the *Web of Science* database using 'nest*' AND 'bird*' OR 'egg*' AND 'bird*' as key words to determine the number of papers published per decade during the 20th and 21st Centuries. Note that 5,375 papers were published between 2010 and 2013 (inclusive) and a further 571 papers have already been published in 2014.

is striking that only papers reporting development and breeding (as a general term) exceeded in number those reporting on nesting biology. Furthermore, the number of papers documenting studies of incubation, laying, and hatching 'held their own' against those describing fledging and recruitment (Figure 1.2). Nesting biology is a broad topic and knowledge of the construction of nests as functional (reproductive) units is relatively poor but slowly improving (Chapters 3 and 4).

1.2 A reluctance to engage?

Our contention does not relate to the *number* of published outputs about nests and eggs. Rather, we argue that ornithologists have been reluctant to engage fully with the study of nesting birds and we attribute this to a number of reasons. The first of these is related to concerns that any activity undertaken by researchers in the vicinity of a nest could influence the data that

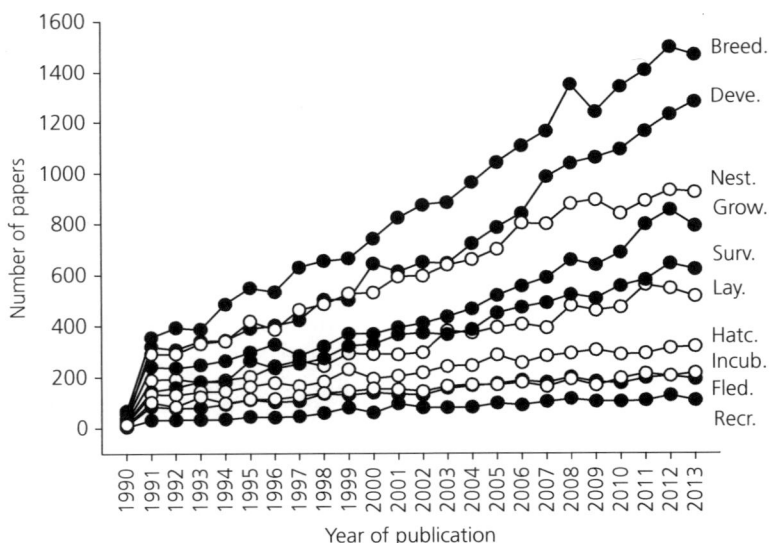

Figure 1.2 The results of a search of the *Web of Science* database using 'bird*' AND various life-history traits as key words to determine the number of papers published per year between 1990 and 2013 (inclusive). Key words used (with focal life-history trait[s]): Recr.—'recruit*' (recruitment); Fled.—'fledg*' (fledging); Incub.—'incubat*' (incubation); Hatc.—'hatch*' (hatch[ing]); Lay.—'lay*' (lay[ing]); Surv.—'surviv*' (survival); Grow.—'grow*' (growth); Nest.—'nest*' (nest-building, nesting and nestling); Deve.—'develop*' (development); and Breed.—'breed*' (breeding). Note that a literature search of papers published in 2014 was not performed as we were only midway through the year at the time of writing.

are generated—this is the basis of the 'uncertainty principle' of Heisenberg (1927) who predicted that the very process of collecting data could exert influence on the nature of the results. Ornithologists have long recognized that this principle could be applied particularly to studies at the nest and they have rightly adapted their nest monitoring protocols to minimize disturbance. Once a nest is discovered and marked appropriately (Ferguson-Lees et al. 2011), ornithologists remain worried that their activities increase its susceptibility to loss from predation by mammals and by both heterospecific (Quinn et al. 2008) and even conspecific birds (e.g. Garvin et al. 2002), and from brood parasitism (Davies 2000). While in some cases such concerns are well-founded (e.g. Anderson and Keith 1980; Pierce and Simons 1986), in others findings are equivocal.

Until recently, the direct effects of our activities at the nest have not been quantified; instead, researchers have 'erred on the side of caution' in their experimental approach. However, Ibáñez-Álamo et al. (2012) carried out a meta-analysis on results from 18 published nest predation studies and found that the activities of researchers did not increase subsequent nest predation. In some orders, such as the Passeriformes, they even found evidence for a positive relationship between human activity at the nest and nest survival (see further discussion in Reynolds and Schoech 2012). Despite these latest findings, we should remain vigilant in designing and executing our nest monitoring protocols to minimize predation risk to adults, eggs, and chicks alike.

Perhaps some of the caution in assessing the 'observer effect' at the nest (Reynolds and Schoech 2012) lies in our inability to identify what to measure. Possible observer effects might range from nest abandonment (resulting in exposure and starvation of nestlings) to subtle changes in nest attendance by adults that result in alterations to the physiological state and nutritional condition of nestlings. For example, Rensel et al. (2010) found that nestling stress levels, as determined from measures of the avian stress hormone corticosterone, were positively correlated with the proportion of time that female breeding Florida scrub-jays (*Aphelocoma coerulescens*) (Figure 1.3) spent away from the nest. Developmental conditions, especially as they relate to exposure to corticosterone, may have profound and chronic effects upon a bird's physiological and behavioural phenotype (Schoech et al. 2012). Note that throughout this book we follow Gill and Wright (2006) in our use of scientific and common nomenclature of avian taxa.

Such considerations of the adverse consequences of nest disturbance are timely as they coincide with those related to other ornithological activities, such as mist-netting (Spotswood et al. 2012) and blood sampling (Arnold et al. 2008; Brown and Brown 2009). Meticulous planning of future studies, and reflection on past studies, promote the adoption of experimental approaches that are effective in acquiring robust data while also promoting the welfare and long-term survival of birds. Indeed, many researchers already seek permissions from national authorities before ringing breeding birds and chicks, and visiting nests, while local and national ethical reviews take place before licenced procedures (e.g. tissue sampling) can be carried out at the nest. For those who are research-active at nests in the UK, approaches must be justified within the context of the so-called 3Rs, i.e. Replacement, Refinement, and Reduction (Festing et al. 2002). The legislative path in the UK can be somewhat hampered by the rather ambiguous definitions of what constitutes 'a procedure' and thus what requires a Home Office licence, particularly when dealing with embryos (Deeming 2011a). There is no doubt that considerable legislation associated with research at the nest has partly engendered a reluctance to engage, especially within the ranks of amateur ornithologists (see Foreword). We refer the reader to Toms (2011) for further details related to such matters.

1.3 A renewed interest in nests and eggs?

Over the 20th Century there was considerable interest in avian incubation, driven particularly by a need to improve artificial methods for hatching large numbers of poultry eggs. Understanding of the avian egg and embryonic development was greatly enhanced by studies during the 1930s through to the 1960s (see Romanoff and Romanoff 1949, 1972; Romanoff 1960, 1967), particularly for the domestic fowl (*Gallus gallus*). In the late 1960s through to the 1980s the work of Rahn, Paganelli, Ar, and numerous colleagues (Rahn and Paganelli 1981; Seymour 1984; Whittow and Rahn 1984; Rahn et al. 1985; Metcalfe et al. 1987; Deeming and Ferguson 1991a; Tullett 1991) developed a more detailed knowledge of embryonic physiology and egg biology in a wider range of bird species, often in natural nests. One consequence of these extensive studies was perhaps a general view that we had 'cracked' the key elements of avian incubation biology. Whilst interest in nests and incubation behaviour was prolonged and sustained over the century, there was considerable

Figure 1.3 A Florida scrub-jay (*Aphelocoma coerulescens*) female incubates a 3-egg clutch at Archbold Biological Station in south-central Florida. A comprehensive and ongoing study of the nesting biology of this species has resulted in a highly successful programme of research to conserve this US federally listed species that is classified as 'Vulnerable' by BirdLife International (2014). (Photo: A. Korpach. See Plate 3.)

emphasis on the ecology of nesting but few scientific syntheses of *how* birds reproduced (Drent 1975) until reviews by Hansell (2000) and Deeming (2002a). Since that time interest in incubation has extended to reptiles (Deeming 2004a) and there has been an increasing use of technology in the study of nesting (Ribic et al. 2012). Use of novel analytical techniques that allow investigation of phylogenetic effects on biological patterns have also been applied to eggs (Deeming et al. 2006; Deeming and Birchard 2008; Deeming 2007a, 2007b, 2008; Birchard and Deeming 2009; Birchard et al. 2013; see Chapters 9 and 11).

Despite ongoing considerations about how we measure the impacts of research activities at the nest, researchers are beginning to appreciate that the nesting biology of birds may considerably shape our understanding of how birds invest time and energy during their breeding attempts. Such considerations may markedly change how we view life-history strategies across species (e.g. Martin and Li 1992) and decision-making during breeding attempts within species (e.g. de Neve and Soler 2002). Both perspectives partly explain the intensity of behaviours such as incubation effort (including nest attentiveness), nest defence, brood provisioning, and even nest construction by adult birds.

Since the publication of Deeming (2002a), there has been a sustained output of published papers reporting on the nesting biology of birds (Figure 1.2). While we

are not suggesting that this is a causative relationship, there may be a number of contributing factors that underlie this trend. For example, so-called 'Citizen Science' (Greenwood 2007; Dickinson and Bonney 2012) has gained momentum with the growing popularity of various established volunteer schemes at both the BTO (e.g. NRS) and the Cornell Lab of Ornithology (e.g. NestWatch) resulting in the public engaging much more in the study of breeding birds (Chapter 17). The harnessing of the power of citizen science to generate data over long temporal and large spatial scales will undoubtedly result in closer associations between ornithological agencies. It is noteworthy that a formal agreement was signed between the BTO and Cornell in 2013 signalling closer collaboration than ever before (see various chapters in Dickinson and Bonney 2012; Chapter 17). We hope that this also signals continued closer associations between amateur and professional ornithologists in this regard.

Arguably, since the publication of Deeming (2002a), rapid advances in the technologies used in video cameras (MacDonald and Bolton 2008) and thermocouples (de Heij et al. 2007), have allowed remote nest monitoring to take place at relatively modest expense to researchers while imposing relatively little disturbance on adult birds in attendance. This has resulted from miniaturization that has made devices much less obvious at the nest and from increased efficiency of batteries reducing frequency of nest visits by researchers to replace them. Technologies are now widely available through commercial suppliers and practical and theoretical information about their effective deployment is available through the scientific literature (e.g. Ribic et al. 2012; Chapter 15) and often from the larger commercial suppliers themselves.

For some of the world's most threatened species, a need for more detailed knowledge about their nesting biology has been clearly identified for their effective conservation (Chapter 16). For example, in the critically endangered kakapo (*Strigops habroptilus*) a programme of supplementary feeding allowed researchers to determine that low breeding frequency was due to nutritional constraint (Elliott et al. 2001). In California condors (*Gymnogyps californianus*) effective breeding to fuel a reintroduction programme was only achieved because of detailed knowledge about the microclimate under which eggs taken from the wild were artificially incubated (Kuehler and Witman 1988). The fundamental importance of food availability to breeding performance of Florida scrub-jays was only revealed through prolonged study of birds at the nest (Figure 1.3). Schoech et al. (2008) clearly demonstrated the potency of food supplementation as a tool to increase both clutch size and chick survival. Increasingly, our understanding of food (Robb et al. 2008) and the key role that it plays in breeding performance is arming conservation biologists in their efforts to translocate populations and to restore native habitats.

1.4 Unexplored gaps in our knowledge

There remain some unexploited areas of ornithology which would certainly be fruitful for future research investigating nests and eggs. For example, we have already acknowledged the nest and egg collections of the NHM Tring, UK (see Section 1.1), but the Western Foundation of Vertebrate Zoology (WFVZ) in Camarillo, CA, USA contains over 18,000 nests and some 225,000 sets of eggs. Although such UK and US collections make major and valuable contributions to scientific research activities, much more use of such collections could be made by researchers (Collar et al. 2003). As an example Russell et al. (2013) asked 140 staff at 112 museums in 37 European countries about the use of their nest collections and discovered that of the only 8% of staff that responded, most reported limited internal and external use of collections for research purposes over the previous decade. Russell et al. (2013) acknowledged that this can be explained partly as a failure in outreach to communicate widely the contents of museum collections.

However, the under-use of such collections can also be explained by limited accessibility to them (but see Russell et al. 2010), and many museums have taken major steps recently to address this issue by starting to database their entire nest collections. It is essential that the gap between workers in the field and museum curators is bridged to ensure that collections are fully exploited for research purposes. It should be noted, however, that museums are generally reluctant to allow researchers to use their materials in a manner that leads to their destruction. Although excess materials can be supplied for research (e.g. Portugal et al. 2014a) and whilst it is an understandable perspective to prevent damage, this can rather restrict the level of engagement between museum collections and researchers.

Fjeldså (2013) suggested that in our research efforts to discover new species, there should be continued collecting of materials to augment museum collections. He described museums as having lost their appetite for collecting in the mid-20th Century when ornithologists, overburdened by regulations governing collecting, turned their attentions to studies of ecology and behaviour of birds in the field. He rightly argued that museum collections still provide much of the data upon which conservation status and descriptions of species in modern field guides are founded. However, interestingly, Fjeldså (2013) provided 'some elements' of a useful description of a new species but they lack any reference to nest or egg traits as being useful descriptions for specimens. For comprehensive discussions about the fundamental role that museums play in research, we refer the reader to Collar et al. (2003).

We are concerned that even the basic breeding biology is not known for some species. For example, while the reproductive strategies of brood parasites such as the common cuckoo (*Cuculus canorus*) have been studied for centuries (Jenner 1788; Darwin 1859; Davies and Brooke 1988; Portugal et al. 2014b), the finer details of how chicks of one well-known genus of brood parasites, the honeyguides (*Indicator* spp.), kill host chicks were only revealed in the last few years (Spottiswoode and Koorevaar 2012). Furthermore, many ornithologists investigate how ecological parameters influence clutch size (e.g. Lack 1947; Ricklefs 1980; Slagsvold 1982), but for many species nest descriptions remain poor despite eggs having been counted within them! In perhaps the most comprehensive study of clutch size to date, Jetz et al. (2008) gathered data on clutch sizes of 5,290 bird species, excluding those of brood parasites and pelagic species. In their analyses they included 'nest type' as an explanatory variable but they only included nest data for 2,816 species (or 53.2% of species for which they had clutch size data). Of course, the authors only had access to sources of primary data that had been published prior to their analyses. For example, for *Handbook of the Birds of the World*, information about nest type was only available for species described up to and including volume 11 (del Hoyo et al. 2006). It is estimated that now nest descriptions could be extended to approximately 4,500 species (W. Jetz, personal communication) of the 9,993 species described in Jetz et al. (2012). However, it is clear that a major shortfall in nest descriptions for many species exists.

Finally, the emerging new field of ethno-ornithology (Tidemann and Gosler 2010) may promise much in furthering knowledge of nests and eggs. It attempts to use traditional ecological knowledge (e.g. information within folklore, nursery rhymes, stories) to augment scientific ecological knowledge about species, thereby expanding our ornithological knowledge (e.g. Sinclair et al. 2010). We envisage that as sources of information contributing to traditional ecological knowledge, such as the Ethno-ornithology World Archive (EWA 2014), expand, so too will our knowledge of the nesting biology of birds, both past and present.

1.5 The structure of '*Nests, Eggs, and Incubation*'

This book builds upon the strong foundations laid down by Deeming (2002a). However, *Nests, eggs, and incubation* is not simply a postscript to it because it is over 12 years since *Avian incubation: behaviour, environment, and evolution* was conceived, written, and published. Understanding of egg biology, nest function, and incubation behaviour has improved in the interim and, of course, during our early discussions there was a strong justification (see Preface) for an updated version of Deeming (2002a). However, some topics in the original book remain valid and relevant and effectively up to date because there has been little substantial progress in further understanding. By contrast, new knowledge has been accrued in many different fields not necessarily covered in the original volume. Therefore, rather than revising Deeming (2002a), we have extended the review of avian egg biology and incubation so as to complement the original, which is now available for readers of this book as a free online resource (see the Preface for more details). The structure of this book is illustrated in Figure 1.4 in which

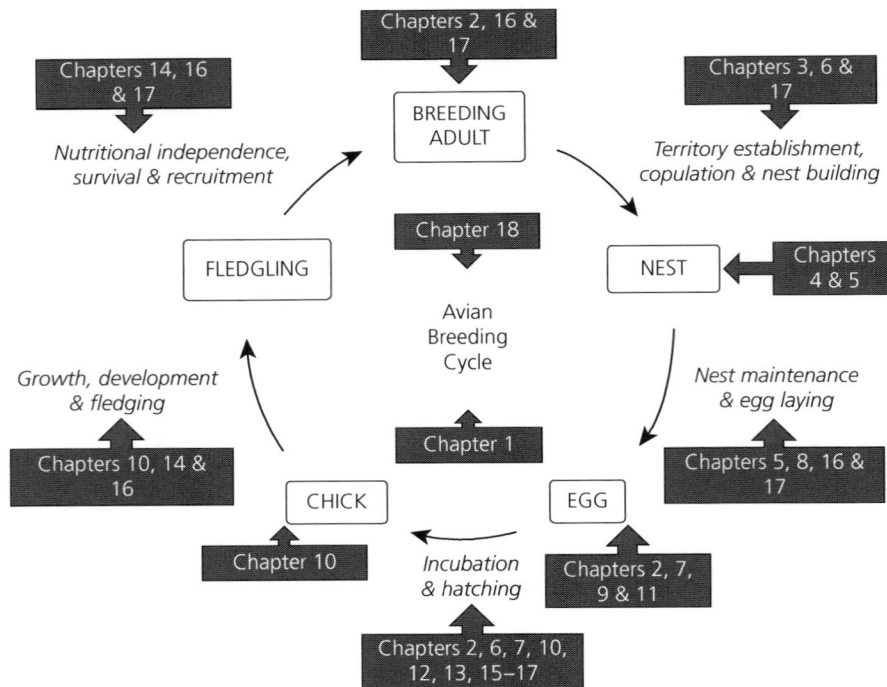

Figure 1.4 A schematic diagram illustrating the overall structure of the book. At its centre is the breeding cycle of a typical avian species with key life stages shown in boxes, biological processes linking stages shown in italicized text and the numbers of relevant chapters referring to each stage and/or process shown in dark grey boxes.

we have shown the avian breeding cycle, its constituent life stages, the biological processes that link sequential stages, and the relevant numbers of chapter(s) of the book that refer to each stage/process. It is clear that the book covers most of the avian breeding cycle except the fledgling stage, which is defined as the period between a young bird fledging and becoming fully independent of its parents (Cox 1996). Among avian life historians 'fledging' is a rather poorly defined unsatisfactory event which has been intensely debated in the literature (e.g. Middleton and Prigoda 2001) and in many species fledging is so transient that little is known about it.

After starting with an updated consideration of the evolution of avian reproductive biology, the book is divided into four broad areas: the nest, the egg, incubation, and the study of avian reproduction. Our understanding of the role of nests has improved greatly since Deeming (2002a) and so these structures are given greater consideration in this book. Other chapters update our understanding of the egg traits and life history as well as incubation as a process, including the currencies of investment paid by breeding birds. We also consider how modern methodologies as diverse as self-contained temperature probes and citizen science help us to gain greater insights into nests and eggs as functional units. The last section of the book includes chapters describing how such basic biological knowledge can be applied to challenges such as defining, and responding to, conservation issues and climate change. We conclude by attempting to define areas that are priorities for future research addressing these key stages of the avian breeding cycle.

Acknowledgements

SJR thanks Laine Wallace for her insightful comments on the chapter.

CHAPTER 2

The fossil record and evolution of avian egg nesting and incubation

D.C. Deeming

2.1 Introduction

The evolutionary origin of birds remains a matter of debate with opinion divided between a non-theropod (Feduccia 1996; Czerkas and Feduccia 2014) and a theropod ancestor (Chatterjee 1997; Dingus and Rowe 1998; Chiappe and Witmer 2002; Currie et al. 2004; Chiappe 2007; Kaiser 2007). The consensus currently favours theropods and so, hereafter, I follow this perspective. Since the earliest evidence of birds (or avian theropods) within the fossil record there have been only a few sporadic reports of evidence of their nests, eggs, or embryos and there have been even fewer considerations of their reproductive biology (Naish 2014). This chapter, therefore, considers the fossil evidence for avian reproduction from the earliest records through to relatively modern times. The fossil records of reproductive biology of Chelonia (Lawver and Jackson 2014) and of extinct non-avian theropods are considered to be relatively well-understood (Grellet-Tinner et al. 2002; Varricchio 2011). For the latter this is because there is fossil evidence of eggs *in situ* within the oviduct (Sato et al. 2005), of eggs and embryos within eggs (Norell et al. 1994; Varricchio et al. 1997), and nests (Dong and Currie 1996; Norell et al. 1994; Clark et al. 1999; Varricchio et al. 1997). Indeed, it has been assumed that avian reproductive biology is directly derived from comparable behaviours in non-avian theropods (Norell et al. 1994; Clark et al. 1999; Varricchio et al. 2008).

However, the idea that the patterns of nesting biology of non-avian and avian theropods are perhaps indistinguishable is not globally accepted and there is evidence to the contrary. The suggestion that male-only incubation, observed in modern-day ratites (Deeming 2002e), has its origins in Mesozoic non-avian theropods (Varricchio et al. 2008) has been seriously undermined by re-analysis of comparative data (Birchard et al.

2013). In addition, water vapour conductance calculated from morphometrics of eggshells from various theropods and a Mesozoic bird is comparable to that of modern crocodilians which bury their eggs and use environmental conditions for incubation (Deeming 2006). This contrasts with extant birds which have open nests and use metabolic heat (Sabath 1991; Deeming 2002b, 2006; Yang et al. 2012). Tanaka and Zelenitsky (2014) challenged the idea that water vapour conductance derived from morphological measurements of the eggshell was a useful way of predicting the humidity conditions within nests. This conclusion was based on an approach that found that conductance values derived empirically by weighing bird eggs were not a perfect match for conductance values calculated from morphometrics of eggshells from the same species. However, the values for conductance were not from the *same* eggs and so cannot be reasonably compared; any differences were almost certainly down to natural variation between eggs. Differences were greatest for small eggs but this probably reflected the difficulty in accurately determining pore density in eggshells (G.K. Baggott, personal communication), which would reduce conductance values. Unfortunately, Tanaka and Zelenitsky (2014) failed to appreciate the clear link between water vapour conductance and nesting environment demonstrated in extant birds (Rahn and Paganelli 1990; Deeming 2011b). Moreover, their relatively small differences would bear little relation to the differences for conductance values for dinosaur eggs and modern birds, which differ by an order of magnitude (Deeming 2006).

There has been speculation about the egg size of *Archaeopteryx* with a couple of estimates (based on the diameter of the pelvic opening) being 14 (breadth) × 25 (length) mm or 18 × 30 mm (Wellnhofer 2009). Unfortunately, these sizes are speculative because no fossil eggs have been attributed to *Archaeopteryx* and

Deeming, D.C., *The fossil record and evolution of avian egg nesting and incubation*. In: *Nests, Eggs, and Incubation*. Edited by: D.C. Deeming and S.J. Reynolds, Oxford University Press (2015). © Oxford University Press. DOI 10.1093/acprof:oso/9780198718666.003.0002 the rationale for egg length estimate is not provided. However, tangible evidence of avian reproduction does exist in the fossil record. This often consists of fragmented eggshells with near-complete or complete eggshells only occasionally being discovered. Rarely do these eggs contain embryos, the ultimate diagnostic feature of any egg specimen that can attribute it unambiguously to a bird. To date, egg–adult associations, either *in utero* or post-oviposition, seen in extinct non-avian theropods (e.g. Varricchio et al. 1997; Clark et al. 1999; Sato et al. 2005) and pterosaurs (Lü et al. 2011), have not been described for any bird species. Evidence of any supposed nesting activity by birds is extremely rare. However, sporadic evidence is available from the earliest stages of avian evolution in the Mesozoic through the Cenozoic and into the relatively modern extinctions during the Pleistocene (Naish 2014). This fossil record should be contrasted with that of the Mesozoic Theropoda and other Dinosauria where evidence of reproduction, in the form of eggs and nests, is relatively common (Carpenter et al. 1994).

Here, I review and update Deeming (2002b) in our current understanding of avian reproduction in the fossil record from the Mesozoic through the Cenozoic to the Pleistocene. Since 2002 there has been a variety of important fossil finds that help to set avian reproduction in a wider biological context. Table 2.1 summarizes the fossil record of avian reproduction providing details of the type of specimen. Many of these reports are discussed in greater detail below. The chapter ends with a consideration of why the fossil record of avian reproduction is so poor compared with other taxa.

2.2 Reproductive behaviour and physiology

Behaviour is rarely preserved in the fossil record but tantalizing glimpses can be seen by morphological characteristics of fossils. In birds it is sexual dimorphism in the appearance of feathers that is often seen as a diagnostic trait of gender classification of fossils. Some examples of *Confuciusornis sanctus* of the Lower Cretaceous of China exhibit two long, central rectrices, i.e. tail feathers, which have been interpreted as indicating display plumage in putative males (Chinsamy et al. 2013). Although morphological evidence did not support this interpretation (Chiappe et al. 2008), later investigation confirmed a specimen lacking such long feathers (Chinsamy et al. 2013), but possessing medullary bone, a characteristic of reproductively active

female birds (Dacke et al. 1993) and a non-avian theropod (Schweitzer et al. 2005).

Incubation behaviour of the earliest birds has been inferred from the relationship between adult body mass and clutch volume in modern birds and in non-avian theropods in the Cretaceous (Varricchio et al. 2008). Both allometry and histology of long bones that demonstrated the absence of medullary bone were used to suggest that 'incubating' non-avian theropods were male. Hence, contrary to McKitrick (1992) who suggested that bi-parental care was the primitive condition for birds, Varricchio et al. (2008) concluded that the male-only incubation behaviour of Palaeognath extant birds was a behaviour derived directly from an ancestral non-avian theropod. However, there are problems with this interpretation. Bone histology reported by Varricchio et al. (2008) was based on only one sample of each species of theropod, and the concept of contact incubation in Cretaceous theropods is not globally accepted (Deeming 2002b, 2006). Moreover, more recent and robust analysis of a larger dataset by Birchard et al. (2013) casts further doubt by showing that incubation behaviour of either the earliest birds or non-avian theropods cannot be inferred from the allometric body mass–clutch size relationships of modern adult birds.

Structures that have been interpreted as fossilized ovarian follicles have been reported from a range of Mesozoic birds, including *Jeholornis prima* and various Enantiornithes from the Jehol Biota of northeastern China (Zheng et al. 2013; O'Connor et al. 2013b). The numbers and relative size of these follicles have been interpreted as indicating the progressive evolution of a modern-style ovarian morphology with sequential follicle maturation and ovulation away from the more synchronous pattern observed in modern crocodilians (Zheng et al. 2013; O'Connor et al. 2013b). Such interpretation has fuelled heated debate resulting in the original hypothesis of Zheng et al. (2013) being challenged (e.g. Mayr and Manegold 2013) and defended (e.g. O'Connor et al. 2013b). Whether the structures indicate ovarian follicles depends on whether the original interpretation is valid; it is interesting that Zheng et al. (2013) suggested that there was only one (left) functional ovary in these Mesozoic birds, although it is unclear how functionality was determined. Such a condition would be typical of most extant birds, with some exceptions (e.g. kiwis [*Apteryx* spp.] and some birds of prey—Kinsky 1971; Johnson 2000). One key point about these specimens is that, as Mayr and Manegold (2013) suggested, there is no evidence of any other soft tissues preserved alongside the (presumably) rather delicate ovarian follicles. More robust tissues such as the heart, kidneys or even feathers are not preserved, thus undermining the argument that the observed structures are yolk follicles.

Intact eggs within the body cavity have been reported for non-avian theropods. The holotype of *Compsognathus longipes* has supposed eggs adjacent to its body cavity (Griffiths 1993), although this is disputed (e.g. Chen et al. 1998). *Sinosauropteryx prima*, a compsognathid from the Jehol Group (Upper Jurassic or Lower Cretaceous) of China, is reported to have two eggs within its body cavity (Chen et al. 1998). An oviraptorosaurian from the Upper Cretaceous of China has also been reported with two fully shelled eggs adjacent to its pelvis (Sato et al. 2005). For avian theropods there are few reports of an egg associated with an adult skeleton. Kaiser (2007) reported a specimen of *Confuciusornis sanctus* that had an egg fossilized in contact with its right foot. This egg had a maximum breadth of 17 mm, which supposedly would fit through the pelvic canal of the adult (Kaiser 2007). Unfortunately, there is no more detailed description of this specimen (B072 of the private Wenya Museum in Jinzhou, China), so further details remain unknown.

Another report is of a femur and lower leg bones in association with two-egg 'clutches' of *Styloolithus sabathi* (Sabath 1991; Varricchio and Barta 2015). These eggs were deemed to have been buried in sediment and Varricchio and Barta (2015) suggested that this indicated an incubation mode comparable to non-avian theropods (e.g. Clark et al. 1999; Fanti et al. 2012). As Deeming (2002b) argued previously for non-avian theropods, such as the oviraptor *Citipati osmolskae* (Clark et al. 1999), it is debatable whether this association of bones with eggs truly reflects contact incubation as practised by modern birds.

Whether *Styloolithus* eggs lacked chalazae, which are found within the albumen of eggs of modern birds, as conjectured by Varricchio and Barta (2015), is difficult to confirm. Romanoff and Romanoff (1949) suggested that the chalazae stabilize the position of the yolk within the albumen when the egg is rotated during egg turning (see Chapter 16), rather than being required to allow eggs to be turned, as was suggested by Varricchio and Barta (2015). Hence, whether these suspensory structures have a key role to play in the evolution of avian eggs is far from clear. I favour caution and suggest that the creation of nests that did not require eggs to be buried, thereby allowing for contact incubation and a higher developmental temperature, was probably a more important step in the evolution of modern avian incubation.

Unequivocal evidence of reproduction for a particular taxon is only available when there is the presence of identifiable bones of an embryo within an egg. In amniotes, in general, this is exceptionally rare in the fossil record (Hirsch 1989), and even more so in birds. The presence of an embryo, and thus the ability to confirm an egg as being of avian origin, has only been found in a few instances during the Mesozoic. The enantiornithine bird *Gobipteryx minuta* from the Late Cretaceous of Mongolia was found to have an embryo within its eggshell (Elzanowski 1981; Kurochkin 1996), but diagenetic alteration meant that the type of eggshell structure was considered difficult to identify (Schweitzer et al. 2002). Furthermore, Schweitzer et al. (2002) reported a partially articulated skeleton of a bird embryo in an egg from the Late Cretaceous of Argentina. The bird was most likely an enantiornithine making this the first unequivocal assignment of a prismatic, trilaminate eggshell structure (typical of modern birds) to a basal avian lineage. A fully articulated skeleton of a well-developed bird embryo was reported from the Lower Cretaceous of China by Zhou and Zhang (2004). This specimen had enantiornithine apomorphies and the presence of feathers strongly suggest that this was a precocial species. Chatterjee (1997) reported two examples of well-developed articulated embryonic skeletons of *Gobipipus reshetovi* from the Upper Cretaceous of Mongolia, which were considered to be included in the Neornithes.

During the Cenozoic, avian-type eggs are less likely to be confused but there are still only a few examples of avian embryos associated with eggshells. Jackson et al. (2013) described embryonic bones within eggs of *Microolithis wilsoni* from the Eocene of Wyoming. It is curious that evidence for embryos in more recently extinct modern-type birds is even rarer. To date, an embryo has been described in high detail for only an *Aepyornis* egg from the Holocene of Madagascar and only following high-resolution X-ray computed tomography (Balanoff and Rowe 2007). Such techniques are likely to help in future studies of fossil finds and as technology improves this should help our understanding of future finds of fossilized eggs or other avian remains.

Other fossil eggs attributed to birds are solely diagnosed on the basis of eggshell structure and there are a number of fossilized eggs from various times in the Cenozoic (Figure 2.1; Table 2.1). For example, Grellet-Tinner and Norell (2002) described a partly intact egg from the Campanian Bayn Dzak locality of southern Mongolia; it had a three-layered eggshell which they attributed to species closer to the Neognath birds

Figure 2.1 Example of a Mesozoic bird egg collected from Upper Cretaceous deposits of Argentina. Scale bar = 1 cm. (Photo: D. Varricchio. See Plate 4.)

than to an enantiornithine bird. Such eggs are attributed to birds mainly on the basis of eggshell type and egg shape. Fossil eggshells attributed to birds on the basis of microstructure are relatively common and range from the Mesozoic to the Holocene in origin (Hirsch and Packard 1987; Mikhailov 1992, 1997a, 1997b; Kohring and Hirsch 1996; Mikhailov et al. 1996; Hirsch et al. 1996; Poinar et al. 2007; Patnaik et al. 2009). Whilst of interest, it is unclear what eggshell alone can tell us about the evolutionary relationships between bird species through geological time. Eggshell microstructure in different taxa will have a phylogenetic signal (Mikhailov et al. 1997b), but I remain unconvinced that it is a robust means to determine phylogenetic relatedness as suggested by Mikhailov (1997a).

Table 2.1 Summary of the fossil record of the various indicators of avian reproduction.

Taxon	Geological era	Location	Specimens	Reference
Mesozoic				
Confuciusornis sanctus	Lower Cretaceous	China	Egg associated with adult	Kaiser (2007)
Enantiornithes	Lower Cretaceous	China	Embryonic remains	Zhou and Zhang (2004)
Neuquenornis volans	Lower Cretaceous	Argentina	Near intact eggs	Fernández et al. (2013)
Jeholornis prima	Lower Cretaceous	China	Ovarian follicles	Zheng et al. (2013)
Enantiornithes	Lower Cretaceous	China	Ovarian follicles	O'Connor et al. (2013a); Zheng et al. (2013)
Neuquenornis? sp.	Upper Cretaceous	Argentina	Embryonic remains in near intact egg	Schweitzer et al. (2002)
Ornithothoraces	Upper Cretaceous	Brazil	Near intact egg	Marsola et al. (2014)
Gobipteryx minuta	Upper Cretaceous	Mongolia	Embryonic remains	Elzanowski (1981); Sabath (1991); Varricchio and Barta (2015)
Nanantius valifanovi [Gobipteryx minuta]	Upper Cretaceous	Mongolia	Eggshell and juvenile remains	Kurochkin (1996)
Styloolithus sabathi	Upper Cretaceous	Mongolia	Eggs	Sabath (1991); Varricchio and Barta (2015)
Gobioolithus minor	Upper Cretaceous	Mongolia	Near intact eggs and eggshells	Mikhailov (1997b); Varricchio and Barta (2015)
Gobipipus reshetovi	Upper Cretaceous	Mongolia	Embryonic remains, near intact eggs	Kurochkin et al. (2013)
Laevisoolithus sochavai	Upper Cretaceous	Mongolia	Near intact eggs	Kurochkin et al. (2013)
Enantiornithes	Upper Cretaceous	Romania	Embryonic remains, near intact eggs	Dyke et al. (2012)
Unattributed	Upper Cretaceous	India	Eggs	Mohabey et al. (1993)
Unattributed	Upper Cretaceous	Mongolia	Near intact egg	Grellet-Tinner and Norell (2002)

continued

Table 2.1 *Continued*

Taxon	Geological era	Location	Specimens	Reference
Cenozoic				
Ornitholithus sp.	Upper Paleocene	Spain, France	Eggshells	Donaire and López-Martínez (2009); Angst et al. (2015)
Metoolithus nebraskensis	Lower Eocene	USA	Near intact egg	Jackson et al. (2013)
Microolithus wilsoni	Lower Eocene	USA	Embryonic remains, partial egg	Jackson et al. (2013)
Lithornis sp.	Lower Eocene	UK	Eggshell fragments	Grellet-Tinner and Dyke (2005)
Presbyornis	Lower Eocene	USA	Eggshell fragments	Leggitt et al. (1998)
Incognitoolithus ramotubulus	Lower Eocene	USA	Eggshell fragments	Hirsch et al. (1996)
Medioolithus sp.	Middle Eocene	Germany	Eggshell fragments	Kohring and Hirsch (1996)
Badistornis aramus	Lower Oligocene	USA	Complete eggs	Chandler and Wall (2001)
Phoenicopteriformes	Early Miocene	Spain	Near intact eggs, nest	Grellet-Tinner et al. (2012)
Struthiolithus	Early Miocene	India	Eggshells	Patnaik et al. (2009)
Trochilidae	Early Miocene	Dominica	Partial egg in amber	Poinar et al. (2007)
Dromornis? sp.	Miocene	Australia	Eggshell fragments	Murray and Vickers-Rich (2004)
Francolinus sp.	Pliocene	Tanzania	Near intact eggs	Harrison (2005)
Numidia sp.	Pliocene	Tanzania	Near intact eggs	Harrison (2005)
Aepyornis maximus	Pleistocene	Madagascar	Embryonic bones	Balanoff and Rowe (2007)
Phoebastria albatrus	Pleistocene	Bermuda	Egg	Olson and Hearty (2003)
Pelecanus occidentalis	Pleistocene	Bermuda	Egg	Olson and Hearty (2013)
Fratercula dowi	Pleistocene	California	Intact eggs	Guthrie et al. (2000)
Genyornis newtoni	Pleistocene	Australia	Eggshell fragments	Miller et al. (1999); Angst et al. (2015)
Dromaius sp.	Pleistocene	Australia	Eggshell fragments	Miller et al. (1999)
Eudyptula minor	Holocene	New Zealand	Near intact egg	Ksepka (2011)
Dinornithiformes	Holocene	New Zealand	Eggs	Anderson (2003); Gill (2006); Huynen et al. (2010)
Dinornithiformes	Holocene	New Zealand	Eggshells	Anderson (2003); Gill (2007); Angst et al. (2015)
Dinornithiformes	Holocene	New Zealand	Nesting material	Wood (2008)
Dinornis gigantea	Holocene	New Zealand	Embryo in egg	Hector (1867); Anderson (2003)
Aepyornis sp.	Holocene	Madagascar	Eggshell fragments	Angst et al. (2015)
Aepyornis sp.	Holocene	Madagascar	Embryo in egg	Balanoff and Rowe (2007)

Egg shape and size have been inferred from fragments of fossilized eggshell in a variety of studies (see Angst et al. 2015). The maximum and minimum curvatures of the shell fragments are considered to correspond to the transversal and longitudinal curvatures, respectively. These values can be used to reconstruct the presumed size of an egg when intact. This technique has been used to predict egg mass in a variety of extinct and extant species (Angst et al. 2015). Eggshell thickness has also been used to predict incubation behaviour of extinct moas (Dinornithiformes), which appear to have reversed sexual dimorphism which was associated with males incubating eggs because the females would have been too heavy to do so without crushing eggs (Birchard and Deeming 2009).

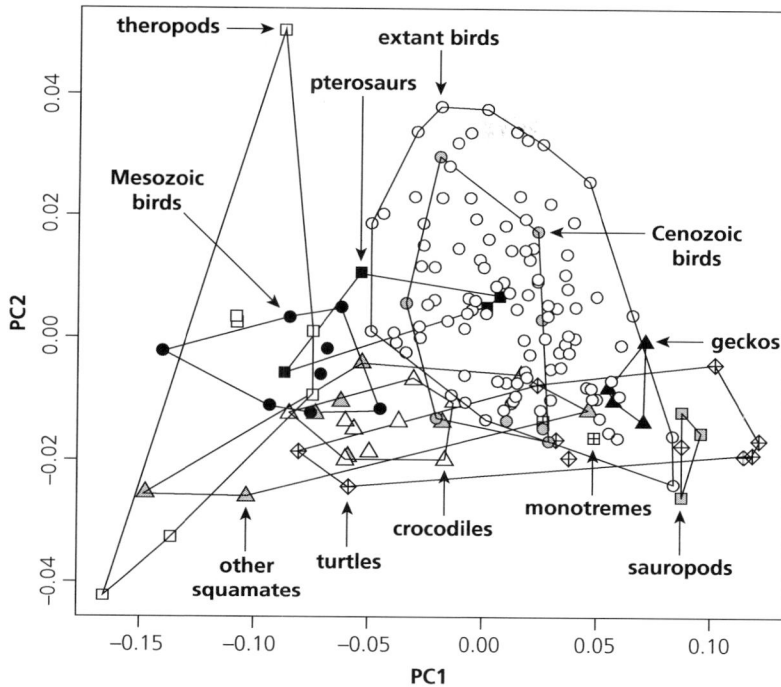

Figure 2.2 Morphospace plot based on the semi-landmark analysis of egg types, with convex hulls delimiting the 11 taxonomic groups in the data set. Briefly, egg outlines were digitally defined, subjected to procrustes transformation and principal components analysis (PCA), prior to semi-landmark analysis. PC1 corresponds to egg shape from elongate to spherical and PC2 corresponds to egg symmetry from symmetrical to asymmetrical. See Deeming and Ruta (2014) for more information.

Deeming and Ruta (2014) investigated evolution of egg shape in extinct and extant oviparous amniotes. Generally, extant avian eggs were significantly less elongated and more asymmetrical than those of extant reptiles and extinct non-avian theropods. Fossil eggs from Cenozoic bird species were indistinguishable from extant avian eggs (Figure 2.2) but eggs attributed to Mesozoic birds were significantly more elongated and less asymmetrical than those of modern birds (Deeming and Ruta 2014). Although more examples of Mesozoic bird eggs are needed to confirm this preliminary analysis, it suggests that the pattern of egg formation in these extinct taxa was somewhat different from that of modern birds.

2.3 The nest environment

Fossil evidence of bird nests is almost non-existent. To date, there is only one report that purports to provide evidence of a stick nest from the Early Miocene lacustrine limestone bed in the Tudela Formation of the Ebro Basin, Spain (Grellet-Tinner et al. 2012). The nest consists of five incomplete eggs and associated plant materials in the form of twigs and leaves attributed to the Fabaceae, although why this diagnosis was made

is unclear. The nest is considered to be comparable to that of a modern grebe (Podicipedidae) but the eggs are attributed to a paleoflamingo on the basis of eggshell characteristics. This specimen was interpreted as representing eggs incubated on a floating nest on a sub-oxic and oligohaline endorheic lake (Grellet-Tinner et al. 2012). Whilst the association between the eggshells and the plant material is suggestive of a nest-like structure, its description as a paleoflamingo is open to question. Eggshells of flamingos have characteristic microglobular covers (Mikhailov 1997a), which are absent in the published images of eggshells of extant and extinct flamingos. A microglobular cover was present on eggshells attributed to *Palaelodus* of the Miocene of France (Mikhailov 1997a). It is unclear why this inorganic cover should have been lost from what seems like an otherwise well-preserved eggshell (Grellet-Tinner et al. 2012). Moreover, extant flamingos construct nests from pillars of mud that raise their eggs above the substrate (Johnson and Kikkawa 2003) rather than constructing floating nests. Even though the nest resembles that of a modern grebe, there is very little fossilized plant material and the associated eggshell does not resemble that of a grebe (see Mikhailov 1997a). Whilst of interest, the specimen does suggest some association between nest materials and eggs but it is difficult to share the same confidence in its origin as do Grellet-Tinner et al. (2012).

Numerous finds of both adult bones and eggshells of *Presbyornis* have been reported at a Lower Eocene site in Wyoming (Leggitt et al. 1998). The presence of numerous eggshells at three different locations within the Fossil Basin site was interpreted as representing shoreline nesting sites (Leggitt et al. 1998) but the evidence seems rather circumstantial.

Nest material is also reported in the much more recent sediments of several rock shelters that measure up to 2 m deep, are dated up to 3,500 years old, and are attributed to a variety of species of moas of the Holocene of New Zealand (Wood 2008). Plant materials, particularly clipped twigs from a variety of species, were apparently processed by adult moas and Wood (2008) drew comparisons between the nests of extant Australasian ratites and these recently extinct moas.

Evidence of more general nest sites of birds is equally rare. Dyke et al. (2012) reported a supposed colonial nest site for an enantiornithine bird from the Late Cretaceous (Maastrichtian) of Transylvania, Romania. This interpretation was based on accumulation of eggshells, some seven near-intact eggs, and the presence of large and small (considered to be embryonic) bones of an enantiornithine bird that were not found unambiguously within an eggshell. These fossil specimens were interpreted as having succumbed to a flooding event but, again, this is but one interpretation and is inconclusive.

An indirect measure of the nest environment can be determined from eggshell conductance (Deeming 2006). Avian eggs incubate in a relatively dry environment compared with that of reptiles and this is reflected in a relatively low eggshell conductance—higher values indicate that eggs are incubated in more humid conditions, which in modern taxa is associated with burial (Deeming 2006). To date, conductance values are available for few avian eggshells irrespective of species.

Fernández et al. (2013) reported eggshell conductance for a Cretaceous ornithothoracine bird egg using a predicted value based on egg size. Unfortunately, despite inferences within this report, such values provide little empirical basis for determination of conductance of these eggshells. Only data for eggshell thickness, pore radius, and pore density from the specimens can be used to calculate conductance values (Deeming 2006), which could then be compared to the predicted values based upon assumed egg mass. To date, the only examples are for the eggshell of *Gobiopteryx* from the Cretaceous of Mongolia (Sabath 1991; Deeming 2002b, 2006) and *Ornitholithus* of the Palaeocene of Spain (Donaire and López-Martínez 2009). In the former case the actual conductance was markedly higher than that predicted based on egg mass of modern birds, implying that eggs had to be buried during incubation (Deeming 2002b, 2006). I have little confidence in the conductance value of the large egg of *Styloolithus* provided by Varricchio and Barta (2015) because this was calculated without the use of comparable values for the

Figure 2.3 Scanning electron micrograph showing the pore canal system in *Ornitholithus* eggshell, grouped together in deep fissures on the valleys between the tubercle ornamentation. Scale bar = 3 mm. Reproduced from Donaire and López-Martínez (2009) with permission from Elsevier.

Gobiopteryx egg because pore density and dimensions were not reported in the former (Sabath 1991).

Eggshells of *Ornitholithus*, which are considered to be from an egg weighing ~ 3 kg and attributed to an unidentified gigantic bird, have extremely low conductance values that are 15–20% of those required for an extant bird's egg of equivalent size (Donaire and López-Martínez 2009). Conductance was attributed to the nest environment being xeric but the authors failed to appreciate that such an eggshell conductance would be incapable of supporting the oxygen requirements of embryos within 3-kg eggs—such conductance estimates are problematic. The pore structure in *Ornitholithus* is unusual as pore radius is narrow (5–10 μm) compared to that of a common ostrich (*Struthio camelus*) egg of half its presumed mass (~ 50 μm; Christensen et al. 1996) and there are deep narrow fissures in the eggshell that appear to extend through it (Figure 2.3; Donaire and López-Martínez 2009), even if the authors suggested that this was not the case. This feature of eggshell anatomy is important because these fissures are long (~ 0.5– 0.75 mm), and if they approximate to a density of one per mm², eggshell conductance increases dramatically. Further research is required to confirm the role of these fissures but they could represent a unique solution to the problem of gas exchange in very large eggs.

2.4 Improving our understanding of reproduction in extinct birds

The fossil record allowing study of reproduction in extinct theropod species is very sporadic and hence difficult to interpret with any certainty. For avian-theropod species the fossil record is exceptionally poor and whilst I expect it to improve as the number of specimens increases, it is puzzling why the fossil record is currently so depauperate. After all, evidence of breeding in many Mesozoic dinosaurs is relatively common (e.g. Carpenter et al. 1994) but evidence for breeding in Mesozoic or Cenozoic birds is almost non-existent. I attribute this disparity to a likely fundamental difference between reproduction of modern birds and their avian and non-avian theropod antecedents.

Extant birds are characterized by their use of contact incubation between the parent and egg within a nesting environment which is relatively open, low in humidity and high in oxygen. With the exception of the megapodes (Booth and Jones 2002), eggs of all species are not buried in a substrate and many are laid in open nests either on the ground or very often well above the ground. By contrast, modern reptiles typically bury their eggs in a substrate irrespective of whether the eggshell has a significant calcium carbonate layer or not (Schleich and Kästle 1988; Packard and DeMarco 1991). This means that the possibility of fossilization will be markedly different for reptiles because their eggs are already buried and much more likely to be fossilized than those of birds laid and incubated in open environments. Many bird eggs are also generally quite small and thin-shelled, which will reduce their chances of preservation. Contrast this with the large, thicker-shelled eggs of dinosaurs that were buried and so were more likely to be fossilized (Deeming 2006). The gas conductance of Mesozoic bird eggshells suggests that they too were buried (Deeming 2006), so I would predict that future fossil discoveries of avian eggs will be more common from the Mesozoic than from the Cenozoic. Whatever the future holds there does seem to be an increasingly popular view that mode of reproduction may have played a key role in the survival of Neornithes across the Cretaceous–Tertiary boundary (Deeming 2002b, 2006; Varricchio and Barta 2015).

The study of evolution of avian reproductive biology is dynamic but it is unclear what new specimens will be revealed in the future. Of more certainty is that our understanding of the evolutionary processes shaping avian reproduction is relatively poor and can sometimes be prone to over-interpretation. However, our knowledge can only improve as we study new fossils and apply new methods of analysis to existing fossils of bird eggs and nests. As the fossil record of reproduction in theropods in general improves, we should be able to examine evolutionary relationships between various taxa in greater detail. However, we need to exercise care during interpretation of findings to ensure that the most parsimonious explanation is achieved.

Acknowledgements

I am grateful to Marcello Ruta, Jim Reynolds, and Helen James for their constructive comments on this chapter. Many thanks go to Dave Varricchio for providing the photograph of the avian egg from Argentina.

CHAPTER 3

Nest construction behaviour

S.D. Healy, K.V. Morgan, and I.E. Bailey

3.1 Introduction

'I believe, in short, that birds do *not* build their nests by instinct' (Wallace 1867). It may have taken some time but one of Alfred Russel Wallace's lesser known ideas finally appears to be bearing fruit. There is increasing evidence that birds build nests that are more variable than once thought and that they will change the material they use to build their nests, depending on their experience. Therefore, birds do not rely entirely on instinct to build their nests. Not only would we argue that the accumulating evidence shows that nest building by birds is not all instinctual but also that the recent data on nest-building behaviour and decision-making are beginning to help us to answer a question that remains, perhaps surprisingly, unanswered, which is why birds build the nests they do.

Birds are one of the best-studied taxa in the world, so it is not surprising that, for the approximately 9,000 species of bird worldwide, there is a plethora of descriptions of the nests they build (e.g. Ferguson-Lees et al. 2011). Although many fewer, there are also multiple descriptions of how different species go about building their nests. In Table 3.1 we provide examples of species for which there are detailed descriptions of the behaviours used by different species when building their nests (for more examples see Hansell 2000; Goodfellow 2011). Importantly, for our purposes, it is difficult to know whether descriptions of nest-building behaviours are typical for a species as usually they have been described only for just one or two individuals per species. In general, birds may either sculpt a nest by removing material from a site or assemble a nest by adding material to a site (Hansell 1984). During nest assembly, material may be piled up, moulded, stuck together, interlocked, sown or woven (for more detail and examples see Hansell 1984). Detailed experimentation investigating how birds know which building techniques to use is rare. Hence, we have little information on how they make choices of materials or whether birds copy the nest-building process and material choice of a parent. We know nothing about whether birds can learn how to manipulate building materials. From recent work, however, it is becoming clear that not only do birds within a species vary in their building behaviour but individuals can be flexible in that building behaviour, variation that can lead to significant variation in the nest produced. These data are increasingly showing that learning has an important role to play in both building dexterity, choice of nest materials and the making of a number of other nest-building decisions. These topics are covered in this chapter.

3.2 Background

Unlike his rather more famous ideas concerning natural selection, the view of Wallace (1867) that it was simply implausible that all of the information required for appropriate nest building was instinctive has languished for over 100 years; it remains both scientific and popular belief that a species' knowledge about nest design, construction, location etc. was all 'in' their genes. We think it most likely that nest-building behaviour is rather more like most other behaviours in that it is shaped both by genes and experience. Here, then, we begin by describing the historical background to the understanding of avian nest-building behaviour and then go on to examine more recent data that begin to provide some of the first evidence for acquisition by birds of nest-building knowledge.

Classical methods to examine the causal basis of a behaviour involve two approaches: 1) deprive the juvenile bird of access to adults so that it has no opportunity to see/experience the behaviour performed by experienced individuals. This is usually done by hand-rearing young birds, sometimes coupled with the isolation of each bird from same-aged conspecifics; and

Healy, S.D., Morgan, K.V., and Bailey, I.E., *Nest construction behaviour*. In: *Nests, Eggs, and Incubation*.
Edited by: D.C. Deeming and S.J. Reynolds, Oxford University Press (2015). © Oxford University Press.
DOI 10.1093/acprof:oso/9780198718666.003.0003

Table 3.1 Bird species for which there is a detailed description of their nest-building behaviour.

Species	Reference
American robin (*Turdus migratorius*)	Howell (1943)
Magpie goose (*Anseranas semipalmata*)	Davies (1961)
Crested oropendola (*Psarocolius decumanus*)	Drury (1962)
Village weaver (*Ploceus cucullatus*)	Collias and Collias (1962b)
Goldcrest (*Regulus regulus*)	Thaler (1976)
Firecrest (*Regulus ignicapilla*)	Thaler (1976)
Rufous-fronted thornbird (*Phacellodomus rufifrons*)	Thomas (1983)
White-winged snowfinch (*Montifringilla nivalis*)	Ambros (1989)
Brown cacholote (*Pseudoseisura lophotes*)	Nores and Nores (1994)

2) restrict the bird from any opportunity whereby it might 'try out' the behaviour. This can be difficult but in the case of nest building it can be achieved by ensuring that the bird has no access to material that may constitute appropriate nest material (this may, in practice, be a little more difficult than it sounds, see below). At a later date, one then assesses whether, in the presence of suitable nest material and, possibly a potential mate, the bird attempts to use the material to form a nest and, if so, what shape that nest takes. Wallace (1867) wrote of such deprivation experiments, describing birds that were raised in cages from eggs, which even when given appropriate nest-building material did not produce a nest that was characteristic of their species. He also suggested that the 'fair' testing of what the birds actually did was to 'turn out a pair of birds so brought up into an enclosure covered with netting, and watching the result of their untaught attempts at nest making.'

Early deprivation experiments with hand-reared rose-breasted grosbeaks (*Pheucticus ludovicianus*) and American robins (*Turdus migratorius*), which had had no experience of nesting materials, resulted in birds that were totally unable to produce a nest when presented with suitable material at adulthood (Scott 1902, 1904). Yet, other species appear to be more capable when deprived in an apparently similar way: domestic Atlantic canaries (*Serinus canaria*), for example, without early experience with appropriate nest-building materials apparently were later able to build a nest of species-typical form (Hinde 1958). Canaries with

either restricted access, i.e. material was provided for only 30 minutes per day, or deprived entirely of material were also observed to perform odd behaviours, including plucking their own feathers, unlike those birds that had had continuous access to building materials (Hinde 1958).

There is more than one reason why lack of early experience may result in first-time builders attempting to build a nest that is not species-typical or failing to build a nest at all. The first of these is that, rather than not 'knowing' what nest to build, the bird lacks the ability to manipulate material in appropriate fashion. That dexterity and practice of manipulating material makes a contribution to the shape of the final nest, and is affected by deprivation, comes from both observational data and deprivation experiments performed on captive, hand-raised village weavers (*Ploceus cucullatus*). Collias and Collias (1964) described a range of actions that young birds raised by their parents had to practise before being as capable as an experienced adult in tearing strips off suitable plants. Those actions, which included where to perch, which part of the grass to begin tearing, which direction to tear, how much to tear off, and so on, all appear to be behaviours that might be required for successful building.

Collias and Collias (1964) also described how young birds, whether deprived of suitable materials or not, begin mandibulating objects from a very early age. In the birds deprived of suitable materials this mandibulation sometimes took the form of weaving their own feathers or those of other birds (Collias and Collias 1964). The plucking of their own feathers by nest material-deprived canaries may also have occurred *in lieu* of learning the mechanical components required for nest building (Hinde and Harrison Matthews 1958). It is possible that it was the deprivation of materials suitable for mandibulating that caused young birds deprived of usual nest-building materials (palm strips or giant reed grass) to be less capable of ripping off strands than were undeprived conspecifics.

The possibility that young weavers may also learn about nest building (whether material choice, method of weaving or both is not clear) from watching the behaviour of nest-building adult males also came from observational data collected by Collias and Collias (1964). They described the establishment by these immature birds of 'play colonies' where they attempt to build nests (Collias and Collias 1964) and young male weavers will also visit nests of adult males as they build, often removing material. It is still unclear, however, why the young males do this and whether they remove the material from the nest of a related male who may be more tolerant of the learning of a young relative. Thus far, these observational data form the only evidence that social learning may play a role in some component of nest-building behaviour, perhaps even in shaping the specifics of the nest structure itself (but see below for a role for social learning in nest-building decision-making).

That early experience has an impact on nest-building decision-making was also provided by a series of experiments conducted by Sargent (1965). These were designed to test the specific suggestion of Wallace (1867) that young birds while in the nest and when recently fledged, have the opportunity to examine the detail of their natal nest and the materials from which it was built. As the young birds forage near to the nest in the early post-fledging days, Wallace (1867) also suggested that this provided an opportunity to compare available material with that contained within the nest. Sargent (1965) considered that there were three different sources of information that might be both available and used by a young bird learning how to build a nest: 1) the materials and the construction of the nest; 2) the substrate immediately surrounding the nest; and 3) the wider habitat in which the nest is located. He included these two latter manipulations because there had been reports of small groups of individuals choosing unusual nesting substrates over multiple generations. For example, western yellow wagtails (*Motacilla flava*) were observed nesting in shrubs rather than on the ground (Schiermann 1939) and redheads (*Aythya americana*) were seen to be nesting on marshland rather than in emergent vegetation over water (Hochbaum 1955). More evidence (including some experimental data) that apparent innate preferences for nest habitats could be modified by experience (Klopfer 1962, 1963) is provided by detection that the natal philopatry of birds following 'transplantations' of eggs or young from one place to another was to sites in which they were reared rather than to those in which they hatched (Välikangas 1933; Löhrl 1959).

Sargent (1965) manipulated each of the three sources of information in turn by providing nest-building captive zebra finches (*Taeniopygia guttata*) with strands of burlap of one of three colours (green, brown, or red) with which to build the nest. The birds also had access to a plywood box or a wire-mesh canary nest cup in which to build, which was located either inside the birds' cage or outside the cage in an enclosed extension accessible from their cage. Sargent (1965) compared the choices birds made when building their first nests with the features of their natal nest. Although birds raised in nests of each of the three colours preferred brown

burlap, the preference for brown of those birds that had been raised in green nests was much less than it was for birds in the other two groups. That natal nest colour need not affect material colour preference, however, was evidenced by birds raised in red nests: they preferred brown burlap as strongly as did the birds raised in brown nests.

Rearing experience also appeared to affect habitat choice: zebra finches of Sargent (1965) preferred to nest in the habitat type in which they were raised although their choice of nest location was not apparently so flexible. All birds preferred to nest in cups rather than boxes, although the strength of this effect seemed variable, depending on the combination of experience and the way in which the options were presented in the test conditions, i.e. at first nesting. The effect of habitat usually overwhelmed the effects of preferences for substrate type.

Sargent (1965) set an early high standard in experimental design, not least in his thorough exploration of sources of information the birds might use in their nest-building decisions. He also differentiated between experience gained from hatching and nestling periods to that gained during the fledgling stage. Although he did not test all the possibilities, Sargent (1965) provided some compelling evidence that experience gained during the fledgling stage has a significant impact on habitat preferences at first nesting. He also looked at the choices of birds building their second nests. Although birds appeared to choose options much as they had for their first nests, they tended to choose more 'natural' or species-characteristic nest materials and situations. This latter result may have helped to shape his ultimate conclusion that for nest building, experience played little role relative to innate predispositions. We have described the work of Sargent (1965) in detail not only because his thorough experimental exploration of the assertions of Wallace (1867) regarding the role of instinct, or lack thereof, on nest-building behaviour is a model of experimental design, but also because his data were the last substantial examination of the influence of experience, i.e. learning and memory, on nest-building decision-making.

By the mid-1960s the work of Collias and Collias (1964) on behaviours such as the weaving of weaverbirds involved in the building of a nest and that of Sargent (1965) on the decisions involved in nest building had both produced evidence of experience-dependent effects. It is perhaps ironic, then, that both studies probably helped to cement the then widespread, and now trenchant, view that nest building by birds is 'instinctive', 'innate' and entirely 'genetic' (each of these labels is used). This is because those data do not allow exclusion of a significant genetic contribution to the observed building behaviours and decisions. Evidence of the hormonal control of nest building by various species, i.e. red-billed quelea (*Quelea quelea*; Butterfield and Crook 1968), Atlantic canaries (Warren and Hinde 1959; Hinde and Steel 1972), pigeons (common pigeons, *Columba livia*, Cheng 1973; ring doves, *Streptopelia risoria*, Martinez-Vargas and Erickson 1973), and zebra finches (Rochester et al. 2008), which showed that testosterone, oestrogen, and progesterone all contribute to motivating birds into nest building, only added further weight to this view. As did the outcome of administering one or other of these hormones, which may lead to the initiation of nest building in the mate of the sex that is not the usual nest builder (Steel and Hinde 1972).

3.3 Habitat choice

Contrary to the assumption that there is little or no flexibility in the behaviour of nest building or in the structure that results, there is one aspect of nest building for which there soon came overwhelming evidence that birds will often modify their behaviour with experience. As Sargent (1965) showed with his experiments, alongside the data from others before him, birds will change the location of their nests based on their experience of the success, or not, of their own previous nest-building attempts. Predation is thought to be the key factor in whether or not birds raise their young successfully (Ricklefs 1969; Martin 1995; Lima 2009) and there is evidence that birds move their nest sites in response to a realized predation risk to their eggs and nestlings (Shields 1984; Sonerud 1985; Haas 1998). Following depredation of a nest a bird may disperse for the next attempt (Howlett and Stutchbury 1997) or for the following season (Dow and Fredga 1983; Doligez et al. 1999). Sometimes birds will both disperse and change nest-cover type (Greig-Smith 1982). Sometimes birds do not go so far as to leave their territory and will reuse a nest site but not in the same season (Styrsky 2005).

Birds may also choose nest sites that reduce the perceived predation risk to their eggs and nestlings. For example, Siberian jays (*Perisoreus infaustus*) respond to taped calls of corvids (egg and nestling predators) by moving to build their nests to safer but cooler sites (Eggers et al. 2006). Moreover, migratory passerines did not settle in a habitat sprayed with water containing American mink (*Neovison vison*) urine and faeces to simulate an increased mammalian predator presence (Forsman et al. 2013). In some cases the apparent predation risk originates from the presence of humans: black redstarts (*Phoenicurus ochruros*) shifted their nests deeper into nest cavities after researchers had disturbed them (Chen et al. 2011).

Predation is not, of course, the only hazard for nest-building parents and birds also choose their initial nest location, and may change it subsequently, so as to mitigate against environmental challenges to their eggs and offspring (see further details in Chapter 5). For example, dabbling ducks (*Anas* spp.) in west-central Montana build their nests so that the surrounding vegetation does not block the sun in the morning but offers shade in the afternoon when the temperatures are considerably higher (Hoekman et al. 2002). Whether these birds learn to do this is not yet clear.

At least some birds can respond to variation in more than one factor causing nest failure. Pinyon jays (*Gymnorhinus cyanocephalus*), for example, will move their nests away from exposed sites (these are warmer due to higher exposure to the sun) only after a nest has failed due to predation. Those birds that first nest in a concealed site will only move to an exposed site when their concealed nest has failed due to the impact of snowfall (Marzluff 1988).

Finally, some species respond to positive memories rather than just by avoiding their previous failures. Mountain bluebirds (*Sialia currucoides*), when given a choice of nestbox type, prefer to return to nest in box types in which they have previously successfully raised a brood (Herlugson 1981).

Success or failure estimates may also be gained from observing heterospecifics' choice of nestbox, even when the characteristics that are copied or avoided appear to be arbitrary. Seppänen and Forsman (2007) showed that European pied flycatchers (*Ficedula hypoleuca*) and collared flycatchers (*Ficedula albicollis*) will choose nestboxes with either a white triangle or square painted around the nest box entrance based on whether the boxes are occupied, or not, by great tits (*Parus major*), blue tits (*Cyanistes caeruleus*), or coal tits (*Periparus ater*). A follow-up study showed that this preference appeared to be based on the success of the tit parents as the more offspring their box contained, the more likely the flycatchers were to choose a box of a similar type (Seppänen et al. 2011). These flycatchers continue to acquire information regarding nesting success of conspecifics through inspecting conspecific nests after a simulated predator visit (Thomson et al. 2013). A more commonplace use of social information is seen in black-legged kittiwakes (*Rissa tridactyla*), which nest on cliffs that vary in their quality across time. New birds will recruit to nest on the cliffs that had been the most productive the previous year. Furthermore, the success of surrounding birds seems to have a greater impact on cliff choice than does a bird's own reproductive performance in any one year, with birds being much more likely to disperse from the cliff on which they had a poor reproductive event when others on the same cliff also bred poorly (Danchin et al. 1998).

Slagsvold et al. (2013) used cross-fostering to show that some part of the preference for nestbox type in tits, at least, is due to imprinting. Although blue tits reared in a small nestbox by blue tit or great tit parents

preferred small over large nestboxes when they nested themselves, blue tits cross-fostered by great tits in a large nestbox much preferred to nest in a large nestbox. Great tits fostered to blue tits also preferred the size of the nestbox in which they were raised.

3.4 Material choice

Typically, although not without exception, birds' nests are layered structures consisting of (an) outer shell(s) of coarser material and (an) inner layer(s) of

Figure 3.1 Various components of nest building by male cape weavers (*Ploceus capensis*): (a) an early stage of building; (b) a male destroying a nest he had been building; and (c–h) six nests built by the same bird showing variation in nest construction. (Photos: I.E. Bailey. See Plate 5.)

finer, lining material (Hansell 2000). The function of the outer part of the nest is to act predominantly as a container to prevent eggs and birds from falling out and perhaps also to provide some protection against adverse effects of climate and predation (Dixon 1902; Collias and Collias 1984; Hansell 2000). The lining of the nest probably provides further insulation against the climate (Mainwaring et al. 2014a), may in some instances provide cushioning to prevent eggs from moving around and cracking against each other or against harder nest materials, and may also aid in nest hygiene (Winkler 1993; McGowan et al. 2004; Dawson et al. 2011; Suárez-Rodríguez et al. 2013; Biddle et al. 2015).

Although nest structure may be roughly species-specific (Ferguson-Lees et al. 2011), there may also be considerable intraspecific variation in nest composition and morphology (Allen 1878; Brewer 1878; Cresswell 1997; Walsh et al. 2010; Britt and Deeming 2011; Álvarez et al. 2013; Chapter 4; Figures 3.1 and 3.2). There is an untested suggestion that there is greater intraspecific variation in the constitution and structure of nests built in vegetation above the ground or adherent to walls or cliffs than are nests built by ground-nesting species (Nickell 1956). In some cases, regional variation in intraspecific nest morphology may simply reflect the abundance of different materials in the local habitat (Álvarez et al. 2013). In other instances, individuals appear to be more selective about their nest materials. For example, killdeers (*Charadrius vociferus*) prefer to use white material to build their nests even when in limited availability (Kull 1977). Other species add pieces of particular species of aromatic plants to their nests. The variability in preference for plant species does not seem to reflect plant species' abundance (Gwinner and Berger 2008; Dykstra et al. 2009; Mennerat et al. 2009a), but rather they have other functions (see Chapters 4 and 7).

It is still unclear what causes variation in preference for plant materials by birds. Like Sargent (1965), others have since suggested a role for imprinting (Muth and Healy 2013). However, material preferences may be based on genetic predispositions, reflect innate responses to climate conditions, or be due to a bird learning the structural or functional features of the material. This may be either through experience of manipulating the material, or associating the outcome of using one material or another with subsequent reproductive success (Muth and Healy 2011, 2014; Bailey et al. 2014).

There are a number of observations from free-living birds in which nest composition and structure vary

Figure 3.2 Six nests built by different male zebra finches (*Taeniopygia guttata*). These males built these nests when they had been provided with four different types of string (all 15 cm long), string that varied in flexibility from stiff to very flexible. Although the males each had access to similar quantities and types of nesting material (i.e. stiff string), these nests demonstrate variation in the nests built by different zebra finch males. (Photo: K.V. Morgan).

intraspecifically with the climate in which the nest was constructed as influenced by latitude (Kern and van Riper 1984; McGowan et al. 2004; Mainwaring et al. 2012, 2014b; Chapter 4) or altitude (Kern and van Riper 1984). These data do not, however, allow us to differentiate between a number of possible explanations for why birds choose different materials: 1) materials differ in their availability as habitat changes; 2) birds have a population-specific genetic predisposition for choosing different materials; or 3) birds learn which materials 'work' best for them. We also do not yet know whether this variation enhances reproductive success.

At least some species can also learn through experience which materials are best for nesting. For example, zebra finches will learn to avoid nest material if it has structural properties that make it difficult to use for building (Bailey et al. 2014; Figure 3.3). They will also learn to choose nest material of a non-preferred colour once they have successfully produced a brood from a nest of that colour (Muth and Healy 2011; Figure 3.4). Some preferences are more difficult to explain. Although zebra finches often strongly prefer material of one colour over another, preferences are not repeatable across populations. Sargent (1965) found that zebra finches much preferred brown over green material while those studied by Muth and Healy (2011) much preferred green over brown when offered a choice.

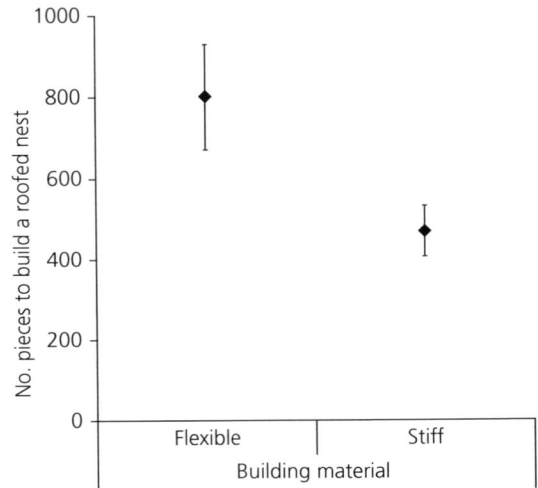

Figure 3.3 The mean number of pieces of either flexible or stiff string male zebra finches (*Taeniopygia guttata*) use to build nests. Modified from Bailey et al. (2014).

Muth et al. (2013) also found that while zebra finches strongly preferred blue over yellow or red nest material, they did not have any colour preferences when they were offered a choice of food of the three different colours.

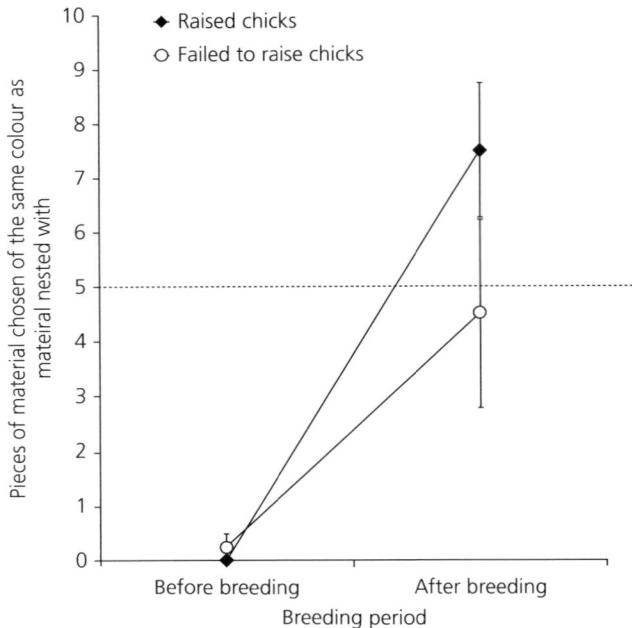

Figure 3.4 In an experiment in which male zebra finches (*Taeniopygia guttata*) built their first nest using material of a colour they did not prefer (preference before breeding), the birds that raised chicks subsequently switched their preference to this initially non-preferred colour (filled symbols). Birds that failed to raise chicks did not switch their preference for material colour (open symbols). Modified from Muth and Healy (2011).

3.5 Nest composition

3.5.1 The outer shell

The composition of the outer shell of the nest may be made almost entirely of a single material type or may be a composite structure made of multiple material types (Hansell 2000). The most basic of nests are shallow scrapes on the ground, such as those built by least terns (*Sterna antillarum*; Stucker et al. 2013), little ringed plovers (*Charadrius dubius*; Kubelka et al. 2014), and common ostriches (*Struthio camelus*; Hanley et al. 2013). In these cases nest-site and material selection are much the same thing as little or no material is moved from elsewhere and added to the nest. Substrate type and shelter are important to such nesting decisions; black-tailed gulls (*Larus crassirostris*) that lay eggs in those locations concealed by vegetation have low rates of egg loss (Lee et al. 2010). Camouflage partly explains this success because eggs are a better match to their background allowing higher egg survival than in areas with low vegetation and poor concealment. Under experimental conditions Japanese quail (*Coturnix japonica*) can select among locations based on the colour of the substrate, as they will lay their eggs in locations that are the best possible match to the colouration of their eggs (Lee et al. 2010; Lovell et al. 2013; Chapter 11). This suggests that the birds learn the colour of their own eggs and choose the nest substrate that will best camouflage them. Alternatively, birds exhibit a preference for nest-substrate colour and the colouration of their eggs that are both genetically determined and genetically linked.

Numerous species build a nest with the outer part of the nest formed from only one material type. Some of the structurally simplest of these are mound or platform nests of Adelie penguins (*Pygoscelis adeliae*), which build unlined mounds of gravel (Williams 1995), or James's flamingoes (*Phoenicoparrus jamesi*), which build mounds of mud (Johnson et al. 1958). At the other end of the continuum of structural complexity, the hanging basket nests built by weaverbirds (Ploceidae) have an outer shell built entirely of long and strong vegetation, usually reed, grass or sedge although some species use strips of palm frond or fine twigs (Crook 1963). The village weaver prefers to build using long green material, a preference that seems to be at least partially learnt. In captivity younger birds are more accepting of less appropriate building materials, such as tooth picks and raffia, than are older more experienced nest builders (Collias and Collias 1962a, 1964). This preference for long fine material can be fatal

when the weavers build with synthetic materials such as string, which can entangle both adults and young alike (Oschadleus 2012).

Why these species-specific preferences develop, or have evolved, is not clear. Although the assumption is that they are fixed, there are instances in which it appears that nest-building birds assess the structural properties of the materials with which they choose to build (Bailey et al. 2014). This is especially so when birds use human-produced materials. For example, although barn swallows (*Hirundo rustica*) typically build a cup-shaped nest out of balls of mud, they will also build with a mix of cement, sand and gravel (Al-Safadi and Al-Aqas 2006). While the usual interpretation would be that this choice is accidental, it is also possible the birds learn that the cement dries harder and stronger than do other materials and so choose it preferentially. Similarly, the addition by common house martins (*Delichon urbicum*) of polysaccharides delivered in their saliva to the clay with which they build their nest increases the compressive strength of the clay. Nest strength may be an important building criterion for these birds and they may learn how to enhance that component of their nest (Silva et al. 2010).

Many species build a nest with an outer shell containing a mix of materials, which may vary widely. For example, the structural components of nests of many small species, such as hummingbirds (Trochilidae), are built using moss and spiders' webs (Calvelo et al. 2006). Spider's web looped around the moss seems to act like Velcro® might to hold the structure together (Hansell 2007). The nests of larger species such as thrushes (*Turdus* spp.) are, on the other hand, made largely of mud and dried grass, which may in the case of the common blackbird (*Turdus merula*), at least, increase insulation (Mainwaring et al. 2014b).

To date, there is, however, rather little information on whether using a mixture of building materials affects the mechanical properties of nests and that birds choose a combination of materials for this reason. Analysis of the co-occurrence of the most common material types (silk, grass stems, grass leaf, and sticks) used by species of the Fringillidae, Tyrannidae and Corvidae in building their nests did not reveal any distinct categories of material type (Hansell 2000). Such a categorization would have suggested that birds used groups of materials with complementary structural properties. As Hansell (2000) pointed out, however, the lack of such evidence might have resulted from coarse nest dissection protocols rather than whether birds choose an appropriate structural mix of materials (see Chapter 4).

While a mixture of materials could aid a nest's structural properties, it is also possible that using a mixture of nest materials in the outer shell may aid nest crypsis via disruptive camouflage as the different colours and textures of the materials could help to break up the nest outline (Hansell 2000; Bailey et al. 2015). Although there is much evidence that birds respond to predation risk by choosing appropriate locations, which results in improved concealment of either the nest (Westmoreland and Best 1985) or the eggs (Lovell et al. 2013), it is perhaps surprising that there are few data to show that birds choose materials to camouflage their nests. There is some anecdotal evidence such as the addition by long-tailed bushtits (*Aegithalos caudatus*) of a thin, apparently decorative, layer to the outside of their nests, which is composed of pale reflective materials such as lichen, spider cocoons and polystyrene. This may help the nest to blend into its background (Hansell 1996). The addition of carnivore scat to nests by common waxbills (*Estrilda astrild*) may also play a role in reducing nest predation either by masking the scent of the nest or by repelling nest predators (Schuetz 2005; Chapter 5). To our knowledge, however, the only experimental evidence to date is provided by zebra finches building with coloured paper in the laboratory: the birds preferred to build with material that was a colour match to their nest site than to build with material of a contrasting colour (Bailey et al. 2015; Figure 3.5).

3.5.2 Nest lining

Nest lining consists of material(s) on the inside that differ from those in the outer layer(s) (Hansell 2000). While not all nests are lined, those that are vary greatly in the material used. The song thrush (*Turdus philomelos*) builds a nest with an outer layer of grass, moss and mud but lines the nest with a thin layer of well-decomposed wood or bark (Ferguson-Lees et al. 2011). The nest of the common blackbird, however, which is very similar externally to the song thrush nest, is typically lined with fine grasses (Ferguson-Lees et al. 2011; Mainwaring et al. 2014b). In some nests it may be difficult to determine whether the inner layer fulfils a lining or structural role. For example, in weaver bird nests, the inside of the nest roof may consist of overlapping leaves, which act in much the same way as do roof slates by allowing rainwater to run off. These leaves may also play a structural role by helping to hold the nest walls apart (Crook 1960, 1963).

Many small birds will also add wool or artificial wool-like fibres to their nest linings, the use of which appears to be opportunistic as birds building closer to the source use more artificial material than do those building further away (Surgey et al. 2012). Long-tailed bushtits line their nests only with feathers, which constitute 41% of the overall nest mass. They also respond to variation in air temperature with a smaller mass of feathers being used

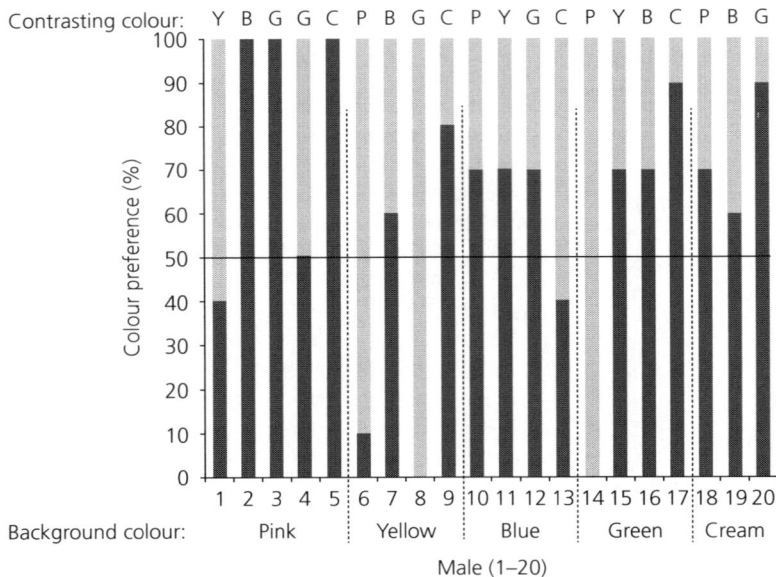

Figure 3.5 In an experiment in which the nest cup and background of the cage were covered with coloured paper, nest-building male zebra finches (*Taeniopygia guttata*) typically preferred to build their nests with materials that matched the colour of the background to nests (dark portion of bars). The colours of background and material were pink, yellow, blue, green or cream and males were presented with materials that either matched the background or were of a contrasting colour (light portion of bars). Modified from Bailey et al. (2015).

in nests built later in the season when it is warmer (McGowan et al. 2004). In an experimental manipulation of the number of feathers in the nests of tree swallows (*Tachycineta bicolor*), chicks from featherless nests grew more slowly and had more parasites than did chicks from feathered nests (Chapter 8). This is possibly because the young birds were colder and had less energy available for growth and immune responses (Winkler 1993). The colour of the feathers chosen by barn swallows appears to be important, as birds prefer white feathers and nests containing more white than dark feathers have a higher hatching success (Peralta-Sánchez et al. 2011). There may be a trade-off between the benefit of feathers for increasing insulation and a cost of increasing predation risk because they increase the conspicuousness of the nest; this idea is supported by the greater incorporation of feathers by hole-nesting than by cup-nesting species (Møller 1984). It is not clear, however, that conspicuousness alone explains any effect of feather lining on predation risk. The addition of breast feathers of common pheasants (*Phasianus colchicus*) to the lining of open cup nests, but not to nests in cavities, increased the predation risk compared to nests without feathers. The addition of fur of European hares (*Lepus europaeus*) to the same nests, on the other hand, had no effect on predation (Møller 1987).

Birds may also trade off the predation risk associated with the conspicuousness of the material with which they line their nests if the materials act as a signal of adult quality to conspecifics. For example, black kites (*Milvus migrans*) add bright white material, often pieces of plastic, to nests. Although decorated nests are more visible and therefore more vulnerable to predation, conspecifics intrude less often into territories containing highly decorated nests than they do into territories with undecorated nests (Sergio et al. 2011).

Birds may also include aromatic substances, usually green plant material but on occasion even material like cigarette butts (Dykstra et al. 2009; Suárez-Rodríguez et al. 2013; Tomás et al. 2013; Chapter 8), when lining their nests. These substances contain chemicals with anti-parasitic and anti-microbial properties that, when added to nests, can improve chick survival (Gwinner and Berger 2008, Suárez-Rodríguez et al. 2013). The choice of species of aromatic materials that common starlings (*Sturnus vulgaris*) and blue tits add to their nests appears to be learnt as birds do not choose herbs relative to their abundance, do not all choose the same herbs, and tend to choose herbs they experienced in their natal nests (Gwinner and Berger 2008; Mennerat et al. 2009a).

3.6 Nest structure

In his review of the roles played by predators on breeding birds, including their nest-building decisions, Lima (2009) noted that there was little evidence that birds changed the structure of the nest they built in response to a perceived risk of predation. Comparative evidence, however, suggests that predation has been a selection pressure not just on choice of where to locate the nest but also in shaping the nest structure itself (Chapter 5). Collias (1997) suggested that those species forced to move to the ground because of competition for nest space would face a greater risk of predation from ground predators. Collias (1997) also proposed that these species would be expected to add a dome to their original cup nest. Indeed, in the babblers (Timaliidae), those species that build on the ground are highly likely to add a dome roof to the basic cup nest, which is not true for those babbler species that build their nests in trees (Hall et al. 2015). Furthermore, it appears that the ancestral state was to build a cup nest off the ground.

Although there are still no data (to our knowledge) that birds will respond to depredation of their own nest by modifying their nest structure, the often long (up to 30 cm) tube entrances to the nests built by some African weavers have often been considered to provide protection from a range of predators, including snakes. This is, in part, because the nest is often pendant from a single relatively fine twig. When Crook (1963) described 62 species of ploceine weavers from Africa, India and the Seychelles, he included anecdotes of snakes slipping off the ends of nest tubes and raptors pulling apart the ends of nest tubes but failing to reach the egg chamber. He concluded, however, that the intraspecific variation in the length of the nest tube was not obviously correlated with variation in predation risk. We suggest that there is more work to be done on explaining nest structure in the weavers.

House wrens (*Troglodytes aedon*) may make structural changes in response to perceived predation risk. Stanback et al. (2013) showed that birds that built in boxes with relatively large entrance holes built up a taller pile of sticks between the entrance hole and the nest cup than did wrens nesting in boxes with relatively small entrance holes. It appears that these birds, at least, can use the size of the entrance hole to assess their vulnerability to possible sources of nest failure, such as predators or brood parasites, and adjust their nest morphology appropriately.

There is increasing evidence for intraspecific structural variation in nests that correlate with variation in weather conditions (Britt and Deeming 2011; Crossman et al. 2011; Mainwaring et al. 2012, 2014b). It is not clear yet whether this variation is due to the responses of individuals to the weather they experience or to natural selection. The construction of a well-insulated nest can reduce both the energetic expenditure required during incubation and the rate at which eggs cool when the parents are away from the nest (Drent 1975; Collias and Collias 1984; van Dijk et al. 2013). It appears that birds usually manage this not by changing the materials they use but by simply adding more material (Crossman et al. 2011; Mainwaring et al. 2014b).

Whether social information plays any role in birds building the nest remains a mystery. It may be that some species modify their nest structure in response to others and the use by great tits of hair to cover their eggs may be such an example. In Finland great tits that have European pied flycatchers as near neighbours fledge fewer offspring than do great tits without such heterospecifics nearby (Loukola et al. 2014). Furthermore, the more offspring in a great tit's nest, the more likely it is that the flycatchers will choose to nest nearby. Loukola et al. (2014) suggested that to reduce the ability of the flycatchers to assess the number of eggs in the nest and therefore to dissuade them from becoming neighbours, great tits cover their eggs with hair. It is unclear, however, whether this functions only as a deterrent to harmful neighbours as great tits will also cover their eggs in areas free from flycatchers.

If there are relatively few data on intraspecific structural variation in nests, there are even fewer concerning what birds do in order to produce that structure. Since the studies by Collias and Collias (1962a, 1964) on weavers, and by Thorpe (1956) on long-tailed bushtits, provided detailed observations on the specific behaviours these birds perform when building their nest, there has been little work to investigate how birds 'know' what structure to build. This is perhaps because detailing nest-building behaviour requires high definition video recordings accompanied by a prodigious effort in their analysis. Determining whether or not birds use a stereotypic sequence of movements when nest building also requires recording of birds building more than one nest. Solitary weavers are a very useful taxon in this regard because they will build multiple nests in a single season. Zebra finches in the laboratory are also useful as they will build 'on command', i.e. as soon as paired and provided with nest material. The logistic issues surrounding the requisite data collection may mean that describing and quantifying the building behaviour of species that build a single nest in each breeding season will prove much more elusive.

Using such detailed video data, however, Walsh et al. (2013) provided a possible explanation for their finding (Walsh et al. 2010) that, for both village weavers (*Ploceus cucullatus*) in Nigeria and southern masked weavers (*Ploceus velatus*) in Botswana, nests built by individual male weavers decreased in size as the season progressed. Walsh et al. (2010) proposed that this was either a response to changing weather conditions or to smaller clutch sizes. The video data showed that the birds also became more dextrous with more building resulting in them dropping diminishing quantities of grass (Walsh et al. 2011; Figure 3.6). Nest-building males may also have woven the grass more tightly as their weaving improved but there are no data to confirm this. The video data also provided an explanation for the lack of repeatability of nest structural characteristics within and across male weavers because individual birds were not stereotypic in the movements they used to build. When building, males also do not follow any obvious behavioural sequence from the beginning to the finishing of a nest as they continue to add new material and to work on older nests. This is the case even when nests are sufficiently complete that they are occupied by adult females and young (Walsh et al. 2013).

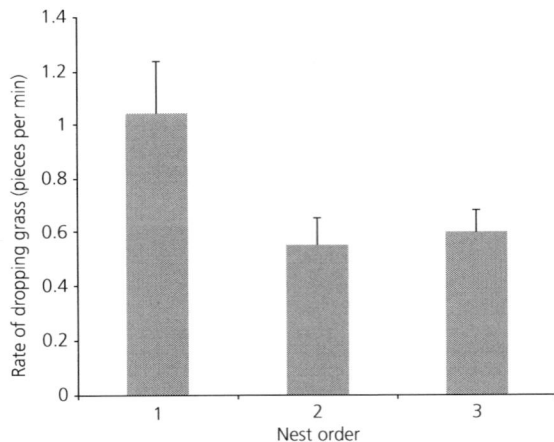

Figure 3.6 The mean rate (± 1 S.E.) at which male southern masked weavers (*Ploceus velatus*, n = 7) dropped pieces of grass during the attachment phase for three successive nests. Modified from Walsh et al. (2011).

3.7 Neural basis of nest building

Concomitant with the recent increase in effort addressing the behaviour and decision-making involved in building nests, work is also beginning on the neural underpinnings of nest building. In zebra finches, at least, activity (as measured by expression of the immediate early gene c-*fos*) increases in several brain regions the more nest-building activity a male undertakes: the anterior motor pathway, the social behaviour network, and the dominergic reward system (Hall et al. 2014). Increased activation, however, is not observed in the posterior motor pathway. The differential gene expression in the anterior but not posterior motor pathway suggests that nest building differs from other motor actions as the anterior motor pathway is specifically involved in motor learning and sequencing (Feenders et al. 2008). More general motor actions are represented by activity in the posterior motor pathway. Gene expression data also suggest that while taking material to the nest and spending time in the nest with his mate is a rewarding behaviour for a nest-building male zebra finch, tucking the material into the growing nest is not. Activation in dopaminergic cells in the ventral tegmental area decreases the more males tuck material into the nest (Hall et al. unpublished data).

The cerebellum, a much-folded brain region involved in complex motor actions, especially manipulative skills, also seems to have a role in nest building. A significant proportion of the variation in the degree of cerebellar folding across species is explained by the complexity of the nest built (Hall et al. 2013; Figure 3.7). How the cerebellum is involved when birds are building their nests is not yet clear. This is just one of the many questions still to be answered with regard to the relationship between the brain and nest-building behaviour.

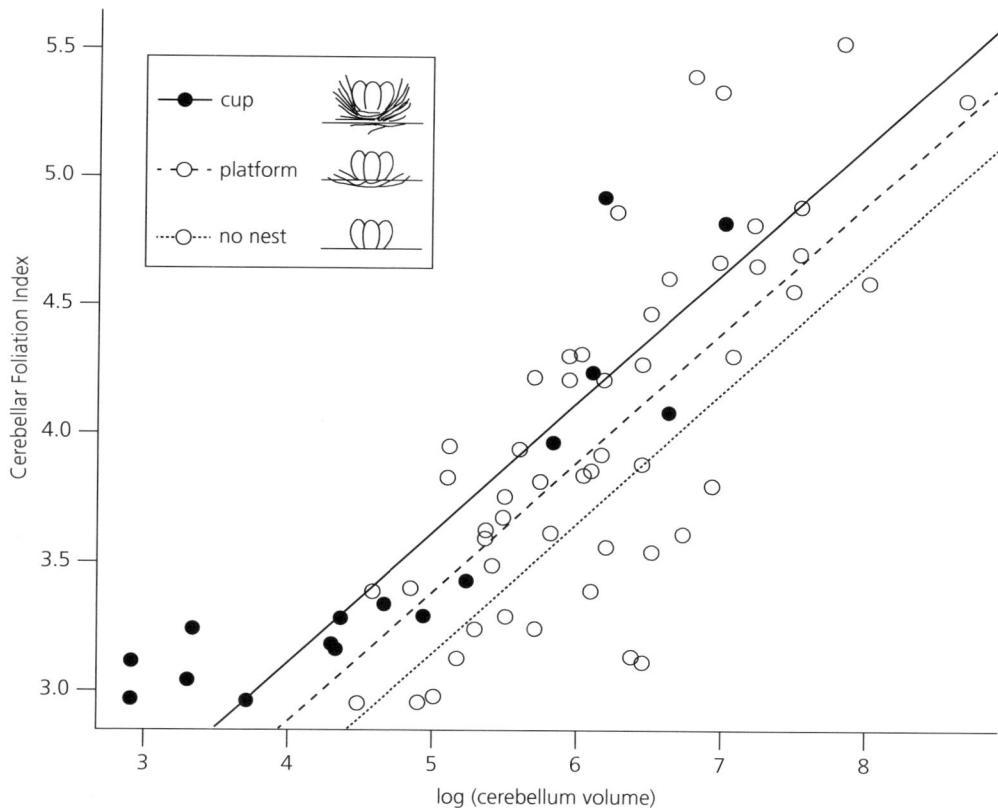

Figure 3.7 Regression lines of cerebellar foliation index on cerebellum volume (log transformed) and CFI for bird species that build a cup nest (filled circles, solid line), a platform nest (open circles, dashed line) or no nest (open circles, dotted line). Modified from Hall et al. (2013).

3.8 Conclusions

For such an important component of a bird's reproductive success, we are still some distance from a full understanding of the behaviours and decision-making of nest-building birds. While there is much evidence showing that birds make decisions over where to locate the nest, modifying their choices with respect to their own experience and with regard to the choices of both conspecifics and heterospecifics, we know much less about the sources of information and decisions birds make over material choice and construction of the nest. It is still unclear, for instance, the extent to which individual birds can use environmental and social cues to fine tune the morphology of their nest to prevailing conditions. It is possible that dependence on learning some or all nest-building attributes is high in some species but not in others. We also need more data on the informational feedback (e.g. building success or reproductive success) that drives whether birds learn about nest design or the value of some materials over others. These are all questions concerning an individual's own experience of nests or nest building but nest building is also a behaviour that might be shaped by observation of others or of the nests of others. We have much to learn about the role of social learning and whether (and if so, what) birds can learn about nest building from watching others build and inspecting the products of their nest-building efforts. We also know nothing at all with regard to the decision-making of birds that build in pairs or groups and how they coordinate their building efforts.

Given the apparent phenotypic similarity between tool use and manufacture in birds and nest building, we suggest that there may be merit in comparisons of the cognitive demands imposed by tool manufacture and those imposed by nest building. Perhaps the first obstacle to overcome is the still trenchant view that while tool manufacture is a product of complex cognition, nest building is entirely genetic. Although we are some considerable distance from fully understanding which aspects of nest building are entirely genetically predetermined and which involve learning from experience, there is increasing evidence that learning plays a role in multiple aspects of nest building. Finally, it may be surprising how little we understand about the structure of the nest itself. For example, how birds might use different materials and combine materials to refine properties of their nest such as insulation, rigidity and waterproofing (see Chapter 4). We will need a combination of both carefully designed field and laboratory experiments with long-term observational data to address these myriad questions.

Functional properties of nests

D.C. Deeming and M.C. Mainwaring

4.1 Introduction

All birds lay eggs but their nests exhibit a great deal of diversity in their size and design, although the various functions of birds' nests have long been overlooked. It is clear that a bird's nest has a number of primary roles which can simply be summarized as a location for incubation of eggs through contact incubation, and as a receptacle for eggs during incubation and nestlings during rearing. Nests may also have other secondary roles related to various components of reproductive biology such as sexual signalling, environmental adjustment, crypsis against predation, and control of parasites (Mainwaring et al. 2014a), many of which are covered in other chapters of this book.

The emphasis of our chapter is not on the functionality of birds' nests *per se*, as this has been reviewed elsewhere (see Hansell 2000; Moreno 2012; Mainwaring et al. 2014a), but about the properties of the nest that allow these functions to be realized. For instance, if the nest has an insulative role then this chapter explores how this is achieved. There are previous considerations of nest construction and function (e.g. Skutch 1976; Collias and Collias 1984; Hansell 2000), but these have tended to be more descriptive, than quantitative, in their approach. Further, recent evidence suggests that rather than being genetically constrained and stereotyped, nest-building behaviours exhibit a great deal of plasticity (Bailey et al. 2014) and thus, this chapter also explores the role of plasticity in nest building and the evolutionary adaptation of nest designs. As we will demonstrate, this is an emerging field of research and we have divided the chapter into three general areas: nest construction, factors affecting the nest microclimate, and the structural properties of nests. The absence of significant research in this area is due, in no small part, to the fact that it is technologically challenging to quantify aspects of nest function, but as technological advances are being made ever more

frequently, we conclude by discussing where future research could be effectively directed.

4.2 Nest construction

Although details of the nest construction period of birds have been reported by a number of empirical studies, their findings have received little synthesis. Nest construction rates are rarely reported in great detail for birds and are difficult to quantify as nests often appear completed several days before eggs are laid in some instances, and are completed after the initiation of egg laying in others. Thus, it is difficult to establish when a nest is complete, with most studies examining nestbox-breeding birds and assuming that the period between the initiation of nest building and the day on which the first egg is laid constitutes the nest-building period. In this way, studies have shown that European pied flycatchers (*Ficedula hypoleuca*) in Spain took an average of 8.6 days to construct a nest with a mass of 24.2 g (3.15 g·day⁻¹), a rate that was not affected by experimentally forcing the females to construct a second nest (Moreno et al. 2008a). Smith et al. (2013) showed that nest construction periods of great tits (*Parus major*) were shorter than for blue tits (*Cyanistes caeruleus*) (16.0 versus 13.0 days, respectively), and were longer earlier in the season (as also shown by Mainwaring and Hartley 2008). Collection of the structural layer (mainly moss) took up a greater proportion of this time for great tits, whilst blue tits took longer to collect nest-lining materials (Smith et al. 2013). Nest building by marsh tits (*Poecile palustris*) in natural cavities was highly variable in its rate and was affected by both prevailing temperatures and the time of year (Wesołowski 2013). Whilst illustrative, these examples are restricted to a few secondary cavity-nesting passerine birds, presumably because of their amenability as 'study systems', yet this does limit our ability to draw any general conclusions about the causes and consequences of the duration in

Deeming, D.C. and Mainwaring, M.C., *Functional properties of nests*. In: *Nests, Eggs, and Incubation*.
Edited by: D.C. Deeming and S.J. Reynolds, Oxford University Press (2015). © Oxford University Press.
DOI 10.1093/acprof:oso/9780198718666.003.0004

nest construction periods. Consequently, there is scope for the better documentation of the duration of nest-building periods as this would enable comparative analyses to investigate whether the duration of nest building is influenced by predation risk and other factors, such as adult body size.

The energetics of nest construction have rarely been considered but the limited number of empirical studies suggest that it is an expensive behaviour (reviewed by Mainwaring and Hartley 2013). Mainwaring and Hartley (2009) found that food supplementation did not influence the duration of nest construction but did result in increased nest mass of blue tits. By contrast, food supplementation led to a significant reduction in the construction time for blue tits but not great tits, and nest size was smaller (Smith et al. 2013). Another study showed that Eurasian magpies (*Pica pica*) built bigger nests when food supplements were provided (de Neve et al. 2004). Thus, although the energetic costs of building a nest are likely to be less than laying and incubating eggs (see Chapter 13), which in turn are likely to be less costly than provisioning offspring (Mainwaring and Hartley 2013), we suggest that future studies should include the energetics of nest construction in any assessment of the energetics of reproduction.

For example, one argument for nest building being energetically costly is that there is evidence of the kleptoparasitism of materials from nests of the cerulean warbler (*Dendroica cerulea*) suggesting that there may be a value placed by other species on particular materials collected by hosts (Jones et al. 2007). By contrast, this may reflect that it is more convenient for kleptoparasites to remove materials previously collected by the target species. Such kleptoparasitism is relatively rarely reported (Slager et al. 2012) but it would be interesting to determine the prevalence of such behaviours in order to examine if kleptoparasitism is due to the costly nature of collecting nest-building materials.

4.2.1 Variation in nest size

Nests are often described and quantified in terms of their exterior diameter(s), depth, and the comparable values within the cup where the eggs are laid (Deeming 2013). Such data allow wall thickness to be calculated and nest mass is also often reported. Interest in these dimensions is growing because of their usefulness for inclusion in analyses that offer both inter- and intraspecific comparisons. Pertinently, the former may provide valuable insights into the evolution of the nest as a structure while the latter may allow a better understanding of the phenotypic plasticity of nest construction as a behaviour pattern.

On the basis of allometric relationships between nest dimensions and adult female body mass, Deeming (2013) modified the concept of the 'bird-nest incubation unit' proposed some years before (Deeming 2002b) to suggest that in larger species nests were simply there to act as platforms where incubation takes place under the feathers of the adult. As avian body size decreases this 'bird as incubator' role changes so the 'bird-nest incubation unit' becomes more relevant because the nest itself has key roles to play in the incubation process. In effect, this means that nests of species of varying body sizes are not directly comparable, perhaps limiting the value of interspecific studies of nest characteristics.

Many species construct a single nest for each breeding season which makes it hard to determine plasticity within an individual. However, an experiment to assess the effects of thermal environment on reproduction of captive great tits showed that nest construction exhibits plasticity (S. Caro and D.C. Deeming, 2013 unpublished data). Nests were removed from pairs of birds that subsequently constructed second and third nests. Nest mass was highest for the first attempt (Figure 4.1); the second nests were significantly smaller (repeated measures ANOVA: $F_{1,7} = 26.70$, $P = 0.001$), but did not differ in mass from the third nests and nor did they show a significant effect of temperature treatment ($F_{1,7} = 2.57$, $P = 0.15$). However, it is prudent to consider that the birds were not allowed to breed successfully and so their previous experience may have affected their willingness to invest in nest building.

From a behavioural perspective, nest construction has long been considered as something instinctive but there is increasing evidence that there is considerable plasticity in nest construction. This concept is fully discussed in Chapter 3 but as an example, male weavers (*Ploceus* spp.) construct several nests in sequence but they exhibit variation as the sequence proceeds: as more nests are constructed, they are shorter and lighter in mass (Walsh et al. 2010).

Nevertheless, empirical studies have generally assumed that such variation occurs and have standardized the stage of reproduction at which nest dimensions are quantified, although some studies (e.g. Britt and Deeming 2011; Mainwaring et al. 2014b) assume that the mass of the nests at the end of breeding is equivalent to that at the start of clutch initiation. Dubiec and Mazgajski (2013) showed that in European pied flycatchers nest mass increased as the breeding period proceeded. Whether this is true for any other species

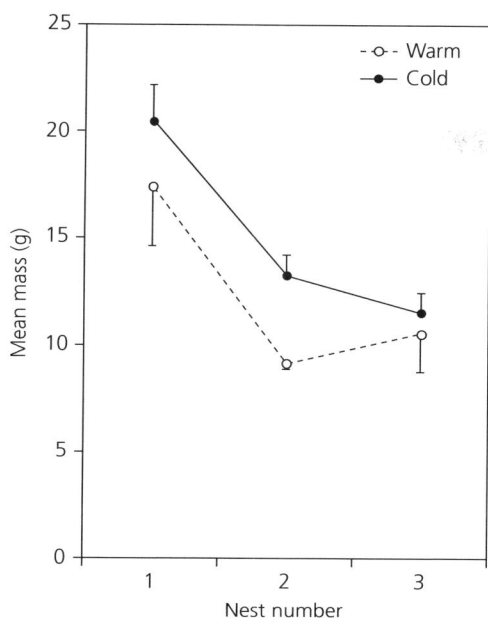

Figure 4.1 The effect of position in a sequence of nests constructed by the same birds on mean (± 1 SE) nest mass of great tits (*Parus major*). Birds were kept in temperature-controlled chambers prior to and during the breeding season with final temperatures of 18°C ('Warm') and 13°C ('Cold'). Sample size for warm and cold = four and five pairs of birds, respectively. (Source: S. Caro and D.C. Deeming, 2013 unpublished data.)

requires further detailed observations from a wide range of species, although further studies that quantify the amount of nest building that is performed after eggs are laid would be useful.

Nests can exhibit intraspecific variation in mass under different circumstances including latitude (Rohwer and Law 2010; Crossman et al. 2011; Mainwaring et al. 2012, 2014b), habitat (Álvarez et al. 2013), altitude (Kern and van Riper 1984), height above the ground (Kern 1984), time of year (Mainwaring and Hartley 2008; Britt and Deeming 2011), and year itself (Britt and Deeming 2011). Lent (1992) showed that nests of grey catbirds (*Dumetella carolinensis*) varied in size with year and within nesting habitat. This was attributed to concurrent variation in nesting substrate or materials and the ability of the birds to modify their nest structure to suit prevailing microhabitat conditions. The wood thrush (*Hylocichla mustelina*) exhibits variation in nest size with the wall width and exterior nest height decreasing as the breeding season progresses whereas cup width increased (Powell and Rangen 2000). Success of nests also appeared to be related to dimensions,

particularly of the nest cup which was smaller in the successful nests (which presumably produced fledged chicks but this was not clear in the report), and in nest height and wall width that were greater. Whether this reflected differences in the thermal properties of nests was unclear (Powell and Rangen 2000). In great tits in Spain nest dimensions positively correlated with various indicators of reproductive success, such as the number of fledglings, but were not influenced by the body size of the building female (Álvarez and Barba 2008). Meanwhile, nest-cup size also varied with laying date in the same population of great tits and nest-cup depth was positively correlated with hatching success but only in one of the two study years (Álvarez and Barba 2011).

Blem and Blem (1994) showed that nest depth and mass positively correlated with nestbox volume in the prothonotary warbler (*Protonotaria citrea*). Eastern bluebirds (*Sialia sialis*) nesting in boxes of different dimensions constructed nests of a greater volume within larger nestboxes (Pitts 1988). At Riseholme Park in Lincoln, UK, a preliminary study showed that blue tits provided with nestboxes with floor areas that were 30% larger than standard nestboxes (which measured 15 cm [l] × 12 cm [w] × 20 cm [h]; du Feu 2003) built nests that filled the nestboxes and weighed on average 39.2 g (SD = 10.8 g, $N = 8$; nests constructed between 2011 and 2014; D.C. Deeming, 2014 unpublished data). Birds breeding in nestboxes that had a floor area 80% larger than standard nestboxes built even heavier nests (mean = 55.2 g, $N = 2$). The impact of this imposed extra nest-building demand on clutch size appeared to be minimal—in the larger nestboxes the mean clutch size was 7.6 (± 2.7) eggs, which was comparable to standard-sized nestboxes but the negative correlation between nest mass and fledging rate approached significance (Spearman's rho = −0.71, df = 5, $P = 0.07$). Thus, extra nest construction activity may have a negative impact on reproductive success of blue tits and is worthy of further investigation.

Nest masses in Table 4.1 indicate that there is variation between geographical location and to some extent between years. At Riseholme Park in Lincoln, UK nest masses of great tits and blue tits were recorded annually from 2008 to 2014 (inclusive; Figure 4.2). There was significant variability in nest mass between years (ANOVA: $F_{6,235} = 2.56$, $P = 0.02$) but there was no effect of species ($F_{1,235} = 0.28$, $P = 0.60$) and there was no significant interaction ($F_{6,235} = 0.51$, $P = 0.80$). Furthermore, for all years there were no significant correlations between nest masses and hatching or fledging successes of blue tits (all r_ss < 0.1, df = 99, all

Table 4.1 The materials used by various open-nesting and cavity-nesting passerine birds to construct their nests at different geographic locations. Values for nest masses and nest components are given in grammes as means (± 1 SE [when known])—indicates that the material was not recorded. Artificial materials refer to human-derived materials such as pieces of string, cloth, various plastics, and cardboard.

Species	Location	Year	N	Nest mass	Mud
Open-nesting species					
Common blackbird (*Turdus merula*)	Northern Scotland, UK	2012	14	211.1 ± 28.2	143.4 ± 23.8
	North-west England, UK	2012	14	205.8 ± 25.0	133.8 ± 22.3
	Western England, UK	2012	15	182.5 ± 24.5	129.7 ± 20.7
	South-west England, UK	2012	14	174.1 ± 15.35	130.7 ± 14.4
	Eastern England, UK	2012	5	193.5 ± 26.7	176.5 ± 24.9
American robin (*Turdus migratorius*)	Southern Ontario, Canada	2009	5	161.3	126.1 ± 19.1
	Northern Manitoba, Canada	2009	5	221.4	168.7 ± 21.3
American yellow warbler (*Dendroica petechia*)	Southern Ontario, Canada	2009	5	4.9	-
	Northern Manitoba, Canada	2009	5	8.4	-
American goldfinch (*Carduelis tristis*)	Southern Ontario, Canada	2009	5	8.9	-
Common redpoll (*Carduelis flammea*)	Northern Manitoba, Canada	2009	5	15.4	-
Savannah sparrow (*Passerculus sandwichensis*)	Southern Ontario, Canada	2009	4	8.7	-
	Northern Manitoba, Canada	2009	5	9.7	-
Willow warbler (*Phylloscopus trochilus*)	Estonia	2004	1	21.8	-
Hedge accentor (*Prunella modularis*)	Eastern England, UK	2008	4	29.1 ± 1.2	-
Hole-nesting species					
Tree sparrow (*Passer montanus*)	Eastern England, UK	2008	5	70.8 ± 10.6	-
Great tit (*Parus major*)	Eastern Spain	2007	11	17.7 ± 1.6	-
	Eastern Spain	2007	10	10.7 ± 0.8	-
	Eastern Spain	2007	10	13.9 ± 1.7	-
	Eastern England, UK	2007	5	17.7 ± 5.6	-
	Eastern England, UK	2008	26	21.5 ± 1.0	-
	Eastern England, UK	2009	20	27.6 ± 1.3	-
	Eastern Spain	2010	6	10.7 ± 4.4	-
Blue tit (*Cyanistes caeruleus*)	Eastern England, UK	2007	5	14.1 ± 1.2	-
	Central Spain	2006	9	38.1	-
	Central Spain	2006	17	24.5	-
	Central Spain	2007	9	22.6	-
	Eastern England, UK	2008	12	26.6 ± 2.5	-
	Eastern England, UK	2009	9	26.8 ± 2.3	-
	North Wales, UK	2012	8	21.7 ± 1.6	-
	North Wales, UK	2013	8	30.6 ± 2.0	-
European pied flycatcher (*Ficedula hypoleuca*)	Central Spain	2006	71	28.9	-
	Central Spain	2006	24	21.8	-
	Central Spain	2007	39	34.5	-
	North Wales, UK	2012	8	22.6 ± 1.9	-
	North Wales, UK	2013	8	27.9 ± 2.1	-
Common redstart (*Phoenicurus phoenicurus*)	North Wales, UK	2013	8	40.6 ± 3.8	-

Moss	Lichen	Grass	Leaves	Twigs	Stems	Roots
8.4 ± 2.0	-	21.5 ± 3.9	13.1 ± 3.2	7.7 ± 3.3	5.6 ± 1.3	1.5 ± 0.4
8.8 ± 2.5	-	17.2 ± 1.8	16.0 ± 4.1	4.9 ± 0.9	4.8 ± 1.1	3.7 ± 1.6
4.5 ± 1.3	-	11.1 ± 1.9	13.3 ± 3.2	6.9 ± 2.0	5.6 ± 0.9	3.3 ± 1.8
3.4 ± 1.8	-	11.1 ± 1.9	13.3 ± 3.2	3.8 ± 1.0	3.8 ± 0.8	2.1 ± 0.6
0.3 ± 0.1	-	19.5 ± 2.6	4.1 ± 2.6	8.0 ± 2.8	-	0.2 ± 0.2
-	4.0 ± 2.4	17.7 ± 2.9	-	11.5 ± 0.4	-	-
-	3.8 ± 2.2	33.3 ± 3.8	-	14.3 ± 0.5	-	-
-	-	3.4 ± 1.5	0.03 ± 0.01	0.03 ± 0.01	-	-
-	-	3.7 ± 0.7	0.08 ± 0.03	0.19 ± 0.13	-	-
-	0.0	4.4 ± 0.3	0.46 ± 0.14	0.7 ± 0.2	-	-
-	0.33 ± 0.17	3.9 ± 0.6	0.17 ± 0.03	7.5 ± 1.1	-	-
0.3 ± 0.1	-	5.7 ± 0.8	0.5 ± 0.3	0.5 ± 0.3	-	-
0.2 ± 0.2	-	5.7 ± 0.4	0.02 ± 0.01	0.4 ± 0.1	-	-
2.3	-	16.2	-	-	-	-
11.7 ± 0.9	-	3.3 ± 1.5	1.1 ± 0.7	6.5 ± 1.2	-	0.2 ± 0.1
-	-	50.0 ± 7.6	0.9 ± 0.3	2.1 ± 0.6	-	-
2.5 ± 0.8	-	-	-	8.6 ± 1.1	-	-
7.9 ± 0.8	-	-	-	0.5 ± 0.2	-	-
8.1 ± 0.6	-	-	-	1.7 ± 0.7	-	-
14.1 ± 5.0	-	0.6 ± 0.2	0.2 ± 0.1	0.8 ± 0.4	-	-
11.9 ± 0.9	-	0.8 ± 0.2	0.2 ± 0.1	1.7 ± 0.4	0.8 ± 0.4	-
12.8 ± 0.8	-	1.0 ± 0.3	0.2 ± 0.1	1.7 ± 0.4	1.1 ± 0.4	-
6.6 ± 2.6	-	-	-	2.2 ± 1.4	-	-
10.8 ± 1.2	-	1.6 ± 0.4	0.1 ± 0.1	0.1 ± 0.01	-	0.0
24.0	-	2.3	-	-	-	-
14.7	-	1.8	-	-	-	-
13.6	-	1.7	0.2	-	-	-
14.6 ± 1.8	-	3.8 ± 0.5	0.4 ± 0.1	0.1 ± 0.1	0.2 ± 0.1	-
15.9 ± 1.4	-	3.6 ± 0.5	0.2 ± 0.1	0.2 ± 0.1	0.9 ± 0.6	-
14.2 ± 1.8	-	2.7 ± 0.4	0.3 ± 0.2	0.1 ± 0.1	-	0.04 ± 0.04
16.0 ± 1.9	-	0.2 ± 0.1	2.6 ± 0.6	0.1 ± 0.04	-	-
-	-	8.4	5.0	-	-	-
0.2	-	14.5	2.9	-	-	-
-	-	3.6	2.3	-	-	-
1.7 ± 0.9	-	9.7 ± 1.8	8.9 ± 1.3	1.3 ± 0.3	-	0.8 ± 0.3
2.7 ± 1.2	-	7.8 ± 0.7	11.6 ± 1.3	0.6 ± 0.1	-	-
8.5 ± 0.8	-	11.6 ± 1.6	6.3 ± 2.0	1.0 ± 0.6	-	-

continued

Table 4.1 *Continued*

Species	Bark	Flowers	Soft plant material	Feathers	Hair
Open-nesting species					
Common blackbird (*Turdus merula*)	-	-	-	-	-
	-	-	-	-	-
	-	-	-	-	-
	-	-	-	-	-
	-	-	-	-	0.2 ± 0.2
American robin (*Turdus migratorius*)	-	-	-	-	-
	-	-	-	-	-
American yellow warbler (*Dendroica petechia*)	-	-	1.1 ± 0.2	0.01 ± 0.01	-
	-	-	3.4 ± 0.4	0.45 ± 0.11	-
American goldfinch (*Carduelis tristis*)	-	-	2.1 ± 0.3	0.001 ± 0.001	-
Common redpoll (*Carduelis flammea*)	-	-	2.3 ± 0.3	0.34 ± 0.07	-
Savannah sparrow (*Passerculus sandwichensis*)	-	-	-	-	-
	-	-	-	-	-
Willow warbler (*Phylloscopus trochilus*)	1.3	-	-	1.7	-
Hedge accentor (*Prunella modularis*)	-	-	-	0.01 ± 0.01	0.8 ± 0.4
Hole-nesting species					
Tree sparrow (*Passer montanus*)	-	-	-	16.8 ± 3.3	0.8 ± 0.2
Great tit (*Parus major*)	-	-	-	0.1 ± 0.02	5.6 ± 1.0
	-	-	-	0.1 ± 0.05	1.5 ± 0.4
	-	-	-	0.1 ± 0.05	1.5 ± 0.8
	-	-	-	0.04 ± 0.03	1.4 ± 0.7
	0.1 ± 0.02	-	-	0.1 ± 0.03	2.2 ± 0.5
	0.1 ± 0.03	-	-	0.1 ± 0.02	2.4 ± 0.6
	-	-	-	0.03 ± 0.03	0.2 ± 0.1
Blue tit (*Cyanistes caeruleus*)	-	-	-	0.7 ± 0.1	0.4 ± 0.2
	1.9	-	-	0.8	8.8
	0	0.5	-	0.5	6.9
	1.8	-	-	0.2	4.7
	0.6 ± 0.3	-	-	0.6 ± 0.2	0.4 ± 0.2
	0.2 ± 0.1	-	-	0.1 ± 0.2	0.4 ± 0.1
	-	-	-	1.5 ± 0.4	2.8 ± 0.9
	1.2 ± 1.0	-	-	1.3 ± 0.2	0.8 ± 0.2
European pied flycatcher (*Ficedula hypoleuca*)	24.7	-	-	-	-
	3.9	-	-	-	0.3
	15.8	0.7	-	-	-
	-	-	-	0.1 ± 0.04	0.1 ± 0.1
	0.1 ± 0.1	-	-	0.3 ± 0.3	0.01 ± 0.01
Common redstart (*Phoenicurus phoenicurus*)	1.4 ± 1.1	-	-	6.9 ± 1.1	0.3 ± 0.1

Wool	Fur	Arthropod silk	Artificial materials[#]	References
-	-	-	-	Mainwaring et al. (2014b); E. Vaughan and
-	-	-	-	D.C. Deeming (2009 unpublished data)
-	-	-	-	
-	-	-	-	
-	-	0.1 ± 0.1	0.1 ± 0.1	
-	-	-	-	Crossman et al. (2011)
-	-	-	-	
-	0.07 ± 0.03	-	-	Crossman et al. (2011)
-	0.09 ± 0.05	-	-	
-	-	-	-	Crossman et al. (2011)
-	-	-	-	Crossman et al. (2011)
-	-	-	-	Crossman et al. (2011)
-	-	-	-	
0.26	-	-	0.07	Elts (2005)
1.3 ± 0.8	-	0.1 ± 0.1	1.9 ± 1.6	E. Vaughan and D.C. Deeming (2009 unpublished data)
0.2 ± 0.1	-	-	0.9 ± 0.6	E. Vaughan and D.C. Deeming (2009 unpublished data)
-	-	-	0.03 ± 0.02	Álvarez et al. (2013); Britt and Deeming
-	-	-	0.8 ± 0.3	(2011); E. Vaughan and D.C. Deeming
-	-	-	2.5 ± 0.6	(2009 unpublished data)
0.5 ± 0.3	-	-	0.9 ± 0.3	
0.7 ± 0.2	2.8 ± 0.4	-	0.1 ± 0.04	
0.4 ± 0.1	3.7 ± 0.5	-	0.1 ± 0.02	
-	-	-	1.7 ± 0.8	
0.5 ± 0.1	-	-	0.1 ± 0.1	Britt and Deeming (2011); E. Vaughan
-	-	-	-	and D.C. Deeming (2009 unpublished
-	-	-	-	data); E. Wilson and D.C. Deeming (2010
-	-	-	-	unpublished data); L. Deacon, D.C. Deeming,
1.2 ± 0.5	1.3 ± 0.4	-	0.01 ± 0.01	P. Coffey, and H. Rowland (2014 unpublished
0.4 ± 0.02	1.8 ± 0.5	-	0.01 ± 0.01	data); Moreno et al. (2009)
0.0	-	-	0.01 ± 0.01	
1.6 ± 0.4	-	-	-	
-	-	-	-	E. Wilson and D.C. Deeming (2010
-	-	-	-	unpublished data); Moreno et al. (2009);
-	-	-	-	L. Deacon, D.C. Deeming, P. Coffey, and
-	-	-	-	H. Rowland (2014 unpublished data)
-	0.03 ± 0.03	-	-	
0.3 ± 0.2	-	-	-	L. Deacon, D.C. Deeming, P. Coffey, and H. Rowland (2014 unpublished data)

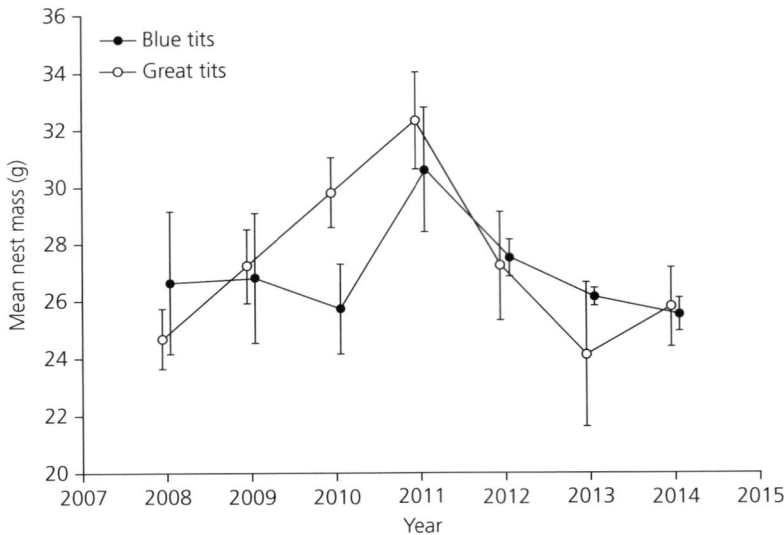

Figure 4.2 Mean (± 1 SE) nest masses of great tits (*Parus major*) and blue tits (*Cyanistes caeruleus*) breeding between 2008 and 2014 (inclusive) at Riseholme Park at the University of Lincoln, Lincoln, UK in nestboxes with bases measuring 15 cm × 11 cm. (Source: D.C. Deeming, 2014 unpublished data.)

$Ps > 0.50$). By contrast, for great tits nest mass was almost significantly positively correlated with hatching success ($r_s = 0.15$, df = 156, $P = 0.06$) and was significantly positively correlated with fledging success ($r_s = 0.20$, df = 156, $P = 0.01$).

The reason for so much variation in nest size within a species (Figure 4.2) is far from clear. Key questions are whether nest and cup are interchangeable in terms of their size, and which of these is more likely to reflect body mass of the incubating bird? Deeming (2013) suggested that the factors determining nest size will be different from those determining cup size. Slagsvold (1989a, 1989b) recognized that the different roles of a nest are crucial in determining the biological significance of cup size. Small, tight-fitting nest cups were more suited to the incubation phase but during rearing of several pulli the cup needs to enlarge to accommodate the growing offspring. Hence, materials such as mosses and lichens, which are readily compressed, are common in nests of small passerines (Hansell 1996). Experimental reduction of nest size reduced clutch size of Eurasian magpies, and impacted on the decision to start incubation (Soler et al. 2001). Lower quality nests also hatched fewer chicks on the first day of hatching and adversely affected chick growth. Interestingly, artificial enlargement of the nest had no significant effect on these parameters. Soler et al. (2001) suggested these results indicated that nest size was a sexual signal. Nest building in the barn swallow (*Hirundo rustica*) was also seen as a reliable sexual signal of the male's willingness to invest in later stages of reproduction (Soler et al. 1998a). This pattern seems to be reflected

in a wide range of European passerines—shared nest building by parents produces bigger nests relative to body size with a greater investment in reproduction (Soler et al. 1998b). Thus, whilst nest construction behaviours play an important role in sexual selection, the number of nests built by birds can also act as a sexual signal. In effect, nests of different species of varying body size are not directly comparable to each other.

4.2.2 Nest construction materials

Nest construction materials are likely to play an important role in determining the reproductive success of birds. However, despite this, we know remarkably little about plasticity in the use of various materials, the use of special materials for nest construction, or the appropriate level of detail for analysis. The materials used by birds to construct nests have been anecdotally described for many species (e.g. Ferguson-Lees et al. 2011) but relatively few studies provide any quantitative empirical data for comparative purposes or otherwise. Thus, we have compiled a list of the materials used to build the nests of several passerine birds (Table 4.1).

Descriptions of nests often rely on single examples or the nest materials are described in relative terms. For instance, a nest of the rufous-fronted thornbird (*Phacellodomus rufifrons*) from Venezuela contained over 3,000 sticks, over half of which measured 1–10 cm in length, ~1,200 were 10–20 cm long, and the total was split 80:20 between non-thorny and thorny sticks (Thomas 1983). However, given that the sample size is just one, we are unsure whether these data are representative of the species. Hansell (2000) described the types of materials used in bird nests, which can be animal- or plant-derived or of mineral origin but are not quantified for any species. For example, mud, either alone or in combination with other materials, is a common component of nests of many bird families (Rowley 1970). It can be used to construct the whole nest, as is the case in the Hirundinidae (swallows and martins) and Phoenicopteridae (flamingos) or it can be used to reduce the size of entrances such as in the Sittidae (nuthatches and wallcreepers) and Bucerotidae (hornbills). Mud and fibre can be used to construct whole nests or be part of the nest lining as in the case of the internal cup of common blackbird (*Turdus merula*) nests (Biddle et al. 2015; Chapter 3). Detailed studies of captive zebra finches (*Taeniopygia guttata*) have recently advanced our understanding of the choice of building materials (e.g. Bailey et al. 2014), yet we still know very little about the function of materials in the nests of wild birds.

This lack of understanding may have arisen because detailed quantitative analyses of nesting materials vary in quality of the information they provide. Kern and van Riper (1984) described the materials used by Hawaii amakihis (*Hemignathus virens*) in terms of species of plants used for some presumably easily identifiable angiosperms but other plant types, i.e. grasses, are less well described. Similarly, species identification of some plant parts (e.g. roots) can be problematic. Hence, lichen-type nests of amakihis comprise 41% grasses and 26% lichens but three angiosperm species constitute approximately 20% of the nest mass (Kern and van Riper 1984). There are racial differences in the materials used in nests of white-crowned sparrows (*Zonotrichia leucophrys*) with nest locations differing between eastern (*Z. l. leucophrys*), mountain (*Z. l. oriantha*), and Nuttall's (*Z. l. nuttalli*) subspecies (Kern 1984). Mountain white-crowned sparrow nests vary in composition (expressed as percentages of total nest mass) according to whether they are built in trees or on the ground (more twigs versus more grasses, respectively), and in the Nuttall's subspecies birds often incorporated leaves into the nest structure from the very bush in which the nest was built. However, our attempt to convert these data into masses was foiled because the mean percentage values for each nest component summed to > 100. Nests of the prothonotary warbler are built within nestboxes and comprised mainly mosses and liverworts (75–80% of dry mass) with particular species dominating the plant species chosen (Blem and Blem 1994). Materials comprising a single willow warbler (*Phylloscopus trochilus*) nest were described by Elts (2005).

There is evidence that nesting materials are not simply used in relation to their availability within the local environment. Antarctic skuas (*Stercorarius* spp.) and kelp gulls (*Larus dominicanus*) in the same habitats differ in the materials used to construct their nests (Quintana et al. 2001). Skuas prefer mosses, particularly *Polytrichum alpestre*, which combined form > 60% of nests by dry mass, followed by pebbles (> 15%) and lichens (almost 10%). By contrast, the gulls preferred to use the grass *Deschampsia antarctica*, which was over 60% of their nests by dry mass, with mosses forming 20%, pebbles ~8%, and limpet (Patellogastropoda) valves ~4%. These differences suggest preferences by each species for specific materials despite their limited availability (Quintana et al. 2001). Elsewhere, van Riper (1977) showed that the use of wool as a nesting material in a variety of passerine species on Hawaii was related to nest complexity with species building more complex nests using more wool. Incorporation of wool seemed to be opportunistic (see also Britt and Deeming 2011) with only one species, the elepaio (*Chasiempis sandwichensis*), seeming to seek this material actively. Consequently, there is reasonable evidence that nesting materials are not used in relation to their availability within the local environment. However, the studies above are observational and so may have been confounded by interspecific competition for the favoured materials. Consequently, further experimental studies are required.

More comprehensive descriptions of the materials used by birds to construct nests are slowly being recorded and reported. To date, the species list is almost exclusively limited to a few European passerines (Table 4.1). The largest species to date, the common blackbird, is reliant on mud but other smaller species appear not to use mud to any great extent despite it being reported as a major nest constituent in many species (Rowley 1970). Also, in contrast to many of the smaller species, the common blackbird does not use animal-derived materials in its nest. We suggest that this may be because blackbirds, as representatives of larger birds, are less likely to use materials such as feathers, mammalian hair, or arthropod silk; this is a subject area where future research could fruitfully be directed. Use of mosses is variable and species-specific. European pied flycatchers do not use moss but tits (Paridae) appear strongly reliant on it for at least 50% of their nest mass (Britt and Deeming 2011; Álvarez et al. 2013; Table 4.1). Breil and Moyle (1976) found that all species of North American passerines used mosses in their nests, with 65 different bryophyte species being recorded. There is significant scope to document further the types and amounts of materials used in nests for both passerines and non-passerines. This would enable us to perform not only interspecific analyses of the use of different nest materials but also to examine the functional significance of various nest materials.

Plasticity in the materials used in birds' nests is shown by several species. For instance, western marsh harriers (*Circus aeruginosus*) nesting in southern Lithuania varied the plant materials in their nests according to neighbouring territories (Stanevičius and Balevičius 2005). Although helophytes tended to dominate nest material choice irrespective of neighbouring habitat, for nests built within sedge-fen and lake-shore meadows, nests were composed of almost 30% marsh plants, which were absent from nests built in forest or reed fen. Nesting location also significantly affected choice of materials by great tits in Spain, with one population nesting in an orange (*Citrus × sinensis*) monoculture relying on sticks to build their nests rather than the more usual moss (Álvarez et al. 2013). However, without a study that experimentally provided the birds with moss to build their nests, we cannot be sure whether the use of sticks was an active choice or whether they were limited to using sticks rather than the more usual mosses.

That local availability of materials is important in choices made by birds when constructing nests was illustrated by a study of four species of tits (Paridae) that used coloured artificial wool-like nest materials provided to birds through dispensers placed in the woodland (Surgey et al. 2012). There were no significant colour preferences between individual females but the amount of material used in the nest lining was a function of distance to the dispenser—the closer the bird was to the source of material, the more was used. There is evidence that the choice of materials can vary according to experience in captive birds—experimental manipulation of coloured nesting material available to captive zebra finches showed that individuals had colour preferences but that those initial preferences could be changed by experience (Muth and Healy 2011). Individual males that built nests using materials of a non-preferred colour would only choose to keep this colour if the reproductive event was successful, whereas those males who suffered a failed breeding attempt changed their colour preference. Thus, the choice of nest material was based on the success of the previous breeding attempt. Moreover, zebra finches prefer blue materials; if raised in a nest of yellow materials the offspring did not construct yellow nests but tended to switch to blue material (Muth et al. 2013). Further experimentation has shown that zebra finches choose coloured

nest materials to match the background colour of their nestboxes (Bailey et al. 2015).

The use of nesting materials is also likely to be influenced by human activities, especially as an ever-increasing area of land is being urbanized across the planet. Use of artificial materials in nests in relation to location was reported by Wang et al. (2009a) for the light-vented bulbul (*Pycnonotus sinensis*). Birds in more urban areas were exposed to more anthropogenic materials (e.g. plastics, cloth) than birds from more rural areas and, accordingly, used more of them in their nests. We still have only a scant understanding of the use of artificial materials within birds' nests and there are some fascinating questions that remain unanswered. For example, are such materials 'mistaken' for natural materials that the species has always used, or are they exploited for the new opportunities they provide? The latter option has implications for the cognition of nest building, which is discussed further in Chapter 3.

We have demonstrated that nest materials are used to modulate the physical environment of the eggs and that they also provide mechanical support, but there are also examples of species using some unusual materials in their nest construction for specialized roles. For example, common waxbills (*Estilda astrild*) incorporate carnivore scat in their nests throughout the construction, incubation, and chick-rearing phases (Schuetz 2005). The presence of such material was shown experimentally to reduce predation rates. Blackstarts (*Cercomela melanura*) construct stone ramparts that weigh nearly 500 g in front of the holes leading to their nesting cavity, which seem not to deter predation, presumably from ground-living mammals, but may act as an early warning system allowing females to escape (Leader and Yom-Tov 1998). By contrast, black kites (*Milvus migrans*) in Spain incorporate variable amounts of white plastic into their nests according to age (Sergio et al. 2011). Compared with young birds, i.e. < 6 years of age, older birds use more materials as a form of extended phenotype and the white plastic supposedly acts as a graduated signal conveying information on the quality of the individual's territory, viability, and social dominance. Trnka and Prokop (2011) reported that great reed warblers (*Acrocephalus arundinaceus*) exhibit a preference for sloughed snake skins, which were woven into the body of the nest but not used in the lining. Experiments indicated that the presence of skins did not influence predation rates and their role may be as a form of sexual signal. Veiga and Polo (2005) also suggested that the use and placement of feathers within the nests of spotless starlings (*Sturnus unicolor*) were influenced by their ultraviolet reflectance and so acted as an intraspecific signal but it was unclear what they denoted.

The presence of ectoparasites in nests has been suggested as the reason why certain materials, particularly some green plant materials, are incorporated into nests and continue to be added after the main construction phase has been completed (e.g. Mennerat et al. 2009a, 2009b, 2009c; Moreno et al. 2009; but see Brouwer and Komdeur 2004 and Chapter 8). Unusually, house sparrows (*Passer domesticus*) and house finches (*Carpodacus mexicanus*) in Mexico incorporated smoked cigarette butts into their nests as a means of reducing the ectoparasite load (Suárez-Rodríguez et al. 2013; Chapter 8). Interestingly, experimental comparison of non-smoked and smoked butts showed that the latter were more effective at deterring ectoparasites. Blue tits on Corsica included a range of aromatic green plants in their nests, which are associated with individual preferences (Mennerat et al. 2009a), and appear to confer some fitness benefits to nestlings (Mennerat et al. 2009b) and may be related to an anti-bacterial role (Mennerat et al. 2009c). However, blue tits in more northerly parts of their European range appear not to include such materials in their nests—whether this reflects localized plasticity in nest construction behaviour or the absence of appropriate plant species remains unclear.

4.2.3 Variation in the number or type of nests

Nest construction is often associated with the individuals incubating at the nest. In species where incubation is shared nest building is also shared and when there is uni-parental incubation, nest construction tends to be by the incubating sex. For example, in the Turdidae nesting in Europe 28 of 30 species have female-only construction of the nests and all 30 have female-only incubation (D.C. Deeming, 2014 unpublished data). In European Sylviidae nest building and incubation can be female-only or bi-parental and only nine of the 48 species deviate from the pattern of the same sex(es) nest building and subsequently incubating. It would be interesting to examine this pattern more widely as it may provide insights into the evolution of incubation behaviour.

However, some species utilize nest construction as an indicator of sexual quality. For instance, male winter wrens (*Troglodytes troglodytes*) and many species of weavers construct a series of nests in order to attract a mate (Skutch 1976; Collias and Collias 1984). In great tits nest mass positively correlates with female quality as measured by plumage colouration (Broggi and Senar 2009).

Meanwhile, the males of some species build multiple nests within their territories, although the exact function of nests that do not contain eggs and nestlings is unclear (Metz 1991). Some studies have shown that those males that build more nests within their territories have greater reproductive success than males with fewer nests (Savalli 1994; Evans and Burn 1996; Evans 1997; Friedl and Klump 2000). Male Australian reed warblers (*Acrocephalus australis*) construct two types of nest—type I are functional incubation units whereas type II are not (Berg et al. 2006). Although supplementary feeding allows males to build more type II nests, implying that their construction is energetically demanding, males do not receive any reproductive benefit and the exact role of nest type II remains unclear (Berg et al. 2006). Another study reported that the experimental addition of nests did not increase the pairing success of male marsh wrens (*Cistothorus palustris*; Leonard and Picman 1987), suggesting that the extra nests may have an anti-predator function whereby the large number of nests dilutes the chances of a predator finding one that contains eggs or nestlings (see Chapter 5). However, our understanding of the function of multiple nests remains uncertain and further research is required.

4.3 Factors affecting the micro-environment of nests

The main function of nests is to provide a location for birds to lay their eggs and in the case of altricial species to raise their nestlings. Thus, the location and design of nests should be expected to create suitable environmental conditions within the nest, as sub-optimal nest microclimates are detrimental to offspring growth and body condition (reviewed by DuRant et al. 2013a). Nest insulation may be important because of its impacts on the energetics of incubation (see Chapter 13), which in turn may influence other aspects of reproductive performance. For example, heating nestboxes, which are an artificial cavity, containing tree swallows (*Tachycineta bicolor*) altered incubation behaviour of females by increasing the time they spent incubating and maintaining higher egg temperatures (Ardia et al. 2009). Pérez et al. (2008) also heated nestboxes containing incubating tree swallows and found benefits to adult female body condition and nestling provisioning rates that improved their growth rates and overall body masses. Dawson et al. (2005a) found that heating nestboxes of tree swallows positively impacted nestling survival. Interestingly, experimental cooling of this species' nests during incubation had adverse effects on nestling body mass and condition and innate immunity (Ardia et al. 2010).

While parents are likely to be able to create suitable microclimates through adaptive changes in their incubation behaviours (Deeming 2011b), such behaviours are still likely to be costly. In this section, we examine the thermal properties of nests and after a brief consideration of the techniques used to quantify them, we explore how environmental conditions, the use of selected materials, and nest type influence nest design. We finish by exploring how these factors influence the humidity within nests.

4.3.1 Thermal properties

Successful hatching depends on the interaction between the eggs and the incubating adult sitting on the clutch—the 'bird-nest incubation unit' (Deeming 2002b). Maintaining egg temperature is, therefore, not purely a function of the presence of the incubating adult but is inter-related with the characteristics of the brood patch, the thermal properties of the egg and the nest (Deeming 2002e). As mentioned above, this idea has been modified to distinguish between smaller species more reliant on the nest for incubation and larger species where the nest is a venue for incubation but the nest itself has no major insulative role (Deeming 2013). We have a relatively good understanding of the thermal relations of eggs under a brood patch (see review by Turner 2002) but the thermal properties of nests have received much less attention (see review by Heenan 2013). However, the thermal environment within nests is not only determined by the structure and/or design of the nest itself, but also the location in which the nest is built.

The location of a nest can also be fundamental in ensuring an appropriate thermal environment. Sidis et al. (1994) showed that nests of Palestine sunbirds (*Cinnyris osea*) are more frequently located such that nest openings face away from the prevailing wind. In addition, fully-shaded nests were commonest and this impacted upon the thermal characteristics of the nest. Shaded nests exhibited generally stable nest temperatures but in the sun the temperature increased significantly. McCracken et al. (1997) showed that nest location can be important in determining nest size of Ross's geese (*Chen rossii*) and snow geese (*Chen caerulescens*). The smaller Ross's geese built larger nests but also habitat was important with more sheltered sites being associated with lighter nests. Changes in wind speed impact on heat loss from nests; Heenan and Seymour (2012) demonstrated that despite physical differences in size, and particularly wall thickness, nests of two Australian passerine species were comparable in the effects of increasing wind speed on their heat loss. Similar results have been obtained when the insulative properties of materials have been tested (Hilton et al. 2004).

In those species where males build multiple nests, females can choose between those nests. Whilst the function of multiple nests is often assumed to be related to sexual selection, some studies show that thermal benefits also influence female choice of nests. The nests of Eurasian penduline tits (*Remiz pendulinus*) vary in size and wall thickness with females appearing to prefer larger nests for incubation (Hoi et al. 1994; Grubbauer and Hoi 1996). Cooling rates of dummy eggs significantly decreased with increasing nest thickness but not with either nest height or chamber volume (Hoi et al. 1994; Szentirmai et al. 2005). Nest size also correlated with increased recess periods during incubation and with higher fledging successes (Grubbauer and Hoi 1996). Nest choice by females seemed to be related to energetics of incubation rather than selection of a mate from the extended phenotype of the nest (Szentirmai et al. 2005).

4.3.2 Measurement of thermal properties

Palmgren and Palmgren (1939) appear to be the first to measure rates of cooling of objects placed in nests. Whittow and Berger (1977) determined thermal conductance values ($W \cdot m^{-2} \cdot {}^{\circ}C^{-1}$), the inverse of insulation, for the nests of Hawaii amakihis by placing a water-filled flask in each nest and heat flux was measured directly as energy required to maintain the water at the same temperature. A similar, steady-state method was used by Walsberg and King (1978a, 1978b) to determine thermal conductance of nests of three species of North American passerine. For 11 species of North American passerine Skowron and Kern (1980) also employed a steady-state approach to measure thermal conductance of nests using a water-filled balloon within nests and thermistors inside and outside of the nest. Many nest characteristics correlated with heat flux and thermal conductance of the nest but the key feature was nest density as measured by light penetration. The same method was used for the nests of wood storks (*Mycteria americana*; Rodgers et al. 1988), and by Kern and van Riper (1984) to show that there was a negative correlation between altitude of nesting and the thermal conductance of the nest of the Hawaii amakihi. Kern (1984) demonstrated that there were racial differences in the physical characteristics of nests of white-crowned sparrows but there was no evidence for a correlation between nest location and thermal conductance in this species. Despite colder and wetter conditions, the sub-Arctic subspecies (*Z. l. leucophrys*) had the smallest nests with the highest thermal conductance, i.e. the lowest insulation. Heenan and Seymour (2011) used a steady-state heating system in the nest cup to extend the recording of nest thermal conductance to 36 species of Australian passerines.

More recent studies have approached measurement of the thermal characteristics of bird nests in a variety of ways (see also Chapter 15). Ar and Sidis (2002) reported cooling rate constants of a steel ball in a common blackbird nest. Lamprecht and Schmolz (2004) used common quail (*Coturnix coturnix*) eggs as a model to measure cooling rates in nests of European passerines and also infrared (IR) thermography to visualize the insulative properties of the nest. Hilton et al. (2004) constructed artificial nests of a variety of natural materials and used domestic fowl (*Gallus gallus*) eggs to determine rates of cooling. An electronic apparatus was used by Pinowski et al. (2006) to mimic heat production of a roosting tree sparrow (*Passer montanus*) and energy requirements to maintain a fixed temperature were measured in nestboxes with and without nests. Air temperatures within the domed nests of zebra finches were significantly higher than the ambient temperatures outside of the nest only during the night, although the roofs may also have served to reduce the amount of direct sunlight that the nest contents are exposed to during the hottest parts of the year as well (Zann and Rossetto 1991). However, new analysis has shown that there was a significant elevation of temperature of ~2–2.5°C due to the presence of the nest whether the birds were either nesting (paired *t*-test: $t_6 = -3.22$, $P = 0.02$) or roosting within the nest ($t_6 = -2.69$, $P = 0.04$). More recently, automatic temperature loggers (*i*Buttons®) (see Chapter 15) have been used to determine the insulatory properties of nests of a variety of species (Mainwaring et al. 2012, 2014b; Deeming and Biddle 2015).

Methods to study the thermal properties of nests have recently improved with the use of IR thermography. This is important as more sophisticated technologies enable us to understand the roles of the adult birds and nest materials in determining nest microclimates. Thermal imaging cameras have been used to locate active nests in the field (e.g. Boonstra et al. 1995; Galligan et al. 2003), which is important in those species where nests are cryptically positioned within vegetation. From a more functional perspective, Lamprecht and Schmolz (2004) placed heated common quail eggs in nests of various songbirds from a museum collection and recorded the way that the nest was warmed using thermography. They concluded that nest insulation varied considerably between species but without any discernible pattern related to, for instance, adult body size, although the bird was not on the nest. Töpfer and Gedeon (2012) used thermography to conduct a preliminary investigation into the heat loss from nests of Stresemann's bush crows (*Zavattariornis stresemanni*) and concluded that the substantial structure offered significant insulation against both heat loss by the adult and eggs but also heat gain from the environment. Boulton and Cassey (2012) used thermography to measure heat loss from eggs of great tits exposed by the incubating bird during incubation recesses. IR thermography was used by Deeming and Pike (2015) to investigate the thermal profile of the surface of great tit and blue tit nests during incubation (Figure 4.3). There was a very steep thermal gradient from the edge of the incubating bird with the surface temperature approaching ambient within 2–3 cm. Deeming and Pike (2015) concluded that the materials lining the cup provided extremely effective insulation.

Tree sparrows nest within nestboxes and utilize a high proportion of feathers in their nests (Pinowski et al. 2006; Table 4.1). Monitoring the energy usage of a heater placed in the open air or within an empty box allowed Pinowski et al. (2006) to estimate the effects of nests within the box on the rate of energy consumption. Compared to the open air with no wind, energy savings with the presence of a box alone amounted to 18% but this doubled to 33–36% for a nestbox containing a complete nest. Rockweit et al. (2012) investigated the factors affecting nest location and structure in spotted owls (*Strix occidentalis*) that construct three different types of nest. Biophysical modelling suggested that the structural configuration of nests influenced egg cooling rates and that this could affect reproductive success. However, varying the structural configuration of nests is not without its problems; both temperature and humidity of the nest chamber may be dampened from fluctuation in ambient conditions but both tend to be generally higher (Chaplin et al. 2002; Lill and Fell 2007; Maziarz and Wesołowski 2013; Mersten-Katz et al. 2013). Therefore, the methods used to quantify the thermal properties of nests have advanced over the years, as more fully discussed in Chapter 15. There is good evidence to show that nests do create suitable microclimates for adults and offspring. However, nest

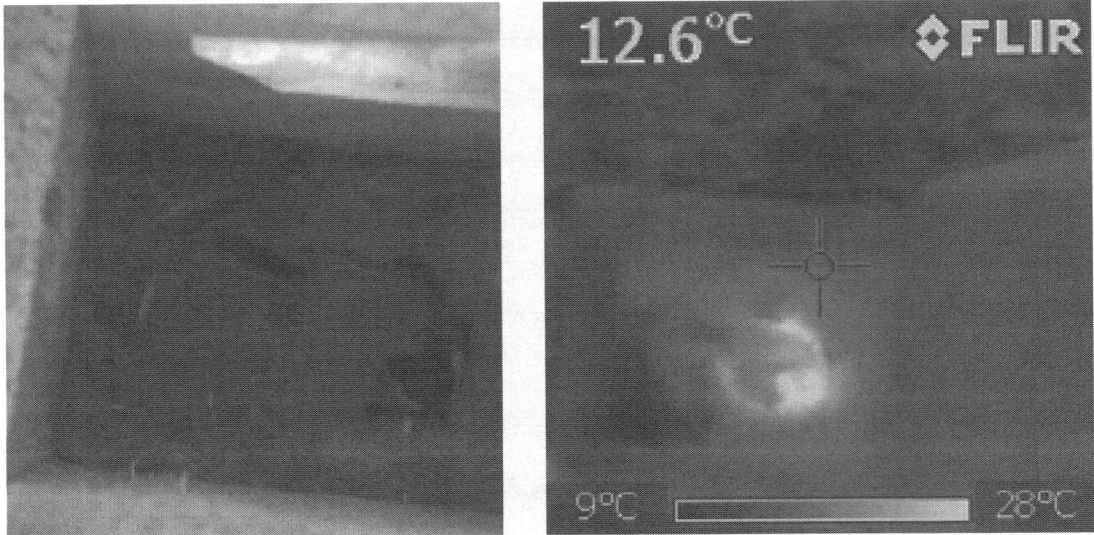

Figure 4.3 Normal (left) and infrared thermal (right) images of an incubating female great tit (*Parus major*) on her nest in Riseholme Park, Lincoln, UK. (Source: Deeming and Pike 2015. See Plate 6.)

building is an energetically costly activity (reviewed by Mainwaring and Hartley 2013) and it is likely that birds are selected to invest such costs during nest construction and by creating suitable thermal conditions for offspring; the initial costs are compensated for by resultant energetic savings in the remainder of the reproductive event.

4.3.3 Thermal properties and environmental conditions

To date, few studies have investigated the interrelationships between bird behaviour and nest characteristics. This is an important omission in our understanding of the location and design of nests as the costs associated with nest construction may be outweighed by the benefits of creating a nest with suitable environmental conditions for offspring growth. Lombardo et al. (1995) showed that eggs within tree swallow nests that were artificially deprived of feathers had longer incubation periods even though there were no significant effects on the incubation rhythm of the females or hatching success. By contrast, hatching success was positively correlated with the mass of moss in great tit nests and the proportion of moss in the nest as a whole (Alabrudzińska et al. 2003). In blue tits, neither hatching success nor the length of incubation correlated

with nest mass (Tomás et al. 2006) although Mainwaring and Hartley (2008) and Britt and Deeming (2011) found that nest mass in this species was negatively correlated with clutch initiation date. The rufous-tailed scrub robin (*Cercotrichas galactotes*) exhibits considerable variation in nest mass (mean mass = 104.81 g with range of 34–226 g) but Palomino et al. (1998) concluded that this was not related to thermoregulation but rather was a sexual signal. Whether this is true is unclear because no measure of construction date was included in the analysis.

Environmental factors affecting nest construction have been observed in a range of passerines. Tiainen et al. (1983) investigated the insulative properties of nests built by wood warblers (*Phylloscopus sibilatrix*), willow warblers (*Phylloscopus trochilus*), and common chiffchaffs (*Phylloscopus collybita*) constructed in Finland. There were strong correlations between ambient air temperature and the insulative properties of the nest for each species but wood warbler nests cooled fastest. This species did not have nests lined with feathers whereas feathers were ~10% of the mass of willow warbler nests and ~14% of common chiffchaff nests. The poorer insulation of wood warbler nests may have contributed to the restricted range of this species to the north.

Mainwaring and Hartley (2008) found that blue tit nest composition changed as the breeding season progressed—the mass of the lining materials showed a seasonal decline. Further investigations have shown that temperature during the seven days leading up to clutch initiation significantly impacts the mass of blue tit (Britt and Deeming 2011) and great tit (Deeming et al. 2012) nests. Colder temperatures are associated with heavier nests, irrespective of the date of clutch initiation (Britt and Deeming 2011; Deeming et al. 2012). Mainwaring et al. (2012) showed that this relationship applied at different latitudes—blue tit and great tit nests built in Great Britain but separated by 5° of latitude had significantly different insulatory properties. This latitudinal effect was also observed in nests constructed by common blackbirds from locations in Britain separated by 7° of latitude (Mainwaring et al. 2014b). In this case the insulatory properties of the nests showed a significant positive correlation with the amount of dry grass in the nest, which is typically found inside the nest cup (Biddle et al. 2015). Long-tailed bushtits (*Aegithalos caudatus*) incorporate many feathers into their nests (41% by mass) and the mass of feathers shows a seasonal decline (McGowan et al. 2004; Figure 4.4). The mass of feathers also had a positive effect on the insulating effect of the nest wall and smaller nests have more feathers, suggesting that this species also adjusts its nest construction behaviour to reflect the prevailing environmental conditions.

In North America, latitudinal effects have been reported for the insulatory properties of Baltimore oriole (*Icterus galbula*) nests (Schaefer 1980). Northern nests from colder climates were less resistant to radiant heat than nests from more southerly locations. Rohwer and Law (2010) compared the morphological and thermal properties of American yellow warbler (*Dendroica petechia*) nests from Churchill, Manitoba and from further south in Elgin, Ontario. This represented a difference of ~14° in latitude, of 5–12°C in temperature, and higher wind speeds further north despite Churchill receiving less precipitation. The Churchill nests were less porous (as measured by light penetration), larger with significantly thicker walls, and contained more feathers and plant materials. This meant that heat loss from these nests was significantly lower than for nests from Ontario. Similar results were observed in nest construction by this and other passerine species from these locations (Briskie 1995; Crossman et al. 2011; Table 4.1). As for the European passerines, latitudinal differences, and associated differences in prevailing temperature, seem to be causing North American passerines to adapt their nest construction behaviours to the prevailing environmental conditions.

There is evidence that different species use different materials to achieve the same common goal of providing a suitable microclimate for the offspring and attending adults. The nests constructed by blue tits, European pied flycatchers and common redstarts (*Phoenicurus phoenicurus*) within the same type of nestbox have been investigated (L. Deacon, D.C. Deeming, P. Coffey, and H. Rowland, 2014 unpublished data). The three species used different nest materials: blue tits mainly used moss, hair, fur, and feathers, compared with leaves and grass for flycatchers, and redstarts used leaves, grass, moss, and particularly feathers. Despite these differences, and that common redstarts constructed significantly heavier nests than the other species, the insulatory properties of the nests (as determined from differences in cooling rates: °C·20s^{-1}) were not significantly different (Table 4.2). They suggested that the different materials all had the same capability of trapping air, which then acted as an insulator. By vacuum packing nests of four passerine species it was shown that removing the air trapped within nest materials significantly decreased the insulatory property of nests as a whole by an average of 20% ± 6%, irrespective of species (Deeming and Biddle 2015). Perhaps in a similar manner European ground squirrels (*Spermophilus citellus*) have a preference for fresh fescue grass (*Festuca* spp.) to build their nests (Gedeon et al. 2010). This seemed to be related to better insulatory properties than dry fescue leaves because it provided a structurally better material to build thicker walls. There are no obvious patterns of variation in insulatory values between small passerine species (Table 4.2). However, whilst there is intraspecific variation that may reflect geographical location or localized temperatures (Mainwaring et al. 2012, 2014b; L. Deacon, D.C. Deeming, P. Coffey, and H. Rowland, 2014 unpublished data), many more studies are required before we can confidently attribute interspecific variation to them.

Another key element of many environments is the extent of precipitation but this seems relatively rarely studied in relation to bird nests. Mainwaring et al. (2014b) found no relationships between characteristics of common blackbird nests from different latitudes. By contrast, for hummingbird (Trochilidae) nests from Patagonia, Argentina there were significant correlations between the amount of precipitation and the amount of plant materials, the species richness of moss and lichens, and the diversity of materials used in nests (Calvelo et al. 2006).

One key element of nest construction behaviour by any species is that it occurs prior to the laying of

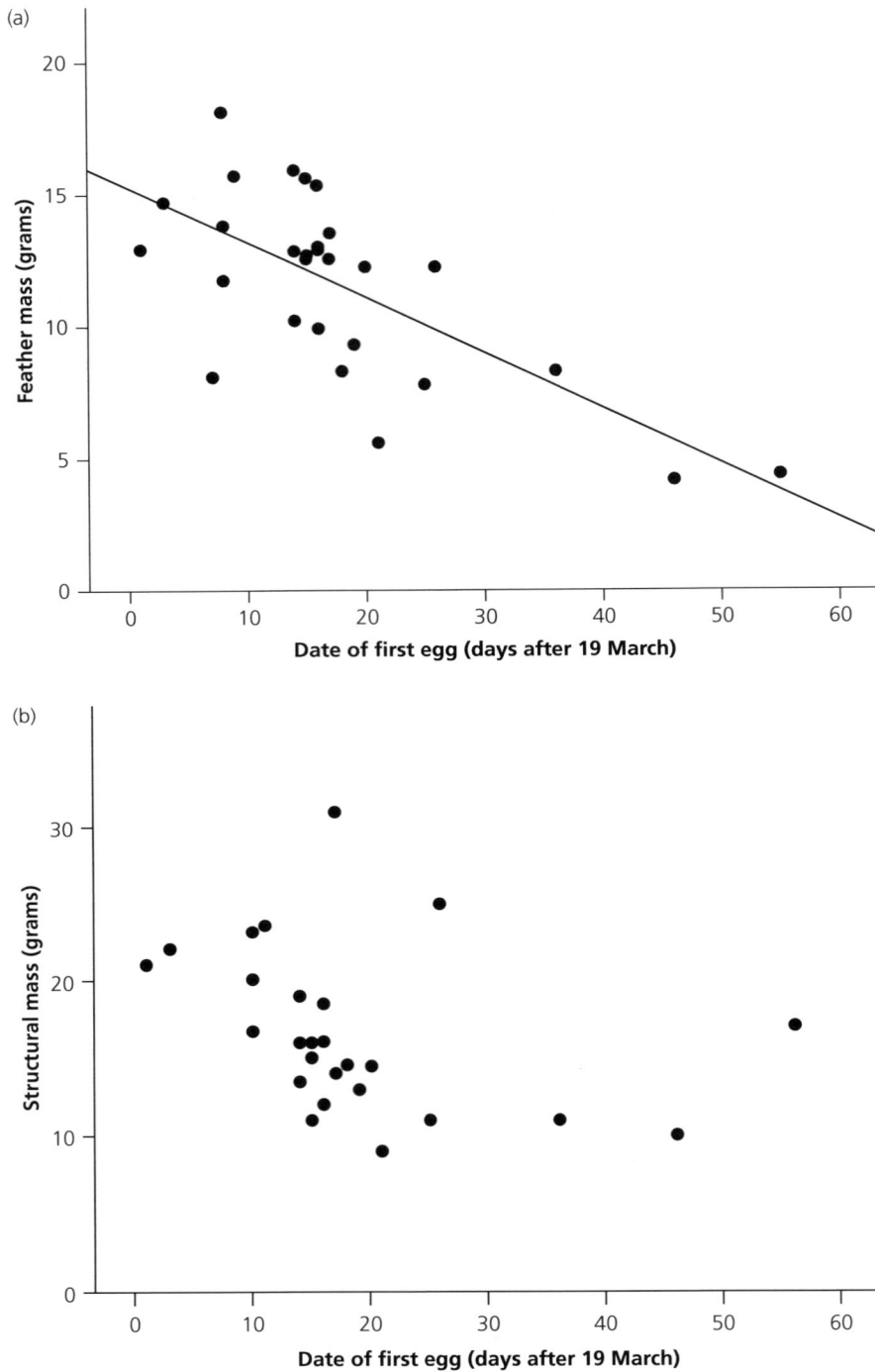

Figure 4.4 The domed nests of long-tailed bushtits (*Aegithalos caudatus*) in Europe are made of structural material and feathers. Whilst the mass of feathers (a) declines as ambient temperatures increase and the need for insulation diminishes, the mass of the structural materials (b) remains unchanged. Figure reproduced from McGowan et al. (2004) with permission from John Wiley & Sons.

Table 4.2 The mean (± 1 SE) insulatory properties of passerine bird nests studied to date. * indicates that the data represent a warming rate of an *i*Button® recorded within each nest. Ranges were not reported for the long-tailed bushtit (*Aegithalos caudatus*).

Species	Insulatory values°C·20s⁻¹ (ranges)	References
Common blackbird (*Turdus merula*)	0.024 ± 0.001 (0.002–0.072)	Mainwaring et al. (2014b)
Great tit (*Parus major*)	0.024 ± 0.001 (0.011–0.039) 0.033 ± 0.004 (0.016–0.053)* 0.028 ± 0.001 (0.021–0.038) 0.029 ± 0.001 (0.019–0.041)	Mainwaring et al. (2012) Deeming and Biddle (2015) S. Caro, L. Biddle and D.C. Deeming (2014, unpublished data for nests constructed under 'cool' controlled temperature conditions, i.e. 13°C) S. Caro, L. Biddle and D.C. Deeming (2014, unpublished data for nests constructed under 'warm' controlled temperature conditions, i.e. 18°C)
Blue tit (*Cyanistes caeruleus*)	0.025 ± 0.001 (0.002–0.037) 0.054 ± 0.003 (0.036–0.080) 0.038 ± 0.002 (0.024–0.050)*	Mainwaring et al. (2012) L. Deacon et al. (2014, unpublished data) Deeming and Biddle (2015)
Long-tailed bushtit (*Aegithalos caudatus*)	0.00270 ± 0.001 0.00207 ± 0.001	McGowan et al. (2004) McGowan et al. (2004)
European pied flycatcher (*Ficedula hypoleuca*)	0.052 ± 0.002 (0.034–0.080) 0.038 ± 0.005 (0.023–0.074)*	L. Deacon et al. (2014, unpublished data) Deeming and Biddle (2015)
Common redstart (*Phoenicurus phoenicurus*)	0.062 ± 0.005 (0.054–0.078)	L. Deacon et al. (2014, unpublished data)
Wood warbler (*Phylloscopus sibilatrix*)	0.040 ± 0.004 (0.026–0.058)*	Deeming and Biddle (2015)

eggs and so clutch size by an individual female may bear no relation to nest properties. It is likely that nest cups are constructed to accommodate the size of the adult bird(s) that will incubate eggs within the structure (Deeming 2013) and that nests are constructed to reflect the thermal requirements of the adult(s) (Britt and Deeming 2011). Hence, if the prevailing temperature is cool, nest walls are better insulated (McGowan et al. 2004; Britt and Deeming 2011; Deeming et al. 2012; Mainwaring et al. 2012, 2014b) not necessarily to maintain egg and/or chick warmth but, rather, as a response by the adult as it constructs the nest. As the breeding season progresses in temperate climates, ambient temperature generally increases and this is reflected in nest construction behaviour. However, the consequences of this pattern of construction for the subsequent clutches and broods have not been fully investigated. Britt and Deeming (2011) suggested that warm temperatures early in the breeding period may produce relatively poorly insulated nests which, if the weather turns cooler, may not provide an appropriate microclimate for young chicks to survive. Mertens (1977) has considered how the thermal properties of the nest and the surrounding microenvironment impact chick survival, finding that high nestbox temperatures are as problematic as low temperatures to great tit broods. Deeming and Pike

(2015) used IR thermography to demonstrate that fledging success in blue tits (but not great tits) was correlated with nest insulation. Increasingly unpredictable weather conditions associated with climate change (see Chapter 6) may impact on breeding success because the nest environment does not offer an appropriate microclimate for survival and growth of nestlings.

4.3.4 Use of feathers

Feathers have high insulatory characteristics in artificial nests (Hilton et al. 2004) and are a key constituent in many bird nests, particularly of small species (Hansell 1995; Hansell and Ruxton 2002). Møller (1984) predicted that there was a conflict between the advantageous insulatory role of feathers and their conspicuousness to predators. It was shown for European passerines that use of feathers in a nest was associated significantly more with early-nesting and smaller species with hole-nests than with open-nesting species that built nests above the ground. Feather use was also associated with more northerly populations of any particular species with a wide geographic range. Møller (1984) also suggested that the insulatory role of feathers could change according to their degree of compression and that behaviours such as 'tremble-thrusting' in various passerine species (see Haftorn 1994) may be responsible for maintaining the insulation of the nest, perhaps by inclusion of air. Feathers in the nest of tree swallows appear not to correlate with incubation duration but seem to have a positive effect on fledgling mass and feather development (Dawson et al. 2011). Liljeström et al. (2009) found that prevailing temperature during nest construction had a significant negative relationship with the number of feathers added to a nest by Chilean swallows (*Tachycineta meyeni*), although this was unrelated to hatching success. Peralta-Sánchez et al. (2010) suggested that feather colour is an important determinant of bacterial load in barn swallow nests, with white feathers perhaps preferred because of their supposed anti-bacterial properties.

While scrape nests often lack a substantial amount of materials, pectoral sandpipers (*Calidris melanotos*) nesting on the Arctic tundra have scrapes with a substantial (~2 cm in depth) lining of feathers or a variety of plant materials (Reid et al. 2002a). The presence of the scrape reduced cooling rates of an artificial egg by 9% compared to the area adjacent to the scrape but the presence of lining materials reduced the cooling rate by 25% (Figure 4.5). Feathers lining the scrape had the best relative insulatory effect with mosses being the worst, whilst wetting the materials significantly reduced their insulatory properties (Reid et al. 2002a). Tulp et al. (2012) also showed that lining of nest scrapes in the high Arctic reduced cooling rates in a variety of shorebird species. There was a body mass effect with smaller bird species having narrow and deep cups with thicker linings than larger species, and the lining was plant-derived (largely the leaves of *Salix* spp.), but reflected the prevailing vegetation in the area where the birds nested. Piping plovers (*Charadrius melodus*) select white pebbles to line their scrape nests, which better reflected the colour of their eggs suggesting a form of crypsis (Mayer et al. 2009). However, the nest scrape as a whole is, as a consequence, more contrasting in colour than the surrounding habitat. Mayer et al. (2009) subsequently demonstrated that the white colouration increased IR reflectance and lowered the temperature within the scrape relative to the surrounding habitat. Elsewhere, the degree of shelter ground-nesting common eiders (*Somateria mollissima*) can provide for their nests can affect nest temperatures under windy conditions; as wind speed increases being sheltered prevented a drop in nest temperature (D'Alba et al. 2009).

4.3.5 Cavity-nesters

While nesting in a cavity can provide a relatively safe location for incubating adults, clutches, and broods from the vagaries of the environment and from the threat of predation (see Chapter 5), the enclosed nature

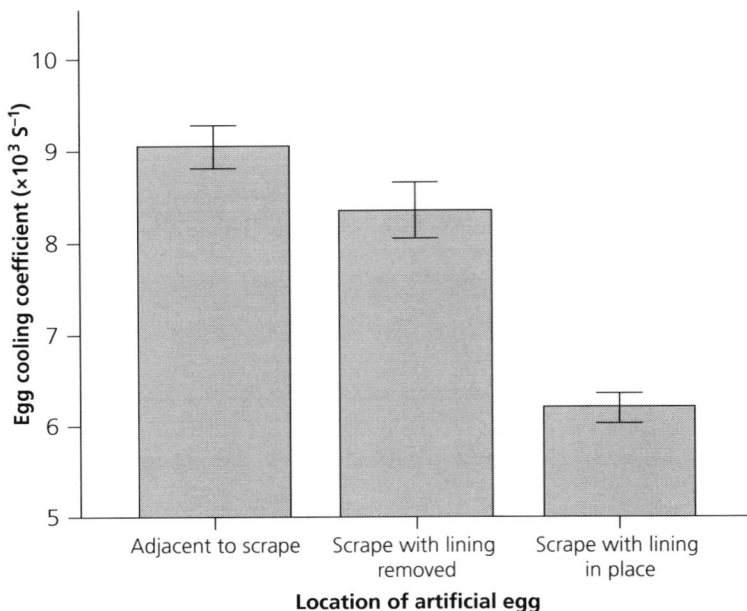

Figure 4.5 Pectoral sandpipers (*Calidris melanotos*) nest on the cool tundra where experiments have shown that nest scrapes that are lined provide the best insulation for eggs. (Adapted from Reid et al. 2002a.)

of a cavity leads to a potential risk of reductions in oxygen (O_2) levels and build-up of carbon dioxide (CO_2). Generally, levels of O_2 are lower and of CO_2 are higher, compared with ambient, but the changes are moderate (1–2%; Birchard et al. 1984; Lill and Fell 2007; Mersten-Katz et al. 2013) and movement within the nest tunnel appears to force the ventilation of the nest chamber (Mersten-Katz et al. 2013). Wesołowski and Maziarz (2012) reported that illumination of nest cavities of great tits and marsh tits is very low and precludes colour vision, which may serve to limit the suitability of cavities to host nests. Avilés et al. (2006) did suggest, however, that this problem could be countered by use of UV reflectance on the eggshell's surface.

4.3.6 Nests and humidity

A key part of avian incubation is mass loss of the egg as water vapour crosses the eggshell, a process controlled by the humidity within the nest in combination with the water vapour conductance of the eggshell (Ar 1991a). Birds do not regulate humidity *per se* but appear to create an environment that raises nest humidity above that of the ambient. Deeming (2011b) demonstrated that nest humidity was a function of prevailing ambient humidity, conductance to water vapour, and nest type. Nest scrapes raise humidity by an average of 5.1 Torr, irrespective of ambient humidity. By contrast, nests with distinct cups had walls that isolated the eggs from the prevailing environment so that, as the ambient air became drier, the difference between nest cup and ambient humidity increased because humidity was much higher inside the nest cup. The nest wall was a significant barrier to loss of humidity with the result that water vapour conductance of eggs had to be relatively high in order to achieve 'normal' rates of mass, i.e. water, loss (see also Portugal et al. 2014a). Deeming (2011b) argued that, because most nest cups are built by birds in cold climates, the nest wall was probably well-insulated against the cold ambient temperatures rather than to retain water. Investigation in a wider range of species nesting in different habitats would further our understanding of these relationships.

In the cavity nests of Syrian woodpeckers (*Dendrocopos syriacus*) and rainbow bee-eaters (*Merops ornatus*) nest humidity and eggshell water vapour conductance are inter-related to produce an appropriate mass loss from eggs during incubation, i.e. ~15% of initial egg mass by external pipping (Ar and Rahn 1980; Lill and Fell 2007; Mersten-Katz et al. 2013). The extent to which nests of Cape Barren geese (*Cereopsis novaehollandiae*) were sheltered also

influences nest humidity with more open nests having lower humidity (Wagner and Seymour 2001).

4.4 Structural properties

A key role for bird nests is the provision of a space that can at least accommodate the clutch and the incubating adult. Another role of the nest for many semi-altricial or altricial species is to contain the brood as they grow to fledging. Given that the combined masses of eggs and adult, or of a brood close to fledging, can be considerable (e.g. clutch mass may represent 130% or more of the mass of an incubating female blue tit; Perrins and Birkhead 1983), it is logical to assume that nests offer some structural support to their contents. An individual's fitness is likely to be significantly reduced if the nest is not strong enough to support the eggs or chicks and there is loss of structural integrity. Studies of the mechanical properties of nests are rare as it is very difficult to quantify such properties accurately. While it may be relatively easy to quantify the tension in materials such as silk and other single materials on their own, it is more complicated for whole structures such as nests that usually are made of a variety of materials. There are two approaches to measuring the structural properties of nests—indirect and direct—and we discuss each of them below.

4.4.1 Nest structure inferred from thermal properties

The first approach to study nest structure used the exponent of the allometric relationship between female body mass and thermal conductance and takes a global, rather than an individual, approach to suggest that nest construction is more concerned with structural integrity than insulatory properties (Heenan and Seymour 2011), but this view has not been broadly accepted (Deeming 2013). Closer examination of the relationship discussed by Heenan and Seymour (2011) indicated that species of larger body size build nests of higher conductance which, inversely, meant that they were less well-insulated than those built by species of smaller body size. However, although data were analyzed using log-transformed values, they exhibited a curvilinear relationship with some of the smallest species, i.e. < 10 g in body mass, having less well-insulated nests than larger species. Heenan (2013) also showed that when plotted against body mass, the area-specific thermal conductance values of Australian nests were higher than those reported by Skowron and Kern (1980) and others. The reason for this difference remains unclear and such relationships require further consideration (Heenan 2013).

The relationships tested by Heenan and Seymour (2011) were between female body mass and thermal conductance but even after log-transformation, the data were not normally distributed (D.C. Deeming, 2014 unpublished analysis). More crucially, if thermal conductance is related to a structural role of the nest then logically this should be tested against nest mass rather than against female body mass. Removal of outliers and regression of the remaining log-transformed data (which were normally distributed—Anderson-Darling tests: all $Ps > 0.05$) shows that there is no significant relationship between nest mass and thermal conductance, i.e. the slope is not significantly different from zero. Over a range of nest masses from 2.7 to 89 g there is no effect on thermal conductance ($N = 23$ species; Figure 4.6). If the approach of Heenan and Seymour (2011) is adopted and adult female body mass is taken as the independent variable, the exponent of the relationship is greatly reduced (Figure 4.6). Therefore, the assertion that nest construction is determined by structural requirements does not appear to be as definitive as Heenan and Seymour (2011) originally proposed.

4.4.2 Directly measured structural properties

The second approach to consider the structural significance of nests is much more direct and involves measurement of the mechanical properties of nest materials (Biddle et al. 2015). Thickness, bending strength, and rigidity were measured for plant materials from various parts of common blackbird nests. Each nest contained a substantial mud cup that incorporated plant materials and was surrounded by twigs. Nest cups were lined with more grass-like materials. The outer layer of twigs was thicker, stronger, and more rigid than materials from the mud cup or the cup lining. This was interpreted by Biddle et al. (2015) as indicating that there was a need for a scaffolding structure during nest construction, which then was superseded by the mechanical properties of the mud cup built subsequently. The materials at the bottom of the cup were thicker, stronger, and more rigid than materials from the top of the cup. Indeed, determination of cup volume using Perspex beads or much heavier dental putty suggests that this layer was compressible. Hence, the bottom of the cup had a mattress-like quality to support the nest contents. These findings reveal that common blackbirds appear to choose materials for particular functions within the nest and that there may be a temporal component to this behaviour. To date, no other studies have considered nest materials in this way although captive zebra finches have demonstrated a preference for stiffer string as they constructed their nests (Bailey et al. 2014; Chapter 3). Woods (1993) reported that loggerhead shrikes (*Lanius ludovicianus*) nesting in dense greasewood (*Larrea tridentata*) bushes used significantly fewer medium- and large-sized sticks in their nests compared to those constructed in more open, sparsely branched sagebrush (*Artemisia tridentata*). The structural support provided by larger sticks seems to be provided by the branches of greasewood. These studies show that it is possible to break down nest structures into their component parts and then to

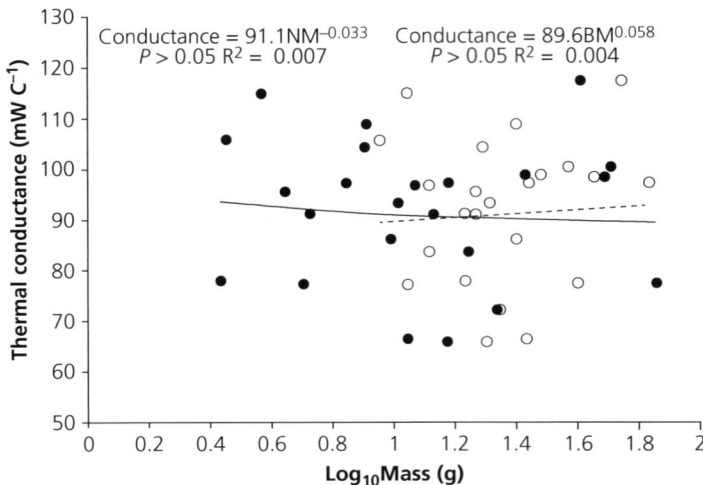

Figure 4.6 Relationships between adult $\log_{10}$ female body mass (o, dashed line, BM) or $\log_{10}$ nest mass (●, solid line, NM), with thermal conductance of the nest wall for 23 Australian cup-nesting passerine species considered in Heenan and Seymour (2011). Equations refer to regression estimates (D.C. Deeming, 2014 unpublished analysis).

Conductance = $91.1NM^{-0.033}$ $P > 0.05$ $R^2 = 0.007$

Conductance = $89.6BM^{0.058}$ $P > 0.05$ $R^2 = 0.004$

make mechanical sense of the whole structure. However, they also illustrate the absence of rigorous quantitative studies of birds' nests and highlight that further studies are required.

Certain materials seem to have key roles in nest construction (Hansell 2000). In particular, mud has been previously considered (Rowley 1970), although a consideration of its mechanical properties is complex (Silva et al. 2010). Spider (Araneae) silk is also commonly used by hummingbirds and by 23 families of the Passeriformes (Hansell 2000). Storer and Hansell (1992) investigated the use of silk within common chaffinch (*Fringilla coelebs*) nests and found that web silk was only used from cribellate spiders but most of the silk was from cocoons. Quantity of cocoon silk was positively correlated with that of lichen in the nests and appeared to be more related to lichen attachment and moss binding. The function of white-coloured lichen and spider cocoons may not be related to the cohesiveness of the nest structure, as Hansell (1996) suggested that the presence of light material on the outer layer of the nest served as a mechanism of concealment by light reflection rather than an attempt to colour match with the nest's local environment. Many species apply lichen to nests when it is absent from branches to which nests are attached. Thus, the mechanical properties of silk to bind and attach small nests to branches and other structures should be relatively easy to study and would be a fruitful area for further research as they involve only a single material in providing tension during nest construction.

4.5 Conclusions and future directions

We conclude that birds exhibit a great deal of plasticity in the materials used in nest construction. This is in contrast to the general perception held a few decades ago, when nest construction behaviours were assumed to be innate and that little variability was possible (see Chapter 3). However, recent studies have convincingly demonstrated that nest-building behaviours are changeable within the course of an individual's lifetime and that individuals use those materials that benefit either themselves or their offspring most. At the interspecific level, there is great variation in the type of nests built and to a lesser extent, the number of nests built per breeding season. At the intraspecific level there is great plasticity in the time taken to build nests, the materials used, and

their final size. Such plasticity often serves to create a suitable microclimate for parents and their offspring, and materials with good insulatory properties such as feathers are key nest components in creating a suitable micro-environment within nests. However, despite such variability in nest construction behaviours, we should not lose sight of the fact that nests are basically structures built to contain parents, their eggs, and in many cases their offspring. Thus, the design of nests is likely to reflect a trade-off between the requirement for a structural platform for the nest contents and for creation of suitable micro-environments for offspring development. Yet, there is no consensus currently regarding the outcome of this trade-off.

Understanding of the functional properties of birds' nests is still in its infancy, with our knowledge generally limited to a relatively small number of studies that focus on hole-nesting passerine species in Europe. Consequently, there are many avenues to explore in the future. First, many more empirical studies are required involving both the commonly studied hole-nesting passerines in Europe and the less well-studied open cup-nesting species both in Europe and in other continents, to increase the generality of the findings acquired to date. Moreover, this would allow us to examine spatial aspects of the functional properties of nests across the world's ecozones. Secondly, such empirical studies could usefully be brought together in meta-analyses that would increase understanding of the determinants of such plasticity in the functional properties of birds' nests. Thirdly, the methods used to quantify the insulatory properties of nests differ between studies and we call for a more standardized approach that will allow more informed comparisons between studies. Also, a comparison of the main methods used to date would be useful so that we can understand the role of methodological inconsistencies in explaining such apparent variation. Finally, given that nests are structural platforms to contain eggs and/or offspring, it would be useful in future work to examine how the requirement to build a nest with a suitable micro-environment for these contents is traded off against parameters considered within this chapter.

Acknowledgements

We thank Mike Hansell and Jim Reynolds for comments that considerably improved previous versions of this chapter.

CHAPTER 5

The influence of predation on the location and design of nests

M.C. Mainwaring, S.J. Reynolds, and K. Weidinger

5.1 Introduction

For many species the risk of predation is ever-present and natural selection favours those individuals that employ behaviours that minimize the risk of predation (Caro 2005). Natural selection exerts strong selective pressures on birds during the breeding season when a predation event not only results in the loss of eggs or the offspring, which reduces reproductive success, but may also result in the death of the attending adult birds. Moreover, the risk of predation can also have a number of non-lethal effects on breeding birds that are probably underestimated compared with the direct effects. Indeed, many birds show an array of behavioural and physiological changes in response to the risk of predation, which often have important influences on the breeding biology of the birds (reviewed by Cresswell 2008). Consequently, there are strong selective pressures on the behaviours of the adult birds (reviewed by Lima 2009; Chalfoun and Schmidt 2012), but also on the location and design of nests. While all birds lay eggs they exhibit much variation in the location and design of nests. However, nests are multi-functional structures and other factors may determine their location and design (reviewed by Mainwaring et al. 2014a; Chapter 4).

In this chapter, we consider the effects of predation on nest location and design. The location of nests is defined in terms of their placement with respect to topography, micro-scale habitat, orientation, height, vegetative cover, and occurrence of other nests. Meanwhile, the design of nests is defined in terms of their building materials, shape, and size. We will also consider how nest ecology is determined by avoidance of predation and how it interacts with other selective factors, such as the requirement to act as a sexual signal, to reduce the impact of parasites or to create a suitable microclimate for offspring growth. Finally, we will indicate areas where further research attention should be focussed.

5.2 Factors affecting nest location

The location of nests varies greatly between species (Skutch 1976; Collias and Collias 1984; Hansell 2000). For example, a variety of seabirds nest on cliff ledges overlooking the world's oceans whilst species such as oilbirds (*Steatornis caripensis*) nest inside caves. A huge range of passerines build their nests in trees, bushes, and vegetation close to the ground, ranging from species such as Australian ravens (*Corvus coronoides*) nesting in trees, Eurasian reed warblers (*Acrocephalus scirpaceus*) nesting in reeds, and grey catbirds (*Dumetella carolinensis*) nesting in woodland understorey vegetation. Elsewhere, a variety of owls, raptors, woodpeckers, and passerines nest in tree cavities. Meanwhile, gamebirds nest in well-concealed locations within tall grasses and other vegetation at ground level whilst waders either build minimal nests or lay their eggs on the bare ground in no nest or just a scrape. These include pectoral sandpipers (*Calidris melanotos*) nesting on arctic tundra and Eurasian dotterels (*Charadrius morinellus*) nesting on alpine plateaux (Hansell 2000).

5.2.1 Topographical features

Birds select nest sites not just according to landscape and vegetation features but also in relation to local topographical features, such as slopes (Newell and Rodewald 2011). Such variation over relatively large scales might be more important at nest sites with little or no vegetation, as in ice-free Antarctic areas. Adelie penguin (*Pygoscelis adeliae*) colonies are positioned at locations within the landscape that provide benign microclimates with low risks of flooding (Moczydłowski

Mainwaring, M.C., Reynolds, S.J., and Weidinger, K., *The influence of predation on the location and design of nests*. In: *Nests, Eggs, and Incubation*. Edited by: D.C. Deeming and S.J. Reynolds, Oxford University Press (2015). © Oxford University Press.
DOI 10.1093/acprof:oso/9780198718666.003.0005

1986, 1989). There is a general trend for ground-nesting birds to select slopes rather than flat areas as places on which to nest. Nests of water pipits (*Anthus spinoletta*) tended to be on east-northeast facing slopes in the Austrian and Swiss Alps where they provided shelter from prevailing winds which, in turn, resulted in greater fledgling production than nests on west-southwest facing slopes (Bohm and Landmann 1995; Rauter et al. 2002). Meanwhile, northern harriers (*Circus cyaneus*) clearly preferred nesting on north-west facing slopes in Scotland although this did not increase their breeding success (Redpath et al. 1998). This may be because the benefits were accrued directly by the adults as was the case in great bustards (*Otis tarda*) that preferentially nested on slopes orientated to the south-east that provided incubating females with protection from the cold north-westerly winds (Magana et al. 2010). Eggs and chicks of Atlantic puffins (*Fratercula arctica*) were less vulnerable to predatory gulls (*Larus* spp.) on slopes as they were less frequently displaced to burrow entrances by panicked take-offs of adults (Nettleship 1972).

Other species prefer to nest on flat areas, although they tend to be those that lay their eggs directly onto bare ground rather than in short vegetation. European golden plovers (*Pluvialis apricaria*) nesting on flat areas had significantly higher rates of survival than conspecifics nesting on slopes, which was probably due to predators being less visible from the latter (Whittingham et al. 2002). Meanwhile, mountain plovers (*Charadrius montanus*) prefer to nest almost entirely on flat areas perhaps as an adaptation to heavy rains since this would reduce the chances of nests being subjected to heavy run-off (Graul 1975).

5.2.2 The orientation of nests

Many studies have quantified the orientation of open cup nests built within short vegetation on the ground. Most of these quantify the orientation of either the nest entrance or the nest cup in relation to a nearby object such as a tuft of grass or a large stone. Studies provide scant evidence that the nest orientation influences their risk of being predated. Gillis et al. (2012) used artificial nests to show that the height of surrounding vegetation was the most important factor determining the location of nests as it simultaneously provided a suitable microclimate for offspring development and protected the nest contents against predators. Creation of a suitable microclimate seems to determine the orientation of open cup nests within short vegetation on the ground. Illustratively, many empirical studies report that birds breeding in relatively cool environments orientate their nests away from prevailing winds (Sidis et al. 1994; Hoekman et al. 2002; Burton 2006; Mallord et al. 2007; Long et al. 2009; Robertson 2009). Birds breeding in relatively warm or even hot environments orientate their nests away from the hot midday sun (Yanes et al. 1996; Hartman and Oring 2003; Tieleman et al. 2008). A systematic review of the subject by Burton (2007) concluded that at lower latitudes in the northern hemisphere, the requirement for shade from the hot sun resulted in the majority of nests facing northwards. Most nests at mid-latitudes face eastwards, probably reflecting a compromise between thermal benefits in the early morning and shade in the afternoon. At higher latitudes most nests face southwards in order to face the sun during the majority of the day (Burton 2007). The limited data available from the Southern Hemisphere conform to these general patterns (Mezquida 2004).

Thus, nest orientation does not appear to be random. When Lloyd and Martin (2004) experimentally changed the orientation of chestnut-collared longspur (*Calcarius ornatus*) nests by removing vegetation and providing artificial shade, nestlings grew more slowly in nests whose orientations had been altered compared with those in control nests that were kept in their original orientations. This confirmed that nestling growth was causally linked to variation in nest microclimate arising from nest-orientation preferences towards the south-east that resulted in high midday temperatures, suggesting that these preferences were not adaptive with regard to nestling growth. We need to consider in greater detail how nest orientation in relation to wind and sun is influenced by the configuration of the local terrain (Sutter 1997; Lloyd et al. 2000). Nest orientation may not always be important in determining nest microclimates or in some cases birds may be prevented from optimal nest orientation. For example, the orientation of saltmarsh sparrow (*Ammodramus caudacutus*) nests is restricted by tidal conditions and prevailing winds that cause grasses to lie in one direction, which constrains the orientation of nest entrances (Greenberg et al. 2006). Moreover, perpendicular orientation of nest openings towards prevailing winds may reduce the dissipation of offspring and adult odours, although this hypothesis remains untested (Conover 2007).

The orientation of nests within tree canopies has been studied to a lesser extent than those on the ground, but similar findings emerge. There is a tendency for open nests to be orientated towards the sides of trees that best shelter them from prevailing winds (Martin and Roper 1988; Viñuela and Sunyer 1992; Summers et al. 2002; Rae and Rae 2013), or excessive sun in warmer environments (Rae and Rae 2013), although not in all cases (Glück 1983). For cavity nests, a review of 16 studies involving 19 species across 11 families revealed that nine studies (56%) found that cavity-entrance orientation was not important during nest-site selection. By contrast, four (25%) found the opposite, and the remaining three (19%) provided evidence both for and against its importance (Rendell and Robertson 1994). Cavity entrances might be simply orientated away from the rest of supporting vegetation rather than to specific compass directions (Zwartjes and Nordell 1998).

Empirical studies provide some, but by no means consensual, support for nests being orientated to avoid prevailing winds or insolation. Such benefits from nest orientation may be indirect because they impact on the behaviour of incubating adult birds. Incubating mountain plovers tend to face away from the sun during the heat of the day and spend much time standing over the nest with raised feathers thus providing shade (Graul 1975). However, it was postulated that such behaviours may result in increased nest detectability to predators (Augustine and Skagen 2014). Incubating Adelie penguins consistently face towards strong winds resulting in reduced levels of vigilance, which makes the

colony more vulnerable to attacks from South Polar skuas (*Stercorarius* spp.) approaching from the opposite direction (Young 1994a). However, our understanding of parental behaviours in relation to the orientation of nests remains poor and further studies could usefully examine if compensatory behaviours occur.

5.2.3 Nesting substrate in ground nests

In those species that lay their eggs directly on bare ground, it has been suggested that background matching reduces the risk of predation (Sánchez et al. 2004; Lovell et al. 2013; Chapter 12). For example, studies of waders nesting on beaches with pebbles have shown that the presence of egg-sized stones may enhance background matching (Cohen et al. 2008; Colwell et al. 2011). Stones may confer additional protection if predators fail to recognize real eggs amidst a backdrop of distracting egg-like objects (Stoddard et al. 2011). Moreover, greater mismatches between the colouration of eggs and surrounding stones resulted in higher levels of predation than nests with close matches (Solís and de Lope 1995; Blanco and Bertellotti 2002; Figure 5.1). Apart from background matching, crypsis

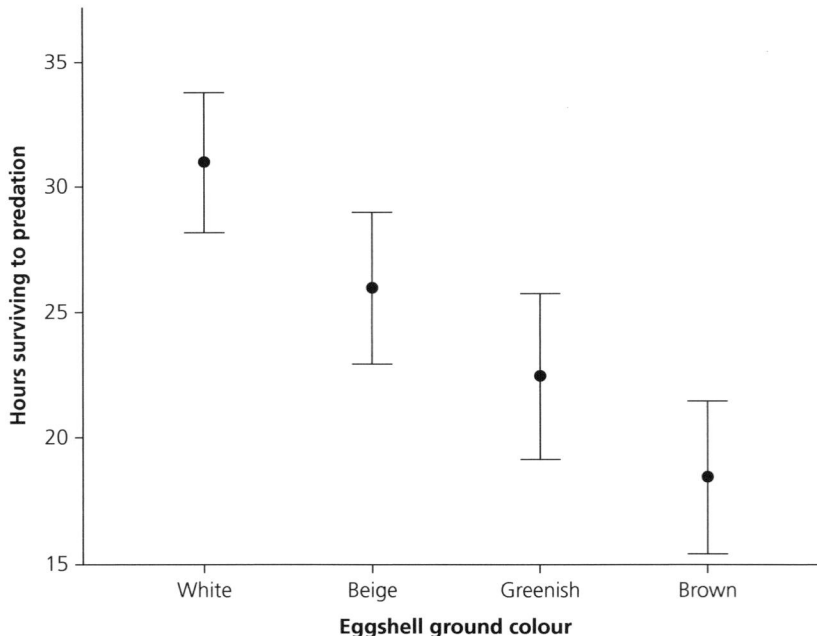

Figure 5.1 The mean number of hours (± 1 SE) that experimental South American tern (*Sterna hirundinacea*) eggs remained unpredated by gulls in relation to their colour along a ground-substrate matching continuum from 'white' eggs (best matching with substrate) to 'brown' eggs (least matching). Sample size was 14 eggs in each colour group. Reproduced from Blanco and Bertellotti (2002) with permission from John Wiley & Sons.

could be enhanced also through visual diversity of egg colouration within a clutch (Lloyd et al. 2000).

There are, however, inevitably trade-offs between maximizing heat reflectance to improve egg micro-climate and minimizing the conspicuousness of nests containing eggs to predators (Mayer et al. 2009; Saalfeld et al. 2012). Both stone curlews (*Burhinus oedicnemus*) and Kentish plovers (*Charadrius alexandrinus*) nest on substrate with sparse amounts of short vegetation. When researchers experimentally removed camouflaging nest material from their nest sites, the birds replaced material rapidly (Solís and de Lope 1995; Szentirmai and Székely 2002). This suggests that such material has a functional role in crypsis and far from nests being simple, they may well be constructed to reduce the risk of predation.

5.2.4 Substrates for burrow and cavity nests

Many species nest in burrows and cavities and the selection of an appropriate substrate may be fundamental to a successful breeding attempt. For example, when they are excavating tunnels in soil banks in which to breed, sand martins (*Riparia riparia*) prefer to burrow into compacted soil to avoid entrance holes collapsing and thereby entrapping eggs, nestlings, or adult birds (Heneberg 2009). European bee-eaters (*Merops apiaster*) prefer to burrow in fine-grained sedimentary substrates while common kingfishers (*Alcedo atthis*) prefer highly compacted substrates (Heneberg 2013).

Hole-nesting species choose to nest in cavities that can either be in living or dead trees, with the latter often showing significant decay and thus containing more potential nesting cavities. Primary hole-nesting species often excavate cavities in dead trees as the wood is softer and hence, easier to excavate. However, secondary hole-nesting species are unable to excavate their own cavities and so rely on cavities that are created by decay over many years or excavated by primary hole-nesting species. Consequently, secondary hole-nesting species are often reliant on such species for breeding opportunities and without them, they must wait for cavity formation through damage from decay or lightning strikes (Cockle et al. 2011). However, studies show that marsh tit (*Poecile palustris*) nests in dead wood and old woodpecker (*Dendrocopos* spp.) holes are predated more frequently than those constructed in living wood and holes of non-woodpecker origin (Wesołowski 2002). Hence, whilst primary hole-nesting species are able to nest in the safest locations as they can excavate their nests at preferred sites, secondary hole-nesting species have to nest where they can find a cavity, which may result in higher rates of nest predation.

5.2.5 Vegetation as nesting substrate

Vegetation provides structural support for nests and camouflage against detection by predators. While the importance of vegetative concealment is well recognized (see section 5.2.9), less attention has been paid to the role of species composition of nest-supporting or concealing vegetation. For example, while birds are well-adapted to concealing their nests in native trees and shrubs, Borgmann and Rodewald (2004) found nest predation rates to often be higher for nests in exotic shrubs than those of conspecifics nesting in native shrubs, although an opposite effect has also been reported (Schmidt et al. 2005). In temperate environments, there is a well-documented pattern of birds choosing to nest in deciduous trees and shrubs with increasing frequency as spring progresses and the amount of vegetation available to conceal nests increases (Morton et al. 1993; Ludvig et al. 1995). Nevertheless, birds choosing nest sites might discriminate even among similar plant species, as in the case of oak (*Quercus* spp.) trees within a mixed-oak forest (Newell and Rodewald 2011). Furthermore, yellow-faced honeyeaters (*Lichenostomus chrysops*) nesting in dioecious plants (*Coprosma quadrifida*) only nested in non-fruiting male plants. When artificial nests were placed in both fruiting and non-fruiting plants, nest predation rates did not differ between them (Boulton et al. 2003). On the other hand, lower predation on artificial nests placed in mistletoe (*Amyema* spp.) compared to eucalypt (*Eucalyptus* spp.) trees might explain the frequent use of this nesting substrate by Australian birds (Cooney and Watson 2008). This highlights the need for manipulative experiments to disentangle the effects of parental behaviour and nest-site traits *per se*.

Some species attempt to minimize the risk of predation by nesting in thorny vegetation which provides both a physical and visual barrier to predators. Thorny vegetation decreases incidents of predation by various snake species on baya weaver (*Ploceus philippinus*) nests (Quader 2006). When the perceived predation risk was experimentally increased using audio playback, female song sparrows (*Melospiza melodia*) built their nests in denser, thornier vegetation (Zanette et al. 2011). These studies suggest that individuals nesting in thorny vegetation are selected for, but further studies would confirm whether this behaviour is widespread.

For those species whose nests are located in more open parts of trees or bushes, patterns and types of branches are also likely to influence nest predation rates, because they may provide crypsis for the eggs, chicks, and attending adults. Tawny frogmouths (*Podargus strigoides*) placed their nests in rough, flaky-barked tree species and on open mid-branch sites with no foliage, where the birds' plumage and posture resembled the colour and form of branches (Rae and Rae 2013). Other studies have shown that nesting success increased with thickness of the supporting branch (Quader 2006; Newell and Rodewald 2011), suggesting that structural support was more important than accessibility to climbing predators. Moreover, a multi-species study provided little evidence for 'branch matching' across 25 species producing lichen-bearing nests (Hansell 1996). As with other aspects of nest-site selection in relation to nest substrate, there are too few studies for us to make any major conclusions.

5.2.6 Factors affecting nest accessibility

Birds attempt to minimize the accessibility of their nest sites to predators by nesting in inaccessible locations such as on cliff ledges. However, such nest sites can potentially be accessed by avian predators and so further selection takes place at such sites to reduce accessibility even further. For example, the risk of predation for thick-billed murres (*Uria lomvia*; Gaston and Elliot 1996) and cape petrels (*Daption capense*; Weidinger 1998) was higher for those birds nesting on wide cliff ledges on which avian predators are able to alight than for those nesting on narrower cliff ledges. A similar pattern of predation emerged in nests of common murres (*Uria aalge*) which were inaccessible to predatory gulls when on narrow ledges, but eggs were exposed to higher risk of accidental displacement during nest defence on narrow ledges (Gilchrist and Gaston 1997).

Weather conditions also influence patterns of predation in addition to cliff topography and in this case when wind speeds were sufficiently forceful, gulls were able to hover adjacent to the nests and take eggs or chicks (Gilchrist and Gaston 1997). Moreover, nests of Antarctic petrels (*Thalassoica antarctica*) that were in more sheltered locations on rocky slopes suffered lower levels of predation but benefits of shelter were traded off against costs of meltwater flowing directly on to nests more frequently (Varpe and Tveraa 2005). Some species nest underground in burrows where accessibility, among other factors, influences the risk of predation. The height of nest entrances was a significant predictor of egg loss in Magellanic penguins

(*Spheniscus magellanicus*; Stokes and Boersma 1998) and the length of tunnels was positively correlated with daily nest survival rates in burrowing owls (*Athene cunicularia*; Lantz and Conway 2009). However, the risk of predation is not the only determinant of nest location in burrowing birds because the characteristics of the soil are also important. Sand martins nesting in burrows on river banks preferentially located their burrows in the upper third of the bank to minimize the threat of flooding and access to ground predators from below (Heneberg 2013). The nests that best survived were also at least 20–50 cm below the top of the bank, preventing inundation of rainwater and attacks from predators from above (Heneberg 2013). Manx shearwater (*Puffinus puffinus*) burrows are located in coastal areas and vary in their susceptibility to flooding. Males that nested in burrows that flooded in one year moved burrows in the following year whereas males that bred in dry burrows in one year tended to breed in the same burrows in the subsequent year (Thompson and Furness 1991). Pallid swifts (*Apus pallidus*) nesting in rock cavities have to balance the risk of flooding in deeper cavities with the risk of overheating in more exposed sites (Penloup et al. 1997). Favourable thermal conditions inside nesting burrows might even relieve parent crab plovers (*Dromas ardeola*) from incubation duties, thus providing benefit in terms of increased foraging and vigilance for predators (Marchi et al. 2015).

Hole-nesting birds suffer lower rates of predation than open cup-nesting species (Martin 1993a) but birds still choose nest locations that offer the most protection from predators with the entrance diameter and wood hardness being among the most important factors (Wesołowski 2002; Paclík et al. 2009). Red-cockaded woodpeckers (*Picoides borealis*) not only choose cavity location, but also modify the surface of the trunk. They excavate resin wells and remove loose bark around their nesting cavities in living pine (*Pinus* spp.) trees so as to resist predatory attacks from climbing snakes trying to access nests (Delotelle and Epting 2007). The pattern of nest-site use by marsh tits reflects security of cavities of different origin (Wesołowski 2002). Cavities situated in living trees and those with small entrances often meet requirements for both predator-resistance and favourable microclimate (Paclík and Weidinger 2007; Maziarz and Wesołowski 2013; Chapter 4).

Nests protected by water are particularly successful because many mammalian predators cannot access them. Survival rates of common pochard (*Aythya ferina*) nests increased along a wetland gradient (Figure 5.2; Albrecht et al. 2006). Nests of prothonotary warblers (*Protonotaria citrea*) were increasingly protected from

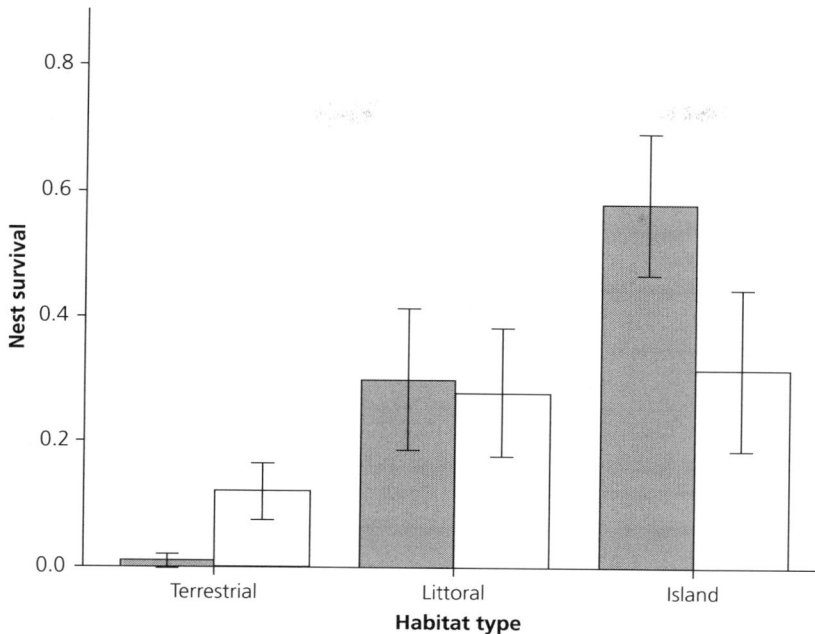

Figure 5.2 The mean survival (with 95% confidence intervals) of natural and artificial common pochard (*Aythya ferina*) nests along a wetland gradient from dry 'terrestrial' areas with minimal water nearby through to wet 'island' areas surrounded by water. Note that natural nests are represented by grey bars and artificial nests are represented by white bars. Nest survival was extrapolated to 32 days, i.e. the period of laying and incubation, from daily survival rates; sample size varied from 20 to 93 nests per group. Reproduced with permission from Albrecht et al. (2006).

raccoons (*Procyon lotor*) by increasing water depth (Hoover 2006). However, Brewer's blackbirds (*Euphagus cyanocephalus*) that nested in bushes located within water traded off their inaccessibility to land predators with reduced levels of nest concealment, because flooded vegetation lost more leaves than that rooted in dry land (Furrer 1975). Nesting over water does not always reduce rates of nest predation and, in fact, a study of large-billed gerygones (*Gerygone magnirostris*) in Australia showed that it may incur increased costs from more frequent flooding events (Noske et al. 2013). Consequently, further studies are required to examine the costs and benefits of nest inaccessibility to predators, particularly within the context of climate change (Chapter 6).

5.2.7 Nest height

The height of nests from the ground represents an obvious physical barrier for those ground predators that are unable to climb. If losses of nests to ground predators are additive to those from avian and climbing predators, we would predict that nest predation rates will decrease with nest height. However, a multi-species comparative analysis of predation rates among vegetation strata that controlled for habitat type showed that the above prediction only held in open grassland habitats (Martin 1993b). However, in woodlands predation was the lowest for ground nests and highest for shrub nests (Martin 1993b). Such broad patterns are less clear in cavity nesters, although within a guild of cavity-nesting species in Arizona, species with lower nest heights had lower nesting success (Li and Martin 1991). Nevertheless, at a within-species level, predation clearly partly drives nest height. Orange-crowned warblers (*Vermivora celata*) on a mainland site exposed to avian and mammalian predators nested at ground level while conspecifics that were introduced to an island where avian predators were absent nested off the ground and away from mammalian predators (Peluc et al. 2008). The above example demonstrates that different classes of predators often exert opposing selective pressures with respect to nest height. If these effects are compensatory rather than additive, the total predation will be independent of nest height (Remeš 2005a). Such predator-specific effects, whether suspected or documented (Söderström et al. 1998; Conkling et al. 2012; Sperry et al. 2012; Vanderwerf 2012), may account for the inconsistent results of empirical studies. Some studies have shown that higher nests are predated less frequently than lower nests (Burhans et al. 2002; Antonov and Atanasova 2003; Yeh et al. 2007), others have found the opposite (Hatchwell et al. 1999; Prokop 2004), and some have shown nest height to have negligible influence on nest predation rates (Hirsch-Jagobson et al. 2012; Latif et al. 2012).

Although the majority of studies suggest some link between predation rate and nest height, predation must surely not be the predominant selective factor determining nest heights in all 'study systems'. For example, more elevated nests are more often usurped by raptors (Prokop 2004) and those nests built in less stable branches are more vulnerable to damage from, and displacement by, strong winds (Bijlsma 1984). Moreover, microclimate in nests located higher and on the periphery of trees and shrubs differs markedly from nests within the *corpus* of vegetation. Under laboratory conditions open cup nests of Australian birds experienced increased heat loss when exposed to windy conditions (Heenan and Seymour 2012). In cooler temperate environments, pinyon jays (*Gymnorhinus cyanocephalus*) nested in more exposed and higher sites during cold weather because solar radiation reduced the energetic costs to nesting females (Marzluff 1988). Nest heights of mountain white-crowned sparrows (*Zonotrichia leucophrys oriantha*) decreased as spring progressed and declining snow cover enabled ground nesting but there was no effect of nest height on predation rates (Morton et al. 1993). Even small differences in height above ground might expose nests to opposing selective pressures. For example, meadow pipit (*Anthus pratensis*) nests placed relatively higher within grass tufts were more vulnerable to predation but incubating females had lower energetic expenditures as the nests were less damp (Halupka 1998). Similarly, the risk of flooding also declines with nest height in large-billed gerygones (Noske et al. 2013), but higher nests of Savi's warblers (*Locustella luscinioides*) and various sparrows (*Melospiza* spp.) are predated more frequently with declines in nest concealment as vegetation becomes more sparse (Aebischer and Meyer 1998; Greenberg et al. 2006). Other species have adapted aspects of their reproduction to the threat of flooding, with one study showing that saltmarsh sparrows time their reproduction to avoid especially high tides, while seaside sparrows (*Ammodramus maritimus*) avoid flooding by nesting in taller vegetation (Gjerdrum et al. 2005; Chapter 15).

5.2.8 Nest concealment by vegetation

We expect nests concealed by vegetation to be predated less because of restricted transmission of visual (Santisteban et al. 2002), auditory (Haff and Magrath 2011) and olfactory cues (Conover 2007) to potential predators. Borgmann et al. (2013) reported seasonal variation in the risk of nest predation associated with foliage density. Hatchwell et al. (1996) reported higher nest predation rates at exposed, rather than concealed, common blackbird (*Turdus merula*) nests, and the same was true in a study of artificial nests in an aviary study (Santisteban et al. 2002).

Studies that experimentally manipulate the vegetative concealment of nests are rare. While some have shown that the experimental concealment of nests influences predation in the expected direction (Stokes and Boersma 1998; Latif et al. 2012), others have not (Howlett and Stutchbury 1996). Such inconsistent findings are currently difficult to explain but, as above, the effects are probably predator-specific and density-dependent and nest concealment may be traded off against other factors. For example, the potential-prey-site hypothesis (Martin and Roper 1988; Martin 1993a) proposes that the probability of a predator finding stationary prey, such as eggs or nestlings, should decrease with increasing numbers of unoccupied potential sites that must be searched. By contrast, the total-foliage hypothesis (Martin and Roper 1988) predicts that predation rate should decrease with the total volume of vegetation containing nests. A habitat manipulation study provided unequivocal support for the former hypothesis as birds chose nest patches containing both greater total foliage and potential nest-site density, but only the latter factor significantly influenced nest predation rate (Chalfoun and Martin 2009). Apart from density of potential nest sites, the effect of nest concealment was modulated by the density of active conspecific nests. Well-concealed nests of Eurasian reed warblers are generally less often parasitized by the common cuckoo (*Cuculus canorus*), which is also an egg predator. However, when the density of available warbler nests was low, cuckoos increased their searching effort to the degree that concealment had no beneficial effect on the rate of parasitism (Jelínek et al. 2014).

Predation on nest-attending adults is potentially an important source of mortality (Reidy et al. 2009). Hence, nest concealment may also be traded off against the degree of visibility that adults afford themselves as they incubate eggs or brood chicks but also remain vigilant against predators (Götmark et al. 1995). Accordingly, great bustards select sheltered nest sites with good visibility of surroundings (Magana et al. 2010), while other ground-nesting birds also prefer generally open habitats (Keyel et al. 2013), even those habitats that are disturbed by a combination of herbivores, fire, and drought (Augustine and Skagen 2014). In rock ptarmigans (*Lagopus muta*) concealed sites were less often detected by predators but were more risky for incubating females that were less able to escape from predators once they had been detected (Wiebe and Martin 1998). Two sorts of evidence support the above correlative data. First, experimentally restricting the view of incubating birds from their nests delayed detection of approaching predators and they were more likely to succumb to predation on covered than on exposed nests (Amat and Masero 2004). Secondly, selection of nest sites might reflect tolerance to stress, where bold individuals should select concealed sites at the expense of lowered chances of escape from predation (Seltmann et al. 2014). Yet another situation where low nest concealment might be an advantage is in the advertisement of its likely low profitability (Galligan and Kleindorfer 2008; see section 5.3.2), which may discourage predators from closer inspection of obvious or already discovered nests (Schmidt 1999).

We have much to learn about the adaptiveness of nest concealment and the lack of progress may be due to simplistic approaches in investigations to date. Most studies use a single metric of habitat or behaviour as a measure of predation risk, but this overlooks a focal species' capacity for compensation in other behaviours (Weidinger 2002). Illustratively, studies have shown that there is sometimes no effect of nest concealment on predation rates at natural nests, but there was a positive effect when such nests were baited with plasticine clutches at which there was no parental activity. Thus, incubating adults probably compensated behaviourally for poor concealment of natural nests (Cresswell 1997; Remeš 2005b). On the other hand, any parental activity may disclose the nest location to potential predators thereby increasing predation rates (Smith et al. 2012), although not all studies have detected such an effect (e.g. Pope et al. 2013). Finally, visual components of nest concealment may also be traded off against non-visual ones, such as olfactory cues (Lehman et al. 2008; Conover et al. 2010). This rather speculative final point reflects our poor knowledge of how concealment acts as an important determinant of nest predation rates in birds. Further research in this area is needed urgently.

5.2.9 Nest sites and density dependence

Nest predators may select for nest-site diversification among co-occurring bird species via density-dependent foraging amongst predators. Hence, nest locations that disrupt the perceived nest density by predators may reduce nest losses from predation (Martin 1988). Indeed, those individuals within a species using nest sites that overlap with co-existing species experienced higher predation rates, thus suggesting nest-site partitioning among species (Martin 1996). However, such an effect is not supported in all studies such as in the case of partitioning of nest sites among vegetation strata (Reitsma and Whelan 2000). Nest predation rates of cavity-nesting common goldeneyes (*Bucephala clangula*) were unaffected by the experimental addition of artificial nests simulating those of ground-nesting mallards (*Anas platyrhynchos*), despite both nests being commonly predated by pine martens (*Martes martes*). Hence, there was no apparent competition between nesting guilds or heterospecifics and no density-dependence in predation risk (Elmberg and Pöysä 2011). Nevertheless, Wada (1994) found that presence of nests of conspecifics determined nest predation rates in Oriental turtle doves (*Streptopelia orientalis*), with heights of neighbouring nests influencing predation risk more than the height of the focal nest itself.

In terms of density dependence the effects on nest predation are equivocal across studies. For example, there was no support for consistent effects of density dependence on predation rates of natural and artificial songbird nests (Schmidt and Whelan 1999). By contrast, predation rates on artificial and natural duck nests were not influenced by density dependence at the level of individual wetlands but there was stronger evidence for it at larger spatial scales (Gunnarsson et al. 2013; Ringelman et al. 2014).

5.2.10 The profitability of nest contents as prey

Nest predation rates may be strongly influenced by the foraging behaviour of predators and the proportion of nests within a search area that are active and thus contain prey in the form of eggs and/or nestlings. Thus, predators might abandon searching for eggs and/or nestlings if the majority of nests within a search area are empty (Schmidt 1999) and so it may be in the interests of birds to decrease the ratio of active:inactive nests. There are three ways in which birds can achieve this: 1) by building multiple nests themselves within a given breeding attempt; 2) by forcing the abandonment of nests of neighbouring conspecifics and heterospecifics; and 3) through the accumulation of old nests.

Several species build multiple nests, including Australian reed warblers (*Acrocephalus australis*) and winter wrens (*Troglodytes troglodytes*). Nests built in addition to those in which eggs are laid are often thought to be part of sexual selection signalling the 'quality' of the builder (de Neve and Soler 2002) but they may also confuse predators (Hansell 2000). Whilst some studies have provided support for the anti-predator hypothesis of multiple nests (Leonard and Picman 1987; Watts 1987; Whitney et al. 1996; Noske et al. 2013), others have not supported this hypothesis (Metz 1991; Cavitt et al. 1999; Berg et al. 2006), even when the number of inactive nests was experimentally increased (Dubois and Getty 2003). For multiple nests to dilute the ratio of active:inactive nests through accumulation, nests must last for more than one breeding attempt. Studies have shown that 74% of Eurasian blackcap (*Sylvia atricapilla*) nests remain undamaged post-breeding and are thus available for potential replacement nesting attempts during the same breeding season (Zieliński 2012). Over 90% of red-winged blackbird (*Agelaius phoeniceus*) nests survive between two successive breeding seasons (Erckmann et al. 1990). However, species such as the hair-crested drongo (*Dicrurus hottentottus*) sometimes destroy their nests to reduce future competition for nest sites (Li et al. 2009). Nevertheless, nests are often sufficiently durable to accumulate at least within a single year, which would predict predation rates decreasing towards the end of the breeding season (Schmidt 1999). This phenomenon is traditionally explained by changes in vegetation concealment and availability, or other prey for potential predators (Borgmann et al. 2013). Explicit tests of this idea are lacking and this field is thus open to further study.

5.2.11 Nest-site reuse

Birds may reuse nests from the previous breeding season or those of heterospecifics (e.g. Kubelka et al. 2014). While this is advantageous in terms of time saving (Cavitt et al. 1999) and having an intimate local knowledge of the breeding territory (Saunders et al. 2012), there may also be some disadvantages. For example, predators may remember nest locations from previous breeding attempts, both within and across seasons (Weidinger and Kočvara 2010) and ectoparasite load may increase over the lifetime of the nest structure (Chapter 8). Open cup-nesting species sometimes reuse old nests, preferably those that were previously successful (Styrsky 2005) or those in safe locations (Yeh et al. 2007). However, other studies reveal that nest reuse is rare (Cavitt et al. 1999), has no effect on nest success (Antonov and Atanasova 2003; Jiménez-Franco et al. 2014), and only occurs when high-quality nest sites are in short supply (Redmond et al. 2007; Cancellieri and Murphy 2013).

Shortage of nest sites is prevalent in hole-nesting species (Hansell 2000; Mazgajski 2009; Cockle et al. 2011; Wesołowski 2012), with excavation of cavities being energetically and temporally expensive for primary hole-nesters (Wiebe et al. 2007). High-quality nest sites and their supply are then restricted in some cases for secondary hole-nesters. From this we can expect frequent reuse of existing tree cavities, in spite of potential costs in terms of ectoparasite exposure (López-Arrabé et al. 2012; Mazgajski 2013). Nestbox experiments suggest that absence of ectoparasites was a more important determinant of nest-site selection than indications of previous successful reproductive events (Stanback and Dervan 2001). However, marsh tits breeding in natural cavities reused holes more frequently after successful reproduction than after failure (Wesołowski 2006). Accordingly, most empirical studies show that there is no clear effect of the presence of old nests on nest-site selection of secondary hole-nesting species (reviewed by Mazgajski 2007), which might be facilitated by rapid decomposition of old nest material (Hebda et al. 2013). Nevertheless, suitability of tree cavities as nesting sites ultimately deteriorates with their age as a consequence of wood decay (Wesołowski 2012) and increasing predation risk due to long-term memory of predators (Mazgajski 2009).

Site fidelity is usually high in long-lived seabirds but even here change of nest sites is more frequent after breeding failure (Ollason and Dunnet 1988; Thibault 1994). Moreover, in colonially breeding black-legged kittiwakes (*Rissa tridactyla*) the breeding success of neighbours may influence site fidelity more than the birds' own breeding experience (Danchin et al. 1998; Bled et al. 2011). Notable in this respect is the nest-site tradition in lesser grey shrikes (*Lanius minor*); most nests in successive years are built in the same or in neighbouring trees, but almost exclusively by different individuals. Hence, nest-site choice in this species is likely influenced by social information gathered from the presence of conspecifics and their breeding success (Krištín et al. 2007; Hoi et al. 2012). In summary, we know very little about the causes and consequences of nest reuse by birds and further studies could usefully examine the relative contributions of predation, parasitism, and conspecific attraction in determining the reuse of old nests.

Apart from reusing old nests, birds may use the presence of old nests of conspecifics as a cue for selecting new nest sites nearby (Safran 2004; Gergely et al. 2009; Ringhofer and Hasegawa 2014). However, high densities of old nests might not indicate suitable habitat as it may result from a high nest predation rate combined with a high rate of renesting. A bird's ability to determine the fates of such old nests may be extremely limited, perhaps explaining why so many studies have failed to find support for old nests acting as such cues (Erckmann et al. 1990; Yahner 1993; Cavitt et al. 1999).

5.2.12 Responses to predators

There is strong selection on the anti-predatory behaviours of breeding birds (Lima 2009). Nest predation is such a critical determinant of fitness that assessment of predation risk is an important component of habitat selection (Mönkkönen et al. 2009). Prior to nest building, birds may assess the local abundance of avian predators through heterospecific eavesdropping (reviewed by Magrath et al. 2014). The local abundance of mammalian predators can be assessed through detection of their excrement in the environment by means of ultraviolet (UV) vision and/or olfaction (Eichholz et al. 2012; Forsman et al. 2013). Birds prospecting potential nest sites can detect chemical cues left by predators inside nest cavities (Amo et al. 2008; Mönkkönen et al. 2009).

Birds not only perceive predation risk but respond to it by choosing safer nest sites (Forstmeier and Weiss 2004; Roos and Part 2004; Eggers et al. 2006; Peluc et al. 2008; Latif et al. 2012) or by spacing out nests (Hogstad 1995). However, they may respond to habitat predictors of predation risk rather than to real presence of predators (Møller 1988). However, in unpredictable environments, breeding birds may lack sufficient cues to determine predation risk accurately but, instead, may have to rely on direct experience of predation (Chalfoun and Martin 2010a). Accordingly, birds can react after their nest contents have been predated by moving to build a nest in a safer location (Marzluff 1988; Haas 1998; Chalfoun and Martin 2010b; Kearns and Rodewald 2013). However, not all studies have found such an effect (Howlett and Stutchbury 1996; Hatchwell et al. 1999; Kershner et al. 2004; Walk et al. 2004).

Humans can also be perceived as predators and disturbance can induce anti-predatory behaviours in birds. Great grey shrikes (*Lanius excubitor*) disturbed at the nest by researchers in one year increased their nest height the following year whereas undisturbed birds nested at the same height across both years (Antczak et al. 2005). Meanwhile, black redstarts (*Phoenicurus ochruros*) located nests in deeper cavities in the subsequent breeding season following human disturbance but nests remained at equivalent cavities across two years when not disturbed (Chen et al. 2011).

Taken together, the above evidence suggests that nest placement is a plastic trait that can develop with individual breeding experience and thus contribute to age-related improvement in breeding success (Horie and Takagi 2012). Although heritability of nest placement might be low (Yeh et al. 2007), nest height was shown to evolve in response to natural selection through novel predation pressure in an island environment (Figure 5.3; Vanderwerf 2012).

5.2.13 Presence of heterospecifics

Some species take advantage of breeding colonially, i.e. closeness and synchrony of breeding with neighbouring conspecifics (reviewed in Caro 2005). Other species sometimes preferentially nest close to other species and gain possible benefits by doing so. By associating with more aggressive species, often not restricted to the same taxa, birds seek protection from predators (reviewed in Quinn and Ueta 2008). In other cases birds simply exploit habitat created by other species or associate with their prey. For example, burrowing owls breed in black-tailed prairie dog (*Cynomys ludovicianus*) colonies where vegetation is kept short (Lantz and Conway 2009). Such interactions can be complex and require extensive research to elucidate their dynamics. For example, while breeding attempts of South Polar skuas (*Stercorarius maccormicki*) close to Adelie penguin colonies fail more often as a result of egg trampling by penguins and predation by conspecifics, they benefit from direct access to penguin eggs and chicks, suggesting that there may be an optimal distance for such heterospecific associations (Hagelin and Miller 1997). Thus, birds often, but not always, accrue benefits from nesting in association with heterospecifics. For example, predation rates on European pied flycatcher (*Ficedula hypoleuca*) nests were higher within, compared with outside, Ural owl (*Strix uralensis*) breeding territories, possibly because some smaller predators that do not directly threaten owl nests were attracted to the owls' breeding territories (Morosinotto et al. 2012).

To nest in close association with protective species birds must adjust location of their nests. Thus, common woodpigeons (*Columba palumbus*) associated with Eurasian hobbies (*Falco subbuteo*) build nests considerably

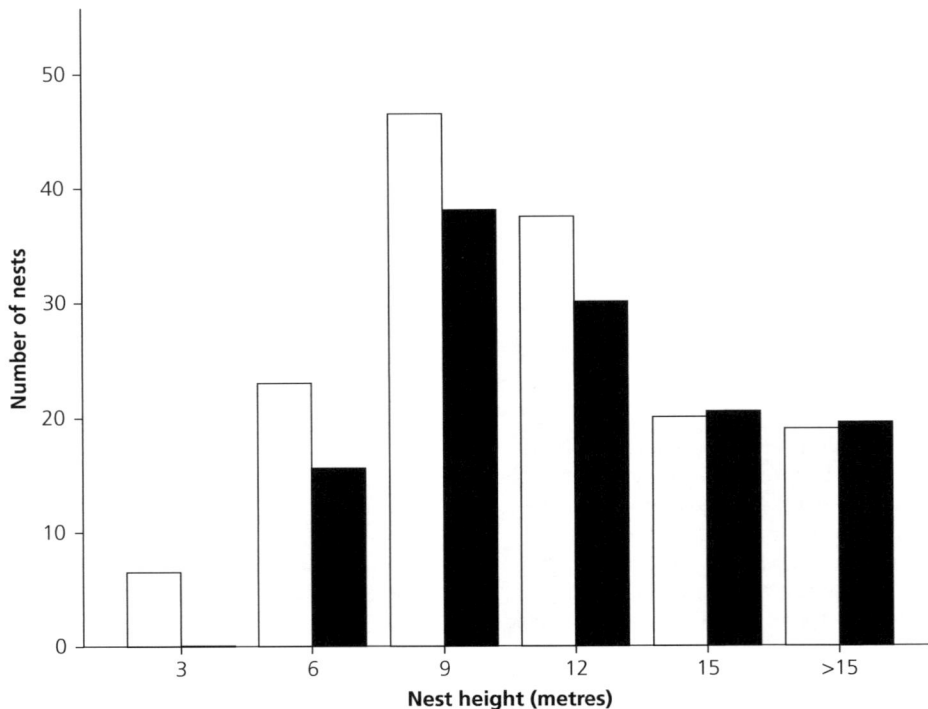

Figure 5.3 The number of elepaio (*Chasiempis sandwichensis*) nests on Oahu, Hawaii, that were predated/abandoned or were successful in relation to their height. The height of sequential nests built by individual birds did not change between 1996 and 2011 (inclusive), but the average height of nests increased by 50% suggesting that predation risk selected for higher nests. Note that predated/abandoned nests are represented by white bars and successful nests are represented by black bars. Reproduced from Vanderwerf (2012) with permission from John Wiley & Sons.

higher and breed in synchrony with them compared to non-associated pairs (Bijlsma 1984). Savannah sparrows (*Passerculus sandwichensis*) nesting within gull (*Larus* spp.) colonies chose nesting microhabitats avoided by gulls (Wheelwright et al. 1997). Azure-winged magpies (*Cyanopica cyanus*) use nest sites that are less concealed by vegetation when nesting in association with Japanese sparrowhawks (*Accipiter gularis*; Ueta 1998). Finally, the costs and benefits of associations are likely to be context-dependent. An experiment with great tits (*Parus major*) and European pied flycatchers showed that birds nested in trees with ants only under increased predation risk, but otherwise preferred trees without ants (Haemig 1999).

5.3 Nest design

In addition to the location of nests, birds can also minimize predation risk through their appropriate design. The three primary ways in which nest design may exert

influence are through: types of nest material used; the basic structure of nests; and the size of nests.

5.3.1 Nest materials

Birds use a wide diversity of nest materials and, whilst their function is comprehensively discussed in Chapter 4, here we discuss some of the more unusual nest materials and their role in minimizing predation risk. For example, some species use snake skin as a nest material, despite snakes being predators of many avian species (Borgo et al. 2006; Clucas et al. 2008). There is mixed empirical support for the hypothesis that snake skin reduces predation risk. Medlin and Risch (2006) found a lowering of predation risk at nests with snake skins but they had no effect in artificial nests (Trnka and Prokop 2011). Some species use mammalian scat as nest material. Some plovers (*Charadrius* spp.) select nest sites that are closer to piles of dung than randomly selected sites (Graul 1975). Burrowing owls routinely collect dry mammalian manure and scatter it around their nesting burrows (Smith and Conway 2011), perhaps to provide olfactory camouflage of eggs and chicks from predators (Levey et al. 2004). Common waxbills (*Estrilda astrild*) regularly use carnivore scat as nest material that lowers predation risk compared to control nests without carnivore scat (Schuetz 2005). The presence of faecal sacs in common blackbird nests had no effect on nest predation rates (Ibáñez-Álamo et al. 2014), but future research should examine the role of olfactory cues in predation events (Conover 2007). This is especially in light of a study of European rollers (*Coracias garrulus*) that confirmed that birds can 'smell the fear' of their offspring (Parejo et al. 2012). In addition, Canestrari et al. (2014) showed that carrion crow (*Corvus corone*) nests parasitized by great spotted cuckoos (*Clamator glandarius*) experienced lower predation due to smelling of repellent secretions from cuckoo chicks.

Feathers are another commonly used nest material and they are usually considered to create a suitable microclimate for offspring development (reviewed by Mainwaring et al. 2014a; Chapter 4). However, the experimental addition of feathers to nests resulted in higher nest predation rates than for control nests (Møller 1987). More broadly, it has been suggested that predation risk is the selective pressure that prevents all species from using feathers. This is because species suffering relatively low nest predation rates, such as hole-nesting species and those nesting above ground level, use feathers as nest material to a much greater extent than species suffering higher levels of predation (Møller 1984).

Many hole-nesting species incorporate green plant material into nests and they normally replenish them daily during the incubation and nestling stages (reviewed by Dubiec et al. 2013). They contain volatile secondary compounds that are biocides of parasites and pathogens (Chapter 8), but they also sometimes play a role in sexual selection (Brouwer and Komdeur 2004; Tomás et al. 2013). Although their incorporation could serve many functions, we know only one study that has examined how (or whether) they influence nest predation: the addition of flowers to artificial nests had no effect on predation outcome (McGuire and Kleindorfer 2007).

During incubation in many bird species no adults are temporarily present at the nest, but some species cover eggs with nest material during incubation recesses. This behaviour is usually explained within the context of an improved nest microclimate for developing embryos but such a hypothesis has received little empirical support (Haftorn and Slagsvold 1995).

Instead, several studies have shown that such behaviour reduces nest predation rates in open (Kreisinger and Albrecht 2008; Fast et al. 2010; Prokop and Trnka 2011) and possibly also in cavity nests (White and Kennedy 1997).

Specific microhabitats, such as tree cavities, represent shared resource for cavity-nesting birds and other non-avian species. Rodents are major predators of cavity-nesting passerines (Adamík and Král 2008a, 2008b; Czeszczewik et al. 2008) and they sometimes even occupy such nests to their own ends to save time and energy in nest construction. However, the role of specific nest materials in such interactions remains unknown. This is seldom considered in studies of open nests but a study of common blackbirds suggested that usurpation of nests occurred by rodents for their own reproduction and as food stores (Tamarit et al. 2012).

Nest materials may also be used to modulate nest humidity, that is, a measure of water vapour retention within nests (Slagsvold 1989b; Deeming 2011b). Although nest materials are usually thought to influence nest microclimate, they might contribute also to olfactory concealment. Indeed, some studies have shown that environmental humidity controls olfactory cues received by predators with the best conditions for predators being cool and humid nights with low air movements (Conover et al. 2010). Accordingly, the predation rates of several birds varied in relation to climatic conditions. The daily survival rates of savannah sparrow nests were highest on cool and wet days (Stauffer et al. 2011). Wet weather conditions resulted in higher nest mortality in mountain plovers as a result of higher activity levels and enhanced olfactory sensitivity of predators (Dinsmore et al. 2002; Dreitz et al. 2012). Thus, predators are more efficient at locating nests under certain weather conditions such as high winds and high rainfall (Webb et al. 2012). However, Mainwaring et al. (2014b) found that nest design of common blackbirds that suffer from high losses from nest predation did not differ in relation to changes in spring precipitation. However, the location and design of nests are likely to vary in relation to weather-induced variation in the behaviours of predators and so further studies should examine these issues in greater detail.

Birds also use a variety of anthropogenic materials as nest materials and as the world is urbanizing (United Nations 2014), we might predict increased incorporation of such materials into nests in the future. Whilst there is a great deal of anecdotal evidence for such materials (see Chapter 4), their functions remain unclear (Igic et al. 2009). Sergio et al. (2011) showed plastics in the nests of black kites (*Milvus migrans*) were involved in intraspecific signalling, but many more studies are required to examine such phenomena in more detail. However, anthropogenic materials are not always advantageous to birds with, for example, adult and nestling great grey shrikes becoming entangled in plastics incorporated in nests (Antczak et al. 2010), effects of which may cause massive mortality in seabirds under some circumstances (Votier et al. 2011). Meanwhile, house finches (*Carpodacus mexicanus*) line their nests with cellulose from smoked cigarette butts as it reduces the number of ectoparasites in nests but effects are not all beneficial. As the number of cigarette butts in nests increased, fledging success and immune responses of nestlings decreased, possibly due to their increased exposure to nicotine and other harmful substances (Suárez-Rodríguez and Macías Garcia 2014).

5.3.2 Nest structures

Nest structures vary greatly across bird species (Hansell 2000; Hansell and Deeming 2002; Chapters 3 and 4). Some nests have false structural components to deceive nest predators. Yellow-rumped thornbills (*Acanthiza chrysorrhoa*) build a lower domed nest where eggs are laid and an upper cup-shaped structure, referred to as the 'false nest'. Artificial nests with false nests had significantly lower predation rates than those lacking such structures (Figure 5.4; Galligan and Kleindorfer 2008), indicating that such nest structures fulfil an anti-predation function. Acadian flycatchers (*Empidonax virescens*) build nests with 'tails' but nest predation

rates were not associated with the prominence of such tails and so their purpose remains unclear (Master and Allen 2012). Although an attractive proposition, structures such as nest roofs provide benefits not only in terms of physical protection from predators, as is the case with the massive nests of magpies (*Cyanopica* and *Pica* spp.; Quesada 2007). Saltmarsh sparrows' domed nests with intact roofs retained eggs during tidal flooding better than nests with experimentally removed roofs (Humphreys et al. 2007), suggesting that domes may protect eggs not just through minimizing nest predation. Consequently, our knowledge is rather incomplete as to how such nest structures influence nesting success. Yet, we are in no doubt that nest predation is one of the major factors selecting for specific nest structures and types.

5.3.3 Nest size

For visual predators nest size must largely determine prey detectability with larger nests expected to be detected more easily than smaller ones (Møller 1990a; Götmark 1992). Some observational studies have shown that larger open cup nests are predated more frequently than smaller nests (Gregoire et al. 2003; Antonov 2004), but such correlative results refer to combined effects of nest size, nest location, and parental behaviour. Nevertheless, experiments with artificial nests have shown that larger nests are predated more frequently when placed in artificial locations (López-Iborra et al. 2004) or when nests of different sizes are

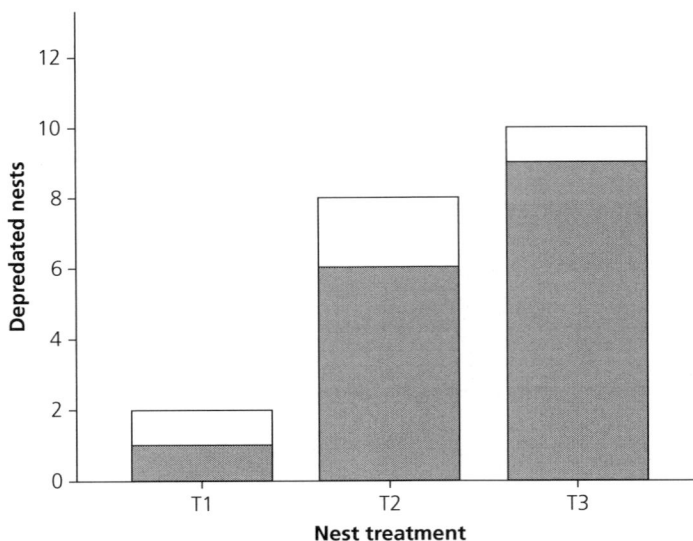

Figure 5.4 The number of yellow-rumped thornbill (*Acanthiza chrysorrhoa*) nests that were predated in relation to the presence of false nests. Thornbills build nests comprising a lower domed section in which eggs are laid and an upper cup-shaped structure which is a false nest. Using artificial eggs, there were 15 nests in each treatment: 'T1'—eggs in domed nests with false nests attached; 'T2'—eggs in domed nests without false nests; and 'T3'—eggs in false nests attached to a dome. Length of exposure: 'Early' (0–16 days) or 'Late' (17–34 days). Early exposure is represented by grey shading and late exposure by white shading. Reproduced from Galligan and Kleindorfer (2008) with permission from John Wiley & Sons.

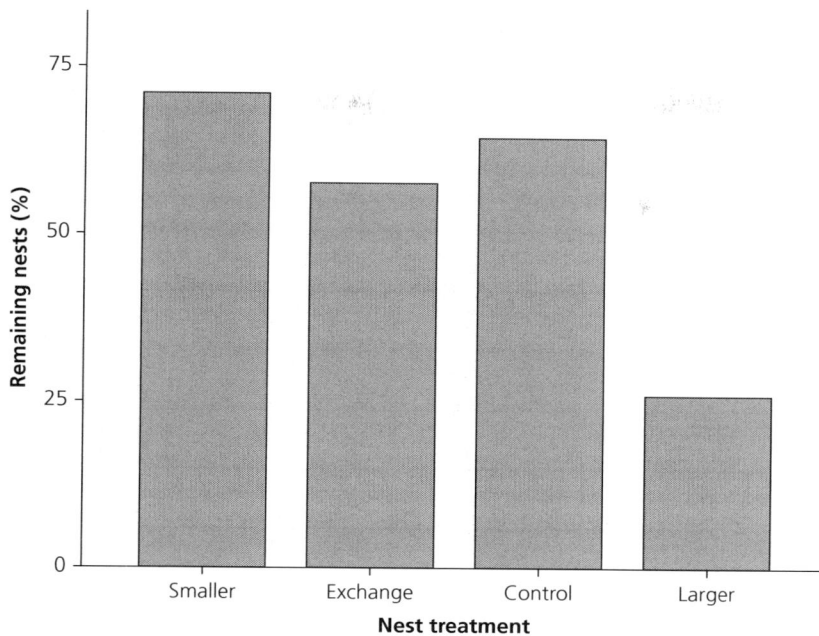

Figure 5.5 The percentage of common blackbird (*Turdus merula*) nests with plasticine eggs that remained unpredated after 14 days of exposure to nest predators in relation to nest size. After they were found, 'smaller' nests were replaced with smaller nests, 'exchange' nests were removed and then replaced at the same location to control for treatment effects, 'control' nests remained unmanipulated, and 'larger' nests were replaced with larger nests. Sample size was 30 nests in each treatment group. Reproduced from Møller (1990a) with permission from John Wiley & Sons.

swapped between original locations (Figure 5.5; Møller 1990a; Biancucci and Martin 2010). Experimental eggs directly attached to branches rather than within nest structures survived better than eggs placed in nests (Götmark 1992). Even for those species that build nests on the ground in the form of stone piles, the presence of more stones might be costly in terms of increased nest predation (Weidinger and Pavel 2013). Reductions in nest size are likely constrained by the need for protective nest structures or balanced by opposing selection on nest size (Fargallo et al. 2001).

The positive relationship between the size of nests and the rate of predation is unsurprising given that larger open cup nests provide larger visual targets for visual predators, while larger nests inside cavities decrease the 'danger distance' between the cavity entrance and nest contents (Mazgajski 2009; Kaliński et al. 2014). However, other studies of open cup-nesting species have found no evidence for nest size influencing nest predation rate (Lent 1992; Hatchwell et al. 1996; Palomino et al. 1998; Herranz et al. 2004; Suárez et al. 2005). In an experimental study Weidinger (2004) studied nests of different species that constructed nests of

different size and exchanged them between locations, showing that nest location was more important than nest size in determining nest predation rate.

Thus, while individual breeding birds may be trading off nest size against location, other factors may also be influencing nest size. The requirement for adequate structural support is the primary selective influence on nest size in Australian cup-nesting species (Heenan and Seymour 2011) although one study showed this may not necessarily always be the case (Palomino et al. 1998). This latter result may be explained by nest size co-varying with microhabitats resulting from physical constraints imposed by the architecture of nest-supporting vegetation (Lent 1992), or from environmental conditions across broad latitudinal gradients (Møller 1984; Mainwaring et al. 2012, 2014b).

Clearly, there is a research opportunity to examine the trade-off between building a sufficiently large nest that is well-insulated against one that is sufficiently small to minimize predation risk (Mainwaring et al. 2014a). Empirical studies to date have generally examined the relationship between nest size and predation rates in a rather oversimplified manner. Future research must examine how nest size interacts with nest location and parental behaviours to advance our understanding of the evolution of nest size of birds.

5.4 Summary and conclusions

Although all birds lay eggs, they show a wide diversity of nest locations and designs ranging from eggs laid on the bare earth by plovers (Charadriidae) through to small open cup nests built by warblers (Sylviidae) and huge stick constructions built by storks (Ciconiidae). Here, we have shown that such diversity is influenced by the risk of predation (Caro 2005). However, nests are multi-functional structures (Mainwaring et al. 2014a) and we discuss how provision of a suitable microclimate for offspring development and, to a lesser extent, their role as sexual signals and in combating parasites and pathogens, also explain such diversity.

We have a reasonable understanding of the determinants of variation in nest location and design, but there are four main avenues where further research is warranted. First, nests are multi-functional structures and studies have generally only considered a single function (e.g. minimization of predation risk) in isolation. For example, many studies have experimentally increased nest size and then monitored nest fate in terms of presence/absence of predation events without considering whether there are confounding variables. Non-building individuals of some species feed offspring more intensely at larger nests in response to the elevated investment of the building bird in the reproductive event. In this case, parental provisioning behaviours may disclose the nest location to predators.

Secondly, a wide range of methodological approaches are currently used to quantify nest location, nest design, and nesting output, and they can introduce bias into research findings. For example, studies focus on those species most accessible to researchers and at sites where birds are present at the highest densities. Moreover, studies have often used artificial nests but we know little about how much they differ from 'natural' nests that they are trying to mimic. For example, there are strong concerns that the absence of parents from artificial nests means that they do not realistically mimic a 'natural' reproductive event. Possibly the most serious weakness of nest predation studies to date is the anonymity of nest predators, but this can be recorded using video technology (Ribic et al. 2012). However, the anticipated anti-predator function of various nest traits is not only traded off against other functions, but the predators themselves might exert opposing selective pressures due to the diversity of their foraging modes.

Thirdly, we are using ever more sophisticated technologies to quantify the size and location of nests but we understand little about the biases associated with such approaches. For example, open cup nests are often quantified in terms of their location within a bush and their size in terms of their outer diameter when measured in a laboratory. However, the vegetation within the bush may be crucial in terms of nest dimensions and function, and is likely to change over the course of the breeding season in temperate regions. Such seasonal adjustments in nest characteristics must be considered and the data analyzed as time-dependent covariates.

Finally, studies should also consider how changing environmental conditions may force birds to alter nest location and design. For example, as ambient temperatures increase in arid regions of the world, nests may need to be located in shaded areas or the parents may need to build further sheltering structures to shade nest contents from direct sunlight. Such aspects are considered further in Chapter 6.

Acknowledgements

We thank Ian Hartley for useful discussions and Charles Deeming for useful comments on previous versions of the chapter.

Nest construction and incubation in a changing climate

M.C. Mainwaring

6.1 Introduction

Birds are among those taxa with oviparous reproduction which means that they must select a nest site in which to lay eggs. Birds' nests vary markedly across species with some constructing elaborate cup-shaped nests whilst others simply lay on the bare ground or on rock on cliff ledges (Hansell 2000; Mainwaring et al. 2014a). One of the most important determinants of nest design is the creation of a suitable microclimate for embryonic development (Ar and Sidis 2002). In passerines embryonic development occurs primarily within relatively narrow thermal limits (34–40°C); prolonged periods of time either below or above these limits result in abnormal embryonic development and death, respectively (Webb 1987; Beissinger et al. 2005; DuRant et al. 2013a). Thermal conditions experienced by embryos during incubation influence their phenotypes after hatching, with both higher and lower than optimal temperatures resulting in nestlings having lower body masses and body condition indices at pre-fledging than those developing under optimal temperatures (Peréz et al. 2008; Ardia et al. 2010). Consequently, birds are under strong selection pressure to situate and construct nests that provide appropriate thermal conditions for incubation and offspring development (DuRant et al. 2013a). Avian nest construction and incubation are both, therefore, inextricably linked to temperature and other environmental conditions, making them sensitive to a changing climate. Global climate change has seen unprecedented temporal changes in climatic variables such as temperature, precipitation, and wind speeds, largely as a result of increasing concentrations of greenhouse gases in the atmosphere from the burning of fossil fuels and land-use changes such as deforestation (Karl and Trenberth 2003; Rosenweig et al. 2008). Moreover, climate change constitutes not only predictable incremental changes in the mean values of climatic variables but also unpredictable and more extreme changes in them that can result in flooding, fires, and hurricanes (Easterling et al. 2000). The effects of both predictable incremental changes and unpredictable extreme changes in environmental conditions are considered in this chapter.

The ecological effects of climate change on breeding birds have been extensively studied (see Stenseth et al. 1999; Walther et al. 2002 and references therein), but empirical research to date has focussed disproportionately on the later stages of reproduction when parents are rearing offspring (reviewed by Dunn and Winkler 2010; Møller 2012). Meanwhile, the early stages of reproduction (e.g. nest construction and incubation) have received less attention with the research to date tending to focus on specific aspects of reproduction such as the breeding phenology of insectivorous birds (e.g. Cresswell and McCleery 2003; Both and Visser 2005; Matthysen et al. 2011; Vedder 2012). In this chapter, I begin by considering how avian nest construction and incubation behaviours may be affected by climate change. I then consider the potential role of both in facilitating adaptive responses to climate change within both ecological and evolutionary contexts before concluding by highlighting those areas that warrant further research. I consider studies performed across all of the world's ecozones and, whilst I focus on birds primarily, I will also use non-avian examples to highlight some specific points.

6.2 Effects of climate change

6.2.1 Nest construction under a changing climate

Climate change has affected, or is likely to affect, nest-building birds in several ways, with threats being

Mainwaring, M.C., *Nest construction and incubation in a changing climate*. In: *Nests, Eggs, and Incubation*.
Edited by: D.C. Deeming and S.J. Reynolds, Oxford University Press (2015). © Oxford University Press.
DOI 10.1093/acprof:oso/9780198718666.003.0006

direct and indirect. The most serious direct threat is the flooding of nests in species that place their nests close to the tideline. Whilst this is a threat to birds globally, the most detailed studies to date have been performed in San Francisco Bay and elsewhere in North America where birds' nests are increasingly susceptible to flooding on spring tides. Studies have shown that more than 85% of saltmarsh sparrow (*Ammodramus caudacutus*) nests were flooded during spring tides (e.g. Gjerdrum et al. 2008; Bayard and Elphick 2011; reviewed by Greenberg et al. 2006), although only 2% of California clapper rail (*Rallus longirostris*) nests were flooded (Schwarzbach et al. 2006). Nesting birds are also extremely vulnerable to rapid increases in maximum high tide levels that result from extreme weather events (Easterling et al. 2000). A long-term study of the internationally important numbers of seabirds breeding on saltmarshes adjacent to the Wadden Sea, bordered by the Netherlands, Germany, and Denmark, and being Europe's largest estuary, found that extreme flooding events became more frequent and increased in their magnitude over the course of the study, between 1971 and 2008 (van de Pol et al. 2010). Tidal effects of climate change constitute both unpredictable extreme changes in flooding events which determine the maximum high tide level and predictable incremental changes in the mean values of water levels which determine the mean high tide level. Whilst the increase in the occurrence of extreme flooding events resulted in the maximum high tide level of the estuary increasing by 0.8 cm per year, increases in the mean high tide level were 0.4 cm per year during the same period, indicating that the occurrence of extreme climatic events increased more than the mean values of climatic variables (van de Pol et al. 2010). The risk of catastrophic flooding subsequently increased for a range of species including common terns (*Sterna hirundo*), pied avocets (*Recurvirostra avosetta*), common redshanks (*Tringa totanus*), common black-headed gulls (*Larus ridibundus*), Eurasian oystercatchers (*Haematopus ostralegus*), and Eurasian spoonbills (*Platalea leucorodia*) over the course of the study. Flooding reduced breeding success of Eurasian oystercatchers below that required to maintain population stability (van de Pol et al. 2010). Sensitivity analyses demonstrated that birds could best adapt by selecting nest sites on higher parts of the saltmarsh but they preferentially selected sites on lower parts that were closer to food supplies on the tideline and afforded better views of potential predators. Therefore, the evidence suggests that lower sites acted as an ecological trap (Schlaepfer et al. 2002) because despite preferentially nesting there, birds experienced reduced fitness;

however, this hypothesis remains to be rigorously tested (van de Pol et al. 2010). Elsewhere, a single extreme storm event on 2–5 June 2012 battered the New England coast of North America, with winds gusting up to 78 km per hour and heavy rain, which, when combined with heavy tide cycles, resulted in many low-lying areas on offshore islands being flooded. This resulted in 22% of herring gull (*Larus argentatus*) nests being destroyed at a breeding colony on an island in the Gulf of Maine (Bonter et al. 2014). There are many other anecdotal accounts of how single storm events can strongly, and sometimes completely, reduce the reproductive output of seabirds. Whilst such events are likely to have always been a part of the life history of the birds, further research is required to examine how both the frequency and magnitude of such extreme weather events, which are predicted to increase in a changing climate (Easterling et al. 2000), affect the population dynamics of seabirds. Such research could usefully examine if breeding birds are more susceptible to the frequency or magnitude of such extreme weather events as, whilst the frequency of an event such as flooding may replenish food sources for seabirds, the magnitude of the flood may destroy many eggs and/ or offspring.

Climate change is likely to have population-scale consequences for those species with temperature-dependent sex allocation such as Australian brush-turkeys (*Alectura lathami*) and the other megapodes (Göth and Booth 2005). In such species, offspring sex is determined by environmental conditions experienced during incubation and whilst a pivotal temperature results in equal numbers of males and females, higher or lower temperatures result in more females or males hatching, respectively (e.g. Hawkes et al. 2007; Refsnider et al. 2013). Although empirical studies of such phenomena are absent for birds, those of reptiles indicate that temperature increases as small as ~1°C can substantially skew sex ratios towards females (Hawkes et al. 2007; Pen et al. 2010). Furthermore, theoretical models predict that temperature increases of as little as ~2°C will result in reptile populations consisting entirely of females with a paucity of adult males resulting in catastrophic population decline (Refsnider et al. 2013). It is reasonable to predict that demographic impacts may also be experienced by the megapodes and studies examining their breeding biology in relation to climate change are urgently needed.

Climate change may also have indirect effects on nest-building birds via changes in the availability of nest sites and some nest materials, and via changes in host–parasite interactions. In Arizona, decreasing winter snowfall over a 22-year period resulted in increased numbers of elk (*Cervus canadensis*) that eat deciduous trees in which species such as orange-crowned warblers (*Leiothlypis celata*) build their nests (Martin and Maron 2012). Although this resulted in declines in breeding bird abundance, these were reversed in just six years following the exclusion of elk from forested areas (Martin and Maron 2012). The availability of some nest materials may also decline. For example, many species use mosses as a material to construct their nests (Breil and Moyle 1976) but climate change is predicted to reduce the range and abundance of bryophytes because higher temperatures increase nutrient concentrations within soils, which in turn enhance the growth of vascular plants that out-compete mosses (Tuba et al. 2012). Moreover, climate change is causing areas of southern Europe to become drier and experience longer droughts (Giorgi and Lionello 2008) and the resulting soil compaction may reduce the availability of mud

for nest building. Illustratively, song thrushes (*Turdus philomelos*) are absent as a breeding species from large areas of apparently suitable habitat in southern Europe (Tomialojc 1992) and their absence is likely to be determined by the low soil moisture levels that make it difficult for them to feed on earthworms (*Lumbricus* spp.) and other invertebrates in the soil (Figure 6.1; Huntley et al. 2007). It has also been suggested that, in the dry conditions, an absence of mud, which the thrushes use to build their nests, accounts for their absence as a breeding species (Tomialojc 1992). Any reduction in the availability of mud will affect numerous species, but particularly the Struthideidae family of mud-nesting birds in Australia. However, it is noteworthy that a study of barn swallows (*Hirundo rustica*) suggested that mud was not limiting as a building material and was, in fact, readily available at ponds and other water bodies (Møller 2006). Interestingly, the same study found that the size of barn swallow nests

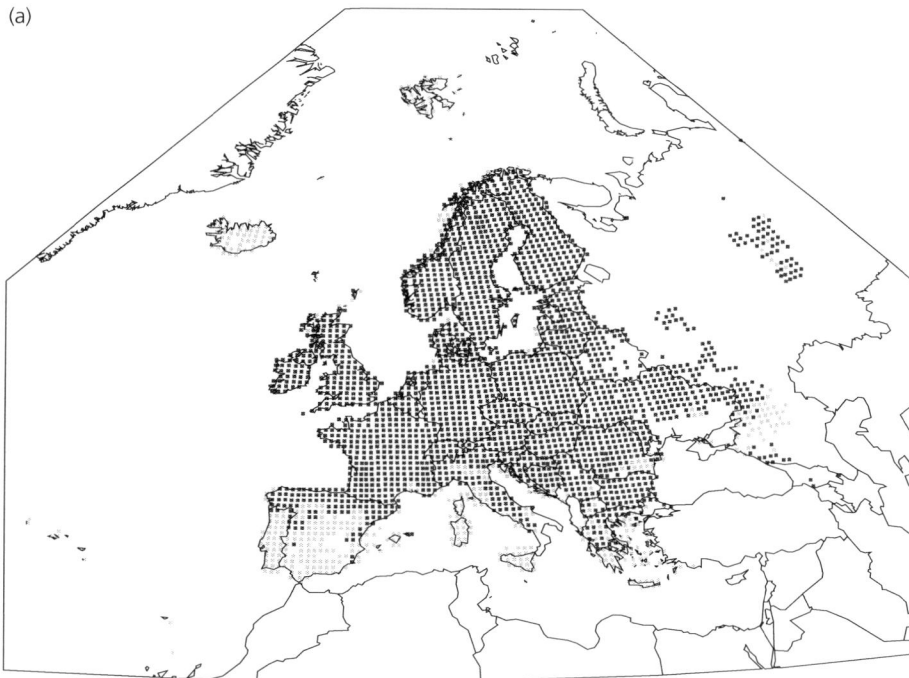

Figure 6.1 The current and future distributions of song thrushes (*Turdus philomelos*) in Europe. Their current distribution is shown as (a) census data from the EBCC Atlas of European Breeding Birds and (b) as predicted by response surface models for the 1961–90 mean climate. (c) shows their future distribution as predicted by response surface models for the mean climate projected for 2070–99 by the Hadley Centre climate model, as further detailed in Huntley et al. (2007). Note that in the maps, each coloured dot represents an area of 2500 km^2 and that whilst white areas have climatic conditions outside the range for which a simulation could be made, black dots represent areas in which birds are simulated as breeding and pale grey dots represent areas in which they are simulated as being absent. Maps reproduced with permission from Huntley et al. (2007). (See Plate 7.)

(b)

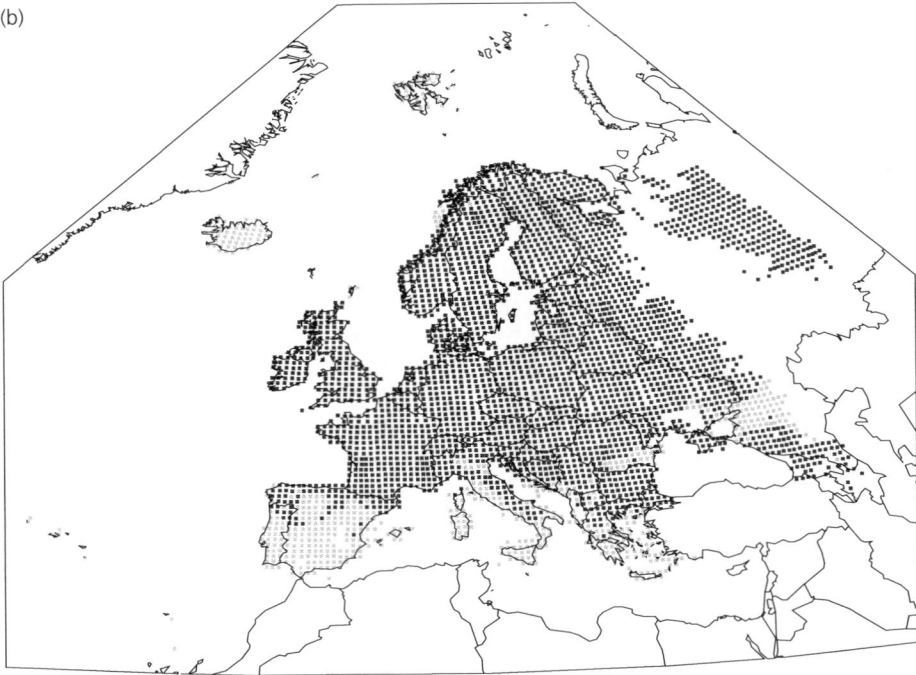

(c)

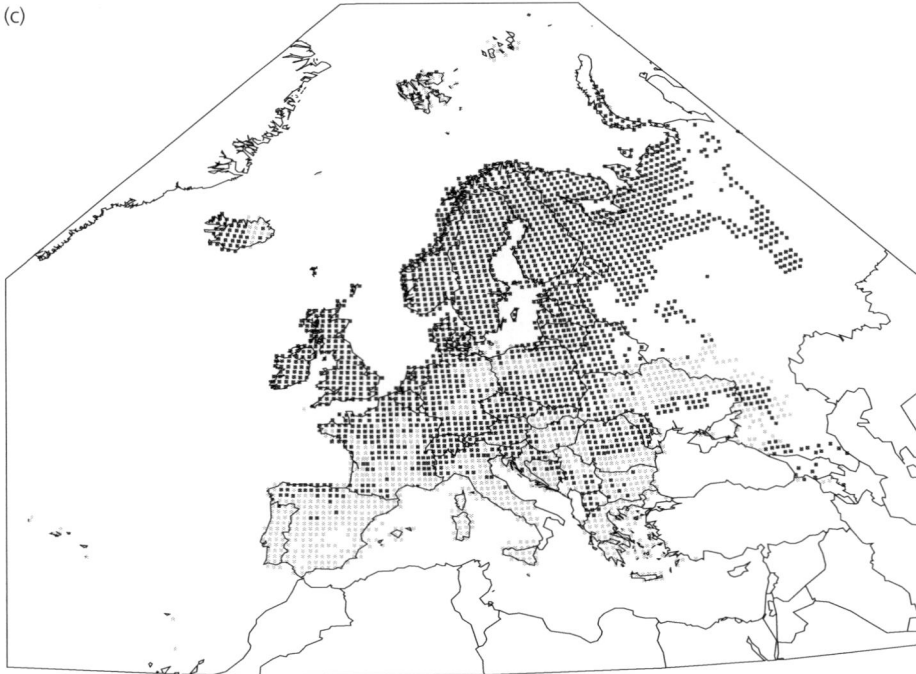

Figure 6.1 *Continued*

decreased between 1977 and 2003, with the outer nest volume and the nest cup volume decreasing by two-thirds and the thickness of nest walls decreasing by half (Møller 2006). Of a number of factors that could have caused such changes, the only significant predictive variable was the tail length of males, a sexually dimorphic secondary sexual character in this species. Barn swallows form socially monogamous breeding partnerships (Turner 2006) and whilst both sexes contribute to nest building, males with longer tails do so to a lesser extent than males with shorter tails. As climate change has led to a temporal increase in the tail length of males, they have invested less in nest construction, and as females have not responded by increasing their own construction efforts, nest sizes have declined (Møller 2006). Consequently, climate change may result in temporal changes in nest sizes in complex ways and further research should examine the determinants of long-term changes in nest characteristics and their fitness consequences.

Climate change may also indirectly affect nest-building birds via changes in host–parasite interactions. Parasites usually have negative impacts upon their avian hosts and so the latter have evolved a range of defences against them (see Chapter 8 for further details). These include nest-building behaviours when birds incorporate green plant material in their nests with volatile secondary compounds having biocidal effects on parasites (Clark 1991). They also incorporate feathers in their nests and these contain bacteria that produce antibiotic chemicals that resist the establishment of more harmful bacteria (Peralta-Sánchez et al. 2010). Although these defences reduce the detrimental impact of parasites, they may be insufficient if the phenology and virulence of parasites alter in a changing climate (Merino and Møller 2010). Empirical studies examining how parasite abundance varies in relation to environmental conditions are scarce. However, the abundance of parasites in tree swallow (*Tachycineta bicolor*) nests was highest within a range of temperatures within one study site at an intermediate temperature of 25°C and lower at both higher (~35°C) and lower (~15°C) temperatures (Dawson et al. 2005b). The abundance of parasites in the nests of great tits (*Parus major*), blue tits (*Cyanistes caeruleus*), and European pied flycatchers (*Ficedula hypoleuca*) was negatively related to ambient temperatures within a single year and within a single study site in Spain (Martínez-de la Puente et al. 2009). Although these studies do not suggest that the abundance of parasites will increase if ambient temperatures continue to increase, studies that examine how the abundance

and virulence of parasites vary with further changes in environmental conditions would be beneficial. Climate envelope models are widely used to predict the current and future distributions of birds in relation to environmental conditions (Huntley et al. 2007; Green et al. 2008) and could be used to predict how parasites could influence birds under future changes in environmental conditions.

6.2.2 Incubation under a changing climate

Unattended eggs in temperate environments usually cool rapidly to a sub-optimal temperature for embryonic development. For example, the temperature of eggs of yellow-eyed juncos (*Junco phaeonotus*) in North America (Weathers and Sullivan 1989a) and of a range of passerine birds in Norway (Haftorn 1988) dropped more than 5°C during the ~15 minute bouts when parents left the nest during incubation. However, the embryos of most species are tolerant to such short-term reductions in temperature and sub-optimal development only occurs when embryos are kept below the optimal temperature of 34–38°C and close to the physiological zero temperature of 24–26°C for prolonged periods of time (Chapter 13; Webb 1987; DuRant et al. 2013a). As ambient spring temperatures in temperate environments are well below such developmental temperatures, climate change is unlikely to disrupt incubating birds in temperate environments. To the contrary, incubating birds may benefit from warmer springs as they will require less energetic investment to maintain egg temperature within tolerable thermal limits (e.g. Haftorn 1979). However, it is prudent to consider scenarios when climate change results in ambient temperatures exceeding thermal conditions for 'normal' embryonic development; birds may be driven to initiate incubation earlier than they presently do in order to prevent the mortality of embryos within early-laid eggs. This is likely to result in increased frequency of asynchronous hatching and subsequently increase the probability of partial brood reduction which, in turn, may result in declines in reproductive success and possibly even population sizes. However, parents are likely to be able to cool eggs in hot temperatures either by shading the eggs from direct sunlight themselves (Walsberg and Voss-Roberts 1983) or by dripping water from plumage directly onto the eggs. Further research into such behavioural adaptations is urgently required.

In contrast, eggs left unattended in tropical and sub-tropical environments are often likely to be well above the physiological zero temperature and sometimes also above the 'normal' incubation temperature. This means that incubating birds in such environments may need to reduce egg temperature by either shading the eggs from direct sunlight themselves (Walsberg and Voss-Roberts 1983), or adding additional material to shade the eggs in their absence at the nest (Szentirmai and Székely 2004). However, we know relatively little about temperature regulation by birds in tropical and sub-tropical environments compared with birds in temperate ones (but see e.g. Stoleson and Beissinger 1999; Martin et al. 2013a); this is a fruitful area for further research. Whilst birds are able to reduce egg temperatures through such behavioural strategies once incubation begins, eggs may heat to lethal temperatures prior to incubation onset. As birds are physiologically constrained to laying just one egg per day (Stoleson and Beissinger 1999), and usually do not begin to incubate their eggs until close to clutch completion (Magrath 1990; Stoleson and Beissinger 1995), early-laid eggs may be exposed to ambient temperatures for several days during which their viability may diminish (Stoleson and Beissinger 1999; Beissinger et al. 2005).

Tree swallows began to incubate their eggs earlier under warmer than cooler ambient temperatures (Ardia et al. 2006a), despite the early onset of incubation being maladaptive as it results in more pronounced hatching asynchrony and partial brood mortality (Magrath 1990). However, empirical studies show that adults need to initiate incubation shortly after the first egg is laid in order to maintain the viability of early-laid eggs. Green-rumped parrotlet (*Forpus passerinus*) eggs that were left unattended for three days or more at temperatures above the 'normal' incubation temperature range were less likely to hatch than eggs left unattended within that range (Stoleson and Beissinger 1999). Pearly-eyed thrasher (*Margarops fuscatus*) eggs were less likely to hatch if they were exposed to either hot (25.8°C) or cold (19.4°C) temperatures for a period of three or more days (Figure 6.2; Beissinger et al. 2005). Malleefowl (*Leipoa ocellata*) eggs that are incubated at temperatures ranging from 32 to 38°C hatched most successfully when incubated constantly at 34°C and less successfully when incubated at both higher and lower temperatures, despite those temperatures being within the normal range for successful incubation (Booth 1987).

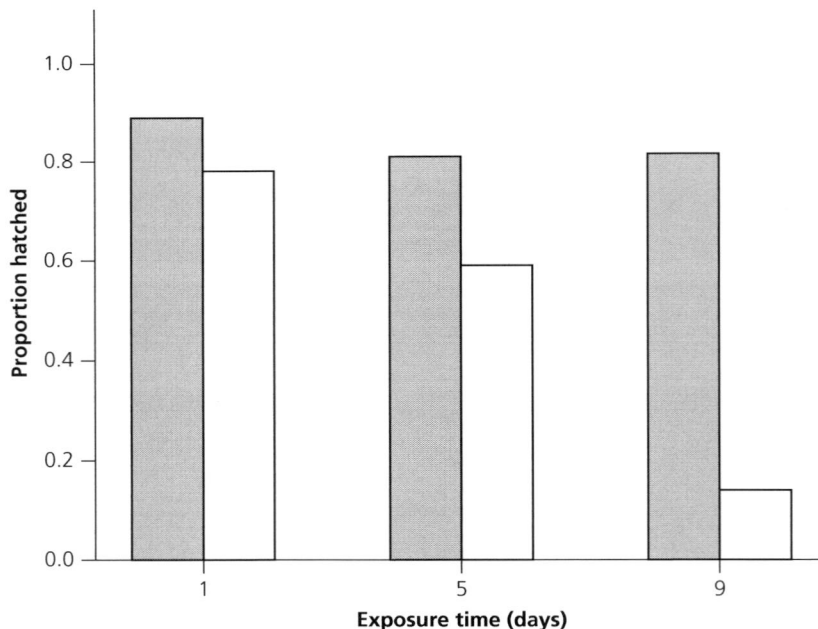

Figure 6.2 Hatching success of pearly-eyed thrasher (*Margarops fuscatus*) eggs that were exposed to ambient temperatures for one, five, and nine days prior to incubation. Note that the grey bars represent eggs that were systematically turned and the white bars represent unmanipulated (control) eggs. Reproduced from Beissinger et al. (2005) with permission from the Ecological Society of America.

Eggs exposed to ambient temperatures in temperate zones typically lose viability more slowly than those from non-temperate zones (Beissinger et al. 2005; Wang et al. 2011). Consequently, eggs left unattended in tropical and sub-tropical environments can quickly lose viability if parents do not initiate incubation earlier (reviewed by Wang and Beissinger 2011), and increasing ambient temperatures are expected to result in an increasing proportion of early-laid eggs failing to hatch. Therefore, studies that examine how parents can respond to these threats via changes in incubation behaviours such as egg turning are urgently required. We know remarkably little about the effects of incubation behaviours on the temperatures experienced by the embryos, and the vast majority of the research to date has been performed on domestic poultry (Deeming 2002f). However, studies have shown that eggs that are turned during the incubation period hatch more successfully than those that are not turned (Figure 6.2; Beissinger et al. 2005), suggesting that further research is required.

6.3 Adaptations to climate change

6.3.1 Adaptations in timing of nesting and location of nests

Nest construction and incubation behaviours offer ways for birds to adapt to climate change. During nest construction birds may adapt via changes in their nesting phenology, nest-site selection, and nest design.

Adaptive changes in the timing of nesting enable birds to breed when environmental conditions are suitable (e.g. when food is most abundant). The reproductive success of insectivorous birds such as great tits and blue tits is dependent on them synchronizing the maximum nutritional and energetic demands of nestlings with the maximum availability of caterpillars (Visser et al. 1998). The caterpillars that form the main part of the nestlings' diet are only abundant for a short time period each spring and insectivorous birds must time their own reproductive event to coincide the period of greatest food demand by their nestlings with the peak availability of their caterpillar food supply. Caterpillars have emerged progressively earlier since the early 1980s and studies in the Netherlands and the United Kingdom have shown that whilst early breeders time their reproduction to coincide with the caterpillars, late breeders exhibit phenological mismatch with food sources, sometimes with severe negative reproductive consequences (Visser et al. 1998; Cresswell and McCleery 2003). However,

spatial heterogeneity in ecological conditions can affect how birds might adapt to climate change. Blue tits on Corsica breed in evergreen and deciduous oak (*Quercus* spp.) woodlands that are separated by just 25 km (Lambrechts et al. 1997). The emergence and peak of the caterpillars are about a month earlier in the deciduous than in the evergreen woodlands and yet the two respective populations breed at the optimal time to match timing of breeding attempts with the availability of caterpillars (Lambrechts et al. 1997). Thus, avian life histories may differ between populations in an adaptive way so that birds can respond to climate change. Irrespective of their location, late-breeding birds reduce the time between the initiation of nest construction and the maximum demand for food by nestlings by building nests more rapidly (Mainwaring and Hartley 2008), by incubating eggs earlier in the period prior to clutch completion, and by laying smaller clutches (Cresswell and McCleery 2003). Thus, adaptive changes in the duration of nest building enable birds to track the phenology of their prey and to respond to it effectively.

Adaptive shifts in nest-site selection may allow birds to adapt to a changing climate; in hot environments birds may adapt by nesting in cooler sites than those selected at random. For example, desert larks (*Ammomanes deserti*; Orr 1970) and greater hoopoe-larks (*Alaemon alaudipes*; Tieleman et al. 2008) in arid environments select nest sites that are shaded from the hot midday sun. However, choosing optimal nest sites is not always straightforward, as illustrated in a study of multi-brooded spotless starlings (*Sturnus unicolor*) breeding in Spain, where spring temperatures increase rapidly throughout the breeding season (Salaberria et al. 2014). In first broods, nestlings benefitted from higher temperatures as they grew longer wings and bills, whereas nestlings in second broods were increasingly lighter and smaller as temperatures increased. Thus, the choice of an optimal nest site switched during the course of the breeding season, yet the availability of suitable nest sites and competition for those sites are likely to impose constraints on that choice (Salaberria et al. 2014). Birds can also exhibit adaptive changes in nest design that may allow those nesting in hot environments to adapt to climate change. However, to date, no empirical studies have tested this idea. Nevertheless, studies of birds nesting in temperate environments show that birds adaptively alter the design and composition of their nests in relation to predictable changes in ambient temperatures (e.g. Chapter 3; McGowan et al. 2004; Crossman et al. 2011). It seems reasonable to expect that birds in hot environments might adaptively respond in the same way and further studies are urgently needed.

6.3.2 Adaptations in incubation

Birds in temperate environments have adaptively changed their incubation behaviours in response to climate change. This has resulted in later breeding by insectivorous birds such as great tits and European pied flycatchers experiencing phenological mismatches between nestling food demands and food availability. Late-breeding birds have attempted to shorten the period between egg laying and hatching by increasing the amount of nocturnal incubation (Lord et al. 2011; Vedder 2012). Such behaviours have resulted in incubation periods becoming shorter over evolutionary timescales (Cresswell and McCleery 2003; Both and Visser 2005; Matthysen et al. 2011) as birds attempt to track their caterpillar food supply (Figure 6.3).

In arid environments incubating birds must keep their eggs cool during the hottest parts of the day. Despite ambient temperatures being above 45°C,

incubating mourning doves (*Zenaida macroura*) kept their eggs at viable temperatures, i.e. 36–40°C, by maintaining contact between them and their bodies, which are held at relatively low core temperatures (Walsberg and Voss-Roberts 1983; Figure 6.4). However, this was achieved through higher levels of evaporative cooling that quickly resulted in dehydration and further studies could usefully examine the consequences of such cooling behaviours. This study suggests that ambient temperatures influence incubation patterns markedly but this contrasts with one that examined the correlates of the onset of diurnal and nocturnal incubation in five cavity-nesting passerines. Neither minimum temperatures, the proportion of time above 24°C, nor seasonality were found to influence significantly the onset of incubation, although it was significantly delayed by increases in both precipitation and wind speed (Wang and Beissinger 2009). As both precipitation (Marvel and Bonfils 2013) and wind speed (Vautard et al. 2010; Young et al. 2011) are predicted to change under future climate change scenarios, further studies could examine how weather conditions influence incubation behaviours.

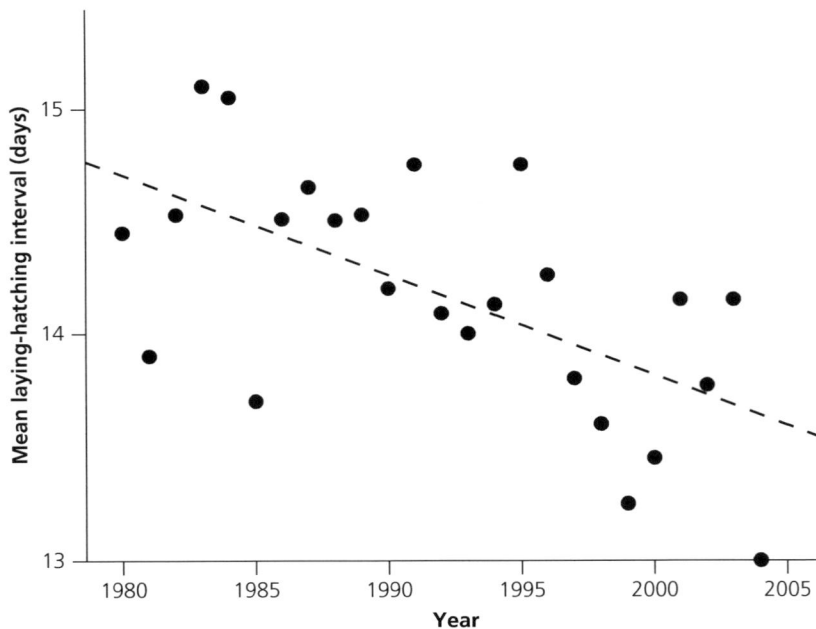

Figure 6.3 Changes in the mean interval between the laying of the last egg and the hatching of the first egg in European pied flycatchers (*Ficedula hypoleuca*) over 24 years in the Netherlands. The interval decreased significantly over the course of the study. Reproduced from Both and Visser (2005) with permission from John Wiley & Sons.

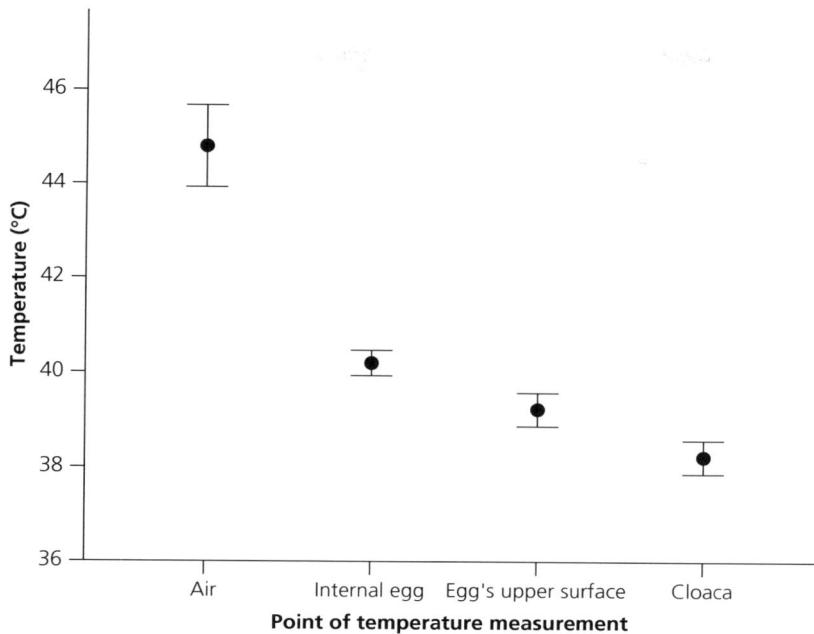

Figure 6.4 The relationship between ambient air temperature, internal and external egg temperatures, and cloacal temperature of mourning doves (*Zenaida macroura*) incubating in extreme heat in the Sonoran Desert, Arizona. Note that values are means ± 1 standard deviation. Reproduced from Walsberg and Voss-Roberts (1983) with permission from Chicago University Press.

6.4 Summary and future research directions

Optimal embryonic and offspring development only occurs within relatively narrow thermal limits and prolonged periods of time spent outside of these limits result in sub-optimal development and sometimes death of young. Therefore, climate change may negatively influence birds during the nest construction and incubation stages in a number of ways. The most serious direct threats to nest-building birds are from inundation of seabirds' nests located just above the tideline and sub-optimal sex allocation in megapodes, whilst the most serious indirect threats include reductions in the availability of nest sites and building materials, and changes in host–parasite interactions. Meanwhile, the most serious threat to incubating birds is the reduced viability of eggs in increasingly hotter environments. However, nest construction and incubation are relatively plastic parental behaviours that may enable birds to adapt to climate change. During nest construction birds can adapt via changes in their nesting phenology, nest-site selection, and nest design;

during incubation birds in temperate and tropical environments can adapt by, for example, changing their nocturnal incubation behaviours and by keeping their eggs cool, respectively.

Our understanding of how climate change influences avian nest construction and incubation behaviours remains in its infancy compared to the chick-rearing stage. Consequently, there are several avenues of research that may prove fruitful. First, I urge more studies to consider how nest construction and incubation behaviours are affected by climate change. In particular, manipulations that alter climatic conditions (e.g. temperature, rainfall, and wind speeds) to which embryos are exposed are likely to be fruitful. For example, studies that cross-foster offspring between nests that are exposed to variable amounts of direct sunlight or precipitation are likely to be informative, as might studies that remove eggs from nests so that embryos are exposed to variable conditions during development. Secondly, research to date has primarily been conducted on species in the Palearctic and Nearctic and further studies from the Neotropics and Afrotropics are required. Thirdly, adaptive changes in nest construction and incubation behaviours provide ways for birds to adapt to climate change in the short term but our understanding of these processes remains poor. Studies should examine the extent to which they might enable breeding birds to adapt to a changing climate in the long term.

Acknowledgements

I thank Charles Deeming and Ian Hartley for many useful discussions, Brian Huntley for providing me with the song thrush maps in Figure 6.1, and Jim Reynolds, Steven Beissinger, and Martijn van de Pol for comments on earlier drafts.

CHAPTER 7

Microbiology of nests and eggs

A. West, P. Cassey, and C.M. Thomas

7.1 Introduction

The avian egg contains a rich medium capable of supporting the growth and metabolism of a highly complex multicellular organism—the developing chick. Microbes, particularly bacteria, are relatively simple by comparison. The range of temperatures, humidity, and resources required to sustain bacterial growth is far broader than for multicellular eukaryotes (Madigan et al. 2010). An egg, incubated at a relatively constant body temperature, filled with a nutrient-rich liquid is, in theory, an ideal environment for microbial growth. The growth rates of microorganisms can be extremely rapid in comparison to an avian embryo, so if provided with the opportunity, they can outcompete the developing bird for its resources, strongly reducing the chances of survival and/or successful hatching.

Microorganisms are capable of contaminating the egg via both vertical transmission, in which the infecting organism is passed between the mother and the egg during its formation, or by horizontal transmission, in which the infection occurs post-laying. There is some dispute over which route is responsible for the majority of egg contamination (Barrow and Lovell 1991; Humphrey 1994; Ruiz-de-Castañeda et al. 2011a, 2011b). Some studies have demonstrated that an increased bacterial load on the eggshell is associated with increased infection rates of the inner egg, and that this in turn reduces hatchability (Cook et al. 2003; Cook 2005a; Soler et al. 2012), although it is also suggested that it is the presence of pathogenic species, rather than the overall quantity of bacteria, which is significant (Ruiz-de-Castañeda et al. 2011b).

Given the abundance of microbes in the environment, preventing them from accessing the egg is impractical. The eggshell provides a physical barrier to microorganisms, but is not completely impenetrable because it must allow the exchange of gases and water

for embryonic survival and development (Paganelli 1991). Various strategies are employed by the incubating bird, and the egg itself, to reduce the chances of microbial infection, including: anti-microbial chemicals; sequestration of molecules essential for microbial metabolism; choosing nest materials with anti-microbial properties; adopting useful symbiotic relationships; and generally making the inner egg an inhospitable place for the growth of anything except the embryo. Following on from a previous review of the microbiology of incubation (Baggott and Graeme-Cook 2002), this chapter discusses these strategies and how they affect microbes in the nest, on the shell, and inside the egg.

7.2 External influences on microbial growth

7.2.1 Humidity, temperature, and incubation

Bacteria are extraordinarily diverse, but can be broadly placed into two key groups: gram-negative and gram-positive. This classification is based on their staining with Gram's stain and relies on diagnostic differences in the properties of bacterial outer cell walls. Compared to gram-positives, gram-negatives generally represent only a small proportion of the bacteria found on eggshells (Seviour et al. 1972) and, subsequently, many of the egg's defences are more effective at targeting gram-positive bacteria. One of the reasons for this may be that gram-positive bacteria are more likely to survive on the egg surface since they are more resistant to desiccation, due to the thickness and chemical composition of their cell walls. Keeping the eggs dry is therefore one of the most basic and effective defences against infection. Infection rates are predictably higher in more humid environments, compared with drier conditions (Cook et al. 2005b; D'Alba et al. 2010a), and have also been shown to increase during periods of higher temperature in cool, montane environments

West, A., Cassey, P., and Thomas, C.M., *Microbiology of nests and eggs*. In: *Nests, Eggs, and Incubation*. Edited by: D.C. Deeming and S.J. Reynolds, Oxford University Press (2015). © Oxford University Press. DOI 10.1093/acprof:oso/9780198718666.003.0007

(Ruiz-de-Castañeda et al. 2011b). Nevertheless, some studies suggest that it is the gram-negatives that are more likely to contaminate the inner egg, particularly bacteria that are non-cluster-forming, and motile, which is thought to enable easier passage through eggshell pores (Cook et al. 2003; De Reu et al. 2006). Notably, *Salmonella enteritidis*, a major cause of human food poisoning, falls into this category.

Although there is evidence that the anti-microbial properties of the albumen are enhanced at higher temperatures (Tranter and Board 1984), incubation of the eggs is a particularly effective mechanism for keeping them dry. In tropical environments, bacterial and fungal infection rates are lower when eggs are incubated immediately following laying (Cook 2005a; Cook et al. 2005b). However, two studies on wood ducks (*Aix sponsa*) found little correlation between the delay in incubation initiation and egg infection by microorganisms (Walls et al. 2011, 2012). In temperate environments, the evidence for the anti-microbial effects of incubation is somewhat conflicting, with studies showing that incubation has an inhibitory effect on the eggshell bacterial load (Ruiz-de-Castañeda et al. 2011a), or no effect (Wang et al. 2011). One study observed an increased bacterial load on incubated mallard (*Anas platyrhynchos*) eggs, compared to unincubated controls, although this was only over the first five days of incubation (Giraudeau et al. 2014). These discrepancies may be attributed to the different species and environmental conditions of the experiments. The differences could also result from incubation promoting the growth of non-pathogenic bacteria at the expense of pathogenic bacteria, although a study using culture-independent techniques which identified almost 1,500 different bacterial taxa found that incubation did not promote the growth of any bacterial species found on the egg, and was uniformly bacteriostatic (Shawkey et al. 2009).

7.2.2 Bacterial assemblages of nests and eggs

Although there is considerable intraspecific variation between the bacterial assemblages of nests, it reveals that assemblages can be sufficiently distinct to identify bird species based on the types of bacteria found in nests. In a study using culture-based methods, 32 bacterial and 13 fungal species were isolated from the nests of two closely related passerine species, blue tits (*Cyanistes caeruleus*) and great tits (*Parus major*). Bacterial species were identified which were only found exclusively in the nests of one species and not the other, allowing the accurate identification of avian species

from the nest microflora (Goodenough and Stallwood 2010). Differences in behaviour and diet were thought to account for the differences in microbial populations. More recently, culture-independent, PCR-based methods have been used to quantify bacterial abundance with up to 10 times greater sensitivity than culture-based methods (Lee et al. 2013). However, caution needs to be exercised when using different identification techniques. A study examining bacterial diversity on eastern bluebird (*Sialia sialis*) feathers showed that culture-dependent and culture-independent identification techniques yielded different results. *Pseudomonas* and *Bacillus* species were the most highly represented in the molecular and culture-based analyses, respectively (Shawkey et al. 2009).

The types of bacteria found on the surface of house wren (*Troglodytes aedon*) eggs were shown to vary during the course of incubation (Potter et al. 2013). Of 20 bacterial genera identified in the study, one was found only in the pre-incubation stage, while eight were specific to early incubation and five to late incubation, likely due to temperature fluctuations over the course of incubation. *Pseudomonas* species, the most common taxa found on nests and eggs, were shown to thrive at all stages of incubation.

An investigation into the influence of brood parasitism on egg bacterial loads found that Eurasian magpie (*Pica pica*) eggs in nests parasitized by great spotted cuckoos (*Clamator glandarius*) carried higher bacterial loads than those in unparasitized nests (Soler et al. 2011b). This difference is most likely due to egg breakage during parasitism, but further highlights the influence of the avian species on the egg and nest microflora. Bacterial loads on the cuckoo eggs were lower relative to the magpie eggs, suggesting that they are better adapted to deal with the high bacterial loads found in parasitized nests (Soler et al. 2011b).

7.2.3 Nest construction

There is evidence that some birds choose their nest materials based on their anti-microbial properties. Studies which examined the properties of the green plants used by common starlings (*Sturnus vulgaris*) in nest construction found that four out of five plant species tested had anti-microbial properties, displaying a range of efficacies and bacterial targets, as well as preventing infestation by lice and mites (Clark and Mason 1985, 1988). More volatile compounds, thought to be responsible for the anti-bacterial activity, were found in the plants preferred by the starlings, compared to others available. The same compounds produced strong electro-physiological responses from olfactory nerves of the starlings, suggesting that the nest-building individuals chose the plants based on odour (Clark and Mason 1987).

Under experimental conditions, male common starlings showed a preference for using plants with anti-bacterial properties in nest construction, in an experience-independent manner (Gwinner and Berger 2008). Bacterial loads were shown to be reduced in the nests of starlings when fresh herbs were incorporated into their construction, leading to a reduction in nestling mortality (Gwinner and Berger 2005). Although the herbs did not have an effect on the number of mites, they did appear to improve the nestling's resistance to their effects. Similar effects were observed in blue tits in which bacterial richness and density were reduced on nestlings when aromatic plants were used in nest construction, particularly in nests with high levels of blowfly (*Protocalliphora* spp.) infestation (Mennerat et al. 2009c). This effect appeared to be exclusive to the nestlings, as no change was detected in the bacterial densities of the parents. In colonial sandwich terns (*Sterna sandvicensis*) a mutualistic relationship was observed between them and the anti-microbial-producing plant, *Artemisia maritima*, where the bird's faeces and uneaten food benefitted the plant, while the plants reduced the abundance of pathogenic *Staphylococcus* bacteria on the eggs, although domestic fowl (*Gallus gallus*) eggs, rather than tern eggs, were used for the experiment (Møller et al. 2013).

Lining nests with feathers has particular advantages, such as thermal insulation (see Chapter 4), but their presence may also influence the total eggshell bacterial load. A comparative study of 24 bird species found that bacterial loads were lower on the eggs of bird species which incorporate feathers into their nest lining (Peralta-Sánchez et al. 2012). In barn swallows (*Hirundo rustica*) the colour of the feathers used in nest construction also appears to be a factor, with un-melanized white feathers carrying a smaller bacterial load than pigmented feathers (Peralta-Sánchez et al. 2014). Barn swallows show a preference for using white feathers for constructing their nests. The study revealed that experimental manipulation of nests to ensure they contained only white feathers decreased the number of eggshell bacteria and increased the chances of successful hatching, compared with nests containing only black feathers. Strangely, white feathers actually supported the growth of feather-degrading, keratinolytic bacteria (specifically, *Bacillus licheniformis*), compared with black feathers. The reduced eggshell bacterial loads were suggested to result from the antibiotic

compounds produced by these bacteria, to aid in their competition for resources.

In another study, *B. licheniformis* isolated from nests and eggs of three waterfowl species inhibited the growth of both gram-positive and -negative bacteria and the fungus, *Pythium ultimum* (Baggott and Graeme-Cook 2002). As well as causing damage to feathers, *B. licheniformis* has been shown to degrade the eggshell cuticle, and may be partly responsible for the observation that mandarin ducks (*Aix galericulata*) show an increase in eggshell conductance over the course of their incubation (Beckett et al. 2003).

7.2.4 Uropygial secretions

Growth of *B. licheniformis* has also been shown to be inhibited by uropygial gland secretions which birds apply to their plumage to protect them from physical damage, and for waterproofing (Soler et al. 2008a). These secretions contain a complex mix of volatile, anti-microbial compounds and oils, which confer activity against fungi and gram-positive bacteria (Jacob and Ziswiler 1982; Shawkey et al. 2003a; Soler et al. 2008a, 2014), and perhaps a wider spectrum of bacteria (Czirják et al. 2013). It is possible that these secretions are transferred from the feathers lining the nest onto the eggs, and exert an anti-microbial influence directly, or that they provide an additional physical barrier to microbial penetration.

European hoopoes (*Upupa epops*) have been observed to apply uropygial secretions directly to their eggs (Martín-Vivaldi et al. 2014). These secretions also harbour antibiotic-producing enterococci bacteria (Law-Brown 2003; Martín-Platero et al. 2006; Martín-Vivaldi et al. 2014), and there is evidence that these bacteria are partly responsible for the anti-microbial properties of the secretions (Soler et al. 2008a; Martín-Vivaldi et al. 2010). Electron microscopy analysis of hoopoe eggs has revealed the presence of crypts on the surface of the eggshell, which are filled with uropygial secretions shortly after laying, retaining more of the substance on the eggshell (Martín-Vivaldi et al. 2014).

Uropygial secretions have also been positively correlated with feather mite population size and diversity (Galván et al. 2006, 2008; Soler et al. 2012). The diversity of the feather mite population has been shown to be negatively related to the eggshell bacterial load, presumably because the mites eat the bacteria, thus demonstrating a true symbiotic relationship between the mites and birds (Soler et al. 2012). Soler et al. (2012) also showed a correlation between the size of the uropygial gland and the eggshell bacterial load, and hypothesized that the link may also result from selective pressure on gland size in species that experience a higher risk of microbial infection of the egg. In house sparrows (*Passer domesticus*) a reduction in the uropygial gland size is observed during an experimentally stimulated immune response, indicating a trade-off between the two modes of microbial defence, although this was only observed in birds of low body mass (Moreno-Rueda 2014).

A recent experiment in mallards, where the bird's access to the uropygial gland was blocked, found no significant effect on the eggshell bacterial load (Giraudeau et al. 2014). This suggests that the effects of uropygial secretions on eggshell bacteria may be species-specific.

7.3 Intrinsic egg defences

7.3.1 The eggshell cuticle

Among its other functions, the eggshell cuticle serves as a waterproofing layer which constitutes one of the most important physical barriers to microorganisms. Several studies have shown that eggs with a reduced cuticle deposition, or an artificially removed cuticle, are at higher risk of infection (Sparks and Board 1984; De Reu et al. 2006). Fungi and bacteria may break down the cuticle layer, decreasing the water resistance of the egg and opening up pores, which increases the risk of infection of the egg contents (Board and Halls 1973; Board et al. 1979).

In Australian brush-turkeys (*Alectura lathami*), the nano-scale structure of the eggshell surface has been shown to be important for preventing microbial penetration of the eggshell (D'Alba et al. 2014). Megapodes are mound builders and bury their eggs, putting them at an increased risk of infection. It seems that despite this their infection rates are relatively low with 86% hatching success observed in a study of 499 eggs (Jones 1988). The surface of these eggs is unusual in that instead of the more common calcium carbonate crystal form vaterite, the surface is composed of calcium phosphate nanospheres. This may provide the surface of these eggs with water-repellent properties approaching super-hydrophobicity. This does not allow water droplets to spread over their surface, inhibiting biofilm formation and minimizing bacterial contact with the egg. Vaterite nanospheres have also been observed on the eggshell cuticles of double-crested cormorants (*Phalacrocorax auritus*), smooth-billed anis (*Crotophaga ani*), hamerkops (*Scopus umbretta*), and emperor penguins (*Aptenodytes forsteri*; Mikhailov 1997a). Future studies may establish whether these nanostructures serve the same purpose as the calcium phosphate spheres. Such research may also help in the design of novel anti-microbial biomaterials.

7.3.2 Eggshell pigments and their light-dependent generation of reactive oxygen species

Key pigments present in the cuticle, protoporphyrin, and biliverdin (Gorchein et al. 2009) may act as photosensitizers. These pigments are haem products that have a photodynamic (light-dependent) bactericidal activity against gram-positive bacteria and are thought to work through light-stimulated generation of reactive oxygen species (Ishikawa et al. 2010).

Porphyrins are also used in photodynamic therapy and photodynamic anti-microbial chemotherapy (reviewed in Dolmans et al. 2003). Ishikawa et al. (2010) showed that roughly crushed eggshells of domestic fowl have a light-dependent anti-microbial effect against gram-positive (*Bacillus cereus* and *Staphylococcus aureus*) but not gram-negative (*Escherichia coli* and *Salmonella enteritidis*) bacteria. The strongest effects were observed for pigmented shells and only weak effects were seen for white shells.

7.3.3 Anti-microbial compounds

In addition to the physical barrier provided by the membrane, shell, and cuticle, various anti-microbial chemicals have been identified in all parts of the avian egg. These compounds display a wide variety of activities and target a broad range of microorganisms (reviewed in Wellman-Labadie et al. 2007). Egg albumen has an alkaline pH, making it inherently inhospitable to many microorganisms (Ng and Garibaldi 1975).

The yolk contains the adaptive immune system's contribution to anti-microbial defence in the form of IgY antibodies which also have various clinical and research applications (reviewed in Spillner et al. 2012). The vast majority of the egg's anti-microbial defences fall into the category of innate immunity. The eggshell bacterial loads of 29 bird species were shown to be lower in those with higher natural antibody and complement levels (Soler et al. 2011a), although many of the anti-microbial compounds which defend the egg do not fall into these categories. The main anti-microbial compounds identified to date are summarized in Table 7.1.

These compounds can be broadly categorized by their activity. Some, such as cystatin, ovalbumin-related

Table 7.1 The major anti-microbial compounds found in the avian egg. This list is not exhaustive and is restricted to the most well-known and characterized compounds. It is likely that many compounds remain uncharacterized.

Compound	Location	Known targets	Mode of activity	Notes	References
Ovocalyxin-32 and 36	Eggshell matrix, vitelline membrane, cuticle	Gram-positives (shown against *Bacillus subtilis*)	Protease inhibitor		Xing et al. (2007) (OCX-32); Gautron et al. (2011) (OCX-36); Rose-Martel et al. (2012)
Avidin	White		Biotin sequestration		Board and Fuller (1974); Wang et al. (2012)
Ovotransferrin	White, eggshell matrix/mammillary layer, cuticle	Gram-positives, gram-negatives, and fungi	Sequestration of iron	Makes up 12–13% of albumen protein	Ibrahim et al. (1998); Ahlborn et al. (2006); Rose-Martel et al. (2012); reviewed in Wu and Acero-Lopez (2012)
Ovocleidin-17/ansocalcin	Eggshell matrix, cuticle	Gram-positives and gram-negatives	Lysis. Binds to peptidoglycan and weakly to chitin and LPS	Ovocleidin-17 in chickens, ansocalcin in geese. Roles in eggshell calcification. Anti-microbial activity is stronger in the presence of Ca^{2+} ions	Wellman-Labadie et al. (2008a, 2008b)
Avian beta defensin protein family	White	Various gram-positives and -negatives and fungi	Lysis via cell membrane damage		Sugiarto and Yu (2004); Hervé-Grépinet et al. (2010)
Gallin (Ovodefensin)	White	*Escherichia coli*	Lysis		Gong et al. (2010); Hervé et al. (2014)
Toll-like receptor 4	White	Gram-negatives (LPS)	LPS binding		Bedrani et al. (2013a, 2013b)
Ovoinhibitor	White, yolk, vitelline membranes, eggshell, cuticle	*Bacillus thuringiensis*	Protease inhibitor		Bourin et al. (2011); Rose-Martel et al. (2012)
Cystatin	White, cuticle	Fungi (*Candida* spp.), bacteria, and viruses	Protease inhibitor		Korant et al. (1985); Kolaczkowska et al. (2010); Bourin et al. (2011); Rose-Martel et al. (2012); Wang et al. (2012),
Lysozyme	White, eggshell matrix, cuticle	Gram-positives' cell walls (peptidoglycan), and fungal cell walls (chitin)	Hydrolytic	Makes up 3.4% of albumen protein	Vadehra et al. (1972); Samaranayake et al. (2001); Rose-Martel et al. (2012)
Biliverdins and porphyrins	Cuticle	Gram-positives	Light-dependent reactive oxygen species generation		Ishikawa et al. (2010)
Ovalbumin	White, eggshell mammillary layer		Protease inhibitor	The major albumen component	Hincke (1995)
Ovomucoid	White	*Streptomyces erythraeus*	Protease inhibitor	Makes up 11% of the albumen protein. Highly allergenic	Nagata and Yoshida (1984); reviewed in Abeyrathne et al. (2013)
beta-NAGase	White, eggshell matrix	Gram-positives (peptidoglycan)	Hydrolytic		Ahlborn et al. (2006)

protein X, ovomucoid, and ovoinhibitor, act by inhibiting proteases which are used by bacteria to break down proteins for their metabolism (Korant et al. 1985; Lu et al. 1993; Bourin et al. 2011; Réhault-Godbert et al. 2013). Avidin is a chelating agent which sequesters biotin, a vitamin required for bacterial growth. Another chelating agent, ovotransferrin, is one of the most abundant egg-white proteins, and sequesters iron, copper, and zinc ions, required for bacterial growth (Baron et al. 2014; reviewed in Giansanti et al. 2012). Ovotransferrin is also known to have anti-bacterial and anti-fungal activities, independent of nutritional deprivation (it is still anti-microbial when saturated with iron), by causing damage to cell membranes (Valenti et al. 1985; Ibrahim et al. 1998, 2000).

Other compounds can bind to bacterial cell wall components, causing lysis. Lysozyme is the most abundant of this type of anti-microbial protein and acts by catalyzing the hydrolysis of the major component of the gram-positive bacterial cell wall, peptidoglycan (Sharon 1967). β-NAGase and ovocleidin-17 also act in this way, by targeting peptidoglycan (Ahlborn et al. 2006; Wellman-Labadie et al. 2008a).

The eggshell itself, and the eggshell cuticle, harbour many anti-microbial proteins, some of which are also found in the albumen, while others are unique to the eggshell. Proteins, such as ovocleidin-17 in domestic fowl and a homologous protein in greylag geese (*Anser anser*) called ansocalcin, have important roles in eggshell formation by catalyzing calcite crystallization. These molecules have anti-bacterial properties resulting from their ability to bind to bacterial cell wall polysaccharides, such as lipopolysaccharides and particularly, peptidoglycan (Wellman-Labadie et al. 2008a). They have more potent activity against gram-positive bacteria than gram-negatives. As zwell as binding to peptidoglycan, ovocleidin-17 has a weak binding affinity for chitin and lipopolysaccharides, which are cell wall components of fungi and gram-negative bacteria, respectively. Bacterial growth is also inhibited by ovocalyxin-32 which is found in the eggshell cuticle of domestic fowl (Xing et al. 2007; Wellman-Labadie et al. 2008b) and ovocalyxin-36 in the calcified part of the shell and shell membrane (Gautron et al. 2007). Experiments have shown that exposing laying fowl hens to lipopolysaccharide results in the production of albumen with a more potent anti-bacterial activity, although the most significant effect was against the gram-positive bacterium *Staphylococcus aureus* (Réhault-Godbert et al. 2013).

It has been hypothesized that in birds laying large clutches, where the first egg in the laying sequence would be exposed to the environment for a much longer period than the last, anti-microbial compounds may be deposited in greater concentrations early in the laying sequence and then decline. This would compensate for the increased infection risk posed to the early-laid eggs by their prolonged exposure to the environment. The current evidence suggests that the situation is far from straightforward. Little variation in the deposition of anti-microbials was seen in eight species with pre-incubation delays between zero and eight days (Shawkey et al. 2008). When this was tested in blue tits, where the first egg can be exposed for up to 13 days, ovotransferrin levels appeared to increase slightly, but in the middle of the laying sequence (D'Alba et al. 2010b). Avidin and lysozyme did not show any correlation with position in the laying sequence. An earlier study of barn swallows showed a decline in lysozyme activity from the beginning to the end of the laying sequence (Saino et al. 2002).

Caspian gulls (*Larus cachinnans*) show the opposite with an increase in albumen lysozyme concentration over the laying sequence, while avidin concentrations decline with the laying order for male-containing eggs, and increases in female-containing eggs (Bonisoli-Alquati et al. 2010). Lysozyme concentrations in these birds were also related to chick phenotype, as a positive correlation was observed between albumen lysozyme concentration and tarsus length at age four. In blue tits, lysozyme levels varied in relation to male quality, with increased levels observed in eggs fathered by more attractive males, based on ultraviolet reflectance of their plumage (D'Alba et al. 2010b). Lysozyme levels also increased when fathers were monogamous rather than polygynous, which could suggest that these female birds invested more resources in certain eggs, based on male quality.

7.4 Conclusions

The work described in this chapter reveals the many strategies that birds can adopt to minimize egg infection and thus ensure chick survival. These strategies range from keeping the eggs dry by controlling nest humidity and the composition of the eggshell, to using anti-microbial compounds produced by the birds themselves, and complemented by additions of nest flora and fauna. The egg's anti-microbial properties make it a potentially useful bio-analogue for biomedical and pharmaceutical applications. For example, fresh eggshell membrane has been utilized to treat cuts, as an anti-microbial tissue adhesive (Ohto-Fujita et al. 2011). Commercial applications may also result from the discovery of novel eggshell properties, such as the hydrophobic, nano-scale surface structures on the eggs of Australian brush-turkeys.

The anti-microbial compounds of avian eggs, their targets, and modes of activity are still being characterized and we predict that this is an ongoing area of considerable research interest and potential. Of particular interest will be how different evolutionary strategies have converged or diverged in response to the many different life-history and environmental selection pressures faced by birds in continually changing environments.

CHAPTER 8

Control of invertebrate occupants of nests

I. López-Rull and C. Macías Garcia

8.1 Introduction

Avian nests accommodate more than just birds. Their protected, moist and warm environment and the regular supply of food makes them useful refuges and convenient breeding habitats for a diversity of invertebrates including bird ectoparasites such as bugs, flies, fleas, ticks, mites, and lice (Figure 8.1). By definition, parasites are harmful to their hosts from which they derive nourishment, protection, and dispersal, while in return causing damage that can be subtle, severe, or even lethal (Price 1980). Such damage can influence the ecology of hosts and act as a selective force shaping their evolution (Loye and Zuk 1991; Clayton and Moore 1997; Newton 2013), giving rise to a suite of host defence mechanisms targeting prevention of infection and, failing that, mitigation of the deleterious effects of parasitism. Defence mechanisms include morphological, physiological, and behavioural adaptations. In this chapter we briefly review the main effects of ectoparasites on bird body condition and fitness. Then, we describe the properties of avian nests that make them suitable habitats for ectoparasites and identify the main invertebrate taxa that are found in nests, distinguishing between those that may visit nests transiently from those that take up more chronic residence within them. Finally, we provide a review of defence mechanisms employed by birds against ectoparasites, emphasizing those involved in preventing or combating nest infestation.

8.2 Ectoparasites and their effects on host fitness

Ectoparasites are defined as those that during any stage of their life cycle inhabit the skin of the host or its outgrowths (e.g. feathers, scales; Hopla et al. 1994). The vast majority of ectoparasites found in avian nests are arthropods, particularly insects and arachnids feeding mostly on blood but also on skin scrapings, exudates, or feathers (see images of representative members of most taxa in Figure 8.1 and a classification in Figure 8.2). Unlike endoparasites that live inside the host (e.g. intestinal worms or blood protozoa), most ectoparasites have free-living stages and direct contact with the host varies from sporadic feeding (e.g. mosquitoes [Culex spp.]) to prolonged attachment (e.g. some ticks [Ixodes spp.]). Some ectoparasites, such as most lice, are host-specific while others, such as some bugs, parasitize a wide range of hosts.

Ectoparasites can reduce host body condition and compromise its survival either directly or indirectly. Their feeding results in tissue damage, anaemia, allergic responses, or secondary bacterial infection. If they feed on blood it can result in inoculation of pathogens for which they are vectors (reviewed in Loye and Zuk 1991; Clayton and Moore 1997). Ectoparasites impose additional costs since the development and maintenance of defence mechanisms in the host are traded off against other metabolic demands (Sheldon and Verhulst 1996; Lochmiller and Deerenberg 2000; Norris and Evans 2000), including the production, maintenance, and display of ornaments. This may reduce fitness if mate choice is driven by ornamental traits where increased body condition can be fundamental to securing a mate (Andersson 1994). Later in the breeding attempt ectoparasites can reduce a parent's ability to incubate eggs effectively, thereby lowering both hatching and overall breeding success (e.g. Møller et al. 1990; de Lope et al. 1993; Oppliger et al. 1994). Infested chicks may grow slowly, attain only low body condition and have a reduced lifetime reproductive success (Møller et al. 1990; Merino and Potti 1995a; Fitze et al. 2004; Puchala 2004), despite having increased energy

López-Rull, I. and Macías Garcia, C., *Control of invertebrate occupants of nests*. In: *Nests, Eggs, and Incubation*. Edited by: D.C. Deeming and S.J. Reynolds, Oxford University Press (2015). © Oxford University Press. DOI 10.1093/acprof:oso/9780198718666.003.0008

Figure 8.1 Some species of ectoparasites commonly found in avian nests: (a) true bug (*Oeciacus hirundinis*); (b) flea (*Ceratophyllus columbae*); (c) hard tick (*Ixodes ricinus*); (d) haematophagous mite *Dermanyssus* sp.; (e) carnid fly (*Carnus hemapterus*); (f) blowfly (*Protocalliphora azurea*); (g) black fly (*Simulium aureum*); (h) mosquito (*Culex quinquefasciatus*); (i) biting midge *Culicoides* sp.; (j) chewing louse (*Laemobothrion tinnunculi*); and (k) louse fly (*Crataerina melbae*). Scale bar in each panel represents 1 mm. (Photos: I. López-Rull [a, b, f, g, h, j and k], G. Tomás [c], S. Merino and E. Pérez Badas [d], F. Valera and M.T. Amat-Valero [e], and J. Martínez-de la Puente [i]. See Plate 8.)

requirements, begging more intensively (Christie et al. 1996; Tripet and Richner 1997; Hurtrez-Boussès et al. 1998), and drawing more parental resources as a result (e.g. European pied flycatchers [*Ficedula hypoleuca*]; Merino et al. 1998).

Through their effect on individual reproduction and survival, ectoparasites can also affect their hosts'

population dynamics (Newton 2013). For instance, in seabird colonies the aggregation of hundreds or even thousands of breeding individuals with relatively long chick-rearing periods favour the build-up of large populations of ticks. In turn, massive tick infestation can prompt large-scale nest desertions as reported for sooty terns (*Sterna fuscata*), Guanay cormorants

Insects
— Flying insects
— Larval infective phase — **Blowflies**** **Muscid flies**** **Flesh flies****
— Adult infective phase — Black flies• *Louse flies******* **Carnid flies**** Biting midges• Mosquitoes•
— Non-flying insects — **Bugs**** **Fleas**** *Lice*△

Arachnids
— **Ticks****
— Mites△

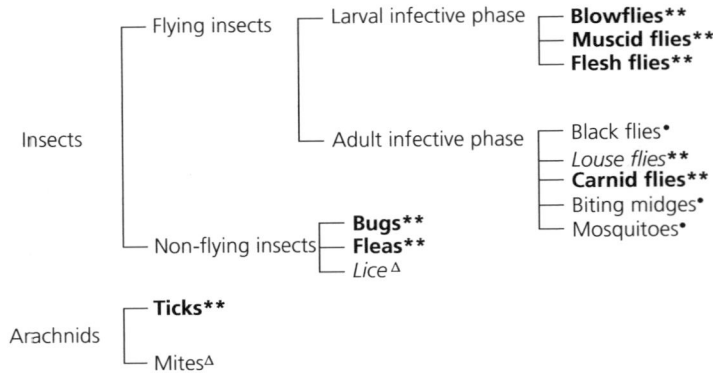

Figure 8.2 Common ectoparasites of nesting birds. Ectoparasites living in the nest material and attaching to the host only for feeding are shown in bold text while those living mostly on the host are shown in italicized text. Symbols according to their feeding habits: **obligate haematophagous; •free-living with only females being haematophagous; and △ some species are haematophagous and others feed on feathers, skin, or skin exudates.

(*Phalacrocorax bougainvillii*), Peruvian boobies (*Sula variegata*), and brown pelicans (*Pelecanus occidentalis*; Feare 1976; Duffy 1983). Ectoparasites can also impact on host demography as was the case in declining populations and local extinctions of many native Hawaiian honeycreepers (Drepanidinae). Birds are now absent from most areas where malaria-bearing mosquitoes (*Culex quinquefasciatus*) are present (Warner 1968; Atkinson et al. 1995; Newton 2013).

Ectoparasites find their hosts by moving through the environment, transferring via direct contact between infested individuals or lying in wait at places frequented by their hosts. By promoting contact for infestation and providing a refuge, nests are key elements in avian host–parasite interactions. However, what exactly makes nests suitable habitats for ectoparasites?

8.3 Nests as refuges for ectoparasites

Most avian species either build a nest or reuse one for successive breeding attempts, and many also use them for roosting. Nests protect eggs and chicks from predators, and allow parents to ensure that temperature and humidity are suitable to promote offspring development (Collias and Collias 1984; Chapters 4, 13, and 16). Non-avian species that would not survive in exposed habitats (e.g. arthropods that are prone to water and temperature stresses) may preferentially survive in the relatively stable micro-environment of an active bird nest where they may also be protected from predation. Since many avian species only use nests during the breeding season, parasites have evolved to synchronize their infective phase with the bird's reproductive phase. Thus, when nest occupancy by birds is seasonal the life cycle of nest-dwelling ectoparasites is coincidental.

Besides protection, nests also provide predictable sources of food. During incubation the temperature requirements of avian embryos are met through direct egg contact by the incubating adult(s) in most avian species (Ar and Sidis 2002) except in the Megapodiidae (Booth and Jones 2002). Therefore, incubating birds are 'sitting targets', both because of their prolonged exposure to nest materials where parasites reside and because many develop a highly vascularized and bare ventral brood patch (Lea and Klandorf 2002) that is readily exposed to hungry haematophagous parasites. After hatching and throughout brooding, nests of altricial species continue to provide warmth and shelter while food remains abundant in the form of nestling tissues (including blood), faeces, and undigested food. Nestlings are particularly vulnerable to parasites because their physiological and behavioural defences are not fully developed (Apanius 1998; Ardia and Schat 2008). Thus, it is precisely during this period when the populations of several ectoparasites reach their highest peak, timing their development with the availability of sufficient food resources (e.g. Heeb et al. 1996; Bize et al. 2004; Valera et al. 2006).

At the end of the breeding season adults and nestlings of many species leave nests while most arthropods burrow into the nest material or its surroundings, or fall off to the ground, and enter diapause. This process allows some insects and acarines to survive harsh conditions sometimes for many years (Tauber et al. 1986). In temperate habitats diapause permits ectoparasites to overwinter as eggs, pupae, nymphs, or adults which in the following breeding season will either parasitize new occupants of the nest or disperse to new nests by either flying or adhering to mobile hosts. Since many avian species often reuse the same nest, parasites remaining in the nest between breeding seasons increase the likelihood of encountering hosts. This particularly applies to avian host species that raise more than one brood per year in the same nest (e.g. colonial hirundines and swifts), or face scarcity of nesting sites (e.g. parids nesting in holes), those for which nest construction spans entire breeding seasons (e.g. ospreys [*Pandion haliaetus*], eagles [*Aquila* spp.], or buzzards [*Buteo* spp.]), or species that breed in traditional colonies such as seabirds. Rózsa et al. (1996) found that colony-breeding rooks (*Corvus frugilegus*) bore more ectoparasites than their close relatives hooded crows (*Corvus cornix*). Rékási et al. (1997) subsequently confirmed this relationship between coloniality and ectoparasite burden in 13 other species: mallards (*Anas platyrhynchos*), northern fulmars (*Fulmarus glacialis*), Manx shearwaters (*Puffinus puffinus*), Wilson's storm petrels (*Oceanites oceanicus*), European storm petrels (*Hydrobates pelagicus*), Leach's storm petrels, (*Oceanodroma leucorhoa*), Eurasian woodcocks (*Scolopax rusticola*), Eurasian collared doves (*Streptopelia decaocto*), common swifts (*Apus apus*), common house martins (*Delichon urbicum*), European pied flycatchers, white-throated dippers (*Cinclus cinclus*), and common reed buntings (*Emberiza schoeniclus*).

8.4 Invertebrate occupants of avian nests

Ectoparasites in avian nests can be: 1) arthropods dwelling mainly in the nest or its surroundings, including wingless (e.g. bugs, fleas, ticks, and nest mites) and winged taxa whose haematophagous stage is confined to the nest (e.g. carnid flies, blowflies, muscid flies and flesh flies); 2) flying insects that live outside the nest and only visit for feeding (e.g. black flies, mosquitoes, and biting midges); and 3) arthropods that primarily reside on the body of the host and are transmitted through direct contact in the nest (e.g. lice, louse flies and diverse groups of mites).

8.4.1 Nest-dwelling ectoparasites

8.4.1.1 True bugs (order Hemiptera, family Cimicidae)

They are obligate haematophagous ectoparasites that only come into contact with the host to feed, usually at night. They overwinter in the nest or its surroundings and can survive periods of up to three years of host absence. Cimicids are hemimetabolous (their development comprises egg, nymph, and adult stages) and undergo five nymph instars that all require a blood

meal to moult into the subsequent instar before becoming sexually mature. Reproduction starts when a host occupies the nest since gamete production requires regular blood meals (Reinhardt and Siva-Jothy 2007). Female bugs lay three to four eggs daily and potentially produce up to 500 eggs during their lives. Bugs disperse adhering to the legs and feet of their hosts (Brown and Brown 2004), and may also switch between host species that are secondary cavity nesters. This is the case of the South American *Acanthocrios furnarii* that is a common parasite of mud nests of rufous horneros (*Furnarius rufus*) but for which primary hosts are species belonging to the Emberizidae, Hirundinidae, Icteridae, Passeridae and Troglodytidae (Di Iorio et al. 2013). The most studied cimicids that infest birds are the swallow bug (*Oeciacus vicarius*) and the house martin bug (*O. hirundinis*) (Figure 8.1a), primarily found in nests of American cliff swallows (*Petrochelidon pyrrhonota*) and common house martins (*Delicon urbicum*), respectively. Bug populations at an active colony of American cliff swallows increase throughout the summer, reaching a peak at approx. the end of fledging when infestations can reach 2,600 bugs per nest and when they can significantly impact on the growth and survival of birds (Brown and Brown 1986; Brown et al. 1995). Experimental increase of *O. hirundinis* nest loads resulted in negative impacts on nestling body condition and survival (de Lope et al. 1993, 1998), while adults nesting in infested nests lost 4% of their body masses on average during the breeding season compared with 0.6% in birds from bug-free nests (Christie et al. 2002). Apart from direct costs, bugs are also vectors of the Buggy Creek virus (Alphavirus) which is a major cause of reduced fitness in house sparrows (*Passer domesticus*) occupying American cliff swallow colonies (O'Brien et al. 2011).

8.4.1.2 Fleas (order Siphonaptera)

They are obligate haematophagous ectoparasites with holometabolous development (comprising egg, larval, pupal, and adult stages). Most fleas dwelling in bird nests belong to the family Ceratophyllidae with the genera *Ceratophyllus* and *Dasypsyllus* being widespread among birds (Figure 8.1b). Females reproduce continuously laying two to five eggs daily throughout their adult lives (Marshall 1981). Eggs are deposited in the nest where larvae develop and pupate (Marshall 1981). Once the host starts to incubate, nest temperatures increase and accelerate flea development causing a first cohort of fleas to develop synchronously (Heeb et al. 1996). Adult fleas spend most of their time in the nest except when they move onto the host to feed. They occasionally disperse onto the fledglings or by jumping onto a passing adult (Humphries 1968), but most remain in the nest until the following breeding season when imagos emerge (Humphries 1968; Tripet and Richner 1999). Horizontal transmission within avian nests is high (up to 72%; Heeb et al. 1996). At the end of the host's breeding season nests may contain hundreds of adult fleas, larvae, and pupae (Harper et al. 1992; Eeva et al. 1994; Heeb et al. 1996). Fleas reduce body condition and survival of nestlings as well as adult breeding success. Their presence delays egg laying by 11 days, increases by 32% the number of nests abandoned, reduces clutch size by 5%, and increases nestling mortality by 32% in comparison with non-infested nests (Richner et al. 1993; Oppliger et al. 1994; Fitze et al. 2004).

8.4.1.3 Ticks (order Parasitiformes)

Those infesting bird nests are obligate haematophagous arachnids (Acari) from the families Argasidae, i.e. soft ticks, and Ixodidae, i.e. hard ticks. Most ticks are generalists and many parasitize mammals as well as birds. The most common genera infesting bird nests are soft ticks *Argas* and *Carios* (formerly *Ornithodoros*) and hard ticks *Ixodes* and *Amblyomma* (Figures 8.1c and 8.3f). Their general life cycle comprises four stages: egg, larva, nymph, and adult. Hard ticks have a single nymph stage while soft ticks can go through up to seven instars, each requiring a blood meal. Longevity ranges from several months to several years (Hoskins and Cupp 1988). Female hard ticks typically oviposit once whereas soft ticks lay eggs following each blood meal. Hard ticks remain attached to their hosts for up to several days whereas most soft ticks feed rapidly and drop off the host to burrow into the nest. When birds leave the nest ticks go into diapause and survive long periods without feeding, becoming long-term residents at breeding sites such as seabird colonies. Ticks may reduce breeding success and even cause nest desertion as was the case in massive infestations by *Carios capensis* in colonies of sooty terns, Guanay cormorants, Peruvian boobies, and brown pelicans (Dietrich et al. 2011 and references therein). Ticks transmit more species of protozoa, rickettsiae, spirochaetes and viruses than any other arthropod taxon (Jongejan and Uilenberg 2004).

8.4.1.4 Mites (orders Parasitiformes and Acariformes)

Bird mites can be divided into those that dwell primarily in or around the nest and those that reside mainly on the body of the host. Mites are a diverse group of arachnids with most bird-associated taxa belonging to either the Parasitiformes, which include both respiratory and nest-dwelling mites (sub-order Mesostigmata), or the Acariformes (mainly the sub-order Sarcoptiformes) to which the greatest diversity of nest-dwelling mites, feather mites and skin mites (infraorder Astigmata) belong (Proctor and Owens 2000). The Trombidiformes or Prostigmata (Acariformes) include a smaller diversity of nest mites. Mites normally pass from egg to larva, nymph, and adult stages. Although not all bird mites are parasitic, those feeding on living avian tissues have a negative effect on their hosts. The best studied nest-dwelling mites are haematophagous (genera *Dermanyssus* and *Ornithonyssus*; commonly termed 'fowl mites') and are mostly found in passerines (Figure 8.1d). Depending on the species, adults live in the nest or on the hosts, but nymph stages dwell in the nest and only climb onto the host for feeding (Proctor and Owens 2000). The haematophagous *Ornithonyssus bursa* require blood meals throughout their lives and cannot survive beyond six weeks without feeding during the reproductive season of their hosts (Powlesland 1977), but can survive eight months in abandoned nests (Møller 1990b). These mites have short generation times but can gather in huge populations of up to 14,000 in one barn swallow (*Hirundo rustica*) nest (Møller 1990b) and 71,000 in a common starling (*Sturnus vulgaris*) nest (Powlesland 1977). When nests are heavily infested, such as through inoculation with 100 mites at the beginning of the host breeding season (Møller 1990b), mites reduce clutch size and brood size at hatching, lower by 5% nestling body condition and reduce fledgling survival by 18%, compared with non-infested nests (Møller 1990b). Mites also reduce adult breeding success; none of the heavily infested nests was reoccupied, compared with 65% occupancy of clean (non-infested) nests. Interclutch intervals were on average 8% longer in infested nests and fewer females laid a second clutch in the same nest if it harboured many mites (nest reuse was 33% in infested versus 77% in non-infested nests). As well as having direct effects upon their hosts, haematophagous mites can transmit pathogenic viruses, rickettsiae and protozoans.

8.4.1.5 Carnid flies (order Diptera, family Carnidae)

They are holometabolous insects that typically possess one pair of functional wings. Among carnids, the most studied species to date is *Carnus hemapterus* (Figures

8.1e and 8.3d), a small fly that feeds on the blood of nestlings and incubating adults of many bird species (Liker et al. 2001). Female flies deposit their eggs in the nest material where three larval stages are completed within 21 days. The pupae undergo diapause and adults emerge the following spring (Marshall 1981). Prolonged diapause has been recorded for this parasite, probably enabling *Carnus* spp. to persist in the nest for several years (Valera et al. 2006). The adult dispersive form has wings that shed after finding a suitable host (Marshall 1981). Adult *C. hemapterus* are present in the nest a few days before incubation starts (López-Rull et al. 2007). In some starlings (*Sturnus* spp.) immaculate eggs often become finely spotted with red-brownish spots (presumably droppings of *C. hemapterus*) as incubation progresses. Therefore, abundance of flies in a nest can be predicted some days after the beginning of incubation by the density of pigment spots on eggshells (López-Rull et al. 2007; Figure 8.3e). *C. hemapterus* may parasitize up to 80% of all the nests in a breeding colony (Fargallo et al. 2001). Their populations increase in size throughout the early nestling period and can exceed 500 flies per nest (Hoi et al. 2010), reaching a peak during the mid-nestling stage of their host (Fargallo et al. 2001; Václav et al. 2008). They become less abundant once nestlings develop feathers (Avilés et al. 2009). Negative correlations between the number of nest-dwelling carnids, and nestling body mass and growth have been recorded (Avilés et al. 2009; Wiebe 2009).

8.4.1.6 Blowflies (order Diptera, family Calliphoridae), muscid flies (order Diptera, family Muscidae; also referred as house flies and botflies), and flesh flies (order Diptera, family Sarcophagidae)

They are holometabolous flies that include free-living adults that typically possess one pair of functional wings and have obligate haematophagous larval stages that dwell in nest material and cause myiasis (infestation of host tissue by dipteran larvae), particularly in nestlings (Figure 8.1f). Typically, female flies lay eggs in nests containing hatchlings. Larvae emerge after one to two days and over a period of seven to 14 days (Little 2008). At the end of the third larval stage they return to the nest material to pupate (Figures 8.3a and 8.3b). Adult flies emerge and they may either mate and then search for nests containing hatchlings, or enter diapause until the following spring (Bennett and Whitworth 1991). While larvae of most species alternate bouts of feeding and hiding within nest materials

(e.g. most *Protocalliphora* blowflies), others develop under the host's skin (e.g. the blowfly *Protocalliphora braueri* and many muscid flies *Philornis*; Little 2008). There are also species (e.g. *Philornis downsi*) that complete their three larval stages in the nasal cavities and then migrate into the nest material, intermittently moving back to the birds for blood meals (Fessl et al. 2006). Unlike blowflies and muscid flies, female flesh flies (e.g. *Wohlfahrtia* spp.) retain the developing eggs until they hatch, and deposit 30–200 larvae directly onto the bird's mucous membranes or fresh wounds from where they penetrate under the skin to feed on the host's underlying tissues. When ready to pupate they leave the host and drop to the ground (Little 2008). Although larval infestation occurs mainly in Passeriformes, they have been recorded on other nidicolous species such as Falconiformes, Strigiformes, Coraciiformes, Piciformes, Columbiformes, Cuculiformes etc. (see review by Little 2008). The most widespread haematophagous larval flies are in the genus *Protocalliphora* (Figures 8.1f and 8.3a), the intensity of its infestation ranging from one to > 1,000 larvae per nest (Whitworth and Bennett 1992). The negative effects on birds are related to blood loss as a result of the feeding larvae, which can consume > 55% of a nestling's blood volume (Gold and Dahlsten 1983) and even cause up to 97% mortality in species such as the pearly-eyed thrasher (*Margarops fuscatus*) infested by *Philornis* spp. (Arendt 1985). Parasitism of *Philornis downsi* larvae on nestlings has led to dramatic population declines; it reached the Galapagos Archipelago and became a major threat to some of its endemic bird species. Its larvae are found in virtually all nests, i.e. 97% out of 177 nests studied, of 12 land bird species examined by Fessl and Tebbich (2002), and a loss of 27% of nestlings was attributed to this parasite. In addition, when infestation is coupled with poor breeding conditions (e.g. inclement weather and food scarcity), reduced fledging success and parental survival have been recorded (Hurtrez-Boussès et al. 1998; Simon et al. 2004). Blowflies also act as vectors of avian tuberculosis (Fischer et al. 2004).

8.4.2 Nest-visiting ectoparasites

8.4.2.1 Black flies (order Diptera, family Simuliidae)

They are dark robust flies (Figure 8.1g) with a wide geographic distribution. Females lay eggs on running water that then sink into the bottom sediments. Simuliids are holometabolous and larvae typically undergo seven moults before pupating. Fully developed adults emerge at the water's surface in a bubble of gas (Currie and Hunter 2008). With the exception of some species that are autogenous, i.e. females can produce at least the first batch of eggs on a blood-free diet, female flies require blood meals to produce eggs. Thus, female diurnal ornithophilic black flies commonly visit nests to feed on adults and nestlings (Bennett 1960). Host specificity varies with some simuliids such as *Simulium annulus* that feeds predominantly on great northern loons (*Gavia immer*) among a few other species, while others are less selective. During years of black fly outbreaks simuliids reduce hatching success by 45%, increase chick mortality by 50%, and increase the number of nests abandoned (at both egg and nestling stages) by 60% (Hunter et al. 1997; Bukaciński and Bukaciński 2000). Simuliids may transmit blood parasites such as filarial nematodes, trypanosomes and protozoa causing avian malaria, such as *Leucocytozoon* and *Haemoproteus* (Anderson 1956; Fallis and Bennett 1961; Votýpka et al. 2002).

8.4.2.2 Mosquitoes (order Diptera, family Culicidae)

They are common visitors to nests where females feed on the blood of adults and nestlings. Female mosquitoes lay their eggs in water or damp soil. Mosquitoes are holometabolous, larvae are always aquatic and they typically undergo four moults before pupating. Development from egg to adult takes five to 40 days depending on the species and temperature (Rueda et al. 1990; Tun-Lin et al. 2000). Adults usually mate within days of emerging after which females seek the blood meals (Figure 8.3g) required to produce eggs (except in autogenous species). This makes them active vectors of blood-borne diseases, including avian malaria caused by *Plasmodium*. In some cases this type of malaria produces no apparent clinical signs but in others it may cause severe anaemia and even death. An extreme example followed the accidental introduction to the Hawaiian Islands of the mosquito *Culex quinquefasciatus* (Figure 8.1h), resulting in the subsequent emergence of malaria that caused the extinction of many endemic honeycreepers (Drepanidinae). Of 41 species and subspecies of Hawaiian honeycreepers, 17 are thought to be extinct and 14 are endangered due to several factors, including vector-transmitted diseases (Atkinson et al. 1995).

8.4.2.3 Biting midges (order Diptera, family Ceratopogonidae)

Biting midges (Figure 8.1i) are amongst the most abundant but least known of the haematophagous insects. They are short-lived crepuscular or nocturnal flies

(Mellor et al. 2000) with a life cycle that is similar to that of black flies except they do not need running water to oviposit. They can lay on saturated soil or animal dung where larvae and pupae develop (Mellor et al. 2000). Typically, adult females require blood meals to produce eggs (when not autogenous) and fecundity is positively related to the amount of blood ingested (Mullen 2002). When nests are highly infested, midges reduce nestling body condition (Tomás et al. 2008). They are also vectors of filarial nematodes (Mellor et al. 2000), of several *Haemoproteus* species (Fallis and Wood 1957; Valkiūnas et al. 2002) and of viruses (Naugle et al. 2004).

8.4.3 Host-dwelling ectoparasites

8.4.3.1 Lice (order Phthiraptera)

Chewing lice (hereafter referred to as 'lice') are small wingless insects with dorso-ventrally flattened bodies classified into two suborders (formerly 'Mallophaga' but now Amblycera and Ischnocera; Figure 8.1j). Amblycera live on feathers and skin, and can make short incursions away from the host or even switch hosts. In contrast, Ischnocera (or 'feather lice') are confined to plumage through specializations that prohibit many species from living away from feathers (Clayton et al. 2008). Lice are obligate parasites that feed on host skin, feathers and, in some cases, blood (suborder Amblycera). Lice have been found on virtually every species that has been examined (Clayton et al. 2008). Some species are strictly host-specific while others can be found on several avian orders. Lice are hemimetabolous with females attaching their eggs to feathers and with all life stages taking place on the host's plumage (Marshall 1981). The timing and duration of each stage vary among species and may be influenced by temperature and humidity (Moyer et al. 2002). Adults live approx. one month with females producing one egg per day (Clayton et al. 2008). Most individuals never leave their host and transmission occurs opportunistically. Lice can cause extensive feather and skin damage. Severe hemorrhagic ulcers in the oral mucosa have been documented in juvenile American white pelicans (*Pelecanus erythrorhynchos*) infested with *Piagetiella peralis*, a species that lives within the pouch of these hosts (Dik 2006), and sublingual fistulas in young masked boobies (*Sula dactylatra*) are proposed to be triggered by such ectoparasites (Hughes et al. 2013). Damage caused by feather lice increases feather breakage (Kose and Møller 1999), and reduces thermoregulatory capacity (Booth et al. 1993), body condition (Potti and Merino 1995), flight performance (Barbosa et al. 2002) and life expectancy (Brown et al. 1995; Clayton et al. 1999). They also delay migration date (Møller et al. 2004) and reduce sexual attractiveness (Clayton 1990; Kose and Møller 1999; Moreno-Rueda and Hoi 2012). Lice can also act as vectors or intermediate hosts of other parasites including some viruses, bacteria and nematodes (Clayton et al. 2008).

8.4.3.2 Louse flies (order Diptera, family Hippoboscidae)

They are obligate haematophagous flies with holometabolous development (Figure 8.1k) that include more

than 150 species (Lloyd 2002). A female fly produces one larva at a time that is retained in her body where it is fed on the secretions of a milk gland located in the uterus. The larva is born, usually into a bird's nest, as a late third instar which pupates immediately. After hatching, winged adult flies emerge and search for a suitable host, thereafter living on its body and feeding solely on its blood several times a day (Coatney 1931). Although hippoboscids do not strictly live in the nest, they pupate on it and are transmitted between parents and nestlings. The effects of hippoboscids on birds include anaemia (Jones 1985), loss of body mass

Figure 8.3 Examples of ectoparasite–bird interactions. Infestations of blowfly *Protocalliphora azurea* showing (a) the base of a blue tit (*Cyanistes caeruleus*) nest, (b) a pupa at higher magnification, and (c) a larva attached to the leg of a nestling of European pied flycatcher (*Ficedula hypoleuca*). (d) A nestling of the European roller (*Coracias garrulus*) infested with the carnid fly *Carnus hemapterus*. (e) The immaculate blue-greenish eggs (left) of some *Sturnus* spp. that become spotted as incubation progresses (right). Recent observations suggest that spots are droppings of *C. hemapterus* living amongst the nest material. (f) A male blue tit with a soft tick attached to his head. (g) A mosquito biting the foot of a domestic fowl (*Gallus gallus*). (h) A larva of a chigger mite *Neotrombicula autumnalis* attached at the eye ring of a red-billed chough (*Pyrrhocorax pyrrhocorax*). (Photos: G. Tomás [a, b, f and g], A. Cantarero [c], M.A. Calero Torrabo [d], ILR [e], and G. Blanco [h]. See Plate 9.)

(Lloyd 2002), and reduced breeding success, as shown in alpine swifts (*Tachymarptis melba*) where parents of parasitized broods produced 26% fewer fledglings than those whose broods were experimentally deparasitized repeatedly during the breeding attempt (Bize et al. 2004). Hippoboscid flies are vectors of several haemoparasites including avian malaria protozoa (Sol et al. 2003), trypanosomes (Baker 1967), and possibly viruses such as West Nile (Farajollahi et al. 2005).

8.4.3.3 Host-dwelling mites (Parasitiformes, Acariformes)

Birds host many groups of mites in their feathers and on their skin. The adults of some skin mites are non-parasitic nest-dwellers, whereas their juvenile stages are subcutaneous parasites (e.g. Hypodectidae). Other groups are strict parasites whose adults and juveniles burrow into the skin, causing lesions on the apterous skin, especially on legs, feet, and around the bill (e.g. Knemidocoptidae or 'scaly leg mites'). Most other groups of skin-associated mites only attach temporarily to the epidermis by their mouthparts (e.g. Trombiculidae or 'chiggers'; Figure 8.3h). Mites feeding on living tissues can damage integument, induce anaemia and in extreme cases necrotizing dermatitis, as observed in Laysan albatross (*Phoebastria immutabilis*) fledglings infested by chiggers (Gilardi et al. 2001). Feather mites (Astigmata) are the most diverse group of arthropods found on birds, with more than 2,500 described species (Proctor 2003). Plumicoles live on the surface of feathers while syringicoles live inside the feather quills. Plumicolous mites are not considered parasites since they feed mainly on uropygial gland oils and pollen and fungi that adhere to the feather barbs, causing no structural damage. By contrast, many syringicolous mites eat the medulla, i.e. pith, of the quills and may thus weaken the feathers. The nature of the interaction between feather mites and birds is a matter of controversy, with some authors reporting detrimental effects on the hosts (e.g. Thompson et al. 1997; Harper 1999; Pérez-Tris et al. 2002), while others report either positive effects on hosts (e.g. Blanco et al. 1997, 2001) or no effect at all (e.g. Blanco et al. 1999; Dowling et al. 2001).

8.5 Bird defences against nest parasites

Apart from external barriers, birds possess physiological and behavioural mechanisms to either repel or fight parasites (Figure 8.4). To be regarded as an adaptation against parasites, a behaviour or physiological function must meet two criteria: 1) investment occurs to the detriment of parasites; and 2) detrimental effects on parasites result in an increase in the host's fitness (Hart 1990, 1997). The main group of physiological defences that constitute the immune response comprises a complex arrangement of interacting cells, tissues and molecules that are widely dispersed throughout the body. By discriminating self from non-self, the immune system controls and prevents parasite infection (Kuby 1998). Complementary to physiological defences, hosts have evolved behavioural ways to avoid, remove, and destroy parasites. Examples of behavioural defences in birds include parasite avoidance, preening, and self-medication (Hart 1997).

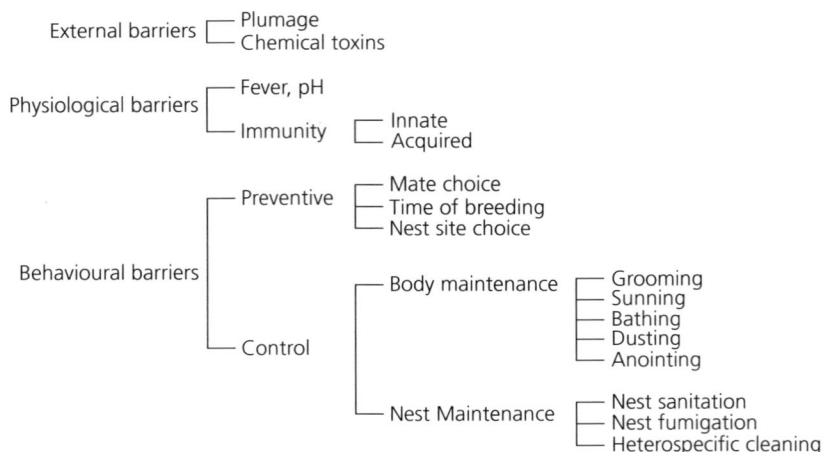

Figure 8.4 Main defence barriers used by birds to prevent, and defend against, ectoparasites.

Individual variation in defences against parasites, environmental heterogeneity and the mechanisms of parasite transmission may all contribute to generate the typically aggregated distribution of parasites, with most hosts harbouring few parasites while some are heavily infested (Hudson and Dobson 1997). Defences against parasites also vary ontogenetically and are normally weaker at early life stages (Apanius 1998). Therefore, nestlings are often heavily infested with ectoparasites while their parents, sharing the same environment, at least for a few hours every day, may not be. Another source of variation in parasite defences is life history. The position of species along the altricial–precocial spectrum (Starck and Ricklefs 1998) is a major determinant of infestation risk early in life. Altricial hatchlings have limited mobility, must be cared for by adults, and lack the ability to evade parasites by leaving the nest, to remove parasites from their bodies, or to mount an effective immune response. Such heightened vulnerability is probably the reason why several parental strategies to reduce parasitism in the nest have evolved. These include physiological adaptations such as transferring maternal immunity to the young via egg components (Chapter 10), as well as behaviours such as selection of nest sites with low risk of parasitism or routine and intense nest sanitation. We will now review bird defence mechanisms against parasites, with particular emphasis on those in the nest.

8.5.1 External barriers

A first line of defence against ectoparasites is plumage. Due to the complex arrangement of interlocking feather barbules, feathers hamper parasite access to skin and therefore blood vessels. However, feathers themselves can be the target of parasites such as keratin-eating lice; these can be thwarted by the presence of melanin pigments that increase feather thickness and resistance to abrasion (Burtt 1986). Altricial chicks are born naked, or only patchily covered with soft down, so they are exposed to skin or haematophagous parasites in the nest and, thus, any defence against them must be preemptive. One of the most striking plumage defences of birds is that exhibited by members of the *Pitohui* and *Ifrita* genera. They store in their skin and feathers alkaloid neurotoxins (similar to those present in poison arrow frogs [*Dendrobates* spp. and *Phyllobates* spp.]) that are acquired through diet (Dumbacher et al. 1992) and are effective against a wide variety of arthropods (Mouritsen and Madsen 1994; Dumbacher 1999). They probably also make birds unpalatable to predators (Weldon 2000). Similarly, the plumage of some seabirds

(e.g. crested auklets [*Aethia cristatella*] and whiskered auklets [*Aethia pygmaea*]) emits a citrus-like aroma that repels ticks and mosquitoes (Douglas et al. 2004, 2005). However, since juvenile vertebrates are generally more vulnerable to dietary toxins than adults, it seems unlikely that defence mechanisms based on neurotoxins or alkaloids are employed by nestlings.

8.5.2 Physiological defences

Birds may either passively 'weather the storm', resisting the effects of parasites and repairing the damage, or actively attack parasites using the immune system. Immune defence can be innate ('non-specific') or acquired ('specific') depending on whether it attacks any foreign invader or whether it is developed in response to a particular pathogen that it can recognize and selectively eliminate. The innate response is the first to be activated upon infection and comprises physiological, inflammatory and phagocytic barriers that not only contribute to fighting non-specific infections but also to the induction and modulation of the adaptive immune system (Juul-Madsen et al. 2008).

Amongst the several processes that constitute the immune response following parasite contact with specialized cells (e.g. macrophages, granulocytes, thrombocytes, and lymphocytes), release of soluble factors (e.g. lysozymes, interferon, complement, and natural antibodies), changes in the pH (e.g. an increase in gastric acidity which may prevent intestinal infections; Kuby 1998) and fever, i.e. generalized increase of temperature, are noteworthy. Fever can potentiate immune response and reduce the availability of blood nutrients, such as iron, which is essential for pathogen development and reproduction (Hart 1988, 1990), and may directly affect pathogen survival (e.g. Hasday and Singh 2000). Neither gastric acidity nor fever appears capable of exerting direct effects on ectoparasites but they may protect birds against the pathogens associated with ectoparasites. The inflammatory response, on the other hand, can reduce the ability of ectoparasites to feed, and even harm them directly. This response is the consequence of a complex sequence of events that can be triggered when a parasite causes tissue damage, promoting recruitment of immunological cells that release several vasoactive and chemotactic factors. They can induce increased blood flow to the area, capillary permeability, and a further influx of lymphocytes and phagocytes (Kuby 1998). These are white blood cells specialized in the ingestion and lysis of pathogens and in the presentation of pathogen antigens to the major histocompatibility complex (MHC) that is subsequently incorporated onto the surface of T-cells, the lymphocytes involved in acquired immunity (Hughes and Yeager 1998). Ectoparasites can find it difficult to extract blood from an inflamed area and the ingestion of quantities of phagocytes may harm their digestive systems (Owen et al. 2010). The nestling's immune system is still maturing (Apanius 1998; Ardia and Schat 2008), and thus chicks depend on their innate immune response and on maternal antibodies (IgG) transferred through the egg yolk to defend against pathogens (Deeming 2002a; Grindstaff et al. 2003; Saino et al. 2003; Müller et al. 2004). They may also be partially protected against ectoparasites if their inflammatory reaction is robust (Owen et al. 2010).

We are not aware of any systematic comparison of the ontogenetic trajectory of the immune system along the altricial–precocial spectrum (Apanius 1998; Ardia and Schat 2008), but we predict that the more altricial a species, the faster its immune system will mature. Altriciality results in nestlings that are 'sitting targets' in an infested environment and are in regular physical contact with potentially infested parents, especially early in the nestling period during intense brooding. Precocial chicks may also encounter high numbers of ectoparasites but more sporadically. We further propose that vulnerability to ectoparasites at the nest may be a major factor promoting/maintaining precociality. Modern bird species demonstrate a diversity of developmental modes from altriciality to precociality, and both were present in dinosaurs although it is unclear which condition was ancestral (see Ricklefs and Starck 1998b and Chapter 2).

8.5.3 Behavioural defences

Behavioural mechanisms against nest parasites may be grouped into two broad categories: pre-emptive and control. The former are those that reduce the risk of nest infestation while the latter remove or destroy parasites already infesting the nest or the nest attendants, or reduce their detrimental effects. We draw this distinction to aid description but it must be borne in mind that by controlling its parasite load, a bird reduces the likelihood of carrying parasites into its nest. Prevention due to pre-emptive behaviours and control behaviours need only result in a low abundance of parasites (one which does not cause a detectable reduction in the host's health) to be deemed successful.

8.5.3.1 Pre-emptive behaviours

The simplest way of dealing with parasites is to avoid contact with them. Selection of a mate and nest

location, and appropriate timing of the breeding attempt, can reduce the probability of nest infestation. Avian mate choice is commonly based on secondary sexual traits. Besides signalling specific identity and sometimes genetic composition, there is substantial evidence that these traits reliably reflect individual quality (Andersson 1994). For instance, fighting parasites and their ill-effects is a costly priority that uses temporal and nutritional resources (e.g. antioxidants) which are therefore not available for the full development of ornaments. Consequently, only those individuals that can avoid and/or combat parasites will develop attractive ornaments and be selected as mating partners (Hamilton and Zuk 1982). Additionally, in several species in which males primarily compete for access to females (e.g. lekking species in polygynous mating systems), the expression of male secondary sexual traits is positively linked to testosterone production which, in turn, may have an immunosuppressive effect. Thus, only individuals in high body condition are able to maintain elevated hormonal levels (and thus produce large ornaments) and mount an effective immune response (Folstad and Karter 1992). By choosing a highly ornamented mate, females may benefit directly through avoiding exposure of themselves and their offspring to ectoparasites. They will also secure a greater contribution of paternal care compared with more heavily parasitized males that may invest less in the breeding attempt. Females also gain indirectly through the inheritance of parasite resistance (Hamilton and Zuk 1982; Andersson 1994).

Timing of reproduction may influence parasite transmission because as the breeding season progresses, ectoparasite populations reach their peak and environmental conditions deteriorate, resulting in larger parasite loads in adults and nestlings through the season (e.g. López-Rull et al. 2010). Large ectoparasite loads have been linked to a reduction in breeding success, as shown in common house martins where increasing numbers of ectoparasites resulted in a 50% reduction in fledging success between first and second broods (de Lope et al. 1993). Laying early in the season could help individuals in lower body condition to reduce exposure to parasitism and hence the cost of fighting it, but it is uncertain whether such individuals may successfully settle in a territory and secure a mate in the face of competition with birds in higher body condition.

Nest-site selection may also influence the risk of parasite infestation through the avoidance of nest sites which already do or will contain large numbers of ectoparasites (Moore 2002). The abundance of ectoparasites at the nest is affected by several characteristics of nest microclimate such as temperature (Dawson et al. 2005b; Martínez-de la Puente et al. 2009, 2010), humidity (Heeb et al. 2000), orientation (George 1959), wind speed (Martínez-de la Puente et al. 2009), and the amount and type of nest material (Mazgajski 2007). Thus, we predict that such factors will influence nest-site selection of birds (see also Chapter 5).

For many species suitable nest sites are scarce (Newton 2013) and birds often occupy previously used nests instead of building new ones. Since old nests are usually infested by ectoparasites that are awaiting the return of their hosts, it has been argued that avoiding old nests may be an effective defence against parasites. However, Mazgajski (2007) found little evidence for this but, instead, found that most species reusing nests have similar breeding performance to those breeding in new ones. Such results suggest the benefits associated with old nests outweigh the costs of parasitism and open the debate of whether nest-site choice is influenced by parasite abundance at all. Old nests signalling successful breeding may indicate the quality of sites (Erckmann et al. 1990; Olsson and Allander 1995) and/or may represent substantial savings in time and energy related to nest building that can be allocated to other activities that help to counteract the negative effects of parasites on offspring. Additionally, old nests may vary in the content of materials that either repel or control ectoparasites, and if this can be assessed by newly arrived birds, it might save them some investment of resources in nest construction without compromising defence against parasitism. A useful rule-of-thumb may be to occupy old nests not heavily infested while avoiding old nests with high parasite abundance. Indeed, birds can assess the degree of infestation as shown in great tits (*Parus major*) for flea burdens and by American cliff swallows for bugs (Loye and Carroll 1991; Oppliger et al. 1994). Also, experimental studies of European pied flycatchers have shown preferential selection of nest sites where old nests have been removed (Merino and Potti 1995b). Interestingly, some bird species like house wrens (*Troglodytes aedon*), common starlings, and European pied flycatchers that reuse nests from previous breeding seasons are known to remove old nest material before laying (Kessel 1957; Merino and Potti 1995b; Pacejka et al. 1996). The particulars of such cleaning behaviours have not been comprehensively studied; the identity of species that do it, the circumstances under which it is done, and the extent of such cleaning remain largely unreported. Crucially, so too do the details of downstream effects on chick and adult survival. In colonially breeding species ectoparasites may accumulate from year to year in the nesting sites, resulting in severe infestations that cause chick mortality, as observed in hirundines, terns (*Sterna* spp.) and kittiwakes (*Rissa* spp.). A seemingly last-ditch defence strategy has been observed in such a context, when sudden and large-scale nest desertions of the breeding site by thousands of breeding pairs occur in response to massive tick infestations (Feare 1976; Duffy 1983; see section 8.2).

8.5.3.2 Control behaviours

Behaviours to control nest-borne parasites include removing and destroying ectoparasites from the body, i.e. body maintenance behaviours, and from the nest, i.e. nest maintenance behaviours. The main behaviour involved in body maintenance is grooming. This is an expensive activity with a metabolic cost of twice the basal metabolic rate (Goldstein 1988) and a temporal cost of between 0.3% (e.g. rockhopper penguin [*Eudyptes chrysocome*]) and 25.4% (e.g. great northern loon) of the daily time budget (Cotgreave and Clayton 1994), comprising 92.6% of all maintenance behaviours. Bird grooming includes preening and scratching (Hart 1997). During preening birds nibble their feathers with their bills to remove ectoparasites and particulate matter, to arrange and re-attach feather barbs and to lubricate the feathers with oily secretions. Effectiveness of preening depends on bill morphology, particularly on the extent of overhang of the maxilla with the mandible (Clayton et al. 2005), as shown by the substantial increase in number of feather lice on common pigeons (*Columba livia*) whose maxillary overhang was experimentally trimmed (an intervention that did not affect feeding efficiency; Clayton et al. 2005). Many studies have shown that preening is an important defence against ectoparasites (reviewed by Clayton et al. 2010). Although most studies dealing with the effectiveness of preening have measured its effects on feather lice numbers, Waite et al. (2012) tested the effectiveness of preening on the more mobile hippoboscid flies and found that birds infested with flies doubled their amount of preening time compared with those without (23.5% versus 11.2% of the time observed). In addition, preening birds killed twice as many flies over the course of the experiment compared with birds with impaired preening, suggesting that preening may also play a role in shaping vector ecology and the evolution of pathogens transmitted by blood-sucking ectoparasites.

Mutual preening, i.e. allopreening, between mates, and between parents and their offspring, is common in birds. Besides promoting social bonds, allopreening may help to control parasites that escape self-preening

(Fraga 1984; Brooke 1985; Murray 1990; Wernham-Calladine 1995). For instance, macaroni penguins (*Eudyptes chrysolophus*) that did not indulge in allopreening had more ticks on their heads and necks than paired birds that did engage in frequent allopreening (Brooke 1985). Since the two focal groups of penguins may have differed in other respects beyond allopreening, this result awaits further experimental confirmation. Scratching involves scraping the skin and plumage with the feet, and may be particularly important to access areas of the body that cannot be self-preened (e.g. head and neck). Birds with deformed or missing feet are usually heavily infested by ectoparasites around these bill-inaccessible regions (Clayton 1991), although, admittedly, birds with such abnormalities may also have a foraging deficit and be in generally low body condition. The pectinate, i.e. comb-like, claw on the middle toe of birds in some orders may have evolved because of its adaptive value in parasite removal. However, to date the only support for this hypothesis comes from a study of barn owls (*Tyto alba*) in which Bush et al. (2012) found that the number of 'teeth' on the pectinate claw was negatively associated with the presence of lice, but no experimental manipulation of this structure has been conducted.

Sunning, i.e. sunbathing, is common in species that expose themselves to solar radiation while adopting a stereotyped posture that maximizes the amount of their plumage that is exposed to the direct sunlight. Blem and Blem (1993) experimentally fumigated violet-green swallows (*Tachycineta thalassina*) and found that they sunned less often than unfumigated (control) birds, suggesting that birds engage in sunning when they feel infested. Sunning can dislodge parasites by killing them directly through overheating, as suggested for lice by Moyer and Wagenbach (1995), or indirectly if the excess heat encourages parasites to move more often in the plumage making them more vulnerable to detection and being preened out.

Despite the popular idea that bathing and dusting constitute defence behaviours against parasites, and in spite of the findings of Moyer and Wagenbach (1995), we are not aware of any study that demonstrates a causal link between natural performance of such behaviours and a reduced parasite load; sometimes poultry are given access to dusting beads containing natural pesticides (Salifou et al. 2013). It has been proposed that dusting discourages infestation by removing an excess of plumage oils on which some ectoparasites feed (Borchelt and Duncan 1974), or that it directly harms parasites as dust blocks their breathing pores and promotes dehydration (Wigglesworth 1944).

However, we are not aware of any study that has ever tested any of these mechanisms. Similarly, anointing has been proposed to constitute a defence against parasites (Clayton and Wolfe 1993), but evidence is inconclusive. During anointing birds rub themselves against aromatic materials or apply them to their plumage. An anointing behaviour common in many passerines is 'anting', in which birds crush and smear ants (Formicidae) or other arthropods on their feathers ('active anting') or allow them to crawl through their plumage ('passive anting'). Interestingly, arthropods used by birds are only those that secrete acids (formic acid in the case of ants), suggesting a possible role for these compounds in killing ectoparasites. While the idea is appealing, no study has provided substantial evidence that anting is an effective parasite defence. The only (unpublished) experimental study that tested this hypothesis found no differences in lice and mite abundances between individuals that engaged in anting and those that did not (as described by Clayton and Wolfe 1993; Clayton et al. 2010). Birds also anoint themselves with citrus fruits, such as lemon (*Citrus limon*) and lime (*C. aurantifolia*). The application of peels, pulp, and juice onto their bodies has also been interpreted as parasite defence, and although volatile compounds present in these fruits repel chewing lice (Clayton and Vernon 1993), ticks (Weldon et al. 2011), and mosquitoes (Weldon et al. 2011) *in vitro*, more studies are needed to confirm the consequences of such behaviours of birds in response to ectoparasite exposure *in situ*.

Birds have evolved several nest maintenance behaviours that reduce the probability of nest infestation and aim to mitigate it once established; these include nest sanitation and fumigation, and heterospecific cleaning. Nest sanitation refers to parental behaviours to clean the nest by removing any object that is neither nest material, nor a viable egg or young (Guigueno and Sealy 2012). Objects removed include eggshells, faeces, dead nestlings, foreign debris (e.g. rings on nestlings) and ectoparasites. Guigueno and Sealy (2012) in a review listed 15 species of passerines that remove parasites from the nest but this small number probably reflects the paucity of studies in the subject area rather than the rarity of the behaviour. Cleaning is a response to nest infestation with experimental removal of parasites from nests of blue tits (*Cyanistes caeruleus*) and European pied flycatchers resulting in females devoting less time to nest sanitation than when nests were heavily infested (Christie et al. 1996; Hurtrez-Boussès 2000; Tripet et al. 2002, Cantarero et al. 2013). Nest sanitation can take up to 17% of the female's daily time budget when nests are infested, but only 5% when they are not (Hurtrez-Boussès 2000), thereby limiting time for other activities such as provisioning nestlings. For example, female great tits reduce their time sleeping from 73.5% of the night-time to just 48.1% in infested nests, performing nest sanitation during the night (Christie et al. 1996).

The incorporation in the nest of materials containing volatile compounds that repel ectoparasites is known as nest fumigation. Many species of Passeriformes, raptors (Falconiformes), and storks (Ciconiiformes) add fresh aromatic plants to their nests at various times between courtship and fledging (reviewed by Dubiec et al. 2013), although it occurs more commonly in species nesting in cavities, i.e. those reusing nests, than in those breeding in open and newly established nests (Wimberger 1984; Clark and Mason 1985). The plants involved are rich in secondary compounds and represent a small, non-random fraction of the species available in the surroundings (Clark and Mason 1985; Gwinner 1997; Petit et al. 2002; Dykstra et al. 2009; Pires et al. 2012). Since secondary metabolic compounds of plants evolved as defences against herbivores (most notably arthropods), a reasonable hypothesis is that bringing to the nest fresh aromatic plants is a defence mechanism of birds against ectoparasites. By incorporating aromatic plants birds may kill or repel parasites and mask the chemical cues that ectoparasites use to find hosts. An additional function may be to enhance offspring immunity. A comparison between passerine species shows that those that add green aromatic plants to their nests are less likely to be infested by *Philornis* spp. (17.4%) than those not adding them (58.3%; Quiroga et al. 2012). However, direct evidence of this phenomenon is mixed: while studies of common starlings (Clark and Mason 1988), Bonelli's eagles (*Aquila fasciatus*; Ontiveros et al. 2008), and tree swallows (*Tachycineta bicolor*; Shutler and Campbell 2007) provide supporting evidence, others have failed to find a negative association between aromatic plant material and ectoparasite load (Fauth et al. 1991; Gwinner et al. 2000; Brouwer and Komdeur 2004; Gwinner and Berger 2005; Mennerat et al. 2008; Tomás et al. 2012). The role of aromatic plants in masking chemical cues also remains unknown. In a laboratory study of chicks of domestic fowl (*Gallus gallus*; Lafuma et al. 2001), aromatic plants repelled mosquitoes and the authors could not rule out that this was due to the masking of the chicks' cues. However, a field study of blue tits found no effect of experimental addition of aromatic plants to nests on the abundance of biting midges and black flies (Tomás et al. 2012). Although convincing experimental evidence of the anti-parasitic role of aromatic

plants in nests is lacking, the subject surely deserves further investigation, particularly because aromatic plants may have positive effects on nestlings' health and development (e.g. Gwinner et al. 2000; Gwinner and Berger 2005; Mennerat et al. 2009b).

Some species in cities line their nests with cellulose fibres from discarded cigarette butts (Suárez-Rodríguez et al. 2013). Filters from smoked cigarettes retain substantial amounts of nicotine and other compounds that may act as arthropod repellents (Wu et al. 1997). The finding that the amount of cigarette butts woven into the nests of the house finch (*Carpodacus mexicanus*) and the house sparrow is negatively associated with the number of nest-dwelling parasites (Suárez-Rodríguez et al. 2013) may be explained by the fact that substances in the cigarette butts repel ectoparasites. When parasites were attracted to heat traps containing smoked or non-smoked cigarette butts, 60% fewer parasites accumulated at the former (Suárez-Rodríguez et al. 2013). While this study supports a role for cigarette butts in repelling parasites, it can only be regarded as an effective anti-parasite defence if cigarette butts are collected and incorporated into nests in response to the presence of parasites, and if this incorporation leads to an increase in host fitness. The latter is likely since the amount of cigarette butts was subsequently found to promote fledgling weight (a predictor of survival) in the house finch, although it also promoted genotoxic damage in chicks (Suárez-Rodríguez and Macías Garcia 2014).

Heterospecific cleaning (*sensu* Hart 1997) refers to instances where one species removes ectoparasites from another. There are only two examples of this involving nest sanitation in birds and the first is the association between oropendolas (chestnut-headed oropendolas [*Zarhynchus wagleri*], crested oropendolas [*Psarocolius decumanus*] and Montezuma oropendolas [*Psarocolius montezuma*]) and yellow-rumped caciques (*Cacicus cela*) with their brood parasite, the giant cowbird (*Molothrus oryzivorus*). Smith (1968) reported that nestling cowbirds remove and consume parasitic botflies (*Philornis* spp.) from their host's offspring, thereby enhancing the fledging success of their hosts. However, this example of cleaning mutualism has not been confirmed experimentally. Gehlbach and Baldridge (1987) described blind snakes (*Leptotyphlops dulcis*) cleaning eastern screech owl (*Megascops asio*) nests. In a wild population of owls, 18% brought at least one live blind snake to their nestlings, whereas all other prey was delivered dead. Blind snakes feed on insect larvae, and chicks in nests with live blind snakes grew faster and experienced lower mortality than those growing in the absence of snakes. The authors proposed that bringing live blind snakes to the nest may be an adaptive strategy to reduce ectoparasite loads. Unfortunately, ectoparasites were not counted and, again, experimental evidence is needed as a priority.

8.6 Conclusions

Birds and parasites are locked in one or many concurrent co-evolutionary arms races centred on nests. The primary reproductive function of avian nests as a central place for the breeding attempt make them suitable places for parasites to thrive. Thus, birds face the challenge of defending themselves and their offspring against parasites while also preserving nest functionality. This is further complicated by the fact that birds often contend simultaneously with several species of parasites, which may call for subtly or markedly different defence strategies.

Factors that increase the likelihood of nest infestation are position on the altricial–precocial spectrum, nest location (e.g. open versus cavity) and reuse of nests both within and across breeding seasons. Yet, our knowledge of the natural history of the interaction between birds and parasites during nesting is far from complete. We know that dispensing with nest materials (as do some colonial seabirds) is associated with the nestlings being afflicted mainly by ticks, but the precise link between nest materials and parasite occurrence and diversity has not been investigated. Parasites synchronize their life cycles with those of their hosts, overwintering in the nests and feeding on hosts once they are breeding. The majority of bird species nest in the tropics where reproduction may occur in most months of the year and where birds often roost in nests. Whether and how nest parasites in the tropics synchronize their life cycles with such avian activities have not been assessed, nor have the countermeasures

that birds adopt to keep themselves free of parasites. Even in well-studied temperate habitats basic questions such as the role of feeding habits on parasite occurrence have not been investigated. While we would predict that species such as predators that bring large food items to the nest would provide greater food substrate to parasitic nymphs than birds that deliver foods discretely into their nestlings' mouths, we do not have sufficient data to substantiate even these generalizations.

The feeding habits of nest-dwelling ectoparasites are harmful to adult and nestling birds (feeding on blood and other tissues) and make them vectors of disease that further threaten the fitness of their hosts. They constitute a selective force that has spurred the evolution of a variety of defence strategies. Such defences may become inefficient under changing conditions, such as when both host bird and parasite species that are vectors to pathogens are introduced to new habitats, or when the native habitat is transformed (e.g. into arable monocultures or cityscapes). The more we learn about these interactions, the better we may predict, or even buffer, the consequences of such changes.

Acknowledgements

We are grateful to Charles Deeming and Jim Reynolds for their invitation to write this book chapter and for the latter's support in improving it through revision. Helga Gwinner, Gustavo Tomás, and Juan A. Fargallo provided valuable ideas and made useful comments on the manuscript. The Entomological Collection of the Museo Nacional de Ciencias Naturales in Madrid provided us with museum specimens to photograph. ILR was supported by a postdoctoral fellowship from CONACYT at the Universidad Autónoma de Tlaxcala. We take full responsibility for all errors and omissions in this chapter.

CHAPTER 9

Egg allometry: influences of phylogeny and the altricial–precocial continuum

G.F. Birchard and D.C. Deeming

9.1 Introduction

The modern study of avian egg biology has been dominated by the use of allometric scaling by egg and female mass. In particular, the work of Rahn and colleagues starting in the early 1970s is notable in this area not only for initially developing many of the relationships between egg structural and functional features including egg size and composition, shell thickness, shell conductance, metabolic rates, and incubation period (Rahn and Ar 1974; Rahn et al. 1975, 1979; Ar et al. 1979; Carey et al. 1980; Paganelli 1980; see summaries by Rahn and Paganelli 1990 and Deeming 2002c). As this area of work developed it also took on an evolutionary perspective. This often meant examining outliers and developing or seeking explanations for differences from the 'avian norm', such as in pelagic seabirds (Whittow 1980), kiwis (*Apteryx* spp.; Calder 1979), and megapodes (Megapodiidae; Seymour et al. 1986). Although the early work on evolutionary questions in avian egg biology was of interest, it often lacked the rigour that more modern evolutionary biology demands. In particular, what was an adaptation versus an evolutionary legacy, i.e. the trait is consistent with phylogeny but not necessarily an adaptation. This changed when statistical procedures accounting for phylogenetic relatedness became more widely available (e.g. phylogenetic contrasts; Harvey and Pagel 1991). At this stage analysis routinely used phylogenetic contrasts to examine trait variation and, thus, often the inference was that after body size was accounted for, phylogeny explained much of the remaining variation in data sets with 'other factors' explaining only small fractions of residual variance (e.g. Sibly et al. 2012). However, there is a potential pitfall in giving phylogeny priority in such analyses—evolutionary convergence(s) can confound analyses and this is likely the case when studying the biology of avian reproduction. To date, no study has examined the effect of phylogenetic constraint on allometric relationships between egg parameters and compared these results with those from analysis of covariance (ANCOVA) determining the effect of avian order (e.g. see Deeming 2007a, 2007b) on the parameter in question. This is pertinent given that some are now questioning the role that allometry plays in investigations of the understanding of evolutionary relationships in birds (e.g. Taylor and Thomas 2014).

Ecological energetics has historically been a subject of much interest in evolutionary biology (Van Valen 1976; Calder 1984; Reiss 1989; Brown et al. 1993). One of the most basic principles of animal energetics is that all organisms must obtain and assimilate energy from the environment and apportion it between self-maintenance, growth, and/or reproduction. One subject of debate is the timing and allocation of energy directed to growth and reproduction because they likely affect fitness (Lack 1968; Brown et al. 1993). Birds have received significant study in this area because it has been assumed that energetic investments in reproduction can easily be inferred from relatively easily collected data (e.g. egg size and number), which have been collected extensively by ornithologists. This allowed clutch mass to be examined within the context of investment packages such as in the trade-off between clutch size and egg mass (Heinroth 1922; Cody 1966; Lack 1968). An interesting and largely unaddressed question in these studies has been whether mass, and particularly egg mass, represents an unbiased estimate of energy in these analyses.

Birchard, G.F. and Deeming, D.C., *Egg allometry: influences of phylogeny and the altricial-precocial continuum*. In: *Nests, Eggs, and Incubation*. Edited by: D.C. Deeming and S.J. Reynolds, Oxford University Press (2015). © Oxford University Press.
DOI 10.1093/acprof:oso/9780198718666.003.0009

The study of egg contents, i.e. yolk, albumen, and shell fractions, and energy has a long history in avian egg biology. Differences in quality related to developmental maturity at hatching were observed as early as Tarchanoff (1874; cited in Carey et al. 1980). The comparative study of egg contents, particularly in the 1980s, created a significant body of data which showed that not only developmental maturity (DM) but also phylogeny were important sources of variation in these data (Ricklefs 1977; Carey et al. 1980; Ar et al. 1987; Sotherland and Rahn 1987). More recently, Deeming (2007a, 2007b) analyzed available data presenting a quantitative picture of how egg contents varied with body mass (BM), initial egg mass (IEM), and DM. What is notable about such information is that it has not generally been well-integrated with egg size in studies of avian reproduction although Dol'nik and Dol'nik (1982) and Vleck and Vleck (1987) are exceptions. Intraspecifically, differences in egg quality, as often indicated by overall egg size despite this not necessarily being a good indicator of quality, have been correlated with increased characteristics indicative of increased hatchling fitness, and this has been a very active area of study (see Williams 2012 for review). However, in general an integration of egg quality into scaling of avian reproductive parameters is lacking. One meaningful way of integrating egg size and quality would be to use egg energy content (EEC). The existing information on egg contents and energy can be used to convert egg mass measurements to energy units and re-evaluate basic reproductive parameters of birds. This approach has been taken only twice previously to our knowledge. Dol'nik and Dol'nik (1982; published in Russian) examined egg and clutch energy relationships of birds by avian order. Vleck and Vleck (1987) used EEC in scaling costs of development.

The avian clade is the only surviving evolutionary branch of the theropod dinosaurs (see Chapter 2). The theropod dinosaurs were oviparous, and some likely built nests and exhibited parental care (Varricchio 2011; Birchard et al. 2013). However, it is questionable whether these avian ancestors ever developed contact incubation as seen in extant birds (Deeming 2002b, 2006; Deeming and Birchard 2008; Chapter 2). The evidence to date is all consistent with precocial young in both the dinosaurian ancestors and the earliest birds (Deeming 2007b; Birchard et al. 2013). However, the phylogenetic distribution of the precocial state indicates that it evolved secondarily from altricial ancestors one or more times in the Neoaves (Figure 9.1; Ricklefs and Starck 1998a; Deeming 2007b).

The evolutionary factors which selected for the evolution of the altricial condition have been a matter of speculation (Ricklefs and Starck 1998a; Deeming 2007b). Ricklefs and Starck (1998b) pointed out in their closing chapter 'We understand so little about the fitness consequences of variations associated with developmental mode that unravelling the special circumstances associated with shifts in developmental mode and the selective forces involved will have to await further research.' It is not clear that our understanding of the origin of different developmental modes has progressed significantly since this statement was made. Transitions along the precocial–altricial continuum in either direction represent significant evolutionary events for birds because they are associated with different energetic and behavioural demands during incubation and post-hatching care by the parent(s). In precocial species the major energetic demands are during laying and incubation. After hatching the young generally feed themselves and are, or very rapidly become, capable of independent thermoregulation. In altricial species the parents invest significantly more energy in feeding their offspring than precocial species when they invest heavily in young until they typically grow to adult size and achieve the capacity to fly. These differences indicate that the relative energy investments of parents in reproduction are more front-loaded in precocial than altricial species. Deeming (2007b) proposed that changes in egg composition could lead to changes in developmental mode. A major implication of the changes suggested would be upfront differences in energy allocation which might result in the evolution of different developmental modes. If fitness is related to energy invested in reproduction (e.g. 'reproductive power' as proposed by Brown et al. 1993) and given energy use is body size-dependent in birds (McNab 2009), it would be expected that a body mass–reproductive energy interaction exists (Sibly et al. 2012) and this interaction would likely be related to DM at hatching in birds.

In this chapter we have two aims. First, we revisit the effects of phylogenetic relatedness on allometric relationships in order to make comparisons with analyses of the effect of avian orders as studied by ANCOVA. Secondly, we review the major parameters, i.e. IEM, clutch mass (CM), clutch size (CS) and incubation period (I_P) (see Table 9.1 for a summary of the most commonly abbreviated terms), and study which determine the energetics of reproduction in the egg-laying and incubation stages of birds. We then examine the impact of using energy rather than mass in these analyses, focussing on how DM and evolutionary

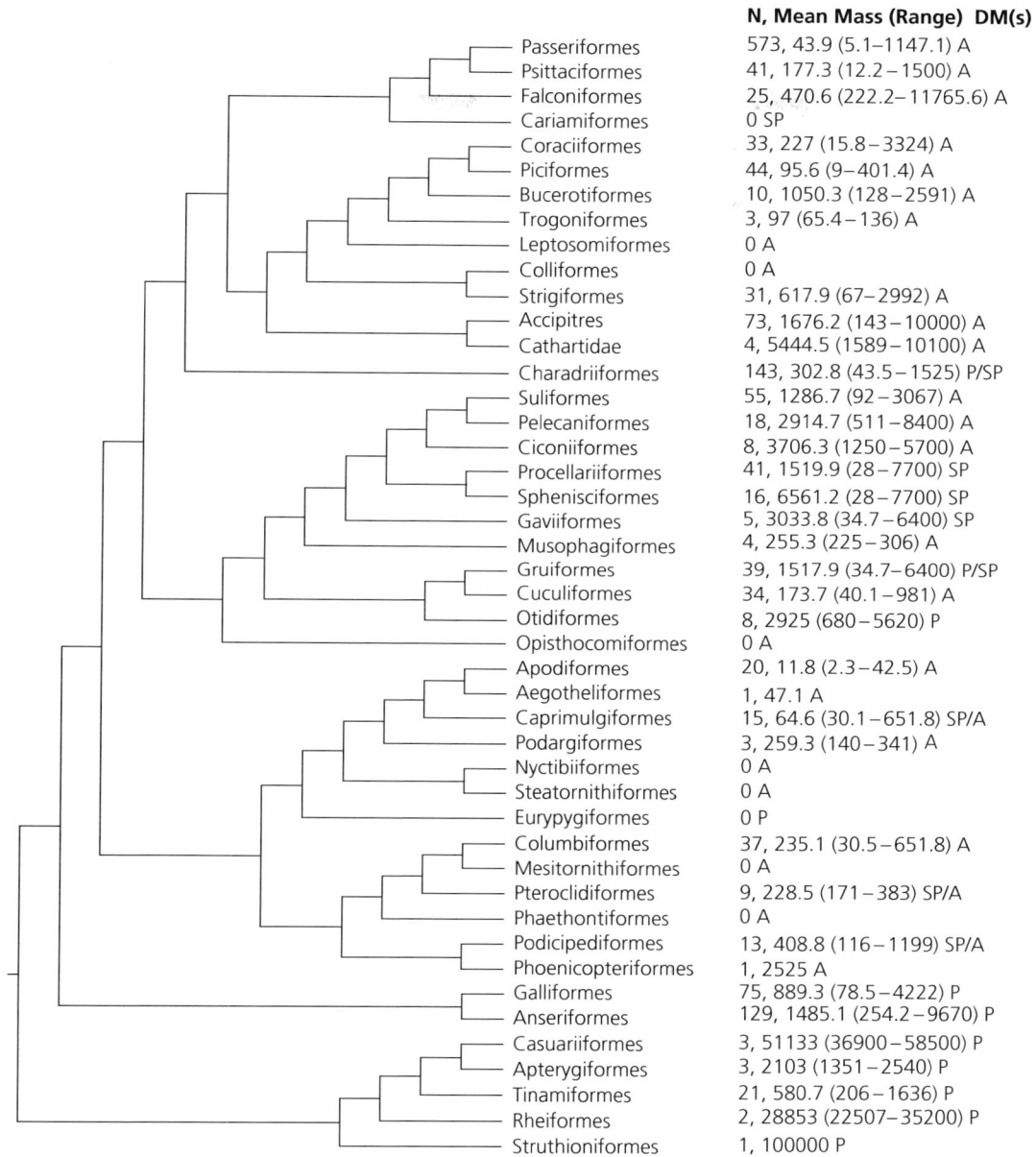

Clade	N, Mean Mass (Range) DM(s)
Passeriformes	573, 43.9 (5.1–1147.1) A
Psittaciformes	41, 177.3 (12.2–1500) A
Falconiformes	25, 470.6 (222.2–11765.6) A
Cariamiformes	0 SP
Coraciiformes	33, 227 (15.8–3324) A
Piciformes	44, 95.6 (9–401.4) A
Bucerotiformes	10, 1050.3 (128–2591) A
Trogoniformes	3, 97 (65.4–136) A
Leptosomiformes	0 A
Colliformes	0 A
Strigiformes	31, 617.9 (67–2992) A
Accipitres	73, 1676.2 (143–10000) A
Cathartidae	4, 5444.5 (1589–10100) A
Charadriiformes	143, 302.8 (43.5–1525) P/SP
Suliformes	55, 1286.7 (92–3067) A
Pelecaniformes	18, 2914.7 (511–8400) A
Ciconiiformes	8, 3706.3 (1250–5700) A
Procellariiformes	41, 1519.9 (28–7700) SP
Sphenisciformes	16, 6561.2 (28–7700) SP
Gaviiformes	5, 3033.8 (34.7–6400) SP
Musophagiformes	4, 255.3 (225–306) A
Gruiformes	39, 1517.9 (34.7–6400) P/SP
Cuculiformes	34, 173.7 (40.1–981) A
Otidiformes	8, 2925 (680–5620) P
Opisthocomiformes	0 A
Apodiformes	20, 11.8 (2.3–42.5) A
Aegotheliformes	1, 47.1 A
Caprimulgiformes	15, 64.6 (30.1–651.8) SP/A
Podargiformes	3, 259.3 (140–341) A
Nyctibiiformes	0 A
Steatornithiformes	0 A
Eurypygiformes	0 P
Columbiformes	37, 235.1 (30.5–651.8) A
Mesitornithiformes	0 A
Pteroclidiformes	9, 228.5 (171–383) SP/A
Phaethontiformes	0 A
Podicipediformes	13, 408.8 (116–1199) SP/A
Phoenicopteriformes	1, 2525 A
Galliformes	75, 889.3 (78.5–4222) P
Anseriformes	129, 1485.1 (254.2–9670) P
Casuariiformes	3, 51133 (36900–58500) P
Apterygiformes	3, 2103 (1351–2540) P
Tinamiformes	21, 580.7 (206–1636) P
Rheiformes	2, 28853 (22507–35200) P
Struthioniformes	1, 100000 P

Figure 9.1 Phylogenetic distribution of developmental mode. Clade names are indicated to the right of the line with P = precocial, SP = semi-precocial, and A = altricial. Clades having more than one developmental mode are indicated by a combination of these codes. Following the clade name are data on the number of species (N), and mean adult body mass (in g, with range within brackets). The phylogeny is from Yuri et al. (2013) and is among the most recently available. It uses their higher taxon names.

history influence the interpretation of these parameters in avian reproduction. In particular, we examine 'whether some regions of the spectrum, particularly that between fully altricial birds and most other birds, are sparsely occupied' indicating 'that there are at least two adaptive peaks or plateaus on the avian developmental landscape' (Ricklefs and Stark 1998a). We examine the consequences of energy use by scaling it with most of the basic relationships considered in studies of avian egg biology. We emphasize phylogeny and,

Table 9.1 A glossary of the commonly used abbreviations and their definitions used in this chapter.

Abbreviation	Definition
AIC	Akaike Information Criterion
ΔAIC	Difference between Akaike Information Criterion values between a pair of models
AICc	Akaike Information Criterion with a correction for a finite sample size
ANCOVA	Analysis of Covariance
CAIC	Comparative Analysis using Independent Contrasts
CEC	Clutch Energy Content (in kJ)
CM	Clutch Mass (in g)
CS	Clutch Size
DM	Developmental Mode (altricial, precocial and semi-precocial)
EEC	Egg Energy Content (in kJ)
FBM	Female Body Mass (in g)
G_{H_2O}	Water Vapour Conductance (in mg $H_2O \cdot$day [d]$^{-1} \cdot$Torr^{-1})
IEM	Initial Egg Mass (in g)
I_P	Incubation Period (in d)
pglm	Phylogenetically constrained linear modelling
Pre-IP V_{O_2}	Pre-internal pipping oxygen consumption rate (in ml $O_2 \cdot$d^{-1})

most importantly, DM because they appear to have the greatest influence on EEC (Ar et al. 1987; Carey 1996; Deeming 2007a, 2007b).

9.2 Effects of phylogeny on allometric relationships

Allometry was used extensively to investigate relationships between egg mass and various other egg characteristics, such as I_P (Rahn and Ar 1974), water vapour conductance (G_{H_2O}; Ar et al. 1974), and pre-internal pipping oxygen consumption (Pre-IP V_{O_2}; Rahn and Paganelli 1990). Bird eggs seem to conform to relatively simple rules associated with factors such as metabolic rate. For example, Pre-IP V_{O_2} scaled with an exponent of ~0.75, which Rahn and Paganelli (1990) interpreted as indicating a standard response of metabolic rate to egg size. One consequence of the intense period of research during the 1970s and 1980s was that our understanding of egg biology dramatically improved. However, it could be argued that research findings were too broadly interpreted and there was a general

acceptance that our understanding of egg biology was comprehensive. This may have resulted in little further research effort to extend broader knowledge of the ecology of avian incubation and oology. Whilst the influence of phylogeny was acknowledged by Rahn et al. (1975), analytical approaches did not accommodate any degree of relatedness. Indeed, Deeming (2002g) indicated that phylogeny had been largely ignored in the study of avian incubation. More crucially, the data sets analyzed by Rahn and colleagues were often heavily biased towards particular avian taxa, such as in study of the relationship between nest humidity and ambient humidity where available data were dominated by Charadriiformes nesting in colder climates (Deeming 2011b).

Since the 1990s the importance of phylogeny in allometric relationships has been realized and statistical techniques have advanced so that controlling for phylogeny in analyses is now routine. Deeming et al. (2006) were the first to approach this issue in their analysis of the relationship between I_P and egg mass in birds and reptiles. Results using comparative analysis using independent contrasts (CAIC; Purvis and Rambaut 1995) revealed that controlling for phylogeny significantly influenced the allometric relationship between IEM and I_P. Using ANCOVA with 'order' as a fixed factor and 'IEM' as covariate, it was clear that it was more accurate to use order-specific regression estimates rather than the single relationship for all birds. A similar conclusion was reached for relationships between both adult female body mass (FBM) and egg components (Deeming 2007a), and IEM and egg components (Deeming 2007b). By contrast, Deeming and Birchard (2007) showed that for analyses of hatchling mass, IEM was a significant covariate but order did not influence the relationship for all species of birds. The importance of avian order was also shown for allometric relationships for eggshell characteristics (Birchard and Deeming 2009) and brood patch temperature (Deeming 2008). As had been previously reported (Carey et al. 1980; Sotherland and Rahn 1987), DM was also important in some of these relationships (e.g. egg composition [Deeming 2007b] and brood patch temperature [Deeming 2008]) but not others (e.g. hatchling mass; Deeming and Birchard 2007).

The CAIC analysis performed by Deeming et al. (2006) and subsequent studies was laborious—the phylogeny had to be constructed from a wide range of literature sources. Moreover, the technique only indicated if independent contrasts were significant rather than providing regression equations that accommodate the effects of phylogeny. However, statistical analyses have advanced rapidly in parallel with the development of phylogenies based upon molecular analyses (e.g. Jetz et al. 2008, 2014). The result is that phylogenetically constrained linear modelling (pglm) can now be used in conjunction with an Aves-wide phylogeny to perform regression analysis (e.g. Sibly et al. 2012). To date, Portugal et al. (2014a) and Tanaka and Zelenitsky (2014) have used a phylogenetically corrected approach to compare IEM and G_{H_2O} but there have been no studies looking at other aspects of egg allometry. Here, we have investigated the effects of controlling for phylogeny on I_P, G_{H_2O} and Pre-IP V_{O_2} based upon egg mass.

9.2.1 Phylogenetically controlled allometry

Two data sets were used in these analyses. The first contained IEM, G_{H_2O} and I_P data for 234 species from 19 different orders while the second contained Pre-IP V_{O_2} and IEM for 90 species from 16 orders. This was to maximize the sample size in analyses. Ratites were grouped together into a single group for these analyses. Analysis involved linear modelling in R (R Core Development Team 2012) to produce standard regression estimates with and without controlling for phylogeny (pglm; Deeming and Pike 2013) as derived from birdtree.org. The data were also analyzed using AN-COVA to test the effect of 'order' as a fixed factor and 'IEM' as a covariate.

The standard regression analyses for the relationships between IEM and I_P, G_{H_2O} and Pre-IP V_{O_2} were comparable to previous estimates on smaller data sets (Ar and Rahn 1985; Rahn and Paganelli 1990; Deeming et al. 2006). However, when phylogeny was controlled for there was no significant change in any of the relationships in both the value of the intercept and the slope, although there was a reduction in the adjusted r^2 value for each relationship (Figure 9.2; Table 9.2). Correcting for phylogeny reduced the slope of the relationship between IEM and I_P and increased the intercept value but neither value differed significantly from the uncontrolled relationship. For both G_{H_2O} and Pre-IP V_{O_2} the change in the regression equation was barely detectable and far from statistically significant (Figure 9.2; Table 9.2). This lack of effect of phylogeny is contrary to that shown by Tanaka and Zelenitsky (2014) who showed larger changes in the slope of the standard and phylogenetically corrected allometric relationships than seen in this analysis but these changes in slopes were not statistically tested.

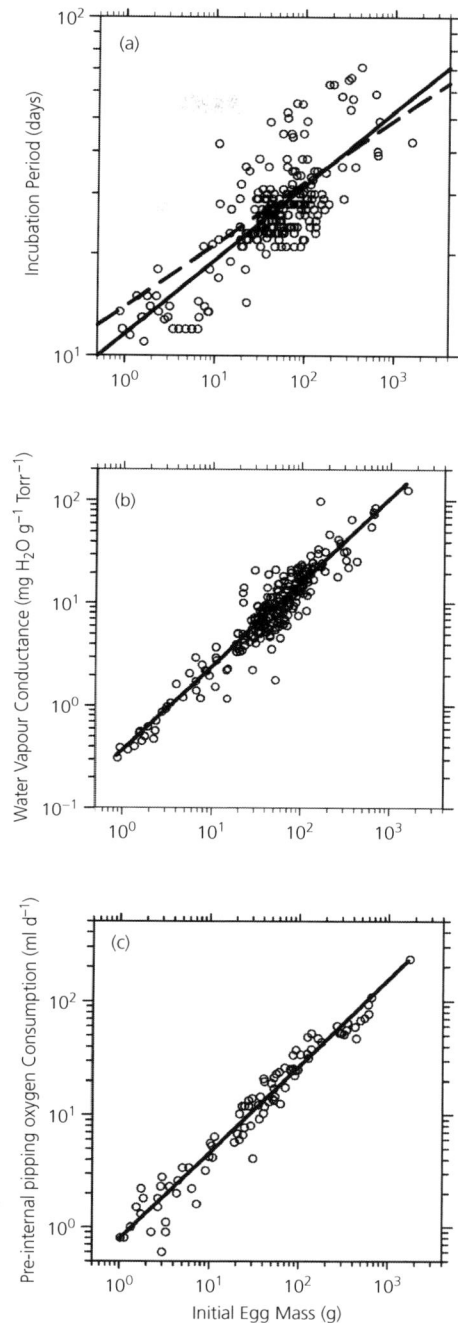

Figure 9.2 Allometric relationships between initial egg mass and (a) incubation period, (b) water vapour conductance, and (c) pre-internal pipping oxygen consumption plotted on $\log_{10}$ scales. Solid lines indicate standard allometric relationships whereas dashed lines (present in all three graphs) indicate regression analysis after controlling for phylogeny.

Table 9.2 Results of linear modelling for the relationships shown indicating the intercept (c) and slope (a) of the relationship: $\log_{10}$ IEM = c + a $\log_{10}$ X where IEM was the initial egg mass (g) and X variables included I_p, the incubation period (days [d]), G_{H_2O} (mg $_{H_2O}\cdot d^{-1}\cdot Torr^{-1}$), and Pre-IP V_{O_2}, the pre-internal pipping oxygen consumption (ml $O_2\cdot d^{-1}$). Regressions illustrated in Figure 9.1 are indicated with a lambda value (bold text) and without a lambda value (lambda = 0) phylogenetic correction. Outputs from the regression analysis, including adjusted r^2, F-ratios, and t-tests (with associated P values) to compare the two intercepts and slopes, are also indicated. In each case the model with the phylogenetic correction provides the best fit with most having negative AICc values.

Dependent variable ($\log_{10}$ Y)	Lambda value	Model AICc	Adj. r^2	F	Intercept	Slope	t-test (intercept)	t-test (slope)
I_p	**0.981**	**−661.92**	**0.322**	**111.8** $P < 0.0001$	**1.145 ± 0.074**	**0.183 ± 0.017**	**1.05** $P > 0.05$	**−1.73** $P > 0.05$
	0	−429.45	0.601	351.5 $P < 0.0001$	1.064 ± 0.020	0.218 ± 0.011		
G_{H_2O}	**0.704**	**−221.17**	**0.719**	**596.5** $P < 0.0001$	**−0.472 ± 0.108**	**0.834 ± 0.034**	**−0.35** $P > 0.05$	**0.36** $P > 0.05$
	0	−191.39	0.885	1794.1 $P < 0.0001$	−0.432 ± 0.033	0.820 ± 0.019		
Pre-IP V_{O_2}	**0.815**	**−109.67**	**0.858**	**528.7** $P < 0.0001$	**−0.105 ± 0.089**	**0.760 ± 0.033**	**−0.02** $P > 0.05$	**−0.13** $P > 0.05$
	0	−100.10	0.949	1611.6 $P < 0.0001$	−0.103 ± 0.032	0.765± 0.019		

Despite this apparent lack of effect of phylogeny on allometry, ANCOVA of the same data set demonstrated that avian order, IEM, and their interaction all had significant effects on I_p (Table 9.3). By contrast, for both G_{H_2O} and Pre-IP V_{O_2}, ANCOVA showed that IEM was a significant covariate but both order and the interaction term were not significant parts of the model.

These analyses demonstrate that controlling for phylogeny has little impact on regression estimates, which is reassuring for our understanding of allometric regressions. However, ANCOVA appears to provide more detail about the influence of phylogeny on these relationships. Whilst IEM is often a significant covariate, there appear to be two categories of relationship with regard to the effects of order. One category,

which includes IEM versus I_p (Deeming et al. 2006; Table 9.3), or IEM versus shell mass, or IEM versus yolk mass or IEM versus albumen mass (Deeming 2007b), and FBM versus brood patch temperature (Deeming 2008), or shell thickness[2] (Birchard and Deeming 2009), has 'order' as a significant fixed factor. By contrast, the relationships between IEM and hatchling mass (Deeming and Birchard 2007), or IEM and G_{H_2O}, or IEM and Pre-IP V_{O_2} (Table 9.3) seem to fit within a second category where 'order' is not a significant fixed factor. Deeming and Birchard (2007) suggested that the first category reflects the nesting and/or feeding ecology and its constraints on variables. By contrast, the second category of variables is more physiologically constrained (Deeming and Birchard 2007). In effect,

Table 9.3 Results of analyses of covariance for $\log_{10}$ I_p with 'order' as a fixed factor and $\log_{10}$ IEM as a covariate, and the interaction term. N represents the sample size of species. Values are F-ratios with associated P values and adjusted (Adj-) r^2s. Orders represented by a single species were excluded from analyses.

Variable ($\log_{10}$ Y)	N	Order	$\log_{10}$IEM	Interaction	Adj-r^2 (%)
IP	228	$F_{13,201} = 4.1$ $P < 0.001$	$F_{1,201} = 28.4$ $P < 0.001$	$F_{13,201} = 3.0$ $P < 0.001$	79.54
G_{H_2O}	228	$F_{13,201} = 1.5$ $P = 0.14$	$F_{1,201} = 76.8$ $P < 0.001$	$F_{13,201} = 1.6$ $P = 0.10$	92.25
Pre-IP V_{O_2}	85	$F_{11,62} = 0.6$ $P = 0.81$	$F_{1,62} = 60.4$ $P < 0.001$	$F_{11,62} = 0.7$ $P = 0.71$	95.30

irrespective of the taxon to which a bird belongs, there is a limit to the structural size it can grow and to the volume of oxygen it can consume. Space precludes further analysis of these relationships for all combinations of variables but what is crucial is the realization that these distinctions are not immediately apparent from simple regression analysis even though they are of great interest to students of avian reproduction.

9.3 Egg mass and its energy content: data and analysis

The basic data set analyzed here is a compilation of FBM, IEM, CM, CS, and I_P (Table 9.1) data carried out by the authors. This compilation involves the same species as Birchard et al. (2013) but with some minor additions. Only FBM was analyzed to avoid issues of sexual size dimorphism and reversed sexual size dimorphism in some species (Cabana et al. 1982; Birchard and Deeming 2009). IEM and CM were converted to energetic measures (kJ), i.e. EEC and Clutch Energy Content (CEC), respectively. Such conversions to energy contents were carried out using conversion coefficients from Ar et al. (1987). Average EECs divided by IEM were determined for each clade and developmental mode. In the case of the Procellariiformes, the data were also sub-divided between large and small eggs reflecting previous findings of Ar et al. (1987) and allowing us to maximize phylogenetic variation and size effects which may result from different FBM-IEM size distributions across clades. Some clades were not represented in Ar et al. (1987). In such cases the conversion coefficient of the most closely phylogenetically related clade with the same level of DM was used. In cases where the combination of DM and phylogeny was in doubt, the general conversion coefficient for the level of DM was used.

Characterization of DM for each clade remains a matter of some debate (Ricklefs and Starck 1998a) so a simplified set of categories (precocial, semi-precocial and altricial) was used. This is consistent with a recently published study on avian reproduction based on egg mass (Sibly et al. 2012) and importantly facilitated accommodating the Procellariiformes and their energy-dense eggs most effectively in our analyses.

Scaling analyses followed Deeming (2007a, 2007b) and Birchard et al. (2013). Statistical analyses were performed using NCSS and the nlwr package in R (R Core Development Team 2012). In discussions of phylogeny we are generally referring to the order level as presented by Sibly et al. (2012). Family-level analyses

were limited to those situations where phylogenetic effects, grades at the family level were indicated, and confounded ANCOVA analysis at the order level. Where non-linear scaling in the log domain was indicated quadratic fits were carried out and the ΔAIC, i.e. the difference (or 'delta') between two Akaike Information Criterion (AIC) values for two models, was used to compare goodness of fit (Burnham and Anderson 2004). In general, we present data with linear fits because size-dependent grades at order and/or family levels appeared to be a major issue in these analyses. Non-linear functions were present in virtually all analyses. This issue is of importance as it limits the utility of using phylogenetic contrasts (Quader et al. 2004).

9.4 Egg mass and egg energy content: results

The basic reproductive unit for birds is the egg. Scaling of IEM as a function of BM was one of the earliest relationships examined and remains one of the largest data sets available through the extensive summaries of Heinroth (1922) and Schönwetter (1960-1971). Figure 9.3a shows a plot of $\log_{10}$ IEM as a function of $\log_{10}$ FBM examining the effect of DM. The plot and regression analysis indicated that a single linear relationship explained most of the variation in IEM (Table 9.4). These results are very similar to those of Rahn et al. (1975) and Dol'nik and Dol'nik (1982). However, Figure 9.3a also indicated that some of the differences in egg mass can be attributed to DM.

Precocial and semi-precocial species generally produce heavier eggs. ANCOVA is consistent with this difference ($F_{2,1507} = 235.5$, $P < 0.00001$) but also showed a significant interaction between FBM and DM ($F_{2,1506} = 33.8$, P 0< 0.00001). Regression analyses indicated that the slope of the relationship for altricial species is greater than that for other DM categories (Table 9.4). This pattern indicates that with increasing FBM the eggs of altricial species approach the size of precocial ones. The ΔAIC for models for all, semi-precocial, and altricial data indicated a quadratic fit was a better supported model (values of 62.7, 16.9, and 11.3, respectively). These results differ from Deeming (2007a) because in part they are based on a different categorization of DM but also there are significantly more species with a significantly greater phylogenetic diversity in the current analyses.

Figure 9.3b shows the relationship between EEC and FBM for the same species as in Figure 9.3a. The

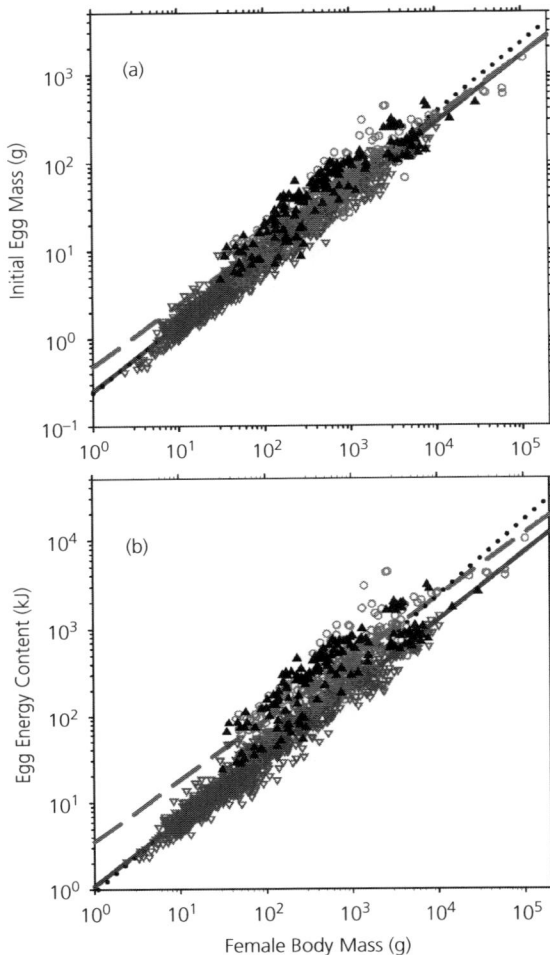

Figure 9.3 (a) Initial egg mass and (b) egg energy content as a function of adult female body mass, all plotted on log$_{10}$ scales. Symbols for developmental mode: open circles (precocial), solid triangles (semi-precocial) and open triangles (altricial). The dotted black line represents the regression for all data in the plot, the solid line is for precocial species, and the long-dashed line is for altricial species. (See Plate 10.)

distribution of data is clearly different from that found for IEM demonstrating the effect of using EEC in such analyses. This finding is consistent with Dol'nik and Dol'nik (1982). This difference in distribution makes interpretation of these data with a single linear relationship questionable, i.e. long-dashed line in Figure 9.3b. The results from ANCOVA ($F_{1,1344} = 1215.0$, $P < 0.00001$) and regression analysis (Table 9.4) showed that different relationships exist for altricial and precocial species. Furthermore, the relationships for

altricial and precocial species have significantly different slopes, as indicated by the significant interaction ($F_{1,1341} = 33.1$, $P < 0.00001$; Table 9.4). The ΔAIC for all data and semi-precocial species indicated a quadratic fit was a better supported model (values of 43.0 and 21.4, respectively). In Figures 9.3a and 9.3b, the precocial orders include the ratites, Tinamiformes, Anseriformes, Galliformes, Charadriiformes, Gruiformes, and Podicipediformes, which are a mix of Paleognathae and Neognathae clades. The presence of the latter three orders indicates an energetic convergence of the Charadriiformes, Gruiformes and Podicipediformes with more ancestral precocial clades.

The semi-precocial clades are generally nested among the precocial ones and are dominated by the Procellariiformes and many of the Charadriiformes, which are characterized by small clutch sizes (typically a single egg) and in some cases long incubation periods (Rahn and Ar 1974; Whittow 1980). The relationship for semi-precocial species cannot easily be compared with those of altricial and precocial species because it is non-linear and phylogenetically split. The Charadriiformes and Procellariiformes have a high EEC relative to the Caprimulgiformes, Gaviiformes, Gruiformes, Podicipediformes, and Sphenisciformes.

Use of energy content demonstrates that from an energetic perspective the difference between altricial and precocial species' eggs is far greater than indicated by analyses using IEM. It is particularly notable that the phylogenetic distribution of precocial species included orders that have both precocial and altricial ancestors. This is consistent with evolutionary convergence. That is, following FBM, DM needs to be accounted for *before* attempting analyses of variation in EEC due to other factors, including phylogeny.

The difference between altricial and precocial species which did not follow phylogenetic lines led to analyses of EEC by order within each DM. When examining order-level regressions for precocial species, it was found that the slope for the Galliformes was significantly greater than for the other orders in the analysis (Figure 9.4a; interaction FBM × order: $F_{5,325} = 2.9$, $P < 0.01$). ANCOVA for precocial orders (with three or more species) without the Galliformes demonstrated significant phylogenetic variation ($F_{5,245} = 56.5$, $P < 0.00001$). Egg energy content from greatest to least was Dinornithiformes > Charadriiformes and Anseriformes > Tinamiformes > Gruiformes and Podicipediformes. Notable in this distribution of orders was that the general pattern was biased toward the clades with the largest eggs also having an egg mass distribution with greater FBMs. Hence, the slope of the general

Table 9.4 Summary of regression analyses for egg, clutch, and incubation parameters examining the effects of developmental maturity on initial egg mass (IEM), egg energy content (EEC), clutch mass (CM), clutch energy content (CEC), clutch size (CS), incubation period (I_p), and developmental mode: all—all developmental modes, pre—precocial, alt—altricial, and spre—semi-precocial. L is from the linear model $\log_{10} Y = a + b \log_{10} X$. Q is from the quadratic model $\log_{10} Y = \log_{10} a + b1 \log_{10} X + (b2 \log_{10} X)^2$. Statistical abbreviations are: sample size (N); standard error (1 SE); coefficient of determination (r^2); probability (P); and the Akaike Information Criterion (AIC).

Variable		N	a	1 SE a	b or b1	1 SE b	b2	1 SE b2	r^2	P	AIC
IEM	L	1,506	−0.6193	0.0113	0.7939	0.0049			0.946	0.00001	1210.1
	Q	1,506	−0.8067	0.0256	0.9866	0.0242	−0.0428	0.0053	0.9484		1147.4
EEC	L	1,506	−0.0322	0.0157	0.8493	0.0068			0.9117	0.0001	2113
	Q	1,506	−0.2820	0.0357	1.1062	0.0338	−1.0570	0.0074	0.9151		2,156
IEM pre	L	326	−0.3182	0.0464	0.7052	0.0165			0.8508	0.0001	314.2
IEM spre	L	162	−0.1252	0.0612	0.6704	0.0218			0.8550	0.0001	182.9
	Q	162	−1.1237	0.2321	1.4206	0.1701	−0.1331	0.0300	0.8711		166
IEM alt	L	1,018	−0.5999	0.0092	0.7604	0.0046			0.9643	0.0001	136
	Q	1,018	−0.6789	0.0230	0.8479	0.0239	−0.0207	0.0056	0.9653		124.7
EEC pre	L	326	0.5513	0.0527	0.7020	0.0188			0.8118	0.0001	400.5
EEC spre	L	162	0.7379	0.0781	0.6479	326			0.7721	0.0001	261.7
	Q	162	−0.6679	0.2919	1.7042	0.2140	−0.1874	0.0377	0.8028		240.3
EEC alt	L	1,018	0.0368	0.0099	0.7630	0.0049			0.9601	0.0001	275.4
	Q	1,018	−0.01807	0.02475	0.8239	0.02565	−0.01442	0.0059	0.9603		271.4
CM	L	1,274	−0.0147	0.0191	0.7612	0.0084			0.8657	0.0001	2242.5
	Q	1,274	−0.1669	0.0445	0.9183	0.0424	−0.0347	0.0092	0.7682		2,230
CEC	L	1,274	0.5722	0.0230	0.8172	0.0101			0.8366	0.0001	2713.9
	Q	1,274	0.3462	0.0535	1.0505	0.0510	−1.0516	0.0111	0.8393		2694.3
CM pre	L	254	0.1375	0.0646	0.8131	0.0230			0.8317	0.0001	371.4
CM spre	L	154	0.2028	0.0572	0.6367	0.0204			0.8655	0.0001	148.4
	Q	154	−1.2476	0.1947	1.7288	0.1430	−0.1940	0.0252	0.9034		99.5
CM alt	L	866	0.0941	0.0165	0.6654	0.0084			0.8789	0.0001	1023.1
	Q	866	−0.0558	0.0426	0.8327	0.0446	−0.0396	0.0104	0.8809		1010.1
CEC pre	L	254	1.0031	0.0705	0.8105	0.0251			0.8048	0.0001	421.2
CEC spre	L	154	1.0684	0.0631	0.6138	0.0225			0.8309	0.0001	178.7
	Q	154	−0.7832	0.1999	2.0080	0.1469	−0.2476	0.0259	0.8947		107.7
CEC alt	L	866	0.7290	0.0163	0.6696	0.0083			0.8833	0.0001	996.8
	Q	866	0.5989	0.0421	0.8148	0.0441	−0.0344	0.0103	0.8848		987.5
CS all	L	1,275	0.6016	0.0189	−0.0305	0.0083			0.0104	0.0003	
CS pre	L	254	0.4105	0.0638	0.1251	0.0227			0.1071	0.0001	
CS spre	L	155	0.2908	0.0674	−0.0227	0.0240			0.0058	0.3474	
CS alt	L	866	0.6893	0.0165	−0.0885	0.0084			0.1143	0.0001	
I_p IEM	L	1,200	1.0587	0.0049	0.2303	0.0039			0.7408	0.0001	
I_p EEC	L	1,200	0.9298	0.0068	0.2114	0.0036			0.7428	0.0001	
I_p IEM pre	L	226	1.0672	0.0204	0.2048	0.0118			0.5718	0.0001	
I_p IEM spre	L	not studied									
I_p IEM alt	L	846	1.0587	0.0049	0.2283	0.0053			0.6908	0.0001	
I_p EEC pre	L	226	0.8915	0.0284	0.2048	0.0111			0.6052	0.0001	
I_p EEC spre	L	not studied									
I_p EEC alt	L	846	0.9128	0.0077	0.2278	0.0052			0.8345	0.0001	

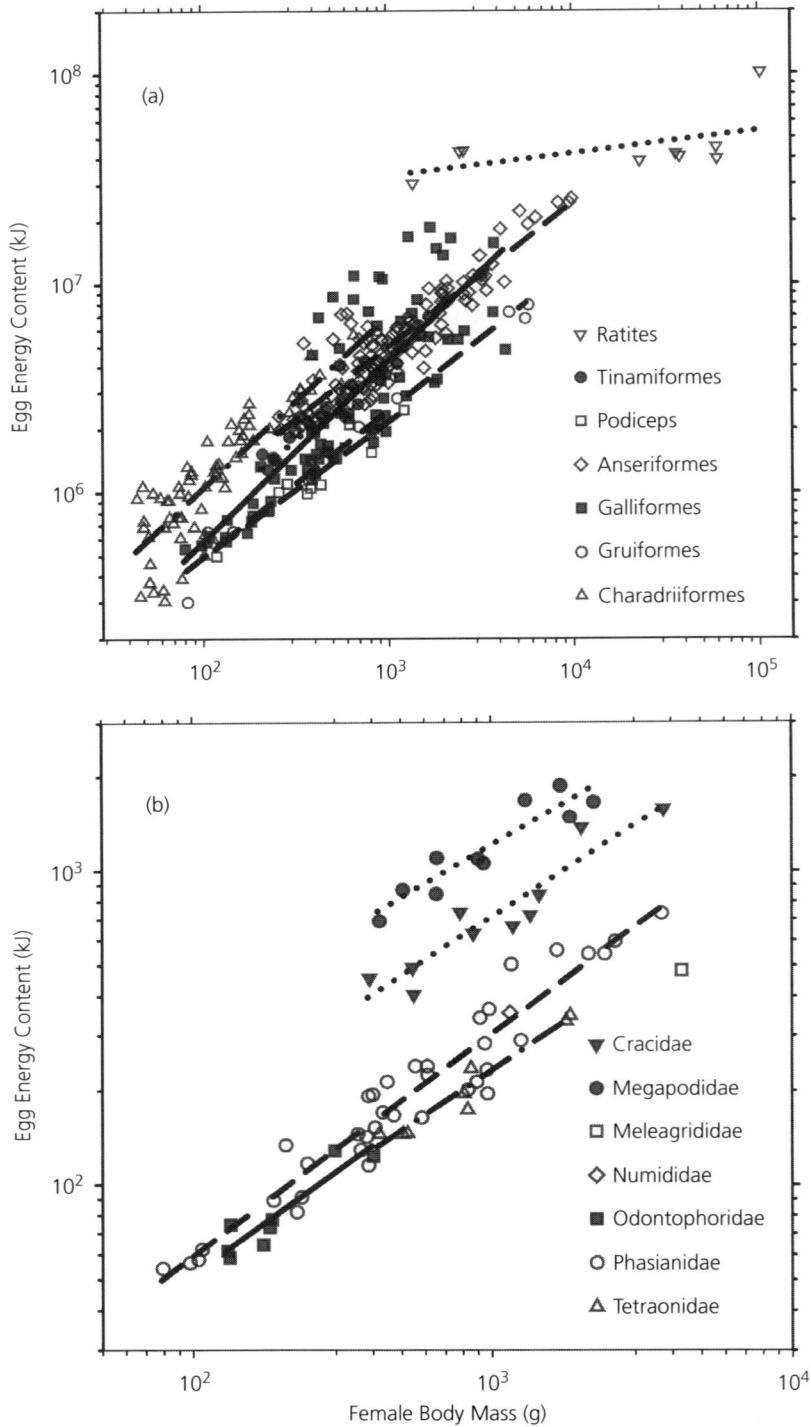

Figure 9.4 Egg energy content of precocial species as a function of female body mass across (a) many orders (different symbols for each), and (b) across families within the Galliformes, all plotted on log$_{10}$ scales. Various solid, dotted and dashed lines indicate regression lines through sets of data points.

relationship for precocial species was biased (greater) because of the existence of grades at the order level.

When data for the Galliformes were plotted by family, the cause of the difference in slopes became evident (Figure 9.4b). The families represent a series of grades having the same slope but different FBM distributions. ANCOVA for families within the Galliformes showed the following variation in EEC: Megapodiidae > Cracidae > Odontophoridae, Phasianidae and Tetraonidae ($F_{4,73} = 181.1$, $P < 0.00001$). Thus, the slope for the Galliformes is greater because the family-level grades had different FBM distributions. It is notable this is a precocial order where precocial to super-precocial hatchlings are represented in different families. An ANCOVA analysis at the family level for all precocial families with three or more species showed that the slopes for families within the Galliformes were not significantly different from those of other orders of precocial birds (interaction FBM × family: $F_{14,308} = 2.9$, $P = 0.80$).

For altricial species many of the same issues existed as in analyses of precocial orders. Using order-level analyses, an ANCOVA indicated phylogenetic differences ($F_{12,976} = 60.5$, $P < 0.00001$). However, as for precocial species, there was a significant interaction between order and $\log_{10}$ FBM ($F_{15,1015} = 4.8$, $P < 0.00001$). A regression analysis was performed on adjusted means from the ANCOVA as a function of the $\log_{10}$ (geometric mean of FBM). This had a positive slope ($F_{1,7} = 7.0$, $P < 0.02$) with the Apodiformes and Passeriformes at the lower end and the Falconiformes and Gruiformes at the upper end of the data distribution. This demonstrated that the slope for EEC for all altricial species (as for precocial species) was biased by grades at the order level. There was also an issue with grades within orders, in particular for the Passeriformes, which is one of the most diverse clades in the analysis. A series of grades existed at the family level with the Tyrannidae and Paridae at the lower end and the Corvidae and Ptilonorhynchidae at the upper end with the largest relative egg size. Thus the issue of grades suggests that it is likely that family-level phylogenetic analyses may be needed due to the confounding nature of grades at the order level.

9.5 Clutch size, clutch mass, and clutch energy content

Clutch size of birds has been a major area of study starting with the early work of Lack (1947) resulting in an emergence of an extensive literature founded on theories of CS and its optimization (Cody 1966; Monaghan

and Nager 1997; Williams 2012). For all birds CS showed a slight but significant decline with FBM (Table 9.4; Figure 9.5). When DM was considered significant differences were found. Clutch size of precocial species increased with FBM whereas for altricial species it decreased (Table 9.4; Figure 9.5). Semi-precocial species generally had small CSs and there was no significant effect of FBM on CS. This result is consistent with previous results (e.g. Jetz et al. 2008). Within the precocial species, order-level differences existed in these relationships with CS of Anseriformes and Gruiformes decreasing with increasing FBM whereas for the Charadriiformes there was no significant effect of FBM. At the family level in precocial species the pattern generally followed that of order. For altricial orders there was a general decrease in CS with FBM across orders and families.

Clutch mass and CEC were analyzed similarly to IEM and EEC above. Clutch mass varied with FBM (Table 9.4; Figure 9.6a) as found in previous studies (Heinroth 1922; Blueweiss et al. 1978; Dol'nik and Dol'nik 1982). In evaluating DM it was found that semi-precocial species showed a non-linear relationship with FBM which made statistical analyses including this group of orders difficult, resulting in us concentrating on the comparison of altricial and precocial species. Clutch mass of precocial species was generally greater than that of altricial and semi-precocial species. An ANCOVA comparing precocial, semi-precocial and altricial species revealed precocial species to have greater CMs than altricial species ($F_{2,1274} = 416.5$, $P < 0.00001$), but there were significant differences in the slopes (interaction: DM × FBM: $F_{2,1274} = 26.4$, $P < 0.00001$). The slope of the relationship for precocial species was significantly greater than that of altricial ones (Table 9.4). Because of this relationship, the CM of altricial and precocial species diverges with increasing FBM with a threshold of approximately 1,000 g above which there is no overlap of data. The linear model is better supported for precocial species. However, for the altricial and semi-precocial data at the greatest FBMs the relationships appear to be asymptotic. This is consistent with the quadratic model being better supported for these two DMs (ΔAICs = 48.9 and 13.0, respectively). The observed non-linearity correlated with the smaller CSs of altricial and semi-precocial species at greater FBMs (Figure 9.5). Furthermore, it was notable that the relationships between FBM and CM of Falconiformes and Procellariiformes, clades whose FBM distribution spans across the 1,000 g threshold, were curvilinear. That is, this non-linear relationship was not just the result of different grades with different size distributions

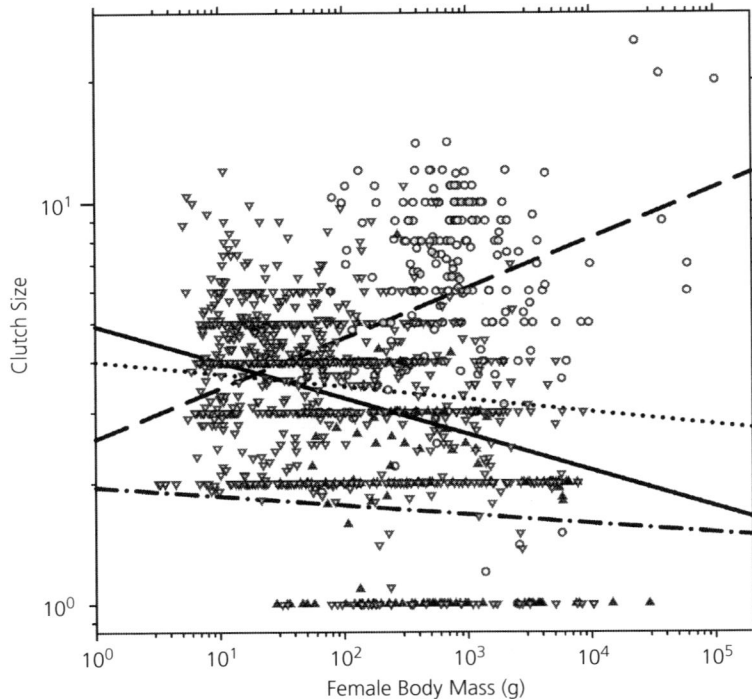

Figure 9.5 Avian clutch size as a function of female body mass plotted on $\log_{10}$ scales. The symbols and lines are as in Figure 9.3.

but it existed within some of the orders in the analysis indicating an effect of size independent of order.

Examination by ANCOVA indicated order-level differences in CM for precocial species ($F_{6,248} = 36.9$, $P < 0.00001$). The descending order of adjusted means was Anseriformes, Dinornithiformes, Galliformes, Tinamiformes, Charadriiformes, Gruiformes, and Podicipediformes. However, as with analyses of IEM and EEC, there were significant differences between slopes of orders (interaction FBM × order: $F_{6,247} = 8.1$, $P < 0.0001$). Similar to previous analyses, plots and regression analysis showed a general positive correlation between CM and FBM indicating the presence of order-level grades, which likely results in an increased slope for precocial species overall.

Order-level differences in CM were also observed among altricial species (ANCOVA: $F_{12,866} = 46.2$, $P < 0.00001$) and differences in slopes also existed (interaction FBM × order: $F_{12,865} = 9.5$, $P < 0.0001$). The descending order of adjusted means was Gruiformes, Ciconiiformes, Strigiformes, Falconiformes, Cuculiformes, Coraciiformes, Psittaciformes, Pelecaniformes, Piciformes, Passeriformes, Caprimulgiformes, Columbiformes, and Apodiformes. As above, the slope for all altricial species was confounded by grades as

indicated by plots and a positive correlation between $\log_{10}$ CM and clade $\log_{10}$ (geometric mean of FBM).

When considering energy contents, analyses revealed that precocial species had a significantly greater CEC than altricial and semi-precocial species (ANCOVA: $F_{2,1274} = 818.2$, $P < 0.00001$), but there was a significant interaction between DM and FBM ($F_{2,1273} = 31.8$, $P < 0.0001$). The overlap between semi-precocial and altricial species was more limited than that for CM (Figure 9.6b). The linear model for precocial species, as with all other analyses of precocial data, is better supported. However, the quadratic model for both altricial and semi-precocial species appears to be better supported (ΔAICs = 9.3 and 71.0, respectively). This is similar to what was observed with CM and is consistent with a decrease in slope of the relationship in CEC–FBM in altricial and precocial species at an FBM above 200–500 g (Figure 9.6b).

Differences between precocial orders were indicated (ANCOVA: $F_{6,248} = 52.6$, $P < 0.00001$), but there were also slope differences among precocial orders (interaction FBM × order: $F_{6,247} = 8.1$, $P < 0.00001$). In this case, quantitatively, the difference was relatively greater than that observed for CM to the point that there was little overlap between precocial and either

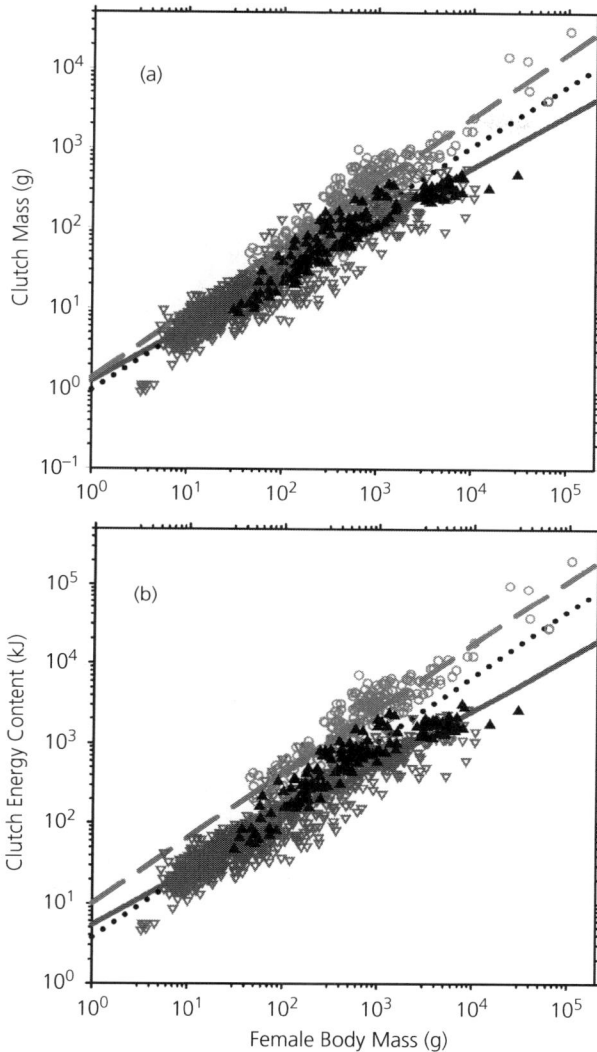

Figure 9.6 (a) Clutch mass and (b) clutch energy content as functions of female body mass plotted on $\log_{10}$ scales. The symbols and lines are as in Figure 9.3. (See Plate 11.)

semi-precocial or altricial species. The precocial orders in sequence from largest to smallest CEC were Anseriformes, Dinornithiformes, Galliformes, Tinamiformes, Charadriiformes, Gruiformes, and Podicipediformes. As for CM, there was a general positive correlation between FBM and clade CEC consistent with a confounding effect of order on the slope of the general relationship for all precocial species.

Order-level differences in CEC were also observed among altricial species (ANCOVA: $F_{12,866} = 42.9$, $P < 0.000001$) and differences in slopes also existed (interaction FBM × order: $F_{12,865} = 8.7$, $P < 0.00001$). The descending sequence of adjusted means by order was Gruiformes, Ciconiiformes, Cuculiformes,

Strigiformes, Falconiformes, Psittaciformes, Pelecaniformes, Coraciiformes, Passeriformes, Caprimulgiformes, Piciformes, Columbiformes, and Apodiformes. This was very similar to that observed for CM. As above, the slope for all altricial species was confounded by a positive correlation between CM and clade FBM.

Procellariiformes were notable for exhibiting some of the greatest IEMs and EECs for their body size (Figure 9.3b). Interestingly, for CM and CEC the interpretation differs. The CMs of Procellariiformes are among the lowest and indeed are lower than most altricial species of similar body size (Figure 9.6a). However, CECs of Procellariiformes are more typical of many other semi-precocial species or 'mid-range' compared to those of altricial species (Figure 9.6b). In other words, the females within this clade apparently invest average energy reserves in their eggs and clutches for their body size. Importantly, this illustrates how the use of energy units can result in very different conclusions than mass units in considerations of relative reproductive investment in birds.

9.6 Incubation period

Incubation period was evaluated for altricial and precocial species as a function of IEM and EEC (Table 9.4). The equation for all data using IEM (Table 9.4; Figure 9.7a) was similar to that found previously in birds (Rahn and Ar 1974). The particularly long I_Ps of some semi-precocial species made comparisons other than that between precocial and altricial species problematic. Incubation in altricial species took longer than precocial species (ANCOVA: $F_{1,1073} = 12.6$, $P < 0.005$) when not including semi-precocial species. However, there was a significant interaction between DM and I_P ($F_{1,1072} = 13.7$, $P < 0.0002$) and Figure 9.7a suggested differences related to DM may not exist.

The use of EEC to scale I_P resulted in a pattern similar to that with IEM (Figure 9.7b), but, as expected, the distribution of some semi-precocial and all precocial species shifted to the right. An ANCOVA of EEC data ($F_{1,1072} = 83.0$, $P < 0.00001$) indicated longer I_Ps for altricial than precocial species. Although there was a significant interaction between DM and FBM ($F_{1,1071} = 14.8$, $P < 0.00001$), Figure 9.7b indicated that altricial species generally take longer to incubate than precocial ones when energy constitutes the unit of analysis.

9.7 Energy versus mass

Qualitative information on egg contents has indicated that significant differences in shell, albumen, yolk mass, and EEC were associated with DM and phylogeny. The integration of this information into the scaling of basic reproductive variables for birds demonstrates that many of the basic scaling rules in avian egg reproductive biology that we have used for the best part of 100 years may require re-evaluation as to their suitability for the hypotheses being tested. For example, morphological and biomechanical hypotheses, incorporating the structure of the reproductive tract and/or pelvis or the function of the eggshell, may require egg mass or volume as appropriate metrics. However, for life-history questions relating to fitness energetic evaluations would

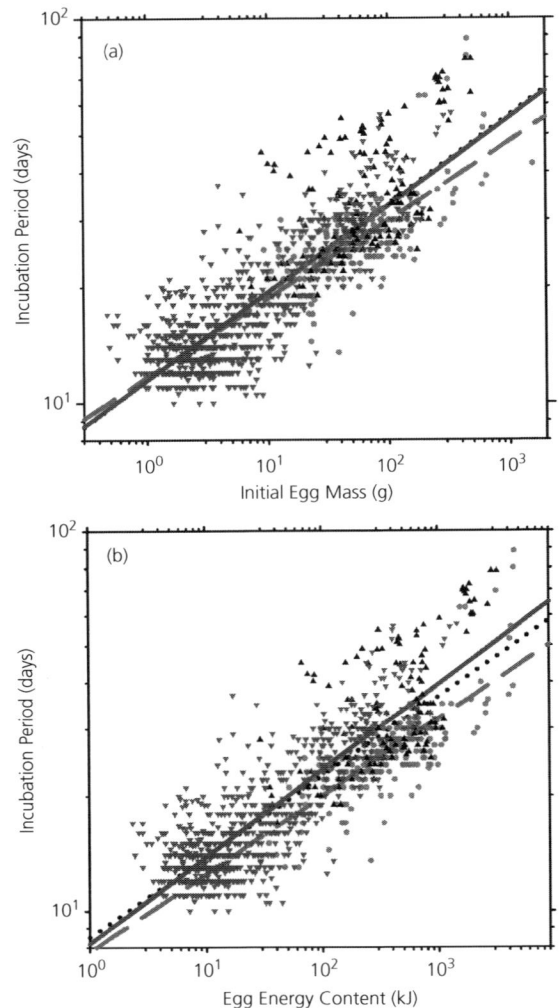

Figure 9.7 Incubation period as a function of (a) initial egg mass and (b) egg energy content plotted on $\log_{10}$ scales. The symbols indicating developmental mode are the same as in Figure 9.3. No regression line for semi-precocial species was calculated. (See Plate 12.)

be more appropriate and insightful given that they offer greater insights into reproductive investment (Williams 2012). At a minimum level researchers should investigate whether equivalent answers arise with mass and energy units for reference when asking many evolutionary questions. As indicated above for the Procellariiformes (see section 9.5), conclusions reached can be markedly different between the two. Pelagic species (dominated by the Procellariiformes) were recently described as having the lowest productivity on the basis of total 'annual mass of eggs' (Sibly et al. 2012). However, when energy units are used the total energy invested in reproduction by pelagic species is significantly greater and falls within the range of many other species with different foraging strategies (G.F. Birchard, 2014 unpublished data).

The variation observed in EEC, CEC, CS and I_P related to both DM at hatching and phylogeny has significant implications for many analyses. First, in the examination of either mass or energy it appears that DM has to be accounted for *before* phylogeny. This is consistent with a phylogenetic distribution which indicates a convergence in characters such as egg mass and energy with DM. In many respects this issue is complicated because of the known variation in the assignment of DM state. The morphological and behavioural characteristics typically used are subject to interpretation (as reviewed by Ricklefs and Starck 1998a). It is likely that adopting measures that are less subject to individual interpretation would be of value (e.g. tissue water content; Ricklefs and Starck 1998a). However, such measures are only available for a limited number of species at present and individual and phylogenetic variation may still result in variation due to classification.

The second major issue indicated by our analyses is at what phylogenetic level analyses should be performed. The issue of grades, found at both the order and family levels, can confound analyses (Martin et al. 2005). Hence, the slope at multiple phylogenetic levels is confounded by grades with different mass distributions. Although order level is most often used, in part this reflects the absence of good phylogenies at within-order, family and genus taxonomic levels. The issue is of significance particularly for examination of the origin of a DM within an order. For example, within the Galliformes these analyses indicate the origin of super-precocial species was a family- (or below) level evolutionary event (Figure 9.4b).

9.8 The evolutionary transition between developmental modes

As noted in the introduction to this chapter, Ricklefs and Stark (1998a) highlighted the lack of information on where in the continuum of traits we could distinguish specific 'spaces' where we could clearly identify presence or absence of the DMs called 'altricial' and 'precocial'. The use of mass rather than energy appears to provide a reason that some of the energetic differences between DMs were not recognized. Examination of the energy-FBM space

(Figures 9.3 and 9.6) indicates several regions which are unoccupied or sparsely occupied by species. First, in comparing egg mass and CM against energy plots (Figures 9.3 and 9.6), it is evident that there are no precocial (or semi-precocial) species below FBMs of 20–40 g. Given the markedly greater EECs and CECs of precocial species, a reasonable hypothesis is that for birds lighter than 20 g it may be difficult to forage for and assimilate sufficient resources to produce eggs containing the energy required for a 'precocial' clutch. This would be consistent with the increasing maintenance costs at small structural sizes which are inferred from the slope of the allometric relationships between body mass and metabolic rate of birds (McNab 2009). However, it is also possible this boundary may be related to other life-history characteristics. For example, the need to hatch a chick of sufficient structural size that it can move within its environment effectively at, or just after, hatching to collect adequate food to grow. The exceptionally small hatchlings that would be produced from eggs of precocial birds of extremely low adult body size may simply lack this ability or have such high mortality rates as to be unable to maintain the viability of a species.

Energy analyses indicate a second region of exclusion, where altricial species appear to be absent. Above 800–1,000 g, particularly for CEC, the relationships for altricial (and semi-precocial) species appear to diverge from those of precocial species. This creates an ever-widening gap between these DMs. Furthermore, there appear to be no altricial or semi-precocial species with body masses above 20,000 g; the largest extant species, the common ostrich (*Struthio camelus*) attains a body mass of 100,000 g and some extinct birds like the moa (*Dinornis robustus*) and elephant bird (*Aepyornis maximus*) were significantly larger than this (Deeming and Birchard 2008). The feature which seems to correlate with this region is that there are no, or never have been, large altricial ground-dwelling species equivalent to the flightless ratites. This may be related to the fact that producing precocial volant species of this structural size would not be possible given the size and energy content of eggs from which they would hatch.

Finally, less obvious, is a more sparsely occupied part of the mass range between 40 and 800 g between the precocial, and the semi-precocial and altricial species (Figure 9.6b). This could correspond to a transitional region between DMs, which is only suboptimal for most species. This region could also be hypothesized as size-specific in which the transition between DMs may occur. This would be consistent with the ease of transition between DM categories being correlated with the magnitude of the energy difference which exists between the two states. This is a body mass range that exhibits the greatest phylogenetic diversity and may represent a 'most common' original 'mass' range for clades, which is a location within the size-energy space of origination and diversification in body size.

9.9 Energy-time interaction

In evaluating the energy results, it is important to note that time is also a factor affecting which evolutionary strategies are energetically possible. Thus, the longer I_ps of altricial compared with precocial species are also significant. First, they suggest that altricial species will likely allocate relatively more energy during incubation for a given body size than precocial species. The other related issue is the sum of incubation and fledging times may be a determining factor in the number of clutches a species is capable of producing in a given year. For example, in altricial species exploiting temporally limited increases in primary productivity, as I_p increases the cumulative total of incubation and fledging times will reach some critical maximum, beyond which it will prove impossible to produce a second clutch. Breeding phenology and its effects on annual productivity are important because CSs of large altricial species are typically small, particularly in comparison to precocial species. Thus, significant life-history changes favouring greater offspring survival are hypothesised in these large altricial species.

9.10 Conclusions

The above hypothesis of size-dependent zones excluding precocial or altricial species of some body masses has important evolutionary implications for birds. This would suggest that, before the evolution of altriciality, birds with a FBM less than 100 g would have been rare or non-existent. Indeed, estimates of the mass of Mesozoic birds suggest that none of these taxa weighed less than 1,000 g (Hone et al. 2008). Body-size evolution has been examined previously in birds (Brown et al. 1993; Mauer 1998) but DM was not considered as a potential factor in these analyses. In particular, the different body mass distributions and energetics across the altricial–precocial continuum raise the question of whether this has also influenced the shape of the body-size distribution of the class Aves?

The work of Rahn and associates in the 1980s and 1990s provided important insights into the biology of the avian egg. They ultimately stimulated others with more evolutionary and ecological interests to work on problems related to the incubation biology of birds. Some of this early work now needs to be revisited and reconsidered in a more detailed manner. As shown here, the use of ever-improving phylogenies in combination with improved phylogenetic statistical methods are providing greater insight into what are the highly constrained characteristics versus the more evolutionarily variable ones. However, as we have demonstrated using these energetically driven investigations, egg biologists must answer evolutionary questions through interpretation based upon appropriately scaled metrics. Also, although phylogeny matters, we must be conscious of possible evolutionary convergences, such as DM, and modify analyses appropriately to account for such effects.

Acknowledgements

We thank Jim Reynolds and Richard Sibly for constructive comments on a previous version of this chapter.

CHAPTER 10

Egg quality, embryonic development, and post-hatching phenotype: an integrated perspective

T.D. Williams and T.G.G. Groothuis

10.1 Introduction

In oviparous taxa such as birds, eggs can be considered 'sealed capsules' that contain all of the nutrients and energy used during embryonic development (Carey 1996) and which only require external heat and oxygen to allow for complete development to hatching. Nevertheless eggs have complex macro- and micronutrient compositions that serve not only as nutrients (resources) but potentially as 'signals', from the mother, that affect embryonic and post-hatching development and which, in turn, potentially affect offspring and adult phenotype and fitness in myriad ways (Williams 2012). Since these egg components are all maternally derived, i.e. transferred from mother to egg during egg formation, they have been considered important pathways for maternal effects, including *non-genetic* contributions of the mother to her offspring, and have become a major focus of life-history studies (Bernardo 1996; Mousseau and Fox 1998a; Badyaev and Uller 2009). Although maternal effects were originally considered to generate troublesome 'noise' in effective breeding programmes, the last 15 years have seen a paradigm shift in viewing such effects as potentially important and flexible 'tools' for the mother to adjust her offspring's phenotype to current environmental conditions, with potentially important consequences for evolution of phenotype. Thus, much of the work on this topic has been conducted at the interface between ecology, evolutionary biology, and physiology within the context of how mothers might 'adaptively' adjust the development and the phenotype of their offspring to match prevailing environmental conditions in order to increase fitness (Mousseau and Fox 1998a; Groothuis and Schwabl 2008; Love and Williams 2008a; Uller 2008).

There have been many reviews of different maternally derived egg components, but these have generally considered variation in single components, or groups of related components (e.g. macronutrients, Hill 1995; hormones, Groothuis et al. 2005; antioxidants, Blount et al. 2000; antibodies, Grindstaff et al. 2003). Nevertheless these reviews have all highlighted the same unresolved, or poorly understood, issues: a) the extent to which, and mechanism(s) by which, females might control or regulate maternal transfer to eggs; b) the extent to which embryos are passive recipients of maternal transfer, or whether they modulate the effects of maternally derived egg components; c) the mechanism(s) by which maternal substances might affect offspring development and ultimately adult phenotype to mediate long-term fitness; but also, d) the relative lack of studies supporting long-term fitness consequences of maternally derived egg components (cf. short-term, single trait studies), especially in free-living birds.

Here, we will focus on these same issues but rather than consider egg macronutrients (e.g. lipid, protein), hormones, antioxidants, or immunoglobulins separately, we attempt to take an integrated approach to consider the effects of overall egg quality on embryonic development and the post-hatching phenotype. This is based on the idea that *if* females use maternally derived egg components to fine-tune offspring phenotype to prevailing environmental conditions, increasing offspring fitness independently of egg size *per se*, then there should be some 'optimum' *combination* of macro- and micronutrients with which females attempt to furnish their eggs (assuming that the different components do not act independently of each other, or affect similar 'targets' in the offspring). Any optimum

Williams, T.D. and Groothuis, T.G.G., *Egg quality, embryonic development, and post-hatching phenotype: an integrated perspective*. In: *Nests, Eggs, and Incubation*. Edited by: D.C. Deeming and S.J. Reynolds, Oxford University Press (2015). © Oxford University Press.
DOI 10.1093/acprof:oso/9780198718666.003.00010

combination of maternal substances might (and probably would) be species-specific, or vary with ecological context even within species. Nevertheless, this simple idea predicts some level of correlation or 'coupling' of the many different egg components by the mother. For example, it would appear counter-intuitive to produce an egg with high immunoglobulin content, perhaps to protect the naïve embryo in an immunologically challenging environment (Boulinier and Staszewski 2008), that also had high yolk testosterone content which might simultaneously cause immunosuppression (see Postma et al. 2014). Similarly, an egg with high levels of *both* testosterone and thyroid hormone would presumably be more advantageous than one with high concentrations of only one of these hormones, given that both have potential growth-promoting effects.

This idea of coupling of multiple, maternally derived egg components (or of their potential uncoupling) has important implications for the unresolved questions that were highlighted above and in earlier reviews (e.g. Hill 1995; Blount et al. 2000; Grindstaff et al. 2003; Groothuis et al. 2005) concerning links between egg quality, embryonic development, post-hatching phenotype, and ultimately consequences for fitness. For example, are different maternally derived egg components correlated (either positively or negatively) within single eggs, or can they vary independently? If different egg components are strongly and positively coupled this would imply that there might be shared mechanisms of transfer from mother to egg. If egg components are negatively correlated this might reflect the fact that they compete for the same transfer mechanism(s), and if different egg components are not correlated this suggests either different mechanisms for independent regulation of egg composition or, more parsimoniously, no control mechanisms for the deposition of these components into the yolk. The same issue is relevant from the embryo's perspective: do mechanisms that allow for modulation of any maternal signal (*sensu* Paitz and Bowden 2013) allow uncoupling of the effects of different egg components, especially where the mother's optimal egg composition does not match that for the offspring? If ecological (or life-history) context is an important determinant of how females might 'manipulate' maternal effects (e.g. Love et al. 2009), are there certain contexts where coupling or uncoupling of different egg components is advantageous? How is this then achieved? This idea of coupling, or uncoupling, of multiple egg components provides the framework for this review. However, we finish with a brief review of the central question of whether egg-mediated maternal effects actually do (or do not) have long-term

fitness consequences in free-living birds, irrespective of potentially interacting effects of different egg components, as this remains a critical gap in knowledge.

10.2 The multivariate egg: are different egg components correlated?

If the absolute amounts of all egg components scale isometrically with egg mass within species this would suggest that egg size itself is a good, i.e. adequate and reliable, measure of egg quality. This would also mean that all egg components are positively correlated and, potentially, transfer from mother to egg occurs through a common, non-specific mechanism. For example, in general larger eggs contain absolutely more shell, dry albumen, and dry yolk; in many species the egg component*$\log_{10}$ egg mass relationship does not differ significantly from $b = 1$ for albumen and yolk, and $b = 0.75$ for shell, i.e. egg component mass is directly proportional to egg mass (Williams 1994). In some passerines and seabirds larger eggs contain disproportionately more albumen, i.e. $b > 1$, and in some precocial species, such as some waterfowl, larger eggs contain relatively more yolk, largely reflecting variation in lipid content. Therefore, different egg components can show positive or negative allometry with egg size depending on taxa (Williams 1994; Hill 1995). Moreover, on average, egg size only explains 57% and 46% of variation in dry albumen and dry yolk content, respectively (N = 11 studies cited in Williams 1994), i.e. there is considerable residual, inter-individual variation in egg composition. Unfortunately, we still know surprisingly little about how the macronutrient composition of eggs might be differentially regulated (Williams 2012).

Most studies of other (micronutrient) egg components have only considered one class of compound (e.g. hormones, carotenoids), or often even only single hormones/carotenoids, in isolation from other egg components. However, Groothuis et al. (2006) explicitly explored the idea of the 'multivariate egg' by analyzing the covariance between androgens, egg mass, antibodies (immunoglobulins), carotenoids, and vitamin E in eggs of common black-headed gulls (*Chroicocephalus ridibundus*). Within clutches, testosterone increased over the laying sequence, egg mass first increased and then decreased, and the other three components all decreased over the laying sequence. The authors suggested that deposition of 'signals', such as steroid hormones, is not costly, but that the other components, being resources, might be costly and reflect maternal depletion. Apart from this general pattern, the only correlation among these egg components across clutches was that clutches with a relatively strong decrease in antioxidants over the laying sequence (produced by mothers of relatively low body masses) also showed a relatively strong increase of testosterone concentrations, perhaps reflecting a compensatory strategy. Groothuis et al. (2006) suggested that 'evolution has not strongly selected for mechanisms that allow the mother to adjust her deposition of pro-immunomodulatory antioxidants and antibodies (e.g. immunoglobulins G, specifically IgG) into eggs to compensate for possible immunomodulatory effects of maternal testosterone'.

Postma et al. (2014) analyzed two components of the multivariate egg: maternally derived yolk IgG and yolk androgens in the great tit (*Parus major*). Within clutches there was no association between these traits suggesting that the two egg components are deposited independently. However, across clutches there was a significant negative relationship between yolk IgG and yolk androgens, suggesting 'that selection has coadjusted their deposition' (Postma et al. 2014). Safran et al. (2008) found no relationship between concentrations or amounts of yolk androgens and carotenoids in barn swallow (*Hirundo rustica*) eggs and suggested each yolk compound may be regulated by different mechanisms. Ruuskanen et al. (2011) also considered variation in multiple egg components, i.e. egg and yolk mass, albumen lysozyme activity, yolk immunoglobulins, yolk androgens, and yolk total carotenoids, in the same sample of European pied flycatcher (*Ficedula hypoleuca*) eggs at the population level. Egg mass and yolk mass were positively correlated across populations, as were yolk testosterone and androstenedione concentrations, lysozyme activity was weakly negatively correlated with egg mass, but there were no other significant correlations between the measured egg components. As an illustration of the uncoupling of different egg components, Figure 10.1 shows the relative difference in absolute egg components for one Norwegian and two Finnish European pied flycatcher populations where population mean egg mass was very similar: 1.69, 1.69, and 1.71 g, respectively. In the two Finnish populations the relative contents of three egg components (IgG, testosterone, and carotenoids) varied in opposite directions, and with markedly different magnitudes (Figure 10.1). Several earlier studies showed that total yolk carotenoid concentration can be positively correlated with vitamin E concentration within collared flycatcher (*Ficedula albicollis*) clutches (Hargatai et al. 2006), and carotenoid, vitamin E, and vitamin A concentrations can be positively correlated within, and, to a lesser extent, among clutches of yellow-legged gulls (*Larus michahellis*; Rubolini et al. 2011). However, there was a negative within-clutch correlation between yolk antioxidants and yolk androgens in lesser black-backed gulls (*Larus fuscus*; Royle et al. 2001) and common black-headed gulls (Groothuis et al. 2006). Similarly, mixed results were reported by Eeva et al. (2011) who reported geographical variation in certain carotenoids (lutein, xanthophylls) but not others (β-carotene and zeaxanthin) in eggs of European

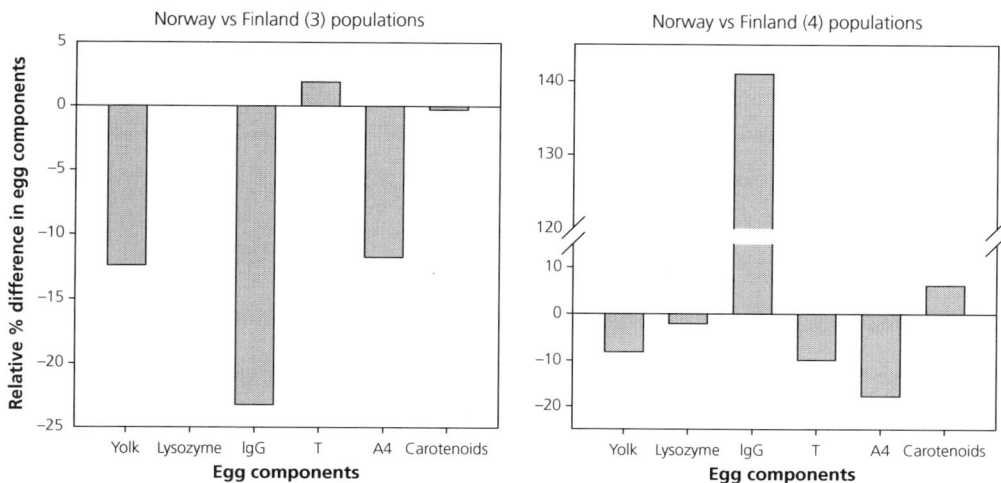

Figure 10.1 Relative (%) difference in absolute egg components (by mass) comparing one Norwegian and two different Finnish European pied flycatcher (*Ficedula hypoleuca*) populations where population mean egg mass was effectively identical, i.e. 1.69, 1.69, and 1.71 g, respectively. Based on data in Ruuskanen et al. (2011).

pied flycatchers, also suggesting that these can be un-coupled within populations. Indeed, the proportions of lutein and xanthophylls were strongly, positively correlated ($r = 0.69$), but were negatively correlated with other carotenoids (r_ss = -0.43 to -0.94).

These studies suggest that different egg components can be uncoupled both within populations, among females, and among populations. This is true even within 'functional groups': if all carotenoids contribute positively to the antioxidant status and immune responsiveness of developing embryos then one would predict that mothers should transfer more of *each* type of carotenoid, perhaps equally, into egg yolks. While this is not the case (Eeva et al. 2011), it also remains unclear if the variation in different egg components reflects biologically meaningful, systematic, regulated variation or non-adaptive variation due, for example, to limits on maternal 'control'. Assuming that this variation is biologically relevant, and reflects the mutual (presumably adaptive) adjustment by females of different egg components that can have opposite effects on offspring phenotype (*sensu* Groothuis et al. 2006 and Postma et al. 2014), this leads to two key questions: a) What are the mechanisms of transfer that might allow a female to 'allocate' multiple resources *differently* to eggs to optimize egg composition and offspring phenotype?; and b) To what extent do mechanisms of embryonic metabolism modulate, uncouple, or even block, the effects of multiple, different egg components?

Before answering these questions we want to highlight a markedly different perspective on this problem

prevalent in the field of toxicology, but one which is rarely acknowledged in the ecological or evolutionary maternal effects literature. There *are* compounds transferred from mothers to eggs that are clearly only likely to have negative, non-adaptive, direct toxic effects or endocrine-disrupting effects, i.e. contaminants or xenobiotics (Fry and Toone 1981; Drouillard and Norstrom 2001; Ottinger et al. 2011). The toxicology literature typically refers to maternal transfer of such compounds as an 'excretory' process whereby laying females 'off-load' some of their own contaminant burden to their eggs during egg production (Bargar et al. 2001; Dauwe et al. 2006), although there can still be systematic variation in transfer (Eng et al. 2013; see Figure 10.2). Many xenobiotics are lipophilic and maternal transfer of contaminants is also considered to be a passive process, with the distribution of toxicants in the egg being dependent simply on the chemical characteristics of the compound (specifically the partition coefficient, log k_{ow}; Ottinger et al. 2011). Concentrations and profiles of contaminants in eggs are considered to reflect tissue contaminant burdens accurately in the female, thereby allowing the use of eggs as bio-monitoring tools for detecting levels, distribution, and long-term trends in environmental contamination (e.g. van den Steen et al. 2009). This view of contaminants being passively 'dumped' into eggs as metabolic waste products differs markedly from that pervading most of the behavioural ecology literature of 'regulated', adaptive, maternal effects of non-contaminant egg components. Passive 'dumping'

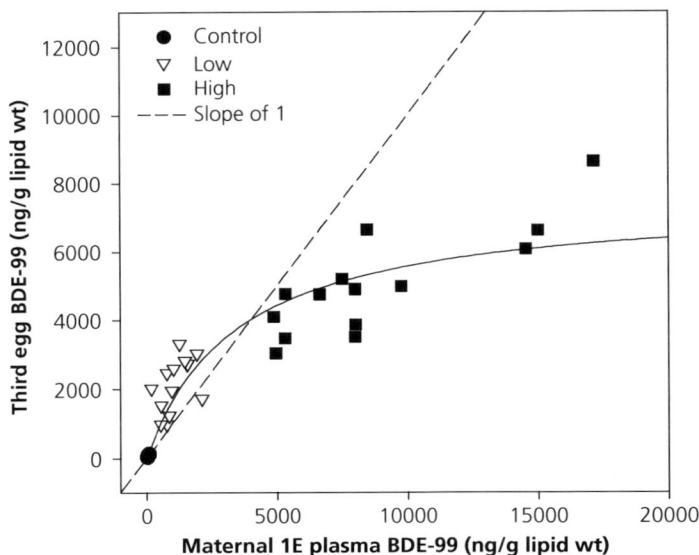

Figure 10.2 Relationship between the lipid-normalized polybrominated flame retardant BDE-99 concentration in the third egg, and the maternal plasma on the day that the first egg was laid (1E) in zebra finches (*Taeniopygia guttata*) in females dosed with 0 ('Control'), 33.7 ('Low'), or 173.8 ('High') ng PBDE·g^{-1} body weight/day during egg laying. Solid line represents the overall fit of an equation based on a saturation binding curve. Reproduced with permission from Eng et al. (2013).

has been considered in only a few papers dealing with other egg components such as hormones and in general we should be careful about assuming too easily that these processes are adaptive; however, we emphasize four points here. First, the use of terminology such as 'active' versus 'passive', 'deposition', or 'control of deposition' can be confusing, as active control can easily suggest a type of conscious decision-making process by the mother. The word 'deposition' suggests a regulatory process and is in contrast to the word 'accumulation' which is more neutral and should perhaps be preferred. Secondly, even when concentrations in the egg of a certain component simply reflect plasma levels of that component, this does not exclude the possibility that maternal regulation of egg composition may be at the level of plasma concentration. Thirdly, even if the latter is not the case, the *passive* accumulation of egg components might still be highly adaptive (Love and Williams 2008a). Finally, for gonadal steroids, in contrast to hormones from other glands, non-adaptive dumping perhaps seems unlikely as the hormones are produced in specialized cells surrounding the ovum, as if evolution has 'designed' the system to provide the eggs specifically with these hormones. It has even been suggested that gonadal hormone production in female birds might be more important for the egg than for the mother's circulation itself (Groothuis et al. 2014; although given that these same hormones play key roles in regulation of gonadal function itself, this provides another reason for this close co-location of gonadal steroids and yolk).

10.3 Do mechanisms of maternal transfer allow coupling/uncoupling of different egg components?

Larger eggs contain, in an absolute sense, more major egg components, i.e. shell, albumen, yolk, and, in general, more maternally derived antibodies and antioxidants (Williams 2012). This suggests that there might be common mechanisms for maternal transfer during egg production: antibodies and antioxidants might be either: a) transported by 'piggy-back' by yolk precursor particles via receptor-mediated endocytosis (MacLachlan et al. 1994; Bortolotti et al. 2003), or during albumen deposition in the oviduct; or b) transferred passively in response to the lipophilic gradient between plasma and lipid-rich yolk (Moore and Johnston 2008; Williams 2012). The extent to which receptor-mediated yolk uptake represents a central physiological pathway for maternal transfer of egg components other

than yolk in response to environmental cues remains largely untested (Müller et al. 2012). In contrast, total amounts of some yolk hormones and albumen antimicrobial proteins can vary independently of yolk or egg size which suggests there are transfer mechanisms for these egg components which are independent of mechanisms underlying yolk formation *per se* (Williams 2012). In fact, in several studies smaller or lighter eggs contain higher concentrations of maternal androgens and even the total amount of these hormones is higher than in larger or heavier eggs, suggesting that hormone accumulation might be compensating for reduced egg nutrients, perhaps by stimulating sibling competition (Groothuis and Schwabl 2002). Furthermore, as Postma et al. (2014) argued, if both yolk immunoglobulins and yolk androgens are 'strategically deposited' within a clutch the lack of a correlation between these egg components would imply that mothers possess mechanisms that, at least within certain limits, allow each of the egg components to vary independently. Unfortunately, despite over 20 years of research we still have a very limited understanding of the physiological mechanisms underlying maternal transfer of any minor egg components (Groothuis and Schwabl 2008; Moore and Johnston 2008; von Engelhardt and Groothuis 2011). This is especially important in the context of whether females can, and do, *adaptively regulate* or 'optimize' egg composition to maximize their, or their offspring's, fitness.

Simple 'passive' mechanisms, i.e. co-transport during yolk formation, can still generate 'adaptive' variation in offspring phenotype and matching of offspring phenotype to prevailing environmental conditions. However, specific mechanisms for selective transfer of different egg components would clearly increase the potential for maternal 'control' of the composition of the multivariate egg. There is some evidence that yolk and plasma *hormone levels* can be uncoupled by selective uptake of hormones into the oocyte, or by barriers to passive diffusion. Groothuis and Schwabl (2008) reviewed correlations between hormone concentrations in the mother's blood plasma and the yolk. Here, we have to make a distinction between gonadal steroids, produced directly in the ovary around the ovum, and other hormones such as corticosterone and thyroid hormones, that are produced in other peripheral endocrine glands. The former can be directly transferred to the yolk but the latter first need to be transported via the circulation. Hence, positive correlations between maternal circulation and yolk levels are more likely for the latter. Although positive correlations have been found between plasma and yolk gonadal steroid levels, these often come from studies in which mothers have been implanted with the hormone (see Table 2 in Groothuis and Schwabl 2008). Such studies have to be treated with caution as they often induced supra-physiological hormone levels in the maternal circulation (but see Williams et al. 2005). Correlations in control birds are equivocal, yielding positive, negative, and neutral results (see Table 1 in Groothuis and Schwabl 2008). As the production of gonadal steroids fluctuates considerably during egg production, making such correlations unreliable, Groothuis and collaborators used two different approaches. First, they selected Japanese quail (*Coturnix japonica*) on maternal yolk hormone deposition obtaining selection lines with high and with low yolk testosterone concentrations. However, plasma testosterone concentrations in the maternal circulation during egg laying did not respond to selection and did not differ between the lines, strongly suggesting independent control of hormonal transfer (Okuliarova et al. 2011). Secondly, a comparative study yielded a positive correlation among species between yolk and plasma concentrations of testosterone (W. Müller and T.G.G. Groothuis, 2014 unpublished data). Together, these data suggest that within species there is some degree of maternal control of hormone deposition in the yolk, but within species-specific limitations. Selective transport of hormones, in either direction across the plasma–yolk boundary, could involve P-glycoproteins (Pgps), membrane-associated transporters belonging to the ATP-binding cassette (ABC) superfamily of proteins, which are known to regulate tissue distribution of endogenous compounds including steroid hormones (Dean and Annilo 2005). Pgps are expressed in birds, although mainly in the intestine (Edelmann et al. 1999; Barnes 2001; Green et al. 2005), but it appears that nothing is currently known about their expression or function in developing follicles in the context of maternal effects. Groothuis and Schwabl (2008) also suggested that steroid-binding globulins could be involved in specific retention and accumulation of passively derived steroids in the yolk to concentrations greater than in plasma, although this idea has not yet been investigated further.

Most studies have found positive correlations between the concentration of specific antibodies circulating in mothers (but not fathers) and the levels of antibodies in eggs or offspring (Grindstaff et al. 2003; Staszewski et al. 2007a; Grindstaff 2008; but see King et al. 2011), although only soon after hatching and not in older chicks (Gasparini et al. 2009). This suggests that antibodies are transferred from females to eggs in proportion to plasma antibody concentrations and there is only very limited evidence to suggest that females can facultatively adjust the concentration of maternal antibodies they transfer to eggs to make 'adaptive' adjustments to egg quality (Saino et al. 2001; Hasselquist and Nilsson 2009). A mechanism for selective transfer of IgG has been identified in birds, involving receptor-mediated endocytosis on the innermost layer of the ovarian follicles by the IgY-Fc receptor (FcRY) although the precise molecular mechanism is currently unknown (West et al. 2004; Murai 2013). However, even if this provides a specific transfer mechanism for IgG there is evidence that uptake of IgG in domestic fowl (*Gallus gallus*) parallels the increase in oocyte mass during the exponential vitellogenic growth phase (Kowalczyk et al. 1985) supporting the idea that IgG uptake is simply coupled to receptor-mediated endocytosis of yolk.

There is clear evidence that plasma concentrations of carotenoids correlate with those in the egg yolk, and feeding maternal lesser black-backed gulls with carotenoids increased carotenoid levels in the egg (e.g. Blount et al. 2002). Interestingly, despite continuous supplementation, carotenoid levels still decrease over the laying sequence as in control birds, suggesting that this laying sequence-specific decrease is not entirely due to depletion of maternal plasma carotenoids. Across avian species, there is a large difference in how richly mothers supply their eggs with carotenoids, and also how they convert them to other carotenoids (Blount et al. 2000), as in the case of androgens (von Engelhardt and Groothuis 2011). The degree of freedom within species for mothers to regulate yolk carotenoid levels independently of their intake, and the potential trade-off between using them for themselves or the embryo is as yet unclear. However, it is notable here that there is some evidence that accumulation of carotenoids can affect other egg traits: limiting carotenoid deposition limits re-laying in adult females after egg removal (Blount et al. 2004).

Finally, variation in, and potential regulation of, anti-microbial proteins (see Chapter 7) must involve secretory mechanisms in the tubular gland cells in the oviduct. Albumen deposition in the oviduct is thought to involve mechanical stress from the yolk descending the oviduct which regulates gene expression in the uterus (shell gland), perhaps with hormonal 'priming' of mechanical sensitivity involving oestrogens or progesterone (Lavelin et al. 2001, 2002; Williams 2012). However, it is currently not known if, or how, this basic mechanism can mediate differential regulation of maternally derived albumen components.

10.4 Does embryonic metabolism allow for coupling/uncoupling of effects of egg components independently of the mother's input?

It has long been recognized that there is variation in embryo energetics, metabolism, and rate and efficiency of development that might modulate the relationship between egg composition and offspring phenotype (Vleck and Vleck 1987; Maurer 1996; Starck and Ricklefs 1998). For example, studies by Burggren and colleagues (e.g. Khorrami et al. 2008; Ho et al. 2011) have documented marked variation among individual embryos in heart rate, oxygen consumption, and haematocrit, even at the same development age. Details of variation in embryo utilization of specific egg components or, with one exception (described below), the mechanisms by which this might be achieved are largely unknown, especially for non-domesticated birds (Deeming 2002d). Thus, the extent to which embryo metabolism might uncouple maternal input to, or optimization of, egg composition remains largely unexplored. Hamdoun and Epel (2007) suggested that cellular mechanisms provide robustness and buffer embryos from the environment and common stressors during development although Robinson et al. (2008) concluded that factor(s) intrinsic to the egg or embryo, which could include 'differences in physiological and immunological components', determine variation in development time (see also Robinson et al. 2014b). Ho et al. (2011) also suggested that the egg yolk environment itself, potentially including yolk hormones, has a significant effect on embryonic body mass, heart rate, and development rate (cf. genotypic differences). However, in zebra finches (*Taeniopygia guttata*) the marked variation in development time of individual

embryos, from 12.8 to 16.0 days (Figure 10.3), is inconsistent with patterns of variation of at least some maternally derived egg components (e.g. yolk androgens; Griffith and Gilby 2013) and is much greater than the reported effect of *manipulated* yolk androgens on hatching time (0.5 days; von Engelhardt et al. 2006).

Developmental effects of different egg components must be mediated by several sequential processes: a) transfer from the maternal circulation to the egg yolk (or albumen) as discussed above; b) transfer from the egg yolk or albumen to the embryonic circulation or tissues; and c) utilization by the embryo. For most yolk components this therefore requires movement from an aqueous compartment, i.e. the mother's circulation, to a lipophilic component, i.e. the yolk, and back to an aqueous compartment, i.e. the embryo, potentially providing many steps for embryonic modulation of maternal input. It is therefore plausible that the specific content or composition of freshly laid eggs is equally, or perhaps even less, important than individual variability in how these are metabolized and utilized by the developing embryo.

Unlike total energy or yolk content, which persists to hatching or even after hatching (Carey 1996), the concentrations of some maternally derived egg components, especially hormones, decline rapidly during early development within a few days of onset of incubation (Elf and Fivizzani 2002; Paitz et al. 2011; but see Eising et al. 2003). Recent work has elucidated details of the embryonic metabolism pathway for yolk hormones (although part of the decline might simply be due to the mixing of yolk with albumen, diluting the concentration; Gilbert et al. 2007). Some studies indicate that maternally derived steroids (testosterone, progesterone, and corticosterone) are already partly metabolized during the first few days of embryo

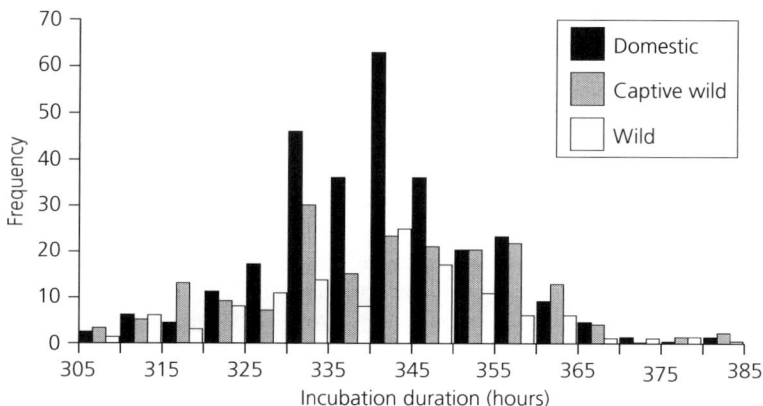

Figure 10.3 Individual variation in development time of embryos in zebra finch (*Taeniopygia guttata*) eggs from three different populations (N = 586). Range = 12.8–16.0 days; bars show frequencies for each interval for each population. Reproduced from Griffith and Gilby (2013) with permission from John Wiley & Sons.

development to water-soluble metabolites *in ovo*, via the sulphonation pathway in zebra finches and common starlings (*Sturnus vulgaris*; von Engelhardt et al. 2009; Paitz et al. 2011; Paitz and Casto 2012). This is a process that is highly conserved among vertebrates and that is 'integral for the regulation of the early endocrine environment of embryos' (Paitz and Bowden 2013). Paitz and Casto (2012) showed that in birds this involves both Phase I modification of steroids by enzymes, such as 5β-reductase, and Phase II metabolism by enzymes, such as steroid sulphotransferases, during early incubation. So, in the context of optimization of the multivariate egg, how is maternal control 'realized in the face of embryonic regulation of the endocrine environment' (Paitz and Bowden 2013)? In mammals the sulphonation of maternal steroids is generally thought to buffer the developing foetus from effects of maternal steroids, especially for steroid-sensitive developmental processes such as sexual differentiation of the brain (Carere and Balthazart 2007). Thus, the sulphonation pathway could inactivate maternal steroids and limit or, at least, modulate the influence of maternal effects on at least some components of offspring phenotype. However, conversion of hormones to water-soluble metabolites could also facilitate transfer of lipophilic hormones from the lipid-rich yolk to the aqueous embryonic circulation, and generate hormone metabolites which themselves are biologically active or which can subsequently be converted back to biologically active forms in the embryo (e.g. oestrone sulphate can be converted back to oestrone by the enzyme steroid sulphatase; Moore and Johnston 2008; Paitz and Bowden 2013). If, or how, hormone metabolites affect offspring phenotype remains unknown but it is clear that embryos are not simply 'passive responders' to the levels of yolk steroids, and potentially other components, present in eggs (Paitz and Bowden 2013). Little is known about potential embryonic modulation of other egg components (e.g. maternal antibodies [MAbs] or antioxidants), although MAbs are rapidly catabolized or inactivated by antigen binding and disappear from offspring within 5–14 days of hatching (Grindstaff et al. 2003; Staszewski et al. 2007b; Bar-Shira et al. 2014).

A crucial unresolved question here is whether the embryo is able to metabolize the maternal signals and resources in a context-dependent manner? Potentially, there can be a conflict between the mother and her offspring as the latter might benefit from more resources than would be most beneficial for the mother to provide. An example is the modification of the competitive position in the sibling hierarchy by means of testosterone in the yolk, and the extreme case of brood reduction in times of food scarcity. It may therefore be in the interest of the offspring not to 'listen' to the mother's signal. However, in this case the offspring must have reliable information about the relevant context such as its place in the sibling hierarchy or food availability before it hatches. No experiments have been undertaken to test this further but for additional discussions see Groothuis et al. (2005) and Müller et al. (2007a).

10.5 How might different egg components affect offspring growth and development to determine adult phenotype?

Adaptive maternal effects not only require mechanism(s) for mothers to (differentially) transfer components to eggs as described above, but they also require those by which offspring respond developmentally or physiologically to these maternally derived components. However, despite the extensive research on maternal effects the underlying mechanisms for putative changes in offspring phenotype remain unknown (Groothuis and Schwabl 2008; Pfannkuche et al. 2011; von Engelhardt and Groothuis 2011). Most work has focussed on maternally derived hormones given the extensive literature showing the critical role of embryonic gonadal hormones on sexual differentiation for at least some behavioural traits in birds (e.g. copulation; Balthazart and Adkins-Regan 2002; Adkins-Regan 2005). Receptors, binding globulins, and enzymes allowing embryos to respond to maternal hormonal signals are expressed by embryos at early stages in development: by day 2–5 of incubation in the domestic fowl, although perhaps slightly later by day 7 of incubation in altricial species (see Figure 4.8 in von Engelhardt and Groothuis 2011). Interestingly, androgen receptors are present in embryos before embryonic production of its own androgens, strongly suggesting that expression of these receptors has evolved for responding to maternal androgens (Godsave et al. 2002).

Maternal hormones have been proposed to affect offspring development in several ways: a) via direct, immediate 'activational' effects on tissue development (e.g. anabolic effects of androgens on muscle tissue) or by pleiotropic effects on reallocation of resources among different physiological systems (e.g. growth at the expense of immune development; Groothuis et al. 2006); or b) via longer-term 'organizational' effects (e.g. on neural architecture in the brain), or other components of the neuroendocrine axes, that lead to differential activation by endogenous hormones later in development and, potentially, at adulthood (von Engelhardt and Groothuis 2011). It has been suggested that maternal hormones might affect hormone-mediated traits by changing hormone sensitivity of the embryo, for example, by up-regulation of hormone receptors, although only one study appears to have tested this idea with *in ovo* hormone manipulation in birds. Pfannkuche et al. (2011) showed that androgen receptor mRNA expression was significantly *lower* in chicks from testosterone-treated eggs 14 days after hatching. This actually suggests another compensatory mechanism to buffer offspring from maternal effects (in addition to conversion of hormones to non-active metabolites, as described above), in this case to avoid potential detrimental effects of high testosterone levels, but it is unknown if these differences persist to adulthood. Several studies have suggested that embryonic exposure to corticosterone can affect hypothalamic-pituitary-adrenal (HPA) axis function in chicks at fledging, although the effect can be sex-specific (Hayward et al. 2006; Love and Williams 2008b). In a recent review on the effects of pre-natal stress and enhanced exposure to corticosterone in birds Henriksen et al. (2011) suggested that maternal stress often affects the egg in a complex manner, making it difficult to disentangle the different pathways. They also suggested that egg levels of corticosterone are often measured in an unreliable way and might be much lower than previously assumed (suggesting the egg is buffered against these hormones as in the case of placental animals) and that the context in which the offspring is tested, i.e. stressful or not, can have a strong effect on the outcome of the study. Furthermore, Crino et al. (2014) showed that although corticosterone treatment of *nestling* zebra finches affected HPA axis function (increased total/free corticosterone responses) up to 30 days post-hatch, there was no detectable treatment effect at 60 or 90 days post-hatch, i.e. at sexual maturity.

Several other developmental processes provide plausible pathways linking maternal effects to offspring phenotype, but these have not yet been studied extensively (or at all) in birds. For example, Haussmann et al. (2012) and Tissier et al. (2014) showed that embryonic exposure to maternally derived oestrogen and corticosterone could uncouple relationships between growth rate, associated changes in oxidative stress, and telomere loss in zebra finches, a species where telomere dynamics in early life can predict subsequent lifespan (Heidinger et al. 2012). Embryonic exposure to corticosterone might increase oxidative stress but reduce telomere length (Haussmann et al. 2012) whereas

oestradiol might increase telomerase activity (Kyo et al. 1999) and down-regulate oxidative stress (Behl et al. 1995). However, the long-term consequences of this potential alternate developmental pathway have yet to be confirmed. Mammalian studies suggest that maternally derived hormones could influence epigenetic programming in the embryo with potential long-term effects on gene expression, development, and phenotype (Crudo et al. 2013; Moisiadis and Matthews 2014). However, to the best of our knowledge this has not been demonstrated in birds (Berghof et al. 2013; Ledón-Rettig et al. 2013).

Although there is currently little information on how maternally derived egg components might affect offspring development, some studies have reported effects on phenotype well beyond the period during which yolk is consumed, such as in the fledging and juvenile stages (e.g. begging, Groothuis et al. 2005; immunity, Müller et al. 2005; neophobia, Tobler et al. 2007; metabolic rate, Tobler and Sandell 2007). Some other studies have demonstrated effects of elevated yolk androgens even into adulthood. They have included studies of competitive behaviour (Strasser and Schwabl 2004; Eising et al. 2006; Partecke and Schwabl 2008), plumage (Strasser and Schwabl 2004; Eising et al. 2006; but see Partecke and Schwabl 2008), ornaments (Rubolini et al. 2006), fertility (Rubolini et al. 2007), and mortality (Groothuis et al. 2005; Schwabl et al. 2012). These studies have suggested that there are organizing effects of the maternal hormones, but whether these represent direct effects or effects mediated indirectly via, for example, body mass that itself causes dominance, or higher fertility, is as yet unclear. Moreover, physiological mechanisms for these long-term effects are also, as yet, unknown. It is well-known that pre-natal exposure to steroid hormones can have organizing effects on sexual differentiation but to what extent yolk steroids affect differentiation within the same sex while not interfering with basic sexual differentiation is, as yet, also unclear (Carere and Balthazart 2007). Groothuis and Schwabl (2008) argued that such interference is unlikely for two reasons: studies on the hormone effect on sexual differentiation in birds have used 100–1,000 times higher dosages of these hormones than those present in the yolk and those used in hormone-mediated maternal effects studies. Moreover, in birds oestradiol is the hormone affecting sexual differentiation in females (the male being the default sex that does not need exposure to gonadal hormones) and oestradiol concentrations are relatively low in egg yolk. Nevertheless, at this stage we reiterate the plea of Groothuis and Schwabl (2008) that 'endocrinologists embark on . . . mechanistic studies and behavioural ecologists adjust their interpretations to accommodate the current [lack of] knowledge of mechanisms'.

10.6 Does ecological context *require* uncoupling of effects of different maternally derived egg components?

One of the most noticeable (or worrying!) conclusions from a review of 20 years of maternal effects studies of birds is the high degree of contradictory or inconsistent results from different studies both among and within species, and for both different and the same traits (see Tables 4.1, 4.2, and 4.3 in von Engelhardt and Groothuis 2011). The most parsimonious explanation for this is that maternal effects are *context-dependent*. In fact, one would *expect* different outcomes of maternal effects in different environmental or ecological contexts, i.e. if maternal effects are a maternal strategy to 'programme' offspring, adaptively adjusting their development to match environmental conditions, then this should occur more, less, or not at all in different contexts (von Engelhardt and Groothuis 2011). This has consequences for both correlative and experimental studies. Patterns of transfer of different maternal components, as well as covariance among components, may vary depending on what is 'needed' by the embryo or chick in a specific environment and, similarly, experimental manipulations *in ovo* will also have context-dependent effects. For example, experimental elevation of yolk concentrations of androgens has both beneficial effects (increased growth and competitiveness) and costs (increased metabolic rate and immune suppression) for the chick. The balance between these costs and benefits may strongly depend on whether the chick is able to bear these costs. This may depend on, for example, the resource (e.g. food) availability (in turn affecting other egg components).

In addition to ecological context, the pattern and magnitude of maternal effects mediated by maternal contributions to egg composition should, theoretically, relate to variation in the life history of different species (Love et al. 2009; Müller and Groothuis 2013) (e.g. whether the species has a long or short lifespan, high or low fecundity etc.). Several comparative studies have found clear relationships between variation in life-history traits and egg concentrations of androgens (e.g. Gorman and Williams 2005; Gil et al. 2007; Schwabl et al. 2007). The potential importance of ecological or evolutionary context for interpretation of the occurrence and significance of maternal effects could most easily be addressed by a meta-analysis of previous studies. Unfortunately, despite the very large number of such studies too few of these report *any* information about the environmental context in the year(s) of the study so such an analysis cannot currently be undertaken (M.S. Müller and T.G.G. Groothuis, 2014 unpublished data).

The importance of considering ecological or evolutionary context is even greater if we consider this from the perspective of the 'multivariate egg', since any single environment is likely to *require* uncoupling of different egg components. As an example, Postma et al. (2014) hypothesized that females should produce eggs with high IgG and low androgen concentrations when parasite prevalence in the local environment is high, but produce eggs with low IgG and high androgen concentrations when parasite prevalence is low. There was (weak) evidence to support this in that covariance of maternal yolk IgG and yolk androgens tended to be different between hen flea (*Ceratophyllus gallinae*)-free and flea-infested nests (*P* = 0.07), mainly due to a significant negative correlation between maternal yolk IgG and yolk androstendione in flea-free nests (Postma et al. 2014). More studies looking at covariance of multiple egg components in known environmental contexts are needed. However, at the very least it is essential that future studies report (and therefore obtain data on) ecological context (e.g. high versus low food years, relative breeding productivity [good versus bad years], high versus low parasite or predator levels etc.).

We would like to highlight two further general points for consideration here. First, in the context of maternal 'programming' of offspring development it is generally assumed that mothers attempt to adjust the development and the phenotype of their offspring to match *prevailing* environmental conditions. This assumes that offspring will encounter, and need to deal with, the same environment as that of the mother during egg production and that this environment is not invariant (otherwise no maternal adjustment would be necessary). This might be the case for the chick stage but not necessarily for long-lasting effects in the adult stage, certainly in species with marked post-natal dispersal. In general, the greater the time-lag between offspring development and reproduction as an adult, the more likely it is that the relevant maternal and offspring environments will be uncoupled. An alternative explanation, however, for long-term effects of egg components might be that the mother applies a bet-hedging strategy in case of unpredictable environments. Different egg components could vary in the extent to which this temporal variation affects how females 'assess' the environment to which she matches offspring phenotype. For example, MAbs will reflect a female's long-term exposure but might not accurately reflect the offspring's future environment, whereas a poor-quality food year might be assessed accurately in the short term, with egg composition adjusted accordingly.

Secondly, it might reasonably be expected that there could be environments or ecological contexts where the optimal maternal strategy is *not* to utilize maternal effects (e.g. to produce eggs with only trace amounts of maternal hormones). The fact that this does not appear to occur would be consistent with a lack of female control over maternal transfer, i.e. mechanisms do not exist to allow females to buffer offspring from circulating hormones, antibodies etc. completely unless a basal level of these substances is necessary for normal development. Unfortunately, this can only be tested by experimentally lowering these substances in the egg, something, which is much more difficult than elevating them by injection. Injection of hormone or receptor antagonists (blockers) for different egg components would also block the effect of these substances when endogenously produced by the embryo; thus, such an experiment would not specifically address the effects of lowering maternal components.

10.7 Do maternally derived egg components actually have long-term effects on fitness?

In many ways all the issues considered above are only of consequence if a mother's optimization of the composition of the multivariate egg, or her *differential regulation* of specific egg components, affects fitness (cf. these egg components simply being 'metabolic waste products' without having fitness effects on the mother or offspring; see above and von Engelhardt and Groothuis 2011). We chose not to focus our entire review on this last question, in part because this remains largely unanswered—even after 20 years of work the majority of studies continue to be conducted on captive and/or domesticated species (e.g. zebra finch, Japanese quail, Atlantic canary [*Serinus canaria*]) and to only consider short-term effects. This multitude of studies has been reviewed by several authors (e.g. Groothuis et al. 2005; Gil 2008; von Engelhardt and Groothuis 2011; Navara 2013) and, therefore, we do not review them again here. Twenty years on from the seminal work of Schwabl (1993) there is a surprising lack of (experimental) field studies correlating yolk hormone

levels with offspring fitness. This contrasts markedly with studies on other maternal traits (e.g. clutch, egg or brood size) where experimental manipulation coupled with long-term fitness measurements has long been the standard approach (e.g. Hegyi et al. 2011; see Williams 2012). Many studies do interpret yolk hormone effects on certain phenotypic traits within a context of *potential* fitness effects based on the fact that these traits have been demonstrated to affect fitness in other studies, but the effects of manipulated yolk hormones on fitness are rarely tested directly *in the same study*. Here, we summarize studies that have directly evaluated effects on embryo/chick survival or long-term, post-fledging effects on maternal or offspring fitness, specifically reproduction and survival, focussing on *free-living* birds. Due to the very limited number of studies looking at more than one class of egg component, we largely consider hormone, antibody, and antioxidant studies separately: for this section of the chapter data are currently insufficient to consider long-term effects in the context of the 'multivariate egg' (highlighting a goal for future studies).

Studies of maternally derived egg antibodies and antioxidants that demonstrate strong, long-term fitness effects are rare. Direct, positive effects of MAbs should be of short duration because antibodies are rapidly catabolized; most MAbs disappear from offspring within 5–14 days of hatching (Grindstaff et al. 2003; Staszewski et al. 2007b). At least in non-domesticated birds, chicks develop their own immune systems and start to produce endogenous antibodies around 10–14 days post-hatch (Grindstaff et al. 2006; Pihlaja et al. 2006), so MAbs should mainly be important in protection of the developing embryo and the newly hatched chick. However, evidence for positive short-term effects of MAbs on offspring phenotype in free-living birds is limited and rather mixed (Addison et al. 2010; Williams 2012). Early studies by Heeb et al. (1998) showed a positive effect on nestling growth rate and recruitment in broods where mothers were exposed to an ectoparasite (hen fleas) during egg laying, and follow-up studies suggested that this maternal effect was due to immunoglobulins transferred via the egg (Buechler et al. 2002; Gallizzi et al. 2008). Longer-term effects of MAbs have also been proposed, largely from studies in mammals, with either a positive or negative effect on development of the offspring's own (endogenous) humoral immune response (Curno et al. 2009; Hasselquist and Nilsson 2009). However, most studies with experimental manipulation of MAb levels have shown no effect on offspring growth, final body size, or survival (Grindstaff et al. 2006; Pihlaja et al. 2006; Jacquin et al.

2012; but see Reed et al. 2006) or inconsistent, short-term results in captive species only (Martyka et al. 2011; Abad-Gómez et al. 2012; Rutkowska et al. 2012).

Many studies on yolk hormones have reported short-term effects on offspring phenotype but several recent studies provide clear examples where these effects *do not* persist to adulthood or reproductive maturity, highlighting the danger of this common assumption (see also Navara et al. 2006; Pitala et al. 2009). Barnett et al. (2011) examined the effect of *in ovo* manipulation of testosterone on provisioning by house wrens (*Troglodytes aedon*). Nestlings hatching from testosterone-injected eggs begged more than control nestlings early in the nestling period (as has been reported in other studies; e.g. Boncoraglio et al. 2006; Helfenstein et al. 2008; Smiseth et al. 2011; Ruuskanen and Laaksonen 2013), but not later in the nestling period. However, treatment had no effect on the levels of parental provisioning or nestling mass gain, nor on hatching success or nestling survival. Similarly, corticosterone treatment of nestling zebra finches affected HPA axis function (increased total/free corticosterone responses) up to 30 days post-hatch, as reported in other studies (e.g. Hayward et al. 2006; Love and Williams 2008b), but there was no detectable treatment effect at 60 or 90 days post-hatch, i.e. at sexual maturity (Crino et al. 2014). Müller and Eens (2009) experimentally manipulated the yolk androgen concentrations of free-living common starling eggs and followed individual males from hatching until the year of first reproduction. Again, there were early effects of hormone treatment (elevated levels of yolk androgens affected the length of the embryonic period), but treatment did not modify the expression of either androgen-dependent or androgen-independent sexually selected male characters, including song phenotype, at adulthood. Müller and Eens (2009) concluded that the results of this study together with the results of a meta-analysis (12 androgen-sensitive traits in six species with an effect size d = 0.184, $P > 0.4$) indicate 'that the effects of yolk androgens on sexually-selected male characters may be comparatively small'.

Nevertheless effects of maternal hormones that persist to adulthood have been demonstrated for traits that could *potentially* affect reproductive fitness under semi-natural or field conditions in some studies (e.g. dominance behaviour, sexual displays, plumage, and metabolic rate; Eising et al. 2006; Partecke and Schwabl 2008; Ruuskanen et al. 2012a). For example, in common black-headed gulls under semi-natural conditions one-year-old birds from androgen-injected eggs performed more sexual displays than those from control eggs, were dominant over the latter, and developed nuptial

plumage earlier (Eising and Groothuis 2003). van Rhijn and Groothuis (1987) had earlier demonstrated that development of nuptial plumage correlated with number of eggs laid in that season. Thus, it is reasonable to assume that yolk androgens *could* affect fitness, although a direct effect of the egg hormone manipulation on egg production and number of chicks that fledge was not tested by Eising and Groothuis (2003). In a field study of great tits it was experimentally demonstrated that parasite load in the nestbox affected yolk hormone levels and that the latter affected natal dispersal of the offspring, suggesting that yolk hormones increase survival through birds avoiding highly infected nest sites, but again a direct effect on survival was not demonstrated (Tschirren et al. 2007). To the best of our knowledge only one study has demonstrated a direct long-term effect on survival under semi-natural conditions in a wild bird species: in the house sparrow (*Passer domesticus*), elevated levels of testosterone in the egg increased survivorship in a sex-dependent manner both in the juvenile and adult stages (Schwabl et al. 2012).

In terms of ecological studies on effects of maternal hormones in free-living birds, Hegyi et al. (2011) sampled eggs of collared flycatchers at 24 nests for yolk testosterone and androstendione by biopsy, hatched eggs in an incubator, and then used a cross-fostering experiment to investigate offspring fitness correlates of natural within-brood androgen patterns while randomizing post-hatching parental effects. There was no relationship between yolk androgen or androstendione and hatching success, or post-hatching chick mortality or growth, and only a weak effect of yolk androstendione on recruitment (based on only N = 7 recruits). Ruuskanen and colleagues published a series of papers (Ruuskanen and Laaksonen 2010; Ruuskanen et al. 2012a, 2012b, 2013) on 'long-term' effects of yolk hormones in *Ficedula* spp. Ruuskanen et al. (2012a, 2012b) represent the *only* example of a large-scale field experiment (N = 240 nests) where yolk hormones were manipulated within the physiological range in free-living collared flycatchers and individual chicks were followed through to recruitment and reproduction. The only significant effect of experimentally elevated yolk androgen was lower local recruitment of male offspring (based on N = 19 individual recruits), with no effect on recruitment of daughters, no effect on any breeding parameters of offspring of either sex, i.e. age at first breeding, timing of breeding, clutch size, hatching or fledgling success, nestling body size and condition, and no effect on the return rate or any breeding parameters of the parents. Furthermore, yolk androgen treatment did not affect the expression of sexually selected plumage ornaments such as forehead and wing patch size, ultraviolet (UV) colouration, or parental feeding rate in either sex. Ruuskanen et al. (2013) studied the long-term effects of experimentally manipulated yolk androgens in male and female European pied flycatchers in a common-garden environment in captivity. Elevated yolk androgen levels increased basal metabolic rate in both females and males in adulthood but had no effect on male melanin colouration or plumage ornaments, timing or speed of moult in either sex, or cell-mediated or humoral immune responses. Thus, despite well-designed studies with initially large sample sizes (even though sample sizes at recruitment were low), this set of studies found significant effects of yolk androgens on only two of 22 or more measured traits (and only three of 44 sex × trait combinations)—a result that would be expected by chance at an alpha threshold of 0.05! Love and Williams (2008a) experimentally manipulated yolk corticosterone in free-living common starlings and demonstrated maternal fitness effects including on cumulative fledgling productivity and local survival, but only within the context of matching/mismatching of maternal effects with the female's own quality (Breuner 2011). Love and Williams (2008a) were unable to assess offspring fitness because of lack of juvenile recruitment in common starlings but further studies of yolk hormones, and other maternally derived egg components, in free-living birds with fitness-related endpoints are clearly needed.

Thus, the many documented short-term effects of maternally derived egg components (see von Engelhardt and Groothuis 2011), without *direct* evidence of fitness consequences, might simply be 'a form of transient activational effect taking place in early life' (Carere and Balthazart 2007), but not permanent organizational effects. However, one could argue that the strongest selection on offspring fitness occurs on survival before adulthood, either in the nest, at independence, or during the bird's first winter. A clear example of a short-term effect with obvious fitness consequences would be where different egg components decrease or increase embryo or nestling mortality (accepting that an increase in offspring mortality might actually increase parental fitness by adjusting brood size or brood sex ratio to the current environment; Groothuis et al. 2005; Love and Williams 2008a). Elevated yolk corticosterone (Love and Williams 2008a) and yolk oestradiol (Williams 1999; von Engelhardt et al. 2004) increase embryo or nestling mortality in a sex-specific manner, i.e. higher male mortality but no effect in females. Similarly, elevated yolk androgens

increased nestling mortality in a sex-specific way in a field study of American kestrels (*Falco sparverius*; Sockman et al. 2008), but increased chick survival in the field in common black-headed gulls (Eising and Groothuis 2003) and common starlings (Pilz et al. 2004). Nevertheless, despite 20 years of research, the relative contribution of maternal effects to variation in life-time fitness, relative to other reproductive traits, remains unclear (Williams 2012), perhaps due to the context-dependent effects discussed earlier.

10.8 Conclusions

In this review we have attempted to consider some different perspectives on the topic of maternal effects via egg quality on offspring development and fitness, and to focus on what we do not know, rather than simply reviewing the canon of previous work. Although we have highlighted many of the same knowledge gaps that have been identified in reviews going back over the last 10 years at least (Blount et al. 2000; Grindstaff et al. 2003; Groothuis et al. 2005; Moore and Johnston 2008; von Engelhardt and Groothuis 2011; Navara 2013), there is some progress in these critical areas of a rather new field of research. We summarize what we see as some of the key issues moving forward by asking the following questions.

To what extent can females regulate and control transfer of hormones, antioxidants, and antibodies to eggs, and what mechanisms do they use? Although the assumption was that the egg reflects circulating hormone levels in the mother, we now know that this may not be true and that the flexibility of the mother to regulate hormone levels independently may be greater than was previously thought, providing more scope for evolution. However, more work is needed and this problem becomes even more acute if we consider that females should be independently regulating, coupling/uncoupling, or optimizing the covariance of different compounds in the 'multivariate' egg.

To what extent does embryo metabolism allow the developing offspring to couple/uncouple, or even ignore maternal input from different egg components? With one exception (Paitz and Bowden 2013) the immediate fate and physiological effects of maternally derived egg components remain largely unknown. However, most of the mechanisms currently identified in embryos (sulphonation, 11β-HSD expression, *downregulation* of receptors) would all appear to protect or buffer the developing offspring from maternal effects which would exacerbate the problem of females potentially optimizing egg composition to programme the offspring's phenotypic development 'optimally'. Whether the embryo can regulate the metabolism or sensitivity to mothers' signals in a context-dependent manner also remains an intriguing but unanswered question.

Is there evidence for changes in offspring development at the molecular, cellular, or physiological level that persist to adulthood, or mechanisms that might mediate such changes, and that are unequivocally linked to early developmental effects of maternally derived hormones? Novel mechanisms (e.g. growth-dependent oxidative stress, telomere dynamics, epigenetic programming) represent exciting possibilities, but these need to be linked explicitly to maternal effects and to long-term fitness effects.

What are the ecological and evolutionary contexts that are critical to how different individuals and species utilize maternal effects, and for the consequences of maternal effects? In addition, the correlation between, and predictability of, parental/offspring environments needs to be considered more often when it is proposed that parents match their offspring's development to the 'prevailing' environment (Burgess and Marshall 2014). Unfortunately, most studies do not consider or report this, and it is imperative that future experimental studies are conducted, and then can be interpreted, within a known ecological, environmental, or life-history context.

How do multiple egg components interact with maternal effects? Studies need to move beyond manipulation and measurement of single egg components, and short-term effects on single, or a few, traits, especially in captive or domesticated birds; we need well-designed studies with experimental manipulation of multiple egg components/maternal effects, with large sample sizes, in free-living populations where individual parents and offspring can be followed in the long term to obtain integrated measures of fitness (e.g. cumulative productivity, future fecundity, survival). In other words, we need the type of experiments that are typically carried out to study other reproductive traits. Recent development of methods to manipulate

non-hormone maternally derived egg components directly (Addison et al. 2010) should facilitate this, but this will require a large-scale collaborative effort, i.e. 'big science', of many research groups to focus efforts on maternal effects.

Why do we readily dismiss a more parsimonious explanation for putative egg-mediated maternal effects? All too often we disregard facts such as: many minor maternally derived egg components are metabolic wastes (as is the case with maternal transfer of contaminants); there has been selection for mechanisms in embryos to attenuate the effects of this unavoidable transfer; and many short-term effects on offspring phenotype (especially in response to experimental perturbation) are simply transient epiphenomena of this unavoidable maternal transfer. However equally, the field is as yet too young to disregard other explanations as discrepancies in the literature due to potential context-dependent effects, and non-significant fitness effects due to too small sample sizes are very likely hampering a synthetic interpretation of current data.

Why after the initial excitement when the topic of egg-mediated maternal effects was placed on the scientific agenda, has momentum not increased in investigating more complex (but no less intriguing!) phenomena? Initial frameworks in behavioural ecology, such as that hormone transfer to the egg would be costly, that mothers face a trade-off between transferring these hormones to themselves versus to their eggs, or that hormones would pose constraints in development, are probably incorrect. The field provides a clear example that we need an integration of mechanistic, ecological, and evolutionary approaches to progress.

Acknowledgements

We would like to thank Ruedi Nager and the editors for helpful comments on earlier drafts of this chapter. This work was supported by a Natural Sciences and Engineering Research Council Discovery grant and an Accelerator grant to TDW.

CHAPTER 11

Egg signalling: the use of visual, auditory, and chemical stimuli

K. Brulez, T.W. Pike, and S.J. Reynolds

11.1 Introduction

The fundamental biological function of avian eggshells is to enclose an incubation environment in which an embryo can develop (Roberts and Brackpool 1994). However, birds have evolved a number of eggshell traits that are capable of transmitting information to a variety of receivers, from incubating parents through to predators and brood parasites. Communication of information consists of three parts: the sender, the signal itself and the receiver. Not all features will have a signalling function but may still be used as a cue by a receiver (e.g. a predator or a brood parasite). A signal is an act, or structure, which has evolved to indicate the quality of a signaller and it functions to change the behaviour of the intended receiver. A cue is a feature of the external world used by a receiver to decide on its future actions (Hasson 1994; Maynard Smith and Harper 1995). A cue can be a signal if it is deliberately communicating that information to the receiver.

Overwhelmingly, research into egg signalling has concentrated on the appearance of the eggshell, namely its colouration and pigmentation (Figure 11.1). Although this chapter will largely focus on this subject, we will also discuss other ways in which eggs may be acting as signalling cues. The first section of this chapter will focus on the acquisition of colouration and will explain the main hypotheses for its functional significance. The second section will discuss how eggs can be used as signals by the reproductive female (and possibly the developing embryo), the nature of signals that eggs are able to convey, and the identity of receivers who interpret these signals.

11.2 The avian eggshell

In order for us to understand how sensory cues in eggshells have evolved, we must first understand the functional properties of the eggshell, the mechanisms of eggshell formation and the acquisition of colouration. The eggshell acts as protection for the developing embryo, excluding most bacteria and resisting pressure of the incubating adult (Board 1982; Board and Sparks 1991; Chapter 7). Pores within the shell allow for gaseous movement between the embryo and the nest microclimate while the shell materials provide a source of calcium to the developing embryo (for a full review of the functions of the avian eggshell see Reynolds and Perrins 2010).

Most of the available information on eggshell formation comes from research into domesticated species such as domestic fowl (*Gallus gallus*), although limited information relating to small passerines suggests that similar processes might take place (Reynolds 1997, 2001; Reynolds and Waldron 1999). Surprisingly, however, given the commercial importance of eggs, little is known about how eggshell pigments are synthesized, mobilized, and deposited.

11.2.1 Eggshell formation

The avian eggshell is a protein matrix lined with mineral crystals, usually of a calcium compound such as calcium carbonate ($CaCO_3$) or apatite ($CaPO_4$; Weiner and Addadi 1991). Within 3 hours of its release from the ovary, the ovum, encapsulated in albumen and immature membranes, reaches the shell gland where it remains for approximately 20 hours, during which biomineralization takes place (Board and Sparks 1991). The various layers of the shell are formed sequentially as the egg rotates within the shell gland (Lavelin et al. 2000). The foundation of the eggshell is provided by the inner and the outer shell membranes. At the onset of eggshell formation, mammillary knobs seed on the outer membrane and crystals grow outwards in

Brulez, K., Pike, T.W., and Reynolds, S.J., *Egg signalling: the use of visual, auditory, and chemical stimuli*. In: *Nests, Eggs, and Incubation*. Edited by: D.C. Deeming and S.J. Reynolds, Oxford University Press (2015). © Oxford University Press.
DOI 10.1093/acprof:oso/9780198718666.003.00011

Figure 11.1 The colour and pigment patterns of eggshells vary greatly between species leading to a range of hypotheses for its possible adaptive significance. From top left to bottom right: Eurasian skylark (*Alauda arvensis*), Eurasian rock pipit (*Anthus petrosus*), meadow pipit (*Anthus pratensis*), European goldfinch (*Carduelis carduelis*), Cetti's warbler (*Cettia cetti*), white-throated dipper (*Cincus cinclus*), hawfinch (*Coccothraustes coccothraustes*), carrion crow (*Corvus corone*), western jackdaw (*Corvus monedula*), common house martin (*Delichon urbicum*), corn bunting (*Emberiza calandra*), yellowhammer (*Emberiza citrinella*), common nightingale (*Luscinia megarhynchos*), common chaffinch (*Fringilla coelebs*), Eurasian jay (*Garrulus glandarius*), barn swallow (*Hirundo rustica*), red-backed shrike (*Lanius collurio*), common grasshopper warbler (*Locustella naevia*), red crossbill (*Loxia curvirostra*), woodlark (*Lullula arborea*), white wagtail (*Motacilla alba*), western yellow wagtail (*Motacilla flava*), spotted flycatcher (*Muscicapa striata*), Eurasian golden oriole (*Oriolus oriolus*), great tit (*Parus major*), house sparrow (*Passer domesticus*), common chiffchaff (*Phylloscopus collybita*), Eurasian magpie (*Pica pica*), dunnock (*Prunella modularis*), Eurasian bullfinch (*Pyrrhula pyrrhula*), goldcrest (*Regulus regulus*), whinchat (*Saxicola rubetra*), Eurasian blackcap (*Sylvia atricapilla*), lesser whitethroat (*Sylvia curruca*), redwing (*Turdus iliacus*), and mistle thrush (*Turdus viscivorus*). Eggshells were made available to the University of Birmingham for a destructive loan from the Natural History Museum, Tring, UK. Eggs have been normalized for size. (Photos: G. Maurer. See Plate 13.)

most directions. When adjacent crystals meet they can only grow outwards and so produce the main palisade layer. Between the palisade columns, narrow pores traverse the eggshell, allowing for gaseous exchange (Rahn et al. 1979; Hunton 2005). Shell formation ceases after deposition of a surface crystal layer (Board and Sparks 1991) and the shell is completed by the formation of a thin outer organic, or inorganic, layer of shell accessory material (Board and Sparks 1991), which is often of a vesicular nature and known as the cuticle (Simons and Wiertz 1966).

Calcium is the most prevalent mineral in the avian body and its homeostasis in birds is controlled by the parathyroid hormone, calcitonin, vitamin D3, and sex hormones (reviewed in de Matos 2008). The increased demand for calcium during eggshell formation is met through augmented intestinal absorption and resorption of calcium from the medullary bone (Klasing 1998). The medullary bone develops in egg-laying females in response to gonadal steroids and acts as a calcium storage compartment required for eggshell formation (Dacke 2000). In the domestic fowl, calcium from the medullary bone can provide as much as 40% of that required for eggshell formation (Dacke et al. 1993), but in smaller species, these endogenous calcium stores are insufficient to meet the calcium demands of eggshell formation (Pahl et al. 1997; Reynolds 2003) and so they must rely on exogenous sources.

11.2.2 How are eggshells coloured?

Eggshell pigments are deposited during the later stages of eggshell formation and consequently occur within the calcitic and cuticular layers of the shell (Poole 1965). Pigments are transferred to the eggshell via surface epithelial cells while in the shell gland (Baird et al. 1975). There are two main types of pigments responsible for the colouration and patterning on eggshells: protoporphyrin IX produces brownish hues, and biliverdin IXα produces blue and green hues (Kennedy and Vevers 1973, 1976; Gorchein et al. 2009). Both protoporphyrin and biliverdin are believed to be derived from a common precursor molecule (Wang et al. 2009b) and are produced during the biosynthesis of blood, most likely in the shell gland (Zhao et al. 2006). The majority (80–87%) of protoporphyrin is located within the calcareous layers of the eggshell while the remainder is located within the cuticle (Samiullah and Roberts 2013). Resulting pigment patterns, known as maculation, are dependent on the quantity and location of the pigment within the eggshell layers. It is believed that

distinct spots on eggshells result from pigment secretion into the eggshell cuticle, while larger patches and streaks are produced from secretion of larger quantities of pigment during the final stages of eggshell formation, possibly whilst the egg is in rotation (Solomon 1987; Sparks 2011). Eggs with pigment present as a pattern rather than as uniform background tend to have the pigment restricted to the shell surface or within various layers of the eggshell (Harrison 1966).

As a product of haem catabolism (Wang et al. 2009b), the synthesis pathway of protoporphyrin is well-documented, but the site(s) where synthesis may occur is still not fully understood (Sparks 2011). However, evidence is growing to show that protoporphyrin is synthesized in the shell gland (Zhao et al. 2006; Wang et al. 2013). Furthermore, whether the synthesis and deposition of protoporphyrin are detrimental or beneficial to the laying female or embryo remains unclear. The accumulation of protoporphyrin within the liver can cause oxidative stress (Afonso et al. 1999), and so the deposition of large amounts of protoporphyrin onto the eggshell may signal a female's ability to cope with intense oxidative stress, or alternatively, may indicate that protoporphyrins have been successfully removed (Moreno and Osorno 2003; Holveck et al. 2010). However, protoporphyrin content in the liver is independent of its content in the shell gland (Zhao et al. 2006), and it may accumulate or deplete in the liver having no influence on what appears on the egg or in the shell gland. Pigment spotting could also affect eggshell permeability (Higham and Gosler 2006; but see Deeming 2011c). Maurer et al. (2011a) argued that neither degree of maculation nor intensity of background colour of an eggshell can affect water vapour conductance, and this is important because insufficient or excessive water loss is detrimental to normal embryonic development (Lundy 1969; Rahn et al. 1977, 1979).

Biliverdin is also a product of haem and is synthesized in the shell gland (Zhao et al. 2006; Wang et al. 2011c). Biliverdin may have some antioxidant properties (Kaur et al. 2003) leading to the hypothesis that it may be costly for females to deposit this pigment onto their eggs (Moreno and Osorno 2003). Some studies have demonstrated positive relationships between biliverdin-based eggshell colouration and egg characteristics, female and/or chick body condition and immune capacity (López-Rull et al. 2008; Moreno et al. 2008b), although other studies have failed to find significant relationships (Cassey et al. 2008; Krištofík et al. 2013).

Although much research has been conducted on eggshells and their pigments, the majority is derived from work on poultry species and, therefore, the focus has been driven by commercial pressures imposed by the human food market (e.g. resistance to breakage during transportation, taste, visual perception). There are two major shortcomings associated with studies of the functional significance of protoporphyrin-pigmented eggs. First, the lack of understanding of the synthesis and mobilization of protoporphyrin (Sparks 2011). Secondly, the failure to recognize the distinction between patterns of protoporphyrin deposition, i.e. maculation, and the absolute content of such pigments in eggshells (Brulez et al. 2014a). Furthermore, eggshell maculation has commonly been quantified through the use of eggshell pattern scoring methods (Gosler et al. 2005; Sanz and García-Navas 2009; De Coster et al. 2012), a method for which repeatability has been questioned and, therefore, must be used with caution (Brulez et al. 2014b). Future studies must consider these shortcomings carefully before carrying out studies that focus on the modes of synthesis and deposition of protoporphyrin onto eggshells.

11.3 A synthesis of hypotheses for eggshell colouration

The great diversity of avian eggshell pigmentation and its possible adaptive significance has fascinated biologists for a long time (see Figure 11.1). Poulton (1890) wrote '…any description of colour and marking [of eggs] will be considered incomplete unless supplemented by an account of meaning and importance in the life of the species'. Kennedy and Vevers (1976) examined the appearance and pigment content of eggshells of 108 avian species and found that nearly half had eggshells containing only protoporphyrin with just over a third containing both protoporphyrin and biliverdin IXα. Ancestrally, avian eggshells were most likely homogenously white and immaculate (Wallace 1889; Kilner 2006), but not necessarily devoid of pigments as even some white, immaculate eggs contain traces of eggshell pigments such as in common woodpigeons (*Columba palumbus*; Kennedy and Vevers 1976). The eggs of extinct paleognath species such as upland moas (*Megalapteryx didinus*) have been shown to contain either protoporphyrin or biliverdin or both pigments (Igic et al. 2010). Therefore, it is highly probable that these eggshell pigments are ancient in origin and highly conserved throughout the diverse radiation of the class Aves (Igic et al. 2010).

The functional significance of eggshell colouration has acquired a range of hypotheses (discussed by Underwood and Sealy 2002; Kilner 2006; Cherry and Gosler 2010; Maurer et al. 2011a). The main hypotheses include thermoregulation (McAldowie 1886; Bakken et al. 1978), crypsis to avoid predation (Wallace 1889; Götmark 1993), and brood parasitism (Newton 1896; Davies and Brooke 1989a). The sexual-selection for the evolution of eggshell colouration (SSEC) hypothesis proposed that pigmentation is based upon females signalling their high body condition to males and thereby inducing more investment by males during breeding attempts (Moreno and Osorno 2003; but see Reynolds et al. 2009). Finally, the structural-function hypothesis (Gosler et al. 2005) proposed that protoporphyrin pigmentation may increase shell strength by acting as a shock absorber (Solomon 1991). These hypotheses are explained in more detail below.

11.3.1 Thermoregulation and gas conductance

Incubation must occur under suitable microclimatic physical conditions within the nest such as appropriate temperatures and water vapour pressures (Ar and Sidis 2002) to ensure that embryos develop and hatch successfully (Drent 1975). Prior to incubation, eggs should neither be too cold to kill the embryo nor too hot to initiate incubation prematurely (Webb 1987; Deeming and Ferguson 1991b). During incubation, eggs vary in core temperature between 34 and 38°C with temperatures above 40°C placing embryos at risk of mortality from heat stress (Bennett and Dawson 1979; Burley and Vadhera 1989). Light-coloured eggs may reflect sunlight and protect eggs from overheating. More than half of the sunlight that falls on eggshells is in the near-infrared (IR) portion of the spectrum. Both protoporphyrin and biliverdin-pigmented eggs reflect more than 90% of light in the near-IR, minimizing heating of the egg by the sun (Bakken et al. 1978). As both of these pigments have low absorbance in this near-IR range of wavelengths, they are unlikely to influence heat gain differentially (Bakken et al. 1978).

By reflecting incident sunlight, lightly pigmented eggs could protect embryos from hyperthermia when adults are away from the nest. Studies using artificially coloured eggs suggest that light-coloured eggs heat more slowly than darker ones (Montevecchi 1976; Bertram and Burger 1981). However, the use of artificially coloured eggs to study thermoregulation can be problematic as these artificial pigments (e.g. paint) likely do not exhibit the same thermo-reflective properties as natural pigments. Naturally pigmented eggs exposed to full sunlight acquired heat more rapidly than eggs in the shade but the rate at which they gained heat did not vary with eggshell colour in either environment (Westmoreland et al. 2007). The authors concluded that differences in reflectivity of eggshell pigments in the visible range (400–700 nm) do not result in different rates of heat acquisition, and therefore did not support the thermoregulation hypothesis.

During incubation, the typical egg loses 15% of its initial mass (Ar and Rahn 1980) due to water loss, which occurs by diffusion of water vapour across the eggshell (Wangensteen and Rahn 1971). Total rate of water loss is determined both by intrinsic properties of the egg (eggshell thickness, egg size, pore density, pore diameter) and micro-environmental factors (e.g. temperature and ventilation of incubating female; Rahn et al. 1979).

11.3.2 Crypsis

It has long been believed that ancestral eggs were white and that all other forms of egg colour and patterning are adaptations to specific micro-environments and function to conceal eggs from predators (Wallace 1889). Amongst the Turdidae, the location of the nest explains some of the variation in their colour and patterning on eggshells. Hole-nesting species are more likely to lay immaculate white eggs while 80% of Turdidae species which have exposed nests lay eggs covered in red or brown speckling (Lack 1958a).

Many studies, through the use of artificially pigmented eggshells, have found no significant difference in predation rates between conspicuous eggs and those mimicking the natural appearance (e.g. Tinbergen et al. 1962; Montevecchi 1976; but see Underwood and Sealy 2002). However, studies using naturally pigmented eggs have found that egg colouration has a significant negative effect on predation rates and hence on nestling survival (Westmoreland and Kiltie 2007; Westmoreland 2008). In the South American tern (*Sterna hirundinacea*), a species that lays eggs with a large variation in background colour, green eggs were depredated less than other colour variants in areas where only mammalian predators, e.g. little grisons (*Galictis cuja* [Mustelidae]), were present. However, in areas where only avian predators, e.g. kelp gulls (*Larus dominicanus*), were present, the rate of artificial nest predation was higher for eggs more conspicuous to the human (*Homo sapiens*) eye than for eggs resembling the nest substrate (Blanco and Bertellotti 2002). In Japanese quail (*Coturnix japonica*), laying females consistently select substrates that maximize camouflage with the degree of maculation of their eggs (Lovell et al. 2013; Figure 11.2). This demonstrates that the selection for crypsis under fluctuating environmental conditions

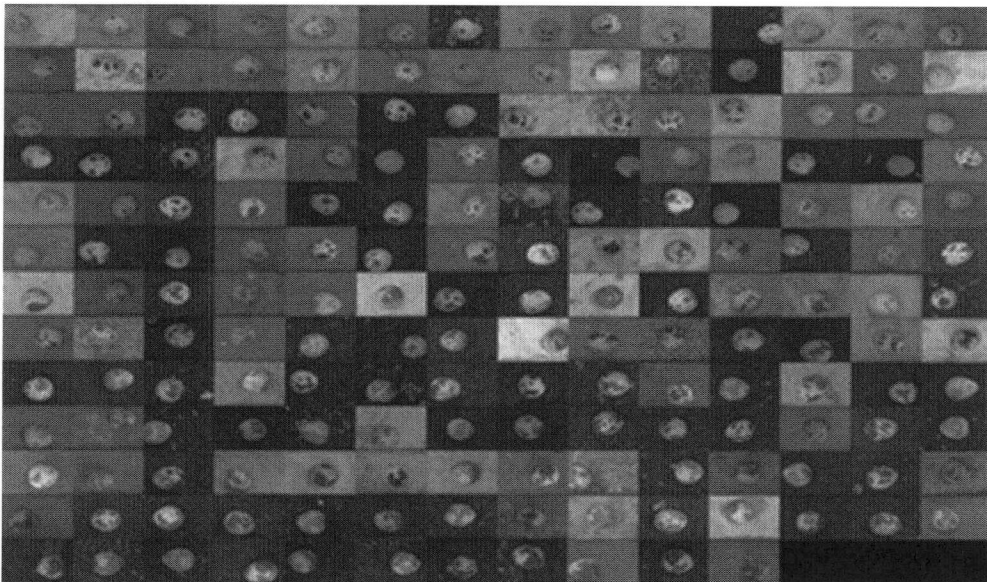

Figure 11.2 Japanese quail (*Coturnix japonica*) females consistently choose laying substrates which best camouflage their eggs and vary the type of camouflage depending on the degree of maculation of the eggs. Heavily maculated eggs were laid on substrates matching the colour properties of the eggs (camouflage through disruptive colouration). Lightly maculated eggs were laid on substrates which best matched egg background colour (camouflage through background matching). Photos are arranged according to level of eggshell maculation, starting with least maculated top left to most maculated bottom right. (Image: P.G. Lovell, G.D. Ruxton, K.V. Langridge and K.A. Spencer. See Plate 14.)

(e.g. variation in predation pressure and nest substrate availability) may be the main evolutionary force driving variation in eggshell colouration (Blanco and Bertellotti 2002).

11.3.3 Egg recognition/brood parasitism

Many species that are potential victims of brood parasites protect themselves by evolving the ability to recognize and reject eggs added to their clutches which are unlike their own (Newton 1896; Davies 2000). Brood parasites may lay eggs in conspecifics' nests (Yom-Tov 2001) or in those of heterospecifics, thereby transferring the costs of parental care to their chosen hosts (Rothstein 1990a; Davies 2000). Unlike facultative brood parasitic species such as black-billed cuckoos (*Coccyzus erythropthalmus*) which are able to provide nests and parental care to their own young, it is imperative for eggs laid by obligate brood parasites such as common cuckoos (*Cuculus canorus*), those species which require the nest and parental care of host species in order to breed successfully, not to be recognized as parasitic eggs by host species. The incurred cost to the host as a result of parasitism has initiated a co-evolutionary arms race between the brood parasite and the host; the latter evolves mechanisms to avoid being parasitized while the former evolves the ability to exploit these newly evolved strategies (Davies and Brooke 1989b; Davies 2000). Strategies to avoid being parasitized can include within-clutch uniformity of egg appearance (Davies and Brooke 1989b), selection for odd-looking last-laid eggs in a clutch (Yom-Tov 1980), recognition of one's own eggs (Victoria 1972; Rothstein 1975), increased nest vigilance (Neudorf and Sealy 1994), and host aggression towards potential brood parasites (Robertson and Norman 1976).

Many species of birds have evolved within-clutch uniformity as well as individual distinctiveness in egg colour and spotting, a combination that facilitates distinguishing between an individual's own eggs and those of a conspecific or a brood parasite (Baker 1913; Davies and Brooke 1989b). Eggshell patterning is genetically female sex-linked in great tits (*Parus major*) breeding in Wytham Woods, Oxford, UK, and is inherited from mother to daughter (Gosler et al. 2000). This allows for the evolution of individual-specific eggshell patterning. Although this has so far only been verified in great tits, it is likely that this hypothesis can be applied to other bird species laying maculated eggshells, although Mahler et al. (2008) found no evidence to suggest eggshell spotting was maternally inherited

in shiny cowbirds (*Molothrus bonariensis*). Nevertheless, it is wise to remember that the findings from Wytham Woods were based on a population that contains almost entirely known, i.e. ringed, birds and has a pedigree that is well-resolved (Savill et al. 2011). Few such longitudinal studies exist to allow the broader applicability of these findings to be tested. Much remains to be investigated, however, because the great tit population of Wytham Woods shows no evidence of brood parasitism from common cuckoos in their recent evolutionary past (Gosler et al. 2000).

The ability to recognize an individual's own eggs has been shown in some studies (e.g. Rothstein 1975, 1978; Bertram 1979; Pike 2011; Yang et al. 2014), but not in others (e.g. Sealy and Lorenzana 1997; Cassey et al. 2009) and so may have evolved in response to specific selective pressures. Some species breeding in colonies, for example, are able to distinguish a conspecific's eggs from their own (Bartholomew and Howell 1964; Buckley and Buckley 1972; Schaffner 1990). Female American coots (*Fulica americana*) combine egg recognition and counting to make decisions based on clutch size, thereby avoiding not only costs of incubating parasitic eggs but also those from wrongly discarding their own eggs (Lyon 2003). Pattern recognition algorithms have shown that species whose eggshell patterns have been mimicked most accurately by brood parasites have also evolved the most recognizable egg patterns (Stoddard et al. 2014).

11.3.4 The sexual-selection for the evolution of eggshell colouration (SSEC) hypothesis

The SSEC hypothesis proposes that egg colour (with an emphasis on blue-green coloured eggs) acts as a sexually-selected trait in females to display their genetic and phenotypic qualities to males as a post-mating selection mechanism (Moreno and Osorno 2003). Female ornaments such as ultraviolet (UV) head colour in female blue tits (*Cyanistes caeruleus*) and tail length in female barn swallows (*Hirundo rustica*; Amundsen 2000 and references therein; Hill 2002), are important in mate choice and paternal care. Moreno and Osorno (2003) argued that the deposition of biliverdin in eggshells may be costly, thereby affecting a female's body condition during an exceptionally stressful phase, i.e. egg laying, and thus acting as an honest signal (Zahavi 1975). Subsequently, males may use this signal to assess a female's body condition and genetic quality through signals reflecting antioxidant capacity, and provide paternal care accordingly.

Relationships have been found between blue-green colouration (BGC) and female body condition at laying (Moreno et al. 2006a), immunocompetence of adults during the nestling period (Moreno et al. 2005), and plasma antioxidant levels (Hanley et al. 2008; Morales et al. 2008). Other studies have found relationships between eggshell colour and nestling health and body condition (Morales et al. 2006; López-Rull et al. 2008; Soler et al. 2008b; but see Stoddard et al. 2012). However, experimental support for a robust association between eggshell colouration and male provisioning effort is mixed (reviewed in Reynolds et al. 2009; Riehl 2011). Blue-green colouration has been positively linked to male parental care in some studies (Soler et al. 2005; Moreno et al. 2006b) but not in others (Krist and Grim 2007; Johnsen et al. 2011). Furthermore, distinguishing between nestling and parental quality can be challenging and must not be ignored when interpreting the results of the aforementioned studies of the association between eggshell colour and nestling health.

The SSEC hypothesis could be extended to include protoporphyrin-pigmented eggs because the accumulation of protoporphyrin within the liver may cause oxidative stress (Afonso et al. 1999; but see Zhao et al 2006). The deposition of large amounts of protoporphyrin onto the eggshell may indicate a female's ability to cope with intense oxidative stress, or alternatively, may indicate that protoporphyrins have been successfully removed (Moreno and Osorno 2003; Holveck et al. 2010). However, Zhao et al. (2006) found that eggshell pigment content was unrelated to that in bile, blood serum, or excreta. Increased protoporphyrin deposition is found in females in lower body condition in some species such as the Japanese quail (Duval et al. 2013), but not in others such as the Eurasian reed warbler (*Acrocephalus scirpaceus*; Krištofík et al. 2013). A study of the northern lapwing (*Vanellus vanellus*) did not show any relationship between protoporphyrin deposition and increased male parental care (Bulla et al. 2012).

11.3.5 Structural-function hypothesis

The molecular structure of porphyrin, the precursor to protoporphyrin, is similar to that of phthalocyanines which are used in solid-state engineering as lubricants, suggesting that protoporphyrin pigmentation may increase shell strength by acting as a shock absorber (Solomon 1991). This led to the structural-function hypothesis proposing that protoporphyrin is deposited into the eggshell for structural strengthening when exogenous calcium is scarce (Gosler et al. 2005). Eggshell pigment spots have greater fracture toughness than unpigmented shell, compensating for reduced eggshell thickness and increasing shell strength (Gosler et al. 2011).

In great tits, protoporphyrin-pigmented spots have been found to demarcate thinner areas of the shell, with darker spots covering thinner areas than paler spots (Gosler et al. 2005), and eggshell calcium content was strongly positively related to local soil calcium content (Gosler et al. 2005). Similar results were found in the Eurasian sparrowhawk (*Accipiter nisus*) breeding in areas contaminated with DDE (dichlorodiphenyldichloroethylene, a metabolite of dichlorodiphenyltrichloroethane, DDT), a pollutant which blocks calcium uptake by the shell gland (Porter and Wiemeyer 1969; Longcore et al. 1971). Shell thinning was positively correlated with DDE concentration and to a greater extent with (internalized) pigmented spots (Jagannath et al. 2008).

11.3.6 Alternative hypotheses

Additional hypotheses addressing the functional significance of eggshell colouration include the blackmail hypothesis of Hanley et al. (2010). This proposed that sexual conflict load (Houston et al. 2005) may be imposed onto males if females lay conspicuous eggs, forcing males to increase parental care. While the appearance of eggs and nests is used as cues by predators and brood parasites, this hypothesis suggests that eggshell cues may have evolved to act as a signal to force males to increase parental care to offset the increased risks due to egg conspicuousness, for example, by increasing male incubation effort, nest vigilance and/ or mate-feeding behaviour. The anaemia hypothesis of De Coster et al. (2012) proposed that as protoporphyrin is derived from blood, an increase in anaemia would result in eggshells with reduced protoporphyrin pigmentation. Moreover, it has further been suggested that pigment spots may act as a defence against bacteria. Protoporphyrin, if activated by light, can reduce bacterial survival on eggshells (Ishikawa et al. 2010; Chapter 7), large quantities of which can increase risk of mortality to the developing embryo (Cook et al. 2003, 2005b).

11.4 Eggs as visual stimuli

For eggs to act as a signal, they must convey information about the quality of the sender to the receiver. This information could signal the quality of the egg, the chick within, or even the quality of the female who laid the eggs. Here, we discuss how eggshell colour, size, and shape can convey information about both egg and nestling quality to a range of receivers.

11.4.1 Does eggshell colouration signal egg quality?

In poultry science, much research has been invested into whether measurement of eggshell colour can be used as a non-destructive method to determine egg quality and has been related to eggshell thickness (Ingram and Homan 2008), eggshell strength (Yang et al. 2009), and hatchability (Richards and Deeming 2001; Baggott et al. 2003; Krystianiak et al. 2005). Eggshell colour is a useful way to monitor stress of laying females in domestic fowl (Mertens et al. 2010). Female Japanese quails are able to regulate eggshell pigment (both protoporphyrin and biliverdin) quantities in response to their body condition while maintaining consistent eggshell colouration. This is believed to optimize camouflage of eggs against the nest substrate and so limit visible changes detectable by predators (Duval et al. 2013).

Eggshells of atypical colouration may possess defective infrastructures which could disrupt their water vapour conductance. Parents may respond to this cue by adapting the incubation environment to accommodate these abnormalities (Higham and Gosler 2006; but see Deeming 2011c). This assumes that females are able to detect such abnormalities in egg structures by inspecting eggs post-laying. Although parents do not actively respond to changes in humidity, they have been shown to modify nest construction behaviour in response to environmental conditions (reviewed in Deeming 2011c). It has been suggested that protoporphyrin pigment spots may either decrease evaporation due to protoporphyrin reflecting strongly in the IR (Bakken et al. 1978), creating 'cold-spots' on the eggshell (Gosler et al. 2005), or directly reduce the permeability of the eggshell by physically blocking pathways for gaseous exchange in great tits (Higham and Gosler 2006) and in blue tits (Sanz and García-Navas 2009). However, Deeming (2011c) argued that the findings from great tits were flawed because of incorrect calculations relating to rates of mass loss and water vapour conductance. Tests under standardized laboratory conditions using eggshells of common black-headed gulls (*Larus ridibundus*) found no difference in water vapour conductance between spotted and un-spotted parts of the eggshell (Maurer et al. 2011a).

Few studies have been conducted on free-living species in this regard and those that exist have found little correlation between eggshell colour and egg quality. In house wrens (*Troglodytes aedon*), eggs which were brighter, i.e. less brown, were heavier and nestlings hatching from brighter eggs were fed at a higher frequency by mothers (Walters and Getty 2010). In collared flycatchers (*Ficedula albicollis*), eggshell BGC and UV colours were not associated with egg volume, eggshell thickness, or pore density (Hargitai et al. 2011). Alternatively, eggshell colour could indicate egg quality by reflecting an egg's anti-microbial properties (Ishikawa et al. 2010; Chapter 7). Eurasian hoopoes (*Upupa epops*) deposit uropygial secretions, associated with anti-microbial activity, onto eggs during incubation, gradually changing their colour from bluish-grey to greenish-brown (Soler et al. 2014). This colour change could be used by males to assess female investment into the egg during incubation and in return allow them to modulate their post-mating care. However, Cassey (2009) states that subtle shifts in hue are not detectable within a cavity nest. The relationship between eggshell colouration and egg quality requires much more empirical research addressing free-living species if we are to make significant inroads in our understanding.

11.4.2 Does eggshell colouration signal nestling quality?

Several studies have been published in this subject area with the amount of BGC of eggshells, produced by biliverdin, reflecting concentrations of maternal antibodies within the yolk (Morales et al. 2006), as well as yolk testosterone (López-Rull et al. 2008). Both yolk testosterone (e.g. Navara et al. 2006) and antibodies (Müller et al. 2007b) are crucial for offspring growth and fledgling success (see Chapter 10). However, experimentally removing the outer layer of pigment (mostly protoporphyrin IX with low concentrations of biliverdin IXα) from eggs laid by common kestrels (*Falco tinnunculus*) did not affect hatching success, nestling body condition, and immunocompetence or the probability of the fledged chick being recruited into the population (Fargallo et al. 2014). Of course, pigment is present throughout eggshells (Sparks 2011; Samiullah and Roberts 2013), and it may be that pigment present within the eggshell structure is more beneficial to the embryo than that on the surface.

The external colour of an eggshell filters the light transmitted through the shell which, depending on the colour, can enhance or protect an embryo's development

Prevent Light

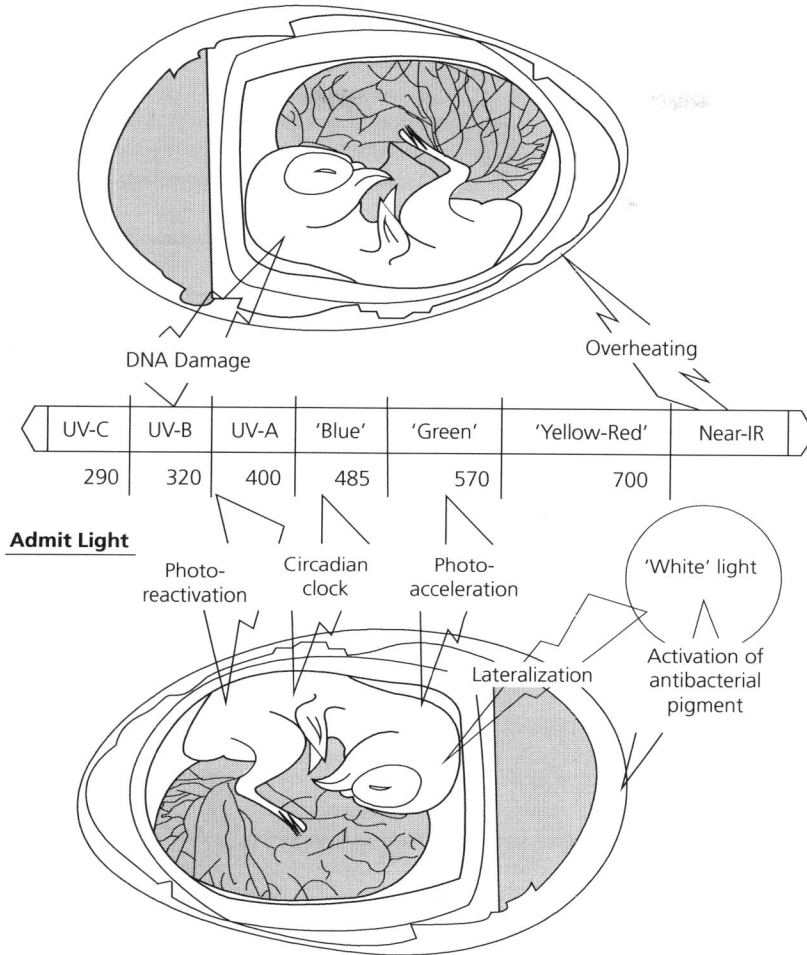

Figure 11.3 Eggshell pigmentation may both protect and enhance the developing embryo by preventing harmful light, i.e. ultraviolet (UV-B) and near infrared (IR), from reaching the embryo and simultaneously allowing beneficial light, i.e. UV-A, blue, green, and white light, wavelengths to enter. Reproduced with permission from Maurer et al. (2011b).

(Coleman and McNabb 1975; Shafey et al. 2004; Maurer et al. 2011b; Figure 11.3). Eggs of cavity nesters tend to be 'brighter' and reflect more in the blue-green spectrum allowing greater light transmission through the eggshell, thereby accelerating embryonic development (Maurer et al. 2011b, 2015). Through a combination of eggshell thickness and colour, eggshells are able to block harmful UV light, potentially protecting the embryo from UV radiation (Maurer et al. 2015). If pigment quantities are limiting, as may be the case if they are required for other (non-signalling) functions, eggshell colouration may be used as a cue by parents about potential

nestling quality, with regards to either embryonic development or exposure to UV radiation.

The SSEC hypothesis predicts that eggshell colouration can be used by males to assess nestling quality and hence allocate parental investment accordingly (Moreno and Osorno 2003). Research has mainly focussed on the relationship between eggshell colouration and parental feeding (Moreno et al. 2004; Sanz and García-Navas 2009) but eggshell colouration might act as a cue in determining rates of courtship feeding, parental incubation behaviour, and intensity of nest vigilance. Bulla et al. (2012) studied bi-parentally incubating northern lapwings and found no significant relationships between eggshell colour and male incubation rates but this and other similar hypotheses with regard to male parental effort should be examined in a wider range of species.

11.4.3 Does egg size signal egg and nestling quality?

Egg size can have a positive impact on nestling survival in the first few days after hatching and has been related to accelerated growth, increased competitive ability and survival (reviewed in Williams 1994, 2012 and Chapters 10 and 14). However, it is not egg size as such, but the associated larger quantities of specific egg components such as carotenoids (Royle et al. 2001; Blount et al. 2002), antibodies (Lung et al. 1996), and lipids (Nager et al. 2000) that promote fitness-related traits in embryos and nestlings such as chick growth and survival to fledging (Chapter 10).

A meta-analysis carried out by Krist (2011) of 283 studies of 162 bird species across 18 orders revealed 1,805 correlations between egg size and offspring quality and concluded that overall egg size is positively related to juvenile survival. Correlations were strongest at the hatchling stage but persisted into the post-fledging stage (Krist 2011). However, Krist (2011) concluded that it remains to be established to what extent these relationships are associated with the quality of the nesting environment (e.g. parental post-hatching care, brood size) rather than direct effects of egg size. Measurements of egg size (including egg mass) appear to be strong predictors of hatchling mass (Deeming and Birchard 2007) but are not associated with other traits, such as nestling immunocompetence in tree swallows (*Tachycineta bicolor*; Whittingham et al. 2007), UV chroma of nestling plumage in eastern bluebirds (*Sialia sialis*; Robinson et al. 2014c), and egg sexual-size dimorphism (Rutkowska et al. 2014), in which eggs containing embryos of a certain sex, usually male, are larger than their counterparts. In red-winged blackbirds (*Agelaius phoeniceus*), an asynchronously hatching species, egg mass was positively related to increased hatchling mass as well as growth rate in both 'core', i.e. those that hatch first, and 'marginal' nestlings, i.e. those that hatch last, but influenced survival through increased body mass only of the latter (Forbes and Wiebe 2010).

Egg size is traditionally considered as an investment by the female rather than as a signal but it may be used as a cue by the male to assess his contribution to parental investment. Males may respond to these cues by increasing mate-guarding and courtship feeding during the egg-laying and incubation stages or could increase nest vigilance and feeding rates during the nestling stage. In glaucous-winged gulls (*Larus glaucescens*), courtship feeding of females by males was positively related to the relative size of the usually smaller third-laid (or 'terminal') egg (Salzer and Larkin 1990). Sexual dimorphism in egg size has been shown in many avian species (Mead et al. 1987; Anderson et al. 1997; but see Rutkowska et al. 2014), but it is not believed that parental reproductive investment is adjusted to either promote or compensate for these differences (Blanco et al. 2003). Nonetheless, egg size has been used as a cue by parents to detect interspecific brood parasitism (Rothstein 1982) and, therefore, has the potential to provide additional cues that may shape reproductive biology.

11.4.4 Does egg shape signal egg and nestling quality?

A spherical egg provides the most resistance to external compressive forces of any egg shape (Bain 1991). It also provides optimal gas exchange between the developing embryo and the nest microclimate (Ar et al. 1974) and is the most conservative in the requirement for shell materials (Gosler et al. 2005). However, despite such theoretical resource and functional considerations, studies have struggled to relate egg shape to fitness parameters of birds. For example, diversity in egg shape was not related to hatching or breeding success of common blackbirds (*Turdus merula*) or great tits (Encabo et al. 2001).

Egg shape may be used as a cue by host to host species to reject eggs laid by brood parasites (Zölei et al. 2012); however, this may not be in direct response to differences in egg shape, but an act of cleaning and sanitizing the nest in which 'foreign' objects are removed (Moskát et al. 2003; Guigueno and Sealy 2012). Egg shape cannot be readily exploited by brood parasites compared with other traits perhaps because of the physiological constraints of egg formation that would mediate changes in egg shape. Of course, there is also a high probability of egg rejection by host species if egg shape is too aberrant (Stoddard and Stevens 2011). There is the possibility that egg shape has evolved due to the physical requirements of incubation rather than as a signal. Optimal egg shape may depend on clutch size for optimal fit under the incubating parent, with eggs in clutches of greater than seven being the most spherical (Barta and Székely 1997; but see Encabo et al. 2001 and Deeming and Ruta 2014).

The structural components associated with egg shape could influence how colouration is perceived by avian observers. Colour is not merely determined by the amount of pigment present but also depends on the structure of the object, its reflective properties (Vevers 1982), and the way in which light is transmitted through it (Prum and Torres 2003). Plumage of many passerines is reflected as white due to the keratin-based structure of feathers and this may alter the visible colouration produced by the pigments therein (Jacot et al. 2010). Furthermore, the UV-blue colour of plumage in some species such as eastern bluebirds is largely due to the composition, size (Prum et al. 1998, 2003), and shape (Shawkey et al. 2003b) of the keratin rods and air spaces in the 'spongy' medullary layer of the feather barbs. It may be possible that the composition, size and shape of the nanostructures (e.g. calcium crystals, pigment molecules) within the eggshell modify chromatic and achromatic colouration. Studies have found negative correlations between pigment concentration (or associated eggshell colouration) and UV chroma (Moreno et al. 2005). Hargitai et al. (2011) concluded that UV reflectance of the shell of collared flycatchers could be due to nanostructural characteristics of the eggshell, i.e. the structure of the calcified layer, or of the eggshell pigments within. Recently, it has been shown that the glossy appearance of the eggs of four tinamou species (*Tinamus major, Eudromia elegans, Nothoprocta perdicaria,* and *Nothura maculosa*) is caused by the surface smoothness created through nanostructural mechanisms within the eggshell cuticle, modifying the underlying background colour (Igic et al. 2015).

Egg shape is known to be variable in some species, e.g. guira cuckoos (*Guira guira*; Cariello et al. 2004) often changing with laying order (black-legged kittiwake [*Rissa tridactyla*]; Coulson 1963). This could be a physiological strategy to maintain eggshell colouration as pigment reserves are depleted during the course of laying. Further research needs to focus on the possible structural colours of eggs and to what extent, if any, females are able to control egg shape. We also need further information about the nanostructures within the eggshell and how they maintain or enhance eggshell colouration for signalling functions when pigment-based colouration is limited or costly to produce.

11.5 What role do other sensory modalities play in egg signalling?

The study of egg signalling has focussed overwhelmingly on visual signals. This is not particularly surprising given the variety of egg colours, shapes, sizes, and patterns of maculation (Figure 11.1), and the relative ease with which these characteristics can be identified and quantified. However, it has been known for several decades that eggs (or more specifically the embryos they contain) can also provide non-visual information in the form of auditory stimuli (Vince 1969), and, more recently, it has also been suggested that they may be a source of chemical stimuli (Ar et al. 2000, 2004; Webster et al. 2015). Avian embryos can perceive and actively respond to their environment during incubation in order to adjust their own development, which may have evolved in response to fitness costs incurred by the embryo when parents favour investment in future over current reproduction (Reed and Clark 2011). The following sections outline how and why embryos may signal their needs to both other embryos within the clutch, and to incubating parents. It should be noted that our understanding of non-visual signals stems from work on a very limited range of species; investigating the generality of these processes, especially in wild birds, and exploring phylogenetic patterns should be a priority of future research.

11.5.1 Eggs as auditory stimuli

Auditory stimuli, including clicks, vocalizations, bill tapping, breathing, heartbeats, limb movements, head-lifting, and beak-clapping, are all known to be produced by embryos during the latter stages of development (Vince 1969), and have been observed across a range of altricial and precocial species (Vince 1966). Many embryonic sounds are almost certainly incidental, with no known functional significance. Clicks, i.e. sharp, metallic sounds, probably caused by air passing over the syrinx (Forsythe 1971), breathing and vocalizations, which are distinct from clicks, their form being under active control (Gottlieb and Vandenbergh 1968), have all been implicated in both embryo-embryo and embryo-parent communication (Brua 2002). Vocalizations are particularly interesting since the embryos of some species are capable of producing a number of distinct sounds (e.g. Gottlieb and Vandenbergh 1968) but we have a poor understanding of their functional significance, and this warrants further investigation.

Communication between embryos plays a role in the maintenance of hatching synchrony in which all the eggs of a clutch hatch synchronously despite incubation and embryonic development beginning prior to the completion of the clutch (Vince 1972; Schwagmeyer et al. 1991). Synchronous hatching is likely to provide benefits to both parents and embryos by coordinating timing of fledging, reducing the chance of predation, or creating an earlier hatch date (Lack 1968; Clark and Wilson 1981; Flint et al. 1994). However, for synchronous hatching to occur, embryonic development in later-laid eggs must be accelerated or retarded in eggs laid earlier in the clutch, or possibly both. Auditory stimuli may play a key role in this process (Vince 1969), although the precise mechanisms are still unclear. In particular, it appears that hatch synchronization can only occur if the eggs are in physical contact with each other during the last 2–3 days of incubation, at least in laboratory-based studies (Vince 1969; Woolf et al. 1976), suggesting that information may be (exclusively or additionally) transmitted between embryos using mechanical stimulation. Concomitant changes in parental attentiveness may also be important (Schwagmeyer et al. 1991), but have received little research attention.

Communication between embryos and parents is most likely to involve vocalizations during the final days of incubation, although parents may also respond to other auditory stimuli (Brua 2002). Several hypotheses have been proposed to explain the purpose of these vocalizations, the information that is being transmitted, and the benefits gained by both parties. Most studies have assumed that embryos are signalling their impending hatch. The traditional explanation is that this signals to parents the need to switch from incubation to brooding behaviours, which may include restructuring the nest, reducing aggression and changing foraging patterns, although the support for this is largely anecdotal. An alternative possibility is that embryos are signalling their viability: vocalizations may therefore act as a strategy to prevent abandonment, or to solicit care (e.g. additional incubation, increased egg turning) from parents whose attention may be directed towards recently hatched chicks (reviewed in Brua 2002).

It has also been suggested that vocalizations may signal the thermal status of the embryo. Since embryos are essentially unable to regulate their own temperature, they are almost entirely reliant on the incubating parent to meet their thermal requirements. Auditory signals, including the frequency of embryonic vocalizations (e.g. Evans 1988; Evans et al. 1994; Brua et al. 1996) have been linked to temperature in a variety of species and may act as a signal to parents to take corrective measures (Brua 2002).

11.5.2 Eggs as chemical stimuli

Birds' eggs are known to release volatile compounds. For example, stoats (*Mustela erminea*) are strongly attracted to the odour of eggs of domestic fowl and Japanese quail (Spurr 1999). However, the mode of action of these compounds and their ecological significance have been almost completely unexplored. One possible role for egg odours may be as a component of parent-embryo and embryo-embryo interactions. Auditory and mechanical stimuli would not be a possible means of communication during the majority of incubation. Instead, volatile chemicals released through the porous eggshell may provide information on the developing embryo, which could influence both parental behaviours and development of other eggs in the clutch. For example, nitric oxide (NO) is produced by developing embryos and emitted at low concentrations (a few parts per billion) from the eggs of several species (Ar et al. 2000, 2004). It has been suggested that NO plays a role in mediating brood-patch development and so may allow the embryo to have some control over its thermal environment. Ar et al. (2000) found that NO emission from eggs was reduced by hypothermia. This is a particularly intriguing mechanism which warrants wider taxonomic investigation, and further work on its adaptive significance.

More recently, gas chromatography-mass spectrometry of volatiles emitted from Japanese quail eggs has shown that there are measurable differences in odour profiles of individual eggs. Composite measures of all the volatile compounds contribute to an egg's unique odour, between fertile and infertile eggs, and also between eggs containing male and female embryos (Webster et al. 2015). These differences are present midway through incubation, and so substantially before the development of auditory stimuli. It is unclear at present whether parents or neighbouring embryos can detect these odours, and the behavioural responses of parents to variation in egg odour remain to be established. However, if egg odours can be detected, it leads to the intriguing possibility that parents may be able to glean further ecologically relevant information from egg volatiles long before this information is available via other sensory modalities. For example, this may include embryonic health (perhaps through volatile markers of oxidative stress caused by the breakdown of lipids by rapidly developing embryos [Surai 2002], or odours resulting from microbial infection) and growth rate. Egg temperature may be signalled through higher concentrations of volatiles being emitted at higher temperatures, while the recognition of eggs from inter- and intraspecific brood parasites may be possible if, for example, eggs exhibit individual- or species-specific odour profiles. Odours may also be used to solicit parental care and allow assessment of maternal quality (Moreno and Osorno 2003).

The lack of progress in this area stems in a large part from our historical under-appreciation of the avian olfactory sense. This is slowly being overturned with recent work showing that olfaction is a well-developed and ecologically important trait in many bird species (Balthazart and Taziaux 2009; Campagna et al. 2012). As such, this almost completely unstudied area is ripe for future research.

11.6 How does the nesting environment affect egg signalling?

The ways in which eggs convey signals are, of course, highly dependent on the nest type and mating system of the focal species. When discussing egg colouration as a visual signal, we must consider the environment in which the receiver is operating. For eggs to function as visual stimuli, the receiver must be able to perceive the egg's signals. Closed- and cavity-nests are poorly illuminated and the difference in light levels between a typical nestbox and normal daylight exceeds 1,000-fold (Reynolds et al. 2009; Figure 11.4). Nests in natural cavities only receive approximately 1% of incoming light in nests of great tits and marsh tits (*Poecile palustris*; Wesołowski and Maziarz 2012) and approximately 4% in nests of collared flycatchers (Maziarz and Wesołowski 2014). The difference in light conditions between ambient daylight levels and those within a cavity nest are considerable and could severely affect vision, especially colour discrimination, for the bird entering the nest cavity (Cassey 2009; Figure 11.4). This begs the question, are any chromatic eggshell signals visible under low light conditions experienced by closed-nesting species?

Studies have shown that nestlings in cavities tend to have brighter mouthparts allowing them to be more visible under low light conditions (Ficken 1965; Kilner 1999). Similarly, eggs in cavities tend to be white, although this may be a response to a lack of predation (Lack 1958a; Kilner 2006) rather than increasing their visibility to adults in attendance. Using models of avian colour vision based on blue tits, a cavity-nesting species, Hölveck et al. (2010) found that it was easier to discriminate between eggshells based on brightness rather than on colour. A phylogenetically controlled study of 98 species of European passerines further showed that eggs laid by hole-nesting species had higher UV reflectance than those laid by open-nesting species (Avilés et al. 2006). However, a further study showed that it is the spectral properties of BGC, rather than UV, of eggshells that allow parents to locate eggs in dark nesting environments (Węgrzyn et al. 2011). Blue-green colouration is the last visible under low-light conditions. Furthermore, brood parasite hosts reject parasitic eggs at more successful rates when common cuckoo eggs of poor mimicry are laid in nests positioned under brightly illuminated conditions (Honza et al. 2011).

Even if individuals are able to observe and receive signals based upon eggshell colour, it is still necessary for the 'target' egg to be compared to others in order to evaluate the strength of the signal. This comparison could be with a bird's own eggs (e.g. for recognizing parasitic eggs), either directly or with a learned template from previous breeding attempts (Lotem et al. 1992), or with eggs laid by other females (e.g. to assess egg or female quality). The former requires the receiver to be able to observe and remember a template of their

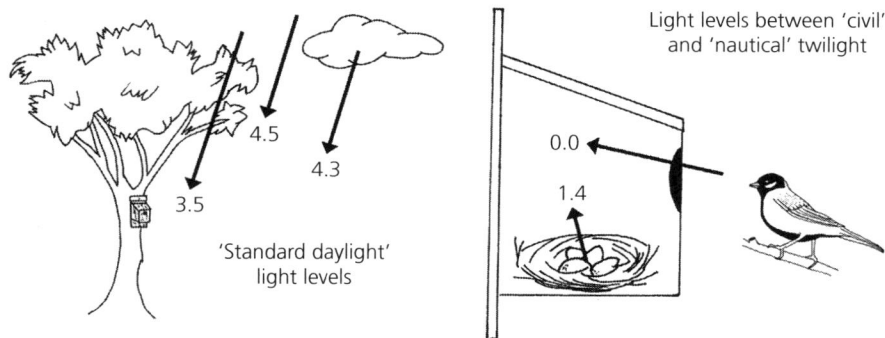

Figure 11.4 Average light levels measured near a standard nestbox on the University of Birmingham campus were 4.5 $\log_{10}$ lux in direct sunlight, 4.3 $\log_{10}$ lux under 75% cloud cover and 3.5 $\log_{10}$ lux under tree canopy (left). These are compared to 1.4 $\log_{10}$ lux from inside the nestbox when the entrance hole is not blocked, i.e. by visiting parent, and 0.0 $\log_{10}$ lux when it is (right). Adapted from Cassey (2009).

own eggs (Lotem et al. 1992), while the latter requires the receiver to visit the egg in question, visit other eggs and nests in order to compare them to the target and remember signals between visits. A comparative study, based upon avian perception models, of eggs laid by species of the superfamily Muscicapoidea predicted that eggs were indistinguishable from those of other clutches of the same species based on background colouration alone (Cassey et al. 2009). The observation and comparison of clutches will most likely only be possible for colonial breeders where eggs are easily visible, or those species which are polygamous where usually the male parent has to make choices about his allocation of parental investment between nests and females. Male blue tits, a polygamous species, were found to visit their own nests while eggs were present and some also visited neighbouring nests, potentially allowing for egg comparison (Holveck et al. 2010).

Some species remember specific outcomes of experience, such as the locations of food caches by Eurasian jays (*Garrulus glandarius*) or nests that have already been parasitized by shiny cowbirds (reviewed in Griffiths et al. 1999). Although females of many species have evolved mechanisms to recognize their own eggs, it remains unclear whether males are able to retain accurate assessments of egg signal strengths between visits. Historically, research has examined egg recognition from the brood parasitism, and specifically the female, perspective. However, males in some species, such as in the Baltimore oriole (*Icterus galbula*; Sealy and Neudorf 1995) and the vinous-throated parrotbill (*Paradoxornis webbianus*; Lee et al. 2005), remove parasitic eggs, suggesting that males can perhaps evolve the ability to remember and compare their own eggs with those of others to determine signal strength. However, much remains to be investigated about the relative strengths of different signals within such study systems.

11.7 What part does the clutch as a whole play in egg signalling?

In many bird species clutch size exceeds one, with more than half of all studied species (Jetz et al. 2008) laying either two or three, i.e. mode of 2.0 and median of 2.8 eggs in a clutch. Therefore, it is likely that observers will take combined cues from multiple eggs within the clutch rather than from individual eggs. For this reason, we contend that consideration of egg signalling should probably focus at the level of the clutch rather than of the egg. Mean clutch size is larger in cavity-nesting species than those nesting in open nests (Jetz et al. 2008). Although this is likely due

to lower predation rates (Lack 1948a; Lima 1987) or limited breeding opportunities (Martin 1993c), within a signalling context a greater clutch size might serve to increase signal strength under low-light conditions. Cavity-nesting species rearing larger broods display greater contrast in brightness between the nestling's skin and the nest background, enhancing nestling detectability by parents (Avilés et al. 2008). It suggests the possibility that females within a species may lay larger clutches to increase egg detectability and hence signal strength. However, it is still early in the formulation of testable hypotheses within this subject area.

To detect eggs laid by brood parasites, host birds may reject foreign eggs which are different from their own through direct comparisons, known as self-referent phenotype matching. This can happen in real time or using a stored template of the hosts' eggs (reviewed in Hauber and Sherman 2001). This involves the comparison of a 'foreign' egg with all others in the clutch and requires females to lay consistently sized, shaped, and coloured eggs. Alternatively, egg arrangement within a clutch may act as a cue to alert host females that their clutch may contain a parasitic egg(s). Both common blackbird and song thrush (*Turdus philomelos*) females that maintained consistent egg arrangements, based on the position and orientation of eggs within the clutch, were more likely to reject foreign eggs (Polačiková et al. 2013). However, experimental manipulations using blackbird clutches found no evidence to support this hypothesis and concluded that females keeping consistent egg arrangements may be more likely to reject parasitic eggs for other correlated reasons (Hanley et al. 2015). Clutch size is further used as a cue to signal the termination of egg laying to the female. In most species, the female relies on the tactile contact between her brood patch and the eggs already laid to trigger a cessation of egg laying (Haywood 1993, 2013). As paternal effort tends to be positively related to clutch size in many species, such as the common starling (*Sturnus vulgaris*; Komdeur et al. 2002), large clutch sizes could be used to force males into greater parental efforts as they may be too large an incubatory task for females alone.

11.8 Can egg signals be used as an environmental cue by humans?

Avian eggs can act as an effective biomonitoring tool (Ormerod and Tyler 1990; van den Steen et al. 2010) due to their high lipid contents that concentrate hydrophobic contaminants (van den Steen et al. 2006). Eggshells in particular are sensitive to persistent organic pollutants, either directly by blocking calcium uptake to the shell gland (Ratcliffe 1970; Lundholm 1997) or indirectly by disrupting the haem biosynthesis pathway and consequently pigment concentrations (Casini et al. 2003). Resident passerine species, such as blue tits and great tits, are particularly effective in monitoring local environmental contamination because of their small territories and foraging areas (Moore 1966; Dauwe et al. 2006).

Understanding eggshell protoporphyrin pigmentation offers the possibility of a non-destructive bio-assay of the health of the egg, the laying female, and/or the environment. For instance, spotting on eggs has been linked to eggshell thickness (Gosler et al. 2005) which is sensitive to environmental calcium availability (Gosler et al. 2005) and pollutants (Eeva and Lehikoinen 1995), and has been identified as an indicator of egg quality (Sanz and García-Navas 2009). Eggshells of Eurasian sparrowhawks with protoporphyrin spots as an internalized layer showed a strong correlation between eggshell DDE content and shell thickness (Jagannath et al. 2008). Egg colour has further been found to be a good predictor of environmental contamination in herring gulls (*Larus argentatus*), although this correlation was found only with UV and blue-green chroma, i.e. biliverdin, but not brown chroma, i.e. protoporphyrin, of eggs (Hanley and Doucet 2012).

Establishing eggshell pigmentation as an environmental monitoring tool would allow us to monitor contaminants such as heavy metals (e.g. lead, cadmium) and their effects. This could be achieved by simply finding nests of free-living birds and documenting pigmentation on their eggs, creating a powerful tool in the monitoring of populations and promoting the effectiveness of conservation actions. In order for this to be successful, however, we must first obtain a much greater understanding of the importance of eggshell colouration as a signal to other bird species before we can take advantage of their signalling function in our own applications.

Acknowledgements

We would like to thank Daniel Hanley for his comments which have greatly improved this chapter. We are grateful to the Natural History Museum, Tring, UK for allowing us to use photographs of their egg collection.

CHAPTER 12

Improvements in our understanding of behaviour during incubation

V. Marasco and K.A. Spencer

12.1 Importance of incubation and overview of incubation behaviours

The understanding of how reproduction and fitness are linked remains a central focus of life-history theory. A key factor is the evolution of different strategies, across different species and among individuals of the same species, in the allocation of efforts of the parents upon breeding. Historically, reproductive effort was thought to be associated with the costs of chick rearing (e.g. Lack 1948a), ignoring the two phases of reproduction required to actually produce the offspring—egg production and incubation (reviewed by Monaghan and Nager 1997). Monaghan and co-workers were the first to demonstrate that the laying of one extra egg by the lesser black-backed gull (*Larus fuscus*) reduced egg quality (Monaghan et al. 1995). Follow-up empirical and experimental work in a variety of bird species has consolidated the idea that the energetic demands of incubation are comparable with, or even higher than, the post-hatching phases during chick provisioning and hence a true understanding of the trade-offs involved in maintaining such behaviours is required (e.g. Ward 1996; Thomson et al. 1998; Visser and Lessells 2001). These fitness costs are crucial for our understanding why parents must optimize the allocation of the energy invested in incubation and the resources invested for their own self-maintenance and survival.

There is a huge diversity in the incubation behavioural strategies across the altricial–precocial spectrum and the classification undertaken by Skutch (1957) remains one of the most detailed (see also Deeming 2002e). Skutch (1957) distinguished between two main incubation patterns: mutualistic sharing by both parents (bi-parental incubation; Williams 1996), or total involvement by one of the parents, the male or the female, for the entire incubation period (andro parental

or gyneparental incubation, respectively; Williams 1996). Shared bi-parental incubation is adopted by approximately 50% of families and is more widespread in non-passerine species (80%) than in passerines; female-only incubation takes place in approximately 37% of families and is more frequent in passerines (62%). Male-only incubation is present in only 6% of families and is almost exclusive to the Paleognathae, such as Struthioniformes and Rheiformes (Skutch 1957; Van Thyne and Berger 1959; Deeming 2002e). Different sub-categories were distinguished within both the bi-parental or uni-parental incubation modes to best capture the interspecific variation that is found, for example, in: 1) the lengths of incubation sessions and recesses, which can vary from a few minutes in female-only incubation of many passerines (Kendeigh 1952; Skutch 1962) to many days in shared incubation of penguins (Williams 1995) and some seabird species (Warham 1990); 2) the gender incubating at night in the bi-parental mode; and 3) taxa where the female is forced to fast such as waterfowl (Afton and Paulus 1992) or is provided with food by its partner, i.e. assisted gyneparental incubation (Williams 1996) as seen in, for example, hornbills and some raptors (Kendeigh 1952; Kemp 1995). Furthermore, Skutch (1976) classified separately: 1) the species in which incubation is carried out by more than two birds at one nest, such as in American bushtits (*Psaltriparus minimus*) or purple swamphens (*Porphyrio porphyrio*); 2) brood parasites, such as the cuckoos (*Cuculus* spp.) in Europe and cowbirds (*Molothrus* spp.) in North America (see Rothstein 1990b for a review on the evolution of brood parasitism); and 3) all the species belonging to the Megapodiidae (Order Galliformes) in which the buried eggs do not require parental heating (Booth and Jones 2002).

A counter-intuitive, but still far from being fully answered question, is why there is large interspecific

Marasco, V. and Spencer, K.A., *Improvements in our understanding of behaviour during incubation*. In: *Nests, Eggs, and Incubation*. Edited by: D.C. Deeming and S.J. Reynolds, Oxford University Press (2015). © Oxford University Press.
DOI 10.1093/acprof:oso/9780198718666.003.00012

variation in incubation rhythms? As incubation is tightly linked to breeding performance (reviewed by Monaghan and Nager 1997), the evolution of different incubation patterns must be a product of selective constraints and life-history strategies. Even more importantly, accumulating empirical work has highlighted that incubation effort can vary widely across individuals, and also between partners (Deeming 2002e), suggesting that intra-brood trade-offs are likely to be much more important than was previously appreciated. Here, we discuss how the resolution of the trade-offs between various components of incubation behaviour are dependent on a variety of both abiotic and biotic factors, as well as interactions between these factors.

12.2 Potential factors affecting incubation behaviour

12.2.1 Abiotic factors

Ambient temperature is commonly thought to influence avian incubation behaviour (e.g. Weathers and Sullivan 1989a; Conway and Martin 2000a, 2000b). For example, Conway and Martin (2000b) found that species with female-only incubation breeding in colder environments can prevent the eggs from cooling below the physiological zero temperature (approximately 26°C; White and Kinney 1974; Drent 1975) by shortening incubation sessions and recesses in comparison with species breeding in warmer environments. Within species, studies under laboratory conditions show that decreased nest or egg temperature increases nest attentiveness and decreases recesses (White and Kinney 1974; Vleck 1981). Interestingly, however, studies of free-living birds show more variable results, with negative, positive, or no correlation between ambient temperature and constancy of incubation. There are also inconsistent results regarding the effects of temperature on other incubation-related behaviours, such as egg turning rates (Walsh et al. 2007; Kwieciński et al. 2009). Some studies suggest a positive relationship between turning rates and ambient temperature, whilst others report no significant relationship at all. Turning of the eggs during incubation, during both day and night even in diurnal species (Shaffer et al. 2014), is paramount in maintaining embryonic development and presenting the chick in the correct position prior to pipping to facilitate successful hatching (Tullett and Deeming 1987; Deeming 1991, 2002e; Wilson et al. 2003). However, there is still some debate as to the

role of abiotic factors in mediating this behaviour. Conway and Martin (2000a) put forward the idea that linear correlation/regression analyses may be inappropriate to examine the links between ambient temperature and incubation as ambient temperature influences both embryonic development and adult metabolism. This conceptual model found support from data for the small orange-crowned warbler (*Vermivora celata*; female-only incubation) in which the lengths of both on- and off-bouts were positively correlated at temperature ranges between 9°C and 26°C, but not at higher and lower temperatures (Conway and Martin 2000a). Similar non-linear patterns have also been identified in the meadow pipit (*Anthus pratensis*, female-only incubation; Kovarik et al. 2009).

These complex relationships may well co-vary and interact with other intrinsic constraints, such as female fat reserves, body condition, physiological status, and also extrinsic ecological factors, such as food availability and the time of the day (Afton 1980; Conway and Martin 2000a; see also section 12.2.2). For example, in the meadow pipit female attentiveness was positively correlated with mate feeding (Halupka 1994). The effects of air temperature on incubation rhythms became apparent only when the influence of the diurnal cycle was filtered out from the data analysis (Kovarik et al. 2009). Additionally, nest temperature can interact with food availability. In northern mockingbirds (*Mimus polyglottos*) manipulation of food availability (by supplementing the incubating females with food) and nest temperature (by placing heaters under their nests) influenced incubation behaviour, but in opposite directions. Food enhanced duration of incubation sessions and enabled females to spend more time in self-maintenance activities (preening), while heat increased the rate of foraging trips and decreased incubation sessions. More importantly, when both food and temperature were increased their effects on incubation schedules counter-balanced each other (Londono et al. 2008).

Rain can also influence incubation behaviour, and probably the effects depend on the severity of the rainfall (Skutch 1962). For example, in northern shovelers (*Anas clypeata*) heavy rain reduced the time spent off the nest (Afton 1980; Cresswell et al. 2003). Other observational studies suggest that open-nesting birds usually show higher nest attentiveness during rain, a common evolved strategy probably to prevent the nest from becoming sodden with eggs cooling as a result (Caldwell and Cornwell 1975; Kovarik et al. 2009). Precipitation has not been shown to affect other behaviours, such as egg turning rates (Walsh et al. 2007).

12.2.2 Biotic factors

Risk of nest predation is an important ecological driver to explain interspecific differences in incubation behaviour (Conway and Martin 2000b; Smith et al. 2012). Conway and Martin (2000b) found that species (female-incubation only) with high nest predation had both long incubation sessions and recesses compared to species with low nest predation. The authors proposed that these different patterns reflect a reproductive strategy adopted to maximize fitness as reducing the daily trips to and from the nest should be associated with a lower probability of attracting the attention of predators (Conway and Martin 2000b; Chapter 5) in a similar manner to that predicted during the post-hatching stages (Skutch 1949; Martin and Ghalambor 1999; Martin et al. 2000). Consistent patterns have been observed within and among ground-nesting shorebird species in which nest predation was positively correlated with the proportion of time the nests were left unattended (Smith et al. 2012). Overall these data suggest that the rate and duration of off-bouts play an important role in determining nest success and that the risk of predation is a key selective pressure contributing to the evolution of behavioural activities in incubating birds, probably acting by reducing conspicuous behaviours and restless movements.

Within species, food availability has been shown to influence the time off the nest and that invested in self-maintenance activities (Cucco and Malacarne 1997; Conway and Martin 2000a; Zimmerling and Ankney 2005). Experiments in which females were supplemented with food showed increased nest attentiveness, shorter incubation periods, and higher hatchability, for example in the blue tit (*Cyanistes caeruleus*; Nilsson and Smith 1988), in the European pied flycatcher (*Ficedula hypoleuca*; Sanz 1996), and in the song sparrow (*Melospiza melodia*; Rastogi et al. 2006). It was suggested that fitness consequences due to food scarcity can be especially severe in species exhibiting gyneparental incubation strategies (Williams 1996), in which the female's hunger levels are thought to determine nest attentiveness (Haftorn 1978; Weathers and Sullivan 1989a; but see Deeming 2002e). Studies addressing this hypothesis showed variable results. For example, in North American passerine species in which only the females incubate, the frequency of mate feeding as well as diet and foraging strategies had a minor contribution in explaining interspecific variation in incubation behaviours (Conway and Martin 2000b). By contrast, a study conducted on a larger scale including both northern hemisphere songbirds (North America) and southern hemisphere species (Australia and New Zealand) reported a strong positive correlation between male incubation feeding and nest attentiveness, and complex interactions were found between incubation feeding rate, nest predation rate, and geographical area (Matysiokova et al. 2011). Despite the importance of these differences, the correlative nature of the data does not allow us to disentangle whether food availability is the causative agent underlying the observed differences in incubation profiles; more experimental work is needed to determine cause and effect.

The stage of incubation is likely to be an important biotic factor influencing nest attentiveness and appears to be dependent on the evolved incubation strategy. For example, in several species of waterfowl (female-only incubation) the proportion of nest attentiveness decreases with incubation time (Afton 1980; Aldrich and Raveling 1983; Yerkes 1998). However, no differences were found in the mallard (*Anas platyrhynchos*; Caldwell and Cornwell 1975), nor in the orange-crowned warbler (Conway and Martin 2000a). In species with bi-parental incubation, nest attentiveness remains generally high throughout the incubation period, but the parental efforts between the incubating adults may differ suggesting that the partners might be able to negotiate their duties and compensate each other (Hawkins 1986; see section 12.4).

There are indications that egg mass and clutch size may also be related to nest attentiveness. A correlative study showed that egg mass and total clutch mass of 421 and 354 species, respectively, correlated significantly with the proportion of nest attentiveness (Deeming 2002e). When the data, however, were split by incubation strategies these factors explained a much larger proportion of variation in species with uni-parental incubation than species with shared incubation (Deeming 2002e). These significant trends have been thought to be linked to changes in egg thermal characteristics, including egg surface temperature and rate of cooling, and with the presence of a brood patch, a structure present in certain species to facilitate heat transfer to the egg (Lea and Klandorf 2002). Birds laying larger eggs (10–100 g) are 'forced' to exhibit high nest attentiveness to prevent rapid cooling of the shell due to the transfer of heat from the surface to the core and to maintain an optimal thermal environment for embryo development (Turner 1987, 1994). Therefore, the thermal conditions of large eggs might have been an important ecological constraint for the evolution of shared-incubation strategies to achieve high nest attentiveness values (Deeming 2002e). Within species, experimental manipulations of egg temperature can alter the behaviour of the incubating bird (Drent 1970; Vleck 1981) and can cause changes in vasodilation of the brood patch (Midtgård et al. 1985). Obviously, egg thermal characteristics will depend on other environmental factors, such as ambient temperature and the degree of insulation of the nest (Conway and Martin 2000b), and also on factors more specifically related to the nest-bird unit. For example, a study of great tits (*Parus major*) showed that during incubation recesses outer eggs within the nest cup are colder and cool faster than the centrally positioned eggs, and importantly, the incubating females are probably able to keep these thermal gradients at optimal values for embryos to develop by moving the outer eggs more frequently than the inner eggs (Boulton and Cassey 2012). They also showed that smaller clutches experienced lower initial incubation temperature and cooled quicker than larger clutches. Hence, also within species, clutch size may be a contributing factor to explain variation in nest attentiveness (Deeming 2002e). Experimental manipulations to enlarge or reduce clutch size would be a suitable way to investigate this hypothesis.

Skutch (1962) suggested that the bird's 'temperament' may be driving incubation behaviour. Today, the recognition in the scientific community of animal personalities offers a more objective framework to test this idea. Personalities, broadly defined as behavioural differences among individuals consistent over time and across different contexts (Sih et al. 2004), have been demonstrated in various vertebrate taxa, including birds (reviewed by Groothuis and Carere 2005). Evidence suggests that such differences could be maintained within populations because of life-history variations. For example, bold individuals have been shown to invest more in reproduction at the expense of their own survival, whereas shy individuals do the opposite (Wolf et al. 2007; Réale et al. 2010). It is therefore possible that at least some components of incubation behaviour may be consistently altered among different personalities. Cole and Quinn (2014) suggested that this might be the case in finding that bold, i.e. fast-exploring, free-living great tits returned faster to nests to incubate eggs after facing an unknown threat than shy, i.e. slow-exploring, females. This translated into bold individuals being more willing to adopt risk-taking strategies and more prone to defend their offspring than the more cautious shy individuals that invested more in their own self-maintenance. It would be interesting to test whether incubation behaviours, such as rate and duration of foraging trips and incubation sessions, differ between bold and shy female great tits and if (and

to what extent) they are related to adult fitness and survival.

Prior reproductive experience or other age-related variables can influence incubation behaviour. Generally, breeding performance of birds tends to increase with age (Harvey et al. 1988; Yerkes 1998), probably as a consequence of more efficient foraging strategies and/or increased reproductive efforts (Curio 1983; Newton 1989). For example, in the southern fulmar (*Fulmarus glacialoides*), adults breeding for the first time lay smaller eggs, and have longer incubation periods and lower hatching success than eggs laid by experienced pairs (Weimerskirch 1990). Age effects in incubation profiles may be more apparent when breeding environmental conditions are poor, i.e. when food is scarce and/or of low quality and weather conditions are severe, or when the social context is unfavourable. For example, young great skuas (*Stercorarius skua*) have reduced breeding success compared to older birds when food is scarce (Hamer and Furness 1991). In the cooperatively breeding Seychelles warbler (*Acrocephalus sechellensis*), breeding performance (including nest quality, incubation periods, and hatching success) improves with age because of experience, but also because of help from the non-breeders (Komdeur 1996). There are also cases, however, in which parental incubation efforts do not seem to be influenced by age or breeding experience, as shown in the savannah sparrows (*Passerculus sandwichensis*; Wheelwright and Beagley 2005). Interestingly, there are some suggestions that early life stress, which can alter many phenotypic traits in later life (reviewed by Schoech et al. 2009 and Henriksen et al. 2011), may be important in determining future investment of adult parents in reproduction, including incubation. An experimental study in a population of captive zebra finches (*Taeniopygia guttata*) found that breeding females, but not males, that experienced stressful conditions as juveniles exhibited reduced incubation efforts compared to control (unstressed) females (Spencer et al. 2010). Of course, we need more studies in different species and especially under wild conditions to test to what extent early-life experience can influence incubation profiles, parental efforts, and possibly fitness outcomes.

In summary, incubation patterns clearly depend on a multitude of factors (Figure 12.1). There is, however, large variation across studies, especially and not surprisingly, in those conducted under natural conditions. It appears that the majority of the studies focussed on discovering the predominant ecological driver of incubation pattern underestimating the importance of the relationships between multiple ecological factors (but see

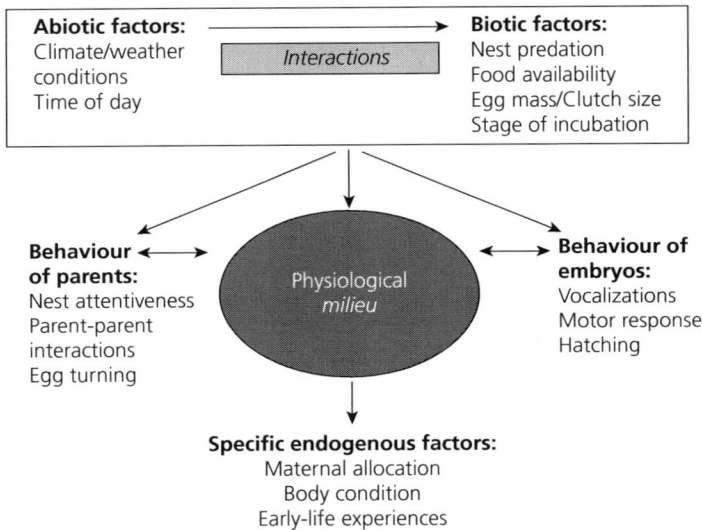

Figure 12.1 Schema illustrating the multitude of ecological factors that influence incubation behaviour of birds.

Londono et al. 2008). Controlled manipulations under captive conditions may be extremely useful to examine the interacting effects of factors such as temperature, food, social context, and parents' prior experience, on relevant components of incubation behaviour. These data could help us to make better predictions of incubation patterns displayed by free-living birds and to make better interpretations of results from field studies. We also need to bear in mind that some incubation behaviours may follow non-linear relationships with environmental and other variables (Conway and Martin 2000b), and may require complex statistical modelling. Recent advances in generalized linear and additive modelling should make it easier to perform statistical analyses of such complex data sets and more effort should be invested by behavioural ecologists to apply these tools.

12.3 Physiological mechanisms mediating incubation behaviour

12.3.1 The importance of the hormonal status of the incubating birds

One of the key factors that can influence incubation behaviour is the hormone *milieu* of the adult bird during reproduction (reviewed by Buntin 1996 and Vleck 1998, 2002). Much of this work has focussed on the hormone prolactin which in birds is usually found at elevated peripheral concentrations in parents during incubation and/or chick rearing (reviewed by Angelier

and Chastel 2009). Seminal studies of domesticated species showed that visual and tactile inputs from the nest, the eggs or the chicks can all induce increases in prolactin secretion (El Halawani et al. 1986; Book et al. 1991). Other stimuli, such as photostimulation and photorefractoriness, can also contribute to raise prolactin (see Sharp et al. 1998 for a review on the hormonal control of prolactin). More recently, a large number of studies have also been accumulated in free-living bird populations that investigate the potential relationships between prolactin and parental efforts in the context of life-history strategies (Vleck 1998; Angelier and Chastel 2009).

Depending on reproductive strategy, the rise in prolactin can be either rapid or slow at the start of egg laying and incubation. For example, in precocial birds plasma prolactin can decline steeply after hatching or remain only moderately high during brooding and drop to a baseline level pre-laying (Criscuolo et al. 2002; Boos et al. 2007). In altricial species, prolactin tends to remain elevated for a longer time after hatching with levels generally comparable to those observed during incubation, and declines to a baseline level only when the young achieve thermal independence (Goldsmith 1991).

Prolactin secretion, however, may also be regulated by an endogenous cycle. For example, in pelagic seabirds, in which the incubating parents spend a long time away from the nest foraging, prolactin levels remain elevated for weeks to months after hatching, possibly to stimulate the continued return of the parents to the colony (Cherel et al. 1994; Vleck et al. 2000; Angelier et al. 2006). Studies of wild turkeys (*Meleagris gallopavo*) and common pigeons (*Columbia livia*) revealed that nest-deprived females were able to restart incubation as soon as they returned to their nests, despite the decline in plasma prolactin concentrations (El Halawani et al. 1980; Lea and Sharp 1989). Whether such behaviour is the consequence of a carry-over effect of the previous prolactin elevation, or whether other physiological inputs (see section 12.3.2) are involved in initiating incubation, are still open questions.

Prolactin concentrations are generally higher in females than in males, but this pattern is intriguingly reversed in species in which only the males participate in incubation, such as in Wilson's pharalopes (*Phalaropus tricolor*; Buntin et al. 1998). These sex differences are likely determined by interactions with steroid hormones. Within domesticated species at least, experimental evidence shows that exogenous prolactin administration induces incubation behaviours, but only in birds that are in a reproductive state or when the treatment is combined with steroids (El Halawani et al. 1986; Youngren et al. 1991). More complex relationships exist, as avian prolactin can be influenced by other hormones such as glucocorticoid stress hormones. Prolactin levels can dampen in response to both acute and chronic stressors (Cherel et al. 1994; Chastel et al. 2005). These data led to the suggestions that corticosterone (the main glucocorticoid in birds) and prolactin have antagonistic effects on parental behaviour (Angelier and Chastel 2009). Some support for this hypothesis comes from studies of Adelie penguins (*Pygoscelis adeliae*) showing that plasma corticosterone elevations increase the rate of nest desertion (Spee et al. 2011; Thierry et al. 2013b)—a behaviour expected to be adaptive as it allows the animal to divert energy from reproduction to immediate survival strategies (Wingfield et al. 1998). Intriguingly, however, in both Spee et al. (2011) and Thierry et al. (2013b) there were still approximately 40% of treated birds in which the elevated corticosterone did not trigger nest abandonment. Based on previous work, it is likely that in those birds prolactin levels were elevated (Spee et al. 2010). Similarly, the experimental inhibition of prolactin secretion in Adelie penguins was not sufficient to induce nest desertion, although it decreased incubation effort (Thierry et al. 2013a). It would be interesting to analyze the independent and combined effects of these two manipulations to examine potential antagonistic effects between corticosterone and prolactin.

In summary, despite the consolidated knowledge that prolactin is a key physiological mediator of the transition from sexual to parental activity (Goldsmith 1991; Sockman et al. 2006), there are still many unresolved issues including: the bi-directional relationships between external cues and endogenous schedules; possible 'carry-over' or 'programmed' effects derived from parents' prior experience; and a lack of clarity concerning whether the relationships between prolactin and components of incubation behaviour are causative or just correlative.

12.3.2 Can the oxidative status of incubating parents influence incubation patterns?

Oxidative stress, defined as the production of oxidative damage to biomolecules (Halliwell and Gutteridge 2007), is proposed as a mechanism mediating the costs of reproduction (Costantini 2014). As incubation efforts are an important component of reproduction (Monaghan et al. 1998; Hanssen et al. 2005), it may be argued that oxidative stress may drive incubation behaviour patterns in birds. Although the data in this context are limited and indirect, there are some suggestions that the capacity of breeding birds to withstand oxidative stress may be linked with at least some components of incubation behaviour. For example, Bize et al. (2008) found that in female Alpine swifts (*Tachymarptis melba*) red blood cell resistance to oxidative stress positively co-varied with clutch size and hatchability. By swapping entire clutches they showed that these significant correlations were dependent on the biological mothers rather than on post-hatching parental care. The recognized link between clutch size and nest attentiveness found among species (see section 12.2.2) begs the question whether incubating parents showing high resistance to oxidative stress may exhibit higher nest attentiveness compared to parents with low resistance to oxidative stress. In blue petrels (*Halobaena caerulea*) circulating antioxidants tend to be lower in individuals in lower body condition (signifying fasting) during incubation (Costantini and Bonadonna 2010), suggesting that nest abandonment might be positively selected to save antioxidants (Costantini 2014). Indeed, due to the long lifespan of blue petrels, it may be safer for the adults to desert their nests during a bad year and to divert resources into self-maintenance until the next breeding attempt. Moreover, in female Audouin's gulls (*Larus audouinii*), plasma antioxidant capacity decreased over the incubation period, and within clutches total antioxidant capacity in the egg tended to decline over the laying sequence (García-Tarrasón et al. 2014). Overall, these correlative data suggest a link between oxidative status and incubation behaviour.

Studies to manipulate parents' oxidative stress or incubation effort will be highly informative in clarifying the role of oxidative stress as a constraint on incubation behaviour.

12.4 Interactions between parents and potential implications for incubation behaviour

Interactions between unrelated individuals are a conserved trait across a large variety of animal species with bi-parental care. On the one hand, it is expected that parents would cooperate together to care for their offspring since this usually enhances the offspring's chances of survival (Clutton-Brock 1991). On the other hand parental care is costly and can increase exposure to predators of the care-providers with negative fitness consequences (Houston et al. 2005). Therefore, while offspring survival is normally dependent on the parents' combined care, each parent can pay the cost individually as a consequence of its own effort (Mock and Parker 1997) and thus, a conflict of interest between the parents can emerge (Trivers 1972).

In some taxa, including birds, the resolution of parental effort decisions has evolved in various forms of behavioural interactions to settle this conflict (reviewed by Lessells and McNamara 2012). Theoretical models predict that each parent should always adopt the best effort given the current effort of its partner—that is, if one parent decreases its level of care, the mate should increase its effort only by a lesser extent and not enough to compensate fully (Chase 1980; Houston and Davies 1985; McNamara et al. 2003). Empirical studies in birds show mixed outcomes, including no compensation, partial compensation, and over-compensation, but also matching responses, which occur when both parents increase (or decrease) their reproductive efforts (reviewed by Johnstone and Hinde 2006). Specifically to the incubation period, the experimental increase of nest temperature in a population of common starlings (*Sturnus vulgaris*; bi-parental incubation) tended to reduce female nest attendance and to increase contribution of the male partner, which resulted in no difference in total nest attendance between the heated and control nests (Magrath et al. 2005). Although these data support the theoretical models well, the extra heat added to both parents also means that the increase in male care could be a consequence of sex differences in the sensitivity to the thermal status of the clutch rather than to the parental contribution of its mate (Magrath et al. 2005). Indeed,

in species with bi-parental incubation, as in the common starling, females have generally a more developed brood patch and, hence, may be more sensitive to nest temperature cues than males (Deeming 2002e). This potential confound was elegantly ruled out by Kosztolanyi et al. (2009) who studied free-living Kentish plovers (*Charadius alexandrinus*), a ground-nesting shorebird, in which the males incubate at night and the females incubate during the day. Kosztolanyi et al. (2009) placed a remote-controlled device under the plovers' nests to manipulate the incubation effort of one parent (either male or female) randomly over two separate days. Similar to the findings of Magrath et al. (2005), they found that an experimental decrease in nest temperature, and therefore, the increased workload of the target parent, led to a decreased effort by its mate. The ability of incubating adults to compensate for their partner's deficits is not a universal rule, however, being rather limited in other species, such as in the Adelie penguin (Spee et al. 2011) and in the lesser black-backed gull (Monaghan et al. 1998).

Between-individual variability in investment can also be sex-dependent, probably due to the asymmetries in brood value and/or role specialization (Trivers 1972; Matysioková and Remeš 2014). Females are usually expected to invest more in parental effort than males. For example, experimental studies in which breeding individuals have been handicapped, for example by removal of flight feathers of great tits (Sanz et al. 2000) and common starlings (Wright and Cuthill 1990), show that males reduce parental effort when handicapped while females maintain their level of effort. Similar effects were observed in breeding pairs of zebra finches under stressful conditions (Spencer et al. 2010). Matching responses appear to occur in semipalmated sandpipers (*Calidris pusilla*) breeding in the high Arctic because incubation bout lengths were positively correlated between pair members and bouts increased in length with incubation time. This was possibly because the breeding pair could negotiate in real time their share in order to match their amounts of care (Bulla et al. 2014).

Overall, there are clear inconsistencies between predictions from theoretical game theory and empirical data, probably due to the omission in the models of factors such as amount, quality, and timing of care (see Bulla et al. 2014 for a detailed discussion). Additionally, early-life experience also appear to be a major factor. Adult female captive zebra finches that were directly exposed to oral doses of corticosterone as nestlings subsequently showed reduced incubation effort, although no effects were seen on nesting success, probably as a result of partial compensation by the male partners (Spencer et al. 2010). Different outcomes may be expected under natural conditions as food scarcity and predator pressure may constrain the ability of the adults to compensate.

12.5 Potential factors affecting behaviour in embryos during incubation

12.5.1 Overview of embryonic behaviours

Although it may seem counter-intuitive, avian embryos exhibit a range of measurable behaviours prior to hatching, including movement, thermoregulatory behaviour, vocalizations, and also learning (Oppenheim 1973; Brua 2002; Rumpf and Tzscenthe 2010; Boulton and Cassey 2012; Chapter 11). However, whilst some behavioural traits have received a large amount of interest from empirical studies, such as movement and vocalizations, others have been scarcely observed or investigated and more work is needed to understand the functional and evolutionary significance of the full range of embryonic behaviours.

Movement is perhaps the most simple behaviour exhibited by developing embryos and the changes in movement patterns over the incubation period were characterized in a seminal work by Oppenheim (1973) investigating the domestic fowl (*Gallus gallus*) embryo. First movements occur after the first fifth of incubation (day 3 in the domestic fowl) and involve stereotyped undulations, followed by uncoordinated jerky movements of all limbs and the corpus, and then coordinated rotatory actions. Activity tends to be in bursts which coincide with increased activity in electromyography (EMG) readings (Bradley and Bekoff 1990). The frequency and duration of these activity bouts increase throughout incubation, peaking midway, after which they abate. Coordinated goal-directed movements begin at day 17–18 in the domestic fowl in preparation for hatching.

Another important function of whole body movements, aside from hatching is the ability of embryos to move within the egg to maintain an optimal temperature for development, in so-called thermoregulatory behaviour (Boulton and Cassey 2012; Li et al. 2014). The ability of an avian embryo to use beneficial strategies to thermoregulate is limited due to the confined nature of the amniotic egg. However, in some species thermoregulatory movements have been observed, particularly with orientation towards warmer parts of the egg (Boulton and Cassey 2012; Li et al. 2014; see section 12.5.2).

As well as general motility and readiness to leave the egg, the avian embryo also exhibits begging behaviour when it produces acoustic signals that stimulate parental behaviour (reviewed in Brua et al. 1996; Brua 2002; Rumpf and Tzscenthe 2010). These vocalizations begin during late incubation, just prior to hatching, which is a very sensitive developmental stage (Vleck and Bucher 1998), and have been implicated in the shift from incubation to brooding behaviour in both parents in several bird species, including little terns (*Sterna albifrons*; Saino and Fasola 1996), American white pelicans (*Pelecanus erythrorhynchos*; Abraham and Evans 1999), and domesticated muscovy ducks (*Cairina moschata*; Rumpf and Tzschentke 2010). They could facilitate imprinting in the post-natal period (Bolhuis and van Kampen 1992).

The production and, more importantly, perception of sounds seem to be consistent across the whole class with precocial species showing hearing sensitivity in the second half of the incubation period, semi-altricial during the last third, and altricial embryos just prior to hatching (Rumpf and Tzschentke 2010). Interestingly, in those species in which much development takes place *in ovo*, i.e. precocial and semi-precocial, their ability to hear and potentially process auditory signals develops prior to that to produce sounds, although this may simply be related to developmental trajectories rather than represent any functional significance.

Sound production is initiated 1–2 days prior to internal pipping in precocial species and during several hours prior to eggshell pipping in altricial species (Vince 1969). The first calls are made irregularly and consist of low frequency single forms. These are then followed by 'clicks', short calls with a high frequency bandwidth, which are the result of respiratory movements. Both of these call types are not produced under syringeal control, but are by-products of other behaviours (Vince 1969; Rumpf and Tzschentke 2010). Sounds under the control of the syrinx are only produced once the embryo has penetrated the inner eggshell membrane and then a repertoire of several species-specific calls develops. It has been suggested that these vocalizations have a functional role in communication between siblings and/or parents. This may be true but we are currently lacking detailed studies. Several studies have suggested that 'clicks' are essential for hatching synchrony in some species (Brua 2002; Reed and Clark 2011) and there is evidence that clicking rates synchronize across a clutch around the time of hatching. For example, in Muscovy ducks embryos were prevented from hearing their siblings' 'click' vocalizations, with the eggs hatching asynchronously as a result (Rumpf and Nichelmann 1993).

The fact that embryos may communicate via simple or even complex vocal signatures that are innate is not perhaps surprising. However, until recently the ability of embryos to learn acoustic traits had not been explored in detail (Lickliter and Hellewell 1992; Lickliter et al. 2006; Markham et al. 2006; Bertin et al. 2009; Columbelli-Négrel et al. 2012). In the superb fairy wren (*Malurus cyaneus)*, which is parasitized by Horsfield's bronze cuckoo (*Chrysococcyx basalis*), it has been shown that embryos can learn a specific call from the incubating parent. Using cross-fostering, Columbelli-Négrel et al. (2012) were able to disentangle the learned and innate components of this call. Playback experiments showed that adult birds responded more to the calls produced by conspecific chicks in their nests compared to cuckoo chicks. This parent-specific 'password' could help females to detect parasitic eggs, but it also elegantly demonstrated learning in an altricial embryo.

Plastic changes in the brain that facilitate learning in any animal are normally associated with being awake and hence the ability to awaken (or sleep) is an important constraint on the ability of developing embryos to learn cues that could enhance their survival once they have emerged from the egg. This prerequisite was discovered in avian embryos by Balaban et al. (2012). Advances in functional imaging techniques, such as positron emission tomography (PET) scans, have allowed studies to quantify neural activity and embryonic behaviour in the same animal. Brief exposure of domestic fowl embryos to a conspecific adult vocalization significantly increased higher brain activity, particularly in the rostral pontine and mesencephalic nuclei, compared to non-vocal sounds (Balaban et al. 2012). This study also recorded correlated patterns of activity in the brainstem and telencephalon/cerebellum during the last fifth of embryonic life that were comparable with conspecifics awake during the postnatal period, suggesting the embryos spend at least part of their time awake.

12.5.2 Biotic and abiotic factors affecting embryonic behaviour

The sensory environment within which an embryo finds itself can have a significant effect on the rate of hatching and hatching behaviours (Sleigh and Casey 2014). Ambient temperature can have significant effects on the avian egg during incubation due to the effects on parental incubation behaviour (see section 12.2.1).

Any time when the egg is left unattended provides a potential cooling of the egg and also of the embryonic environment. Experimental placement of eggs of the common quail (*Coturnix coturnix*), the domestic fowl, the common pigeon, and the Pekin duck (*Anas platyrhynchos* var. *domestica*) in either constant heat (38°C) or lateral heating applied directly to the egg surface (39.5°C), resulting in a temperature gradient within the egg, caused movement from the centre of the egg to the edge near the shell in each species (Li et al. 2014). This study demonstrated the intrinsic motor response of embryos to their environment.

While some have found that a reduction in ambient temperature, which equates to the nest temperature in some cases, reduces vocalization behaviour (e.g. Evans 1992), others (e.g. Budgen and Evans 1999; Gräns and Altimiras 2007; Poisbleau et al. 2013) found no such discernible effects of temperature on vocalizations. Brua et al. (1996) found an increase in rate of vocalizations once egg temperature dropped, which was thought to facilitate increased parental care in black-necked grebes (*Podiceps nigricollis*). In this species embryos also responded to egg turning by parents by increasing vocalization rates, also potentially facilitating more care (Brua et al. 1996). Whilst the overall frequency or duration of calling may not be affected, and this may be due to parental compensation in increasing nest attentiveness (see section 12.4), more research should target the potential structural changes in vocalizations which could provide more information for parents on the needs of the developing embryo under varying environmental conditions.

It has been suggested that certain call types may have different meanings with distress calls characterized by descending frequencies and comfort calls by ascending and descending frequency cascades (Rumpf and Tzschentke 2010). However, the functional significance of these call types has not been rigorously examined. Two studies have used a finer scale of analysis when investigating the effects of ambient temperature on vocalization behaviour. When the ambient temperature is sub-optimal, embryonic budgerigars (*Melopsittacus undulatus*) responded by producing longer calls without changing overall activity (Berlin and Clark 1998). In the little grebe (*Tachybaptus ruficollis*) call complexity decreased with increasing ambient temperature (Brua et al. 1996).

12.5.3 Physiological mechanisms

It is recognized that there is maternal allocation of more than merely nourishment in the form of yolk and albumen into the avian egg (see Chapter 10). In particular, the concentration of many steroid hormones, such as corticosterone or testosterone, have received substantial recent attention and have been shown to affect several traits, both during incubation and in later life. However, one candidate mechanism for mediating *in ovo* vocalizations is that of testosterone.

Circulating levels of androgens, testosterone, in particular, have been shown to increase the intensity of chick begging behaviour in the post-natal period in European pied flycatchers (Goodship and Buchanan 2006). In addition, pre-natal exposure to androgens results in elevated begging displays during the post-natal period and aggressiveness in adulthood (Schwabl 1996; Eising et al. 2006). There are also significant sex differences of the response to testosterone experienced in the egg in the zebra finch, with female begging intensity being elevated, but males being unaffected (von Engelhardt et al. 2006).

The potential role of testosterone in preparing embryos for life outside the egg has recently been extended to mediating behaviours within the egg. Boncoraglio et al. (2006) have shown that injection of testosterone into yellow-legged gull (*Larus michahellis*) eggs increases the amplitude of embryonic vocalizations prior to hatching, with no effects on call structure or frequency of calling bouts. Of course, it could be debated whether 'louder' is equivalent to 'better' in this situation, as more noise could increase predation risks. While more research in this area is needed, this study was undertaken in the field, allowing natural behaviours to be studied despite no predation rates being reported.

In addition to mediating vocal production, testosterone can also influence vocal learning *in ovo*. Bertin et al. (2009) experimentally increased yolk testosterone concentrations prior to incubation in northern bobwhite (*Colinus virginianus*) eggs whilst providing a template conspecific assembly call for a brief period in the day. Post-hatching chicks that had experienced elevated testosterone levels were more likely to respond to the familiar assembly call than unmanipulated controls. The effect of testosterone was to effectively double the learning speed of the embryos as the stimulus was presented at half the frequency of previous studies of this species (Lickliter et al. 2004). Several other hormones, such as corticosterone, have effects on learning in the post-natal period (Hodgson et al. 2007). However, the effects of such ligands on *in ovo* learning by embryos have not been tested and we urge future work in this area.

12.6 Conclusions

To summarize, it is important to emphasize that variation in incubation profiles depends on how parents allocate their time between incubation and other non-breeding activities. The resolution of these trade-offs needs to be considered throughout an individual's lifespan. This is a key concept to aid our understanding of how benefits and costs of reproduction are shared between parents and current offspring. Whilst several studies have showcased the multitude of ecological factors that can influence incubation profiles, we still have a limited knowledge of how these factors are inter-related. Furthermore, we still do not know enough about how partners influence each other in relation to their incubation patterns and the implications of the resolution of these conflicting interests on embryonic development and behaviour. We need more work to explore how within-pair variation has impacted on avian incubation strategies across the precocial–altricial spectrum and in the context of different life-history strategies and incubation modes such as in comparisons of long-lived versus short-lived species and those undertaking bi-parental versus uni-parental incubation.

Birds are excellent 'models' to study the physiological mechanisms underlying behavioural differences in incubation as the embryos are physically isolated from the mothers. A considerable amount of work in this area has focussed on steroids (primarily prolactin metabolism in incubating adults and androgen metabolism in the embryos), although work on other hormones is emerging. There are some suggestions that biomarkers other than steroids, such as those tracking oxidative stress, could also play a role in mediating variation in parental investment, which could in turn influence embryonic development and behaviour. This is a new research area that would be interesting to explore using avian models. Overall, a deeper understanding of mechanisms is perhaps easier to achieve in captive populations and it will be helpful to postulate appropriate trade-offs and identify relevant physiological constraints that are likely to occur in natural populations.

Acknowledgement

We thank David Costantini for providing comments on earlier chapter drafts.

CHAPTER 13

The energetic costs of incubation

A. Nord and J.B. Williams

13.1 Introduction

Fifty years ago Kendeigh (1963) focussed attention on the cost of avian incubation by modelling heat loss for an incubating house wren (*Troglodytes aedon*): he concluded from his calculations that incubation by birds is associated with increased energy expenditure. His findings were challenged by King (1973), who instead suggested that the heat requirements for embryonic development could largely be provided by heat generated from basal metabolism of the adult bird. King and co-workers later supported this view in a series of papers where, based on heat loss models, they concluded that the energy expenditure of an incubating bird is some 15–18% below that of resting conspecifics (Walsberg and King 1978a, 1978b). This idea rapidly gained momentum and the prevailing view for the next two decades was that the incubation period of birds is a time of relatively low energy expenditure compared with other reproductive stages. During this time research into avian life history, ecology and evolution was largely biased towards examining constraints on the rearing capacity of parents (reviewed by Monaghan and Nager 1997). However, the view that incubation was a cost-free period slowly started to change in the late 1980s and early 1990s with the publication of several papers showing that experimentally enlarged clutches took longer to incubate (e.g. Coleman and Whittall 1988; Moreno and Carlson 1989; Smith 1989), and that reducing the foraging cost of females on the nest via supplementary feeding decreased the length of the incubation period and increased hatching success (Nilsson and Smith 1988). This was later followed by a study on interspecific variation in the energy costs of incubation by Tatner and Bryant (1993), who found that the daily energy expenditure (DEE) during incubation was significantly lower than during the brood-rearing stage in only five of 17 species that had been studied.

The view that incubation was a period of elevated energy expenditure was supported in the first formal review on the energetics of avian incubation, in which Williams (1996) concluded that: 1) the energy expenditure during incubation is significantly elevated above resting levels of metabolism, at least when birds incubate in temperatures below their thermal neutral zone; 2) the energy cost of rewarming eggs after an incubation recess could be substantial; 3) the energy expenditure during incubation is higher when parents incubate larger clutches; 4) at least for systems in which only the female incubates, there is no evidence to suggest that energy expenditure is lower during incubation than during other reproductive stages; and 5) because of high sustained energy expenditure under simultaneous constraints on energy acquisition, incubation alone can result in significant energy shortage. These ideas were revisited by Tinbergen and Williams (2002) who, based on a larger data set, supported the conclusions reached by Williams (1996).

Research into the costs of incubation has lain quiescent, for the most part, during the 21st Century, with only one study per year being published following Tinbergen and Williams (2002). Thus, there is currently little reason to challenge the perception that incubation is far from a resting period, and that the demands on incubating parents are as important as those during brood rearing in the evolution of avian breeding behaviour and physiology (Williams 1996; Tinbergen and Williams 2002). However, it is still poorly understood how increased energy expenditure during incubation results in (direct or indirect) fitness costs (cf. Reid et al. 2002b). Nor do we know much about the mechanisms by which incubation costs are transferred from adults to their young, typically manifested as increased embryonic mortality (Moreno et al. 1991; Siikamäki 1995) or reduced size and growth of chicks following incubation when demands on parents have been experimentally increased (Heaney and Monaghan 1996;

Nord, A. and Williams, J.B., *The energetic costs of incubation*. In: *Nests, Eggs, and Incubation*. Edited by: D.C. Deeming and S.J. Reynolds, Oxford University Press (2015). © Oxford University Press.
DOI 10.1093/acprof:oso/9780198718666.003.00013

Cichon 2000; Ardia et al. 2010). To shed light on such processes, this chapter discusses the possibility that energy costs of incubation can only be covered by reducing allocation of energy to other somatic functions when resources are limited, and that fitness costs of incubation occur as a result of such constraints acting on other aspects of physiological regulation. We then turn to evaluating evidence for the hypothesis that incubation costs to chicks (which may be observed independently of any effects acting on parents) are mediated via a facultative reduction in incubation temperature in resource-limited environments. This could occur if parents are unable or unwilling to maintain the thermal environment of the nest (which is energetically costly)

(e.g. Tøien et al. 1986) in favour of self-maintenance when ambient conditions deteriorate. Finally, while much research has focussed on the temporal, spatial and interspecific organisation of energy expenditure by incubating birds, we still know little about the importance of, and the variation in, non-energetic (physiological) costs of incubation, and their relationship to (or dependence on) a bird's energy balance. We will exemplify this using data on water balance during incubation in desert-breeding birds. However, we will start by reviewing current knowledge of the energetics of avian incubation, focussing particularly on findings from the last decade. For convenience, Table 13.1 lists abbreviations used in the text.

Table 13.1 List and definitions of abbreviations used in the text.

Abbreviation	Meaning	Definition
AGI	Assisted gyneparental incubation	The female alone incubates eggs but the male assists in incubation by bringing food to the nest
BCI	Bi-parental continuous incubation	Both parents take responsibility for incubation
GCI	Gyneparental continuous incubation	The female incubates without assistance from the male and does not leave the nest to forage during the incubation period
GII	Gyneparental intermittent incubation	The female incubates alone and intermittently without assistance from the male
BMR	Basal metabolic rate	The metabolic rate ($kJ \cdot d^{-1}$) for post-absorptive animals resting in their thermal neutral zone
EE_{warm}	Energy expenditure for rewarming eggs	The energy (kJ) required to rewarm eggs after an incubation recess
FMR_{inc}	Field metabolic rate of incubation	Energy expenditure across all activities during the incubation period ($kJ \cdot d^{-1}$)
FMR_{chick}	Field metabolic rate of chick rearing	Energy expenditure across all activities during the chick-rearing period ($kJ \cdot d^{-1}$)
MR_{inc}	Incubation metabolic rate	The metabolic rate ($kJ \cdot d^{-1}$) when incubating eggs
$MR_{off-nest}$	Incubation metabolic rate for periods off the nest	The metabolic rate ($kJ \cdot d^{-1}$) of incubating parents during recesses from incubation
PZT	Physiological zero temperature	The egg temperature below which no embryonic development occurs
ROS	Reactive oxygen species	Chemically reactive oxygen-containing molecules (e.g. anions and peroxides) that may be harmful to cell components
T_a	Ambient temperature	The temperature in the environment of animals
T_b	Body temperature	The temperature in the body core of an animal
T_e	Operative temperature	The body temperature of an animal if in thermal equilibrium with the environment in the absence of metabolic heating or evaporative cooling
T_{egg}	Egg temperature	The temperature in a single egg during incubation
$T_{egg-unatt}$	Egg temperature of unattended eggs	The temperature of eggs during incubation recesses
T_{inc}	Incubation temperature	The temperature beneath the incubating bird when averaged over the clutch
TEWL	Total evaporative water loss	The rate at which water evaporates from the body ($g \cdot h^{-1}$)
TNZ	Thermal neutral zone	The temperature interval within which body temperature can be maintained without increasing metabolic rate above basal levels

13.2 Energetics of incubation

One of the most heavily debated areas in research on avian incubation is arguably the relative energy costs of incubation versus chick rearing, the latter traditionally having been viewed as the main energetic bottleneck during breeding (cf. Lack 1947). However, incubation is energetically costly, and often as expensive as chick rearing. In this section, we will highlight some of the main findings from previous reviews of incubation energetics (Williams 1996; Tinbergen and Williams 2002), and describe work published during the last decade.

13.2.1 Energy expenditure during incubation in terrestrial birds

We compiled studies on terrestrial birds for which data on field metabolic rate (FMR) are available either for the incubation period (FMR_{inc}), or for both the incubation and chick-rearing (FMR_{chick}) periods (Table 13.2). We then used this information to describe and explore aspects of the energetics of incubation in terrestrial birds. To estimate the energy cost of incubation, we expressed FMR_{inc} as multiples of basal metabolic rate (BMR) and regressed these data on the logarithm of body mass in Arctic versus non-Arctic breeding birds (incubation in the former habitat being known to carry particularly high energy costs; Piersma et al. 2003). We used averages in all analyses when more than one study had been performed on a single species, except when data were available from both Arctic and non-Arctic latitudes (which was the case for the ringed plover [*Charadrius hiaticula*]) where we regarded data from different breeding sites as separate observations (cf. Tinbergen and Williams 2002). FMR_{inc} was significantly elevated above BMR for all species, but the increase in energy expenditure during incubation was 43% higher for birds breeding in the Arctic ($F_{1,32} = 31.3$, $P < 0.001$; Figure 13.1a). Specifically, birds outside Arctic latitudes augmented energy expenditure 2.88 (± 0.13) (mean ± 1 SE) $\times$ BMR, whereas the corresponding increase in Arctic birds was 4.13 (± 0.10) $\times$ BMR. Although it seems likely that challenging weather conditions during incubation at Arctic latitudes could explain at least some of the larger increase in energy expenditure in this group (Piersma et al. 2003), it is noteworthy (as also observed by Tinbergen and Williams 2002) that the only species for which data are available from both Arctic and non-Arctic breeding areas, the ringed plover, showed similar increases in energy expenditure regardless of nesting site (Arctic:

4.0 $\times$ BMR; Piersma et al. 2003; non-Arctic: 4.6 $\times$ BMR; Tatner and Bryant 1993; Table 13.2). This is interesting because the only other wader in this genus in Table 13.2, the non-Arctic Kentish plover (*Charadrius alexandrinus*), does not show elevated energy expenditure during incubation, i.e. 2.3 $\times$ BMR (Amat et al. 2000). The elevated FMR_{inc} in temperate-breeding ringed plovers could potentially be explained if birds did not resume normal incubation behaviour after injection with doubly-labelled water, instead engaging in particularly energy-consuming activities during measurements of FMR_{inc}. For instance, incubating red knots (*Calidris canutus*) expended energy at almost twice the rate during foraging bouts away from the nest compared with when sitting on eggs (Piersma et al. 2003).

In both Arctic and non-Arctic breeding birds, the energy cost of incubation is lower for larger birds over the range of body sizes included in the analysis, i.e. 9.5 to 491.0 g ($F_{1,32} = 5.0$, $P = 0.03$; Figure 13.1a), with the ratio of FMR_{inc} to BMR decreasing by 0.54 for each unit increase in $\log_{10}$ body mass. This could be explained if overall activity during incubation was higher in smaller birds to sustain their inherently higher foraging requirements resulting from increased metabolic intensity and limited capacity to deposit endogenous reserves (McNab 2002; but see Tulp et al. 2002).

13.2.2 Energy costs of incubation in relation to those of chick rearing

For the species listed in Table 13.2 in which both FMR_{inc} and FMR_{chick} were measured, the workload when expressed as multiples of BMR was some 15% higher during chick rearing (3.38 $\pm$ 0.17 $\times$ BMR) than during incubation (2.93 $\pm$ 0.15 $\times$ BMR) (repeated measures ANOVA: $F_{1,21} = 22.8$, $P < 0.001$; Figure 13.1b). However, consistent with results for the ratio of FMR_{inc} to BMR (see section 13.2.1), the difference between reproductive stages tended to be less in smaller species (breeding stage $\times$ $\log_{10}$ body mass interaction: $F_{1,21} = 4.2$, $P = 0.05$). This lends further support to the view that small birds work relatively harder during incubation.

The variation in the ratio of FMR to BMR shown in Figure 13.1b suggests there is considerable variation in energy management strategies between species of similar body size. It seems likely that similar variation could be present also within species in different study populations. For instance, there was no difference between FMR_{inc} and FMR_{chick} in a coastal lowland population of white-crowned sparrows (*Zonotrichia*

Table 13.2 Overview of the species of terrestrial birds for which data on basal metabolic rate (BMR) and field metabolic rate (FMR) are available, with the latter for either the incubation stage (FMR$_{inc}$) or for both the incubation and chick-rearing (FMR$_{chick}$) periods. All presented values are means across studies except for Ns which are sums. 'N/A' means 'Not applicable'. Data on FMR$_{chick}$ for the non-incubating sex in AGI and GII incubation systems are not included (but data averaged over both sexes are). Basal metabolic rate (BMR) was estimated using the phylogeny-corrected equation from Reynolds and Lee (1996). Stage: denotes whether birds were incubating ('I') or feeding chicks (C). System: AGI—female-only incubation with male incubation feeding, i.e. assisted gyneparental incubation; BCI—bi-parental continuous incubation; and GII—female-only intermittent incubation with no male assistance, i.e. gyneparental intermittent incubation.

Species	Stage	System	Body mass (g)	BMR(kJ·d⁻¹)	FMR(kJ·d⁻¹)	FMR/BMR	FMR$_{chick}$/FMR$_{inc}$	N	Reference
Arctic breeding									
Ringed plover (*Charadrius hiaticula*)	I	BCI	57.0	55.0	218.8	4.0	–	3	Piersma et al. (2003)
Red knot (*Calidris canutus*)	I	BCI	142.0	98.2	373.5	3.8	–	6	Piersma et al. (2003)
Sanderling (*Calidris alba*)	I	BCI	59.0	56.2	229.2	4.1	–	4	Piersma et al. (2003)
Purple sandpiper (*Calidris maritima*)	I	BCI	71.5	63.5	271.9	4.3	–	11	Drent and Piersma (1990); Piersma et al. (2003)
White-rumped sandpiper (*Calidris fuscicollis*)	I	GII	39.0	43.2	189.6	4.4	–	1	Piersma et al. (2003)
Ruddy turnstone (*Arenaria interpres*)	I	BCI	108.0	82.6	348.1	4.2	–	16	Piersma et al. (2003)
Dunlin (*Calidris alpina*)	I	BCI	53.0	52.5	210.7	4.0	1.1	11	Piersma et al. (2003); Tulp et al. (2009)
Dunlin	C	BCI	47.2	48.8	237.6	4.9	–	6	Tulp et al. (2009)
Little stint (*Calidris minuta*)	I	GII	28.8	35.7	169.4	4.7	0.9	24	Piersma et al. (2003); Tulp et al. (2009)
Little stint	C	GII	27.5	34.6	159.9	4.6	–	11	Tulp et al. (2009)
Long-tailed skua (*Stercorarius longicaudus*)	I	BCI	336.0	169.7	635.0	3.7	–	2	Piersma et al. (2003)
Non-Arctic breeding									
Common kestrel (*Falco tinnunculus*)	I	AGI	275.0	149.4	324.0	2.2	1.1	N/A	Masman et al. (1988)
Common kestrel	C	AGI	235.0	135.3	346.0	2.6	–	N/A	Masman et al. (1988)
Eurasian oystercatcher (*Haematopus ostralegus*)	I	BCI	491.0	215.9	400.0	1.9	1.8	1	Kersten (1996)
Eurasian oystercatcher	C	BCI	492.3	216.3	708.0	3.3	–	3	Kersten (1996)
Pied avocet (*Recurvirostra avosetta*)	I	BCI	323.4	165.6	425.5	2.6	–	6	Hötker et al. (1996)
Common ringed plover (*Charadrius hiaticula*)	I	BCI	74.8	65.4	301.5	4.6	–	4	Tatner and Bryant (1993)
Kentish plover (*Charadrius alexandrinus*)	I	BCI	40.6	44.4	103.4	2.3	–	15	Amat et al. (2000)
Common sandpiper (*Actitis hypoleucos*)	I	BCI	54.7	53.6	135.5	2.5	1.2	3	Tatner and Bryant (1993)
Common sandpiper	C	BCI	48.5	49.7	156.9	3.2	–	1	Tatner and Bryant (1993)
Blue-throated bee-eater (*Merops viridis*)	I	BCI	35.4	40.7	90.4	2.2	0.9	9	Tatner and Bryant (1993)

continued

Table 13.2 *Continued*

Species	Stage	System	Body mass (g)	BMR(kJ·d⁻¹)	FMR(kJ·d⁻¹)	FMR/BMR	FMR$_{chick}$/FMR$_{inc}$	N	Reference
Blue-throated bee-eater	C	BCI	33.2	39.0	80.3	2.1	–	18	Tatner and Bryant (1993)
Dune lark (*Certhilauda erythrochlamys*)	I	GII	26.4	33.7	88.1	2.6	1.0	6	Williams (2001)
Dune lark	C	GII	25.5	33.0	88.5	2.7	–	6	Williams (2001)
Sand martin (*Riparia riparia*)	I	BCI	14.3	22.9	81.7	3.6	1.1	3	Westerterp and Bryant (1984)
Sand martin	C	BCI	13.0	21.5	90.0	4.2	–	4	Westerterp and Bryant (1984)
Barn swallow (*Hirundo rustica*)	I	GII	20.5	28.7	120.6	4.2	0.9	1	Westerterp and Bryant (1984)
Barn swallow	C	GII	19.2	27.6	108.5	3.9	–	8	Westerterp and Bryant (1984)
Pacific swallow (*Hirundo tahitica*)	I	GII	14.7	23.3	53.2	2.3	1.4	8	Tatner and Bryant (1993)
Pacific swallow	C	GII	14.1	22.7	76.6	3.4	–	6	Bryant et al. (1984)
Common house martin (*Delichon urbicum*)	I	BCI	19.0	27.4	80.2	2.9	1.2	10	Westerterp and Bryant (1984)
Common house martin	C	BCI	18.5	26.9	95.9	3.6	–	56	Westerterp and Bryant (1984)
Tree swallow (*Tachycineta bicolor*)	I	AGI	22.6	30.6	118.9	3.9	1.1	9	Williams (1988)
Tree swallow	C	AGI	19.4	27.8	136.4	4.9	–	11	Williams (1988)
White-throated dipper (*Cinclus cinclus*)	I	AGI	62.8	58.5	204.8	3.5	1.2	25	Tatner and Bryant (1993); Engstrand et al. (2002)
White-throated dipper	C	AGI	57.1	55.1	250.1	4.5	–	7	Tatner and Bryant (1993)
European robin (*Erithacus rubecula*)	I	AGI	20.6	28.8	75.1	2.6	0.8	3	Tatner and Bryant (1993)
European robin	C	AGI	17.9	26.4	58.8	2.2	–	4	Tatner and Bryant (1993)
Great tit (*Parus major*)	I	AGI	21.0	29.2	95.4	3.3	1.0	22	Bryan and Bryant (1999); de Heij et al. (2008)
Great tit	C	AGI	17.5	26.0	97.4	3.7	–	40	Tinbergen and Dietz (1994); Verhulst and Tinbergen (1997); Sanz et al. (2000)
Blue tit (*Cyanistes caeruleus*)	I	AGI	12.0	20.5	61.1	3.0	1.1	4	Tatner and Bryant (1993)
Blue tit	C	AGI	11.0	19.4	66.9	3.5	–	2	Tatner and Bryant (1993)
European pied flycatcher (*Ficedula hypoleuca*)	I	AGI	14.7	23.3	63.0	2.7	1.0	11	Moreno and Carlson (1989); Moreno and Sanz (1994)
European pied flycatcher	C	AGI	12.5	21.0	60.9	2.9	–	1	Tatner and Bryant (1993)

continued

Plate 1 Amos Ar in 2007. (Painting by Amnon David Ar [Amos' youngest son]. See Dedication figure 1)

Plate 2 Amos Ar in 2012. (Photograph by D.C. Deeming. See Dedication figure 2)

Plate 3 A Florida scrub-jay (*Aphelocoma coerulescens*) female incubates a 3-egg clutch at Archbold Biological Station in south-central Florida. A comprehensive and ongoing study of the nesting biology of this species has resulted in a highly successful programme of research to conserve this US federally listed species that is classified as 'Vulnerable' by BirdLife International (2014). (Photo: A. Korpach. See Figure 1.3)

Plate 4 Example of a Mesozoic bird egg collected from Upper Cretaceous deposits of Argentina. Scale bar = 1 cm. (Photo: D. Varricchio. See Figure 2.1)

Plate 5 Various components of nest building by male cape weavers (*Ploceus capensis*): (a) an early stage of building; (b) a male destroying a nest he had been building; and (c–h) six nests built by the same bird showing variation in nest construction. (Photos: I.E. Bailey. See Figure 3.1)

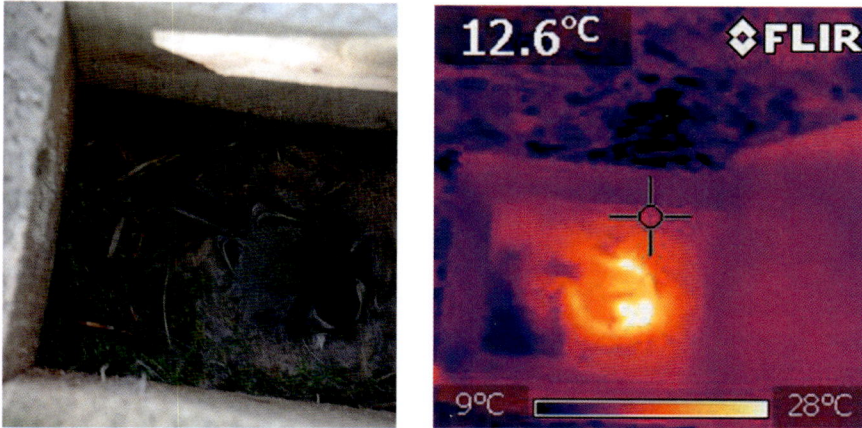

Plate 6 Normal (left) and infrared thermal (right) images of an incubating female great tit (*Parus major*) on her nest in Riseholme Park, Lincoln, UK. (Source: D.C. Deeming and T. Pike, 2014 unpublished data. See Figure 4.3)

(a)

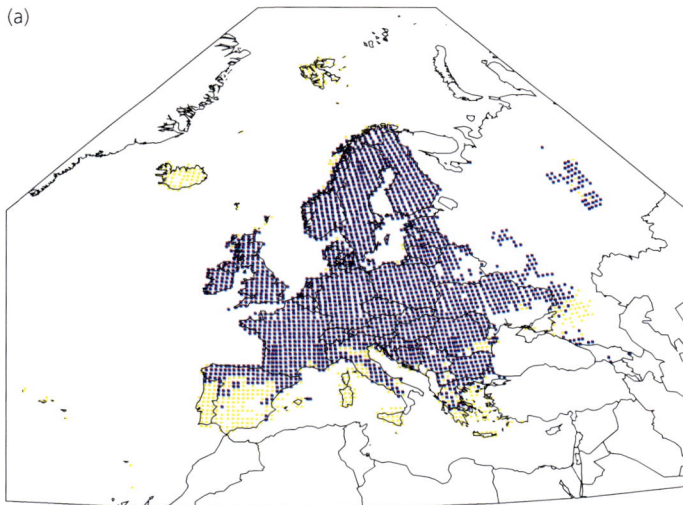

(b)

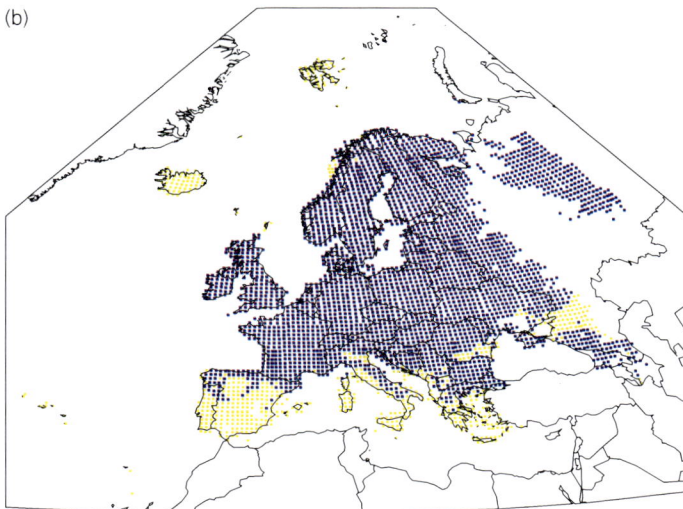

Plate 7 The current and future distributions of song thrushes (*Turdus philomelos*) in Europe. Their current distribution is shown as (a) census data from the EBCC Atlas of European Breeding Birds and (b) as predicted by response surface models for the 1961–90 mean climate. (c) overleaf shows their future distribution as predicted by response surface models for the mean climate projected for 2070–99 by the Hadley Centre climate model, as further detailed in Huntley et al. (2007). Note that in the maps, each coloured dot represents an area of 2500 km^2 and that whilst white areas have climatic conditions outside the range for which a simulation could be made, blue dots represent areas in which birds are simulated as breeding and yellow dots represent areas in which they are simulated as being absent. Maps reproduced with permission from Huntley et al. (2007). (See Figure 6.1)

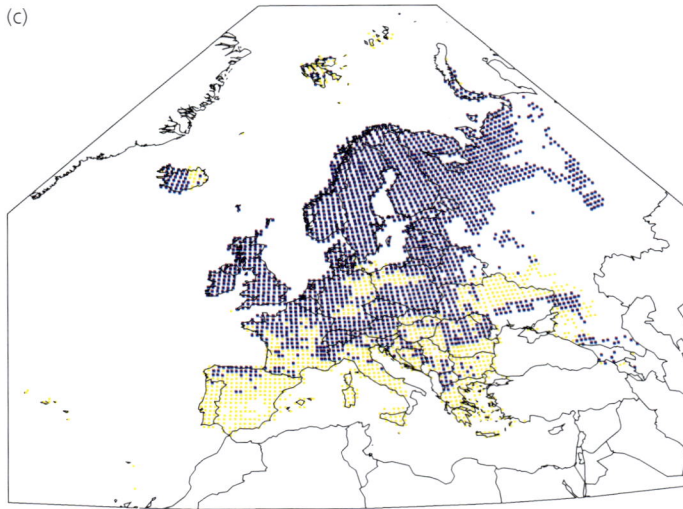

(c)

Plate 7 *Continued*

Plate 8 Some species of ectoparasites commonly found in avian nests: (a) true bug (*Oeciacus hirundinis*); (b) flea (*Ceratophyllus columbae*); (c) hard tick (*Ixodes ricinus*); (d) haematophagous mite *Dermanyssus* sp.; (e) carnid fly (*Carnus hemapterus*); (f) blowfly (*Protocalliphora azurea*); (g) black fly (*Simulium aureum*): (h) mosquito (*Culex quinquefasciatus*); (i) biting midge *Culicoides* sp.; (j) chewing louse (*Laemobothrion tinnunculi*); and (k) louse fly (*Crataerina melbae*). Scale bar in each panel represents 1 mm. (Photos: I. López-Rull [a, b, f, g, h, j and k], G. Tomás [c], S. Merino and E. Pérez Badas [d], F. Valera and M.T. Amat-Valero [e], and J. Martínez-de la Puente [i]. See Figure 8.1)

Plate 9 Examples of ectoparasite–bird interactions. Infestations of blowfly *Protocalliphora azurea* showing (a) the base of a blue tit (*Cyanistes caeruleus*) nest, (b) a pupa at higher magnification, and (c) a larva attached to the leg of a nestling of European pied flycatcher (*Ficedula hypoleuca*). (d) A nestling of the European roller (*Coracias garrulus*) infested with the carnid fly *Carnus hemapterus*. (e) The immaculate blue-greenish eggs (left) of some *Sturnus* spp. that become spotted as incubation progresses (right). Recent observations suggest that spots are droppings of *C. hemapterus* living amongst the nest material. (f) A male blue tit with a soft tick attached to his head. (g) A mosquito biting the foot of a domestic fowl (*Gallus gallus*). (h) A larva of a chigger mite *Neotrombicula autumnalis* attached at the eye ring of a red-billed chough (*Pyrrhocorax pyrrhocorax*). (Photos: G. Tomás [a, b, f and g], A. Cantarero [c], M.A. Calero Torralbo [d], ILR [e], and G. Blanco [h]. See Figure 8.3)

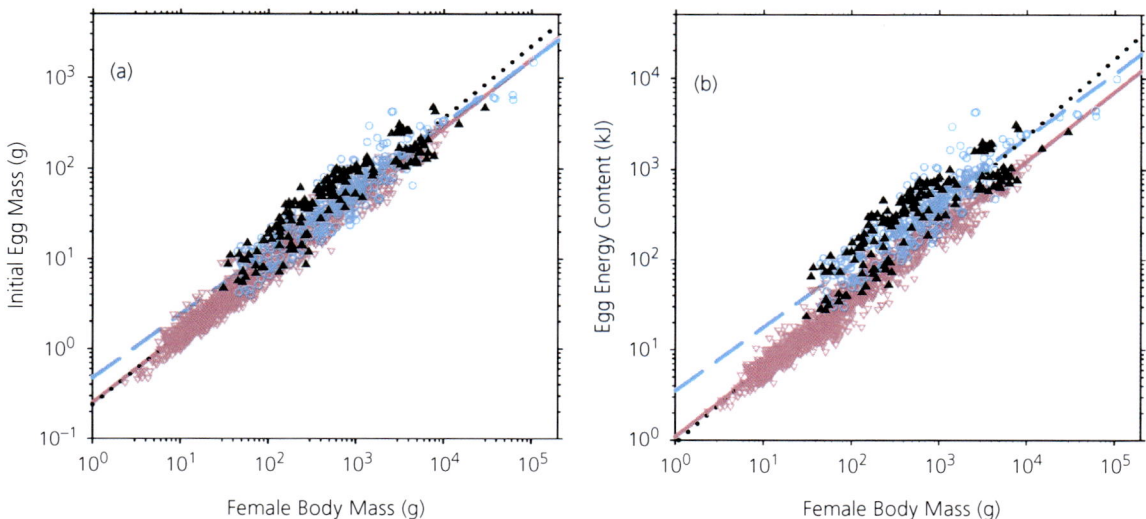

Plate 10 (a) Initial egg mass and (b) egg energy content as a function of adult female body mass, all plotted on $\log_{10}$ scales. Symbols for developmental mode: open blue circles (precocial), solid black triangles (semi-precocial), and open pink triangles (altricial). The dotted black line represents the regression for all data in the plot, the solid line is for precocial species, and the long-dashed line is for altricial species. (See Figure 9.3)

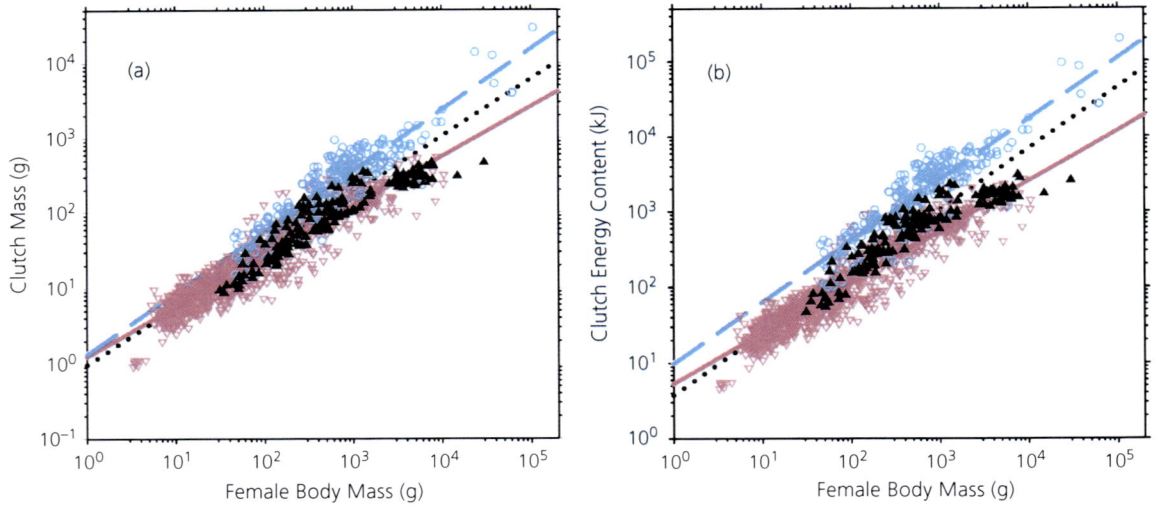

Plate 11 (a) Clutch mass and (b) clutch energy content as functions of female body mass plotted on $\log_{10}$ scales. The symbols and lines are as in Plate 10. (See Figure 9.6)

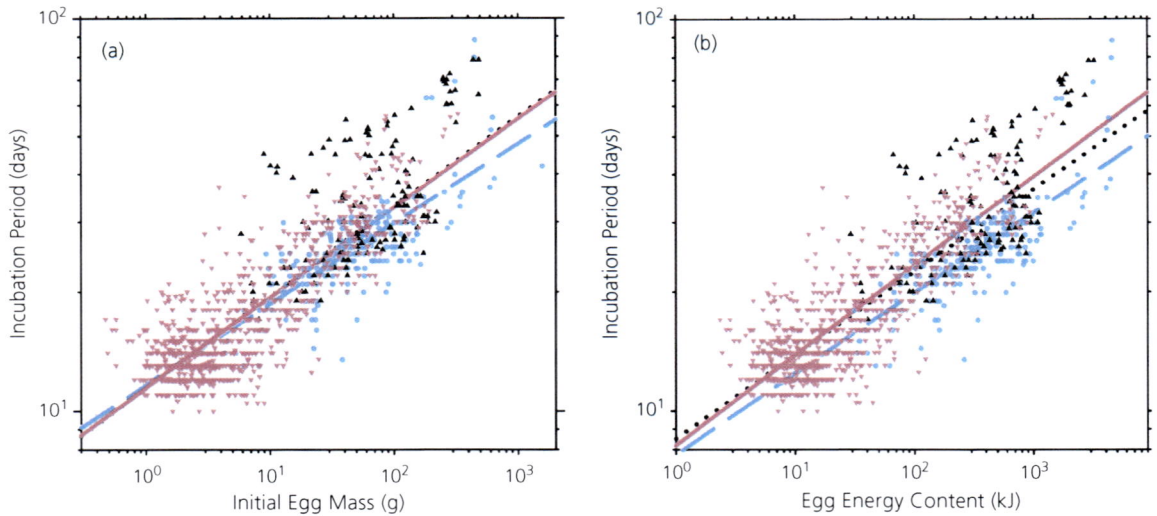

Plate 12 Incubation period as a function of (a) initial egg mass and (b) egg energy content plotted on $\log_{10}$ scales. The symbols indicating developmental mode are the same as in Plate 10. No regression line for semi-precocial species was calculated. (See Figure 9.7)

Plate 13 The colour and pigment patterns of eggshells vary greatly between species leading to a range of hypotheses for its possible adaptive significance. From top left to bottom right: Eurasian skylark (*Alauda arvensis*), Eurasian rock pipit (*Anthus petrosus*), meadow pipit (*Anthus pratensis*), European goldfinch (*Carduelis carduelis*), Cetti's warbler (*Cettia cetti*), white-throated dipper (*Cincus cinclus*), hawfinch (*Coccothraustes coccothraustes*), carrion crow (*Corvus corone*), western jackdaw (*Corvus monedula*), common house martin (*Delichon urbicum*), corn bunting (*Emberiza calandra*), yellowhammer (*Emberiza citrinella*), common nightingale (*Luscinia megarhynchos*), common chaffinch (*Fringilla coelebs*), Eurasian jay (*Garrulus glandarius*), barn swallow (*Hirundo rustica*), red-backed shrike (*Lanius collurio*), common grasshopper warbler (*Locustella naevia*), red crossbill (*Loxia curvirostra*), woodlark (*Lullula arborea*), white wagtail (*Motacilla alba*), western yellow wagtail (*Motacilla flava*), spotted flycatcher (*Muscicapa striata*), Eurasian golden oriole (*Oriolus oriolus*), great tit (*Parus major*), house sparrow (*Passer domesticus*), common chiffchaff (*Phylloscopus collybita*), Eurasian magpie (*Pica pica*), dunnock (*Prunella modularis*), Eurasian bullfinch (*Pyrrhula pyrrhula*), goldcrest (*Regulus regulus*), whinchat (*Saxicola rubetra*), Eurasian blackcap (*Sylvia atricapilla*), lesser whitethroat (*Sylvia curruca*), redwing (*Turdus iliacus*), and mistle thrush (*Turdus viscivorus*). Eggshells were made available to the University of Birmingham for a destructive loan from the Natural History Museum, Tring, UK. Eggs have been normalized for size. (Photos: G. Maurer. See Figure 11.1)

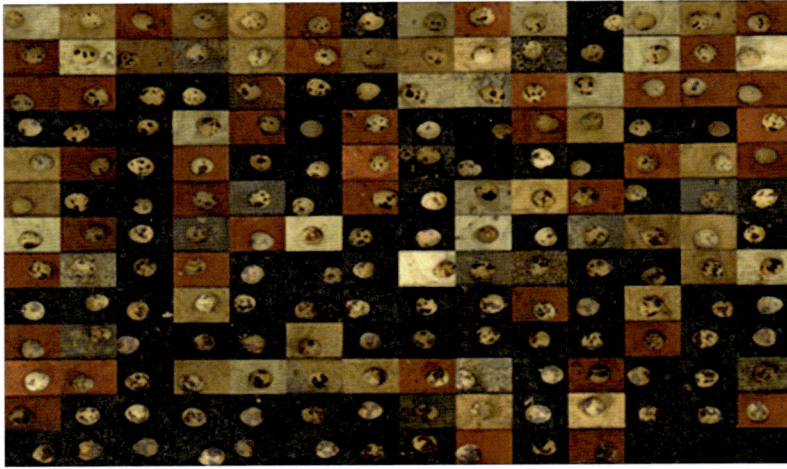

Plate 14 Japanese quail (*Coturnix japonica*) females consistently choose laying substrates which best camouflage their eggs and vary the type of camouflage depending on the degree of maculation of the eggs. Heavily maculated eggs were laid on substrates matching the colour properties of the eggs (camouflage through disruptive colouration). Lightly maculated eggs were laid on substrates which best matched egg background colour (camouflage through background matching). Photos are arranged according to level of eggshell maculation, starting with least maculated top left to most maculated bottom right. (Image: P.G. Lovell, G.D. Ruxton, K.V. Langridge and K.A. Spencer. See Figure 11.2)

Plate 15 (a) A pair of the Critically Endangered Madagascar pochard (*Aythya innotata*). This diving duck species was thought extinct until a small population of approximately 20 birds was discovered on a remote lake in North Madagascar in 2006. Ornithologists recorded many failed breeding attempts at the lake and this prompted conservationists to begin a conservation breeding programme to safeguard birds against extinction. (Photo: S. Dench/WWT.) (b) The poorly maintained 160 km track to the lake precluded egg translocation from wild nests to a rearing facility in a hotel. (Photo: P. Cranswick/WWT.) (c) The incubation tent set-up located under the canopy of the lakeside forest. (Photo: M. Roberts/WWT.) (d) A Brinsea 'Hatchmaker' with accessory incubation equipment and materials. (Photo: M. Roberts/WWT.) (e) Madagascar pochards nest in dense papyrus (*Cyperus* spp.) marsh at the lakeside. (Photo: M. Brown/WWT and NSJ/WWT.) (f) A nine-egg clutch of a Madagascar pochard. (Photo: M. Brown/WWT and NSJ/WWT.) (g) Ducklings and pipped eggs in the 'Hatchmaker'. (Photo: G. Young/Durrell.) (h) Adult drake Madagascar pochard: at the beginning of 2015 the conservation breeding programme comprised 54 ducks. (Photo: P. Cranswick/WWT. See Figure 16.4)

Table 13.2 *Continued*

Species	Stage	System	Body mass (g)	BMR(kJ·d⁻¹)	FMR(kJ·d⁻¹)	FMR/BMR	FMR$_{chick}$/FMR$_{inc}$	N	Reference
Collared flycatcher (*Ficedula albicollis*)	I	GII	15.5	24.1	72.1	3.0	1.0	3	Moreno et al. (1991); Tinbergen and Williams (2002)
Collared flycatcher	C	GII	12.9	21.4	71.7	3.3	–	20	Moreno et al. (1991); Tinbergen and Williams (2002)
Orange-breasted sunbird (*Anthobaphes violacea*)	I	GII	9.5	17.6	66.2	3.8	–	10	Williams (1993a)
Common starling (*Sturnus vulgaris*)	I	BCI	83.2	69.9	197.8	2.8	1.4	10	Ricklefs and Williams (1984); Westerterp and Drent (1985)
Common starling	C	BCI	79.2	67.7	274.6	4.1	–	33	Ricklefs and Williams (1984); Westerterp and Drent (1985)
Northern mockingbird (*Mimus polyglottos*)	I	GII	50.9	51.2	112.4	2.2	1.1	1	Utter (1971)
Northern mockingbird	C	GII	46.7	48.5	122.5	2.5	–	4	Utter (1971)
Yellow-eyed junco (*Junco phaeonotus*)	I	GII	20.5	28.7	66.7	2.3	1.1	6	Weathers and Sullivan (1989b); Tinbergen and Williams (2002)
Yellow-eyed junco	C	GII	18.9	27.3	74.4	2.7	–	7	Weathers and Sullivan (1989b); Tinbergen and Williams (2002)
Savannah sparrow (*Passerculus sandwichensis*)	I	GII	19.5	27.8	79.5	2.9	0.9	26	Williams and Nagy (1985); Williams (1987); Williams and Dwinnel (1990)
Savannah sparrow	C	GII	17.7	26.1	74.2	2.8	–	23	Williams (1987, 1993b); Williams and Dwinnel (1990)
'Nuttall's' white-crowned sparrow (*Zonotrichia leucophrys nuttalli*)	I	GII	29.3	36.1	82.6	2.3	1.0	22	Weathers et al. (2002)
'Nuttall's' white-crowned sparrow	C	GII	28.4	35.3	85.1	2.4	–	22	Weathers et al. (2002)
'Mountain' white-crowned sparrow (*Zonotrichia leucophrys oriantha*)	I	GII	28.6	35.5	98.5	2.8	1.1	31	Weathers et al. (2002)
'Mountain' white-crowned sparrow	C	GII	27.6	34.7	108.4	3.1	–	31	Weathers et al. (2002)

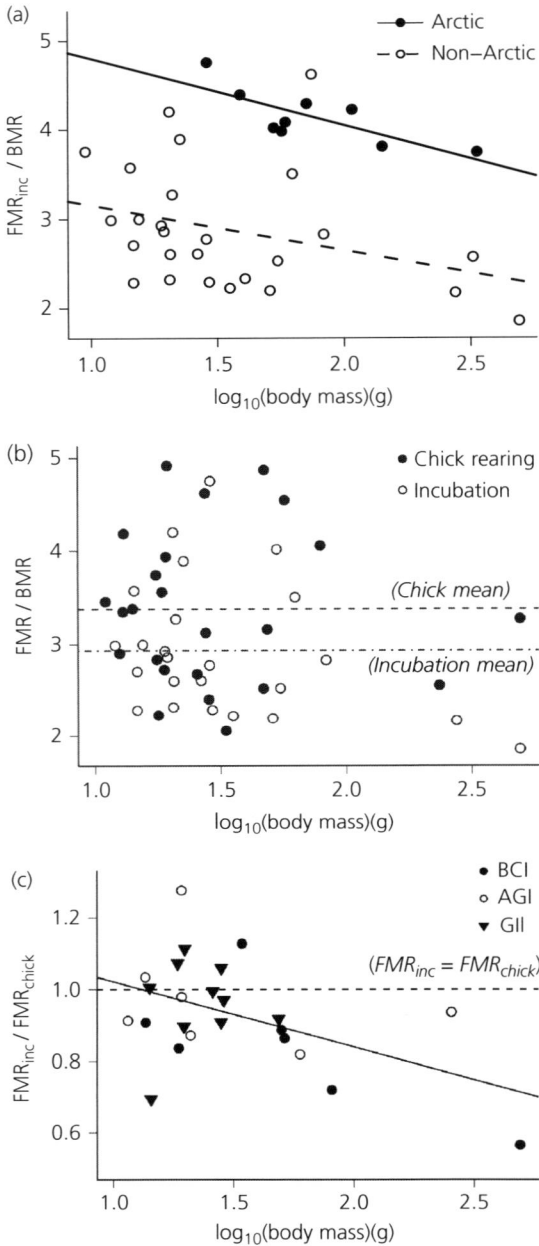

Figure 13.1 The energy costs of incubation in relation to the logarithm of adult body mass in terrestrial birds, expressed as multiples of basal metabolic rate (BMR) (a, b), or relative to the energy cost of chick rearing (c). All illustrations are based on data in Table 13.2. (a) shows the increase in field metabolic rate during incubation (FMR_{inc}, $kJ \cdot d^{-1}$) relative to BMR of birds breeding in the Arctic (Arctic; filled circles, solid line) and at lower latitudes (Non-Arctic; open circles, dashed line). (b) shows the energy cost of incubation (open circles) and chick rearing (filled circles), respectively. The dotted-dashed and dashed lines show means for incubation and chick rearing, respectively. (c) shows the relative cost of incubation, expressed as the fractional difference between FMR_{inc} and field metabolic rate during chick rearing (FMR_{chick}, $kJ \cdot d^{-1}$) partitioned for birds with bi-parental continuous incubation (BCI, filled circles), assisted gyneparental incubation (AGI, grey circles), and gyneparental intermittent incubation (GII, open circles) (for further details see Table 13.1). The fitted line describes the relationship between FMR_{inc}/FMR_{chick} across incubation systems. The dashed line denotes the point where $FMR_{inc} = FMR_{chick}$.

leucophrys ssp. *oriantha*) whereas FMR_{inc} was 14% lower than FMR_{chick} in a high-altitude population of the same species (*Z. l.* ssp. *nuttalli*; Weathers et al. 2002). It is possible that FMR_{chick} was higher in the high-altitude population (where overall FMR was also 24% higher than in the coastal population) because food resources to meet chick demands were scarce in their harsher environment. Clearly, we need more data on FMR_{inc} *and* FMR_{chick} during the same breeding attempt, and for the same species in different populations *and* under different environmental contexts so we can study any such effects in more detail.

13.2.3 Energy costs in different incubation systems

Williams (1996) broadly distinguished modes of incubation behaviour as falling into one of three categories (bi-parental continuous incubation—BCI; assisted gyneparental incubation—AGI; gyneparental intermittent incubation—GII) as defined in Table 13.1. Given the likely between-system variation in predicted constraints on energy acquisition experienced by the incubating parent, it is relevant to ask if this is reflected in the relative energetic costs of incubation. We investigated this by calculating ratios of FMR_{inc} to FMR_{chick} and then regressed these data on incubation system and $\log_{10}$ adult body mass. Absolute means varied in the predicted direction: species with BCI had lower overall relative FMR_{inc} ($0.84 \pm 0.065 \times FMR_{chick}$), whereas those with AGI ($0.98 \pm 0.057 \times FMR_{chick}$) and GII ($0.96 \pm 0.038 \times FMR_{chick}$) had similar relative FMR_{inc} that did not deviate qualitatively from the 1:1 line, i.e. where the energy cost of incubation is equal to the energy cost of chick rearing (Figure 13.1c). These differences, however, did not quite reach statistical significance ($F_{2,20} = 2.8$, $P = 0.09$). In line with this, when measuring FMR in two sympatric species of Arctic-breeding shorebirds, the uni-parental little stint (*Calidris minuta*) and the bi-parental dunlin (*Calidris alpina*), Tulp et al. (2009) could find no difference between FMR_{inc} and FMR_{chick} in either species. These findings are not surprising in view of the small number of species that have been investigated to date, and the general lack of consistent measurements of FMR_{inc} and FMR_{chick} within a single breeding attempt. To this end, it is difficult to predict how general patterns will be affected when more species are studied.

13.2.4 Energetics of incubation in seabirds

Incubation behaviour in seabirds differs from that of most terrestrial birds in that the incubation period is comprised of relatively lengthy bouts of nest attendance followed by prolonged foraging journeys at sea. Thus, the energy costs of incubation in these taxa can be partitioned into the energy expenditure (i.e. MR_{inc}), and that while parents are away from the nest foraging ($MR_{off-nest}$). Tinbergen and Williams (2002) compared MR for these two stages of seabird incubation using data from Table 11.5 in Williams (1996). On average, the workload during foraging trips was considerable, i.e. $4.8 \times BMR$, whereas that of actual egg heating was comparatively low, i.e. $1.7 \times BMR$. Largely due to energetically expensive foraging trips,

FMR_{inc} averaged $3.3 \times BMR$ across both stages (assuming birds spent half their time at sea; Tinbergen and Williams 2002), which is similar to the overall mean for terrestrial birds, i.e. $3.5 \times BMR$. These estimates are in line with more recent results. Black-browed albatrosses (*Thalassarche melanophrys*) expended energy at a rate of $4.0 \times BMR$ during foraging trips, and at $1.3 \times BMR$ during incubation on the nest, i.e. an average of $2.7 \times BMR$ for both activities (Shaffer et al. 2004). Similar values were obtained by Green et al. (2009) who, using heart-rate loggers for a full year in macaroni penguins (*Eudyptes chrysolophus*), estimated $MR_{off-nest}$ as $3.8 \times BMR$, and MR_{inc} as $1.8 \times BMR$. With an average of $2.5 \times BMR$, FMR_{inc} in macaroni penguins was 16% lower than the yearly average (which was $2.9 \times BMR$). However, when partitioning the yearly routines into active and passive periods, MR during incubation foraging trips could not be statistically separated from that during chick rearing (or other active parts of the annual activity budget; Green et al. 2009). The pronounced activity-related variation in the energy cost of incubation in seabirds highlights the need for detailed assessment of the subject's activity budgets in conjunction with measurement of energy expenditure that integrates longer time periods.

13.3 Energetic and thermal consequences of clutch size

The idea that demands of incubation can constrain the evolution of clutch size has been a tenet in ornithology for decades (Kendeigh 1963; Tatner and Bryant 1993; Williams 1996; Monaghan and Nager 1997). Central to this is the notion that larger clutches are energetically more costly to incubate (Thomson et al. 1998), and that effective incubation of such clutches can only be achieved at the expense of parental fitness (Reid et al. 2000a; Visser and Lessells 2001; Engstrand and Bryant 2002; de Heij et al. 2006). Studies that have measured the energy expenditure of a bird whilst sitting on eggs, i.e. MR_{inc}, consistently show that parents augment their metabolic rate when incubating enlarged clutches (Table 13.3). Of the studies that have examined the energetic consequences of clutch size, the mean (± 1 SE) increase in MR_{inc} of $3.4 \pm 0.2\%$ per egg is remarkably consistent regardless of species and whether measurements were made in the field or in the laboratory (Table 13.3). For instance, de Heij et al. (2007) measured MR_{inc} of free-living great tits (*Parus major*) incubating enlarged or unmanipulated clutches in their natural environment during two subsequent years with a mean ambient temperature (T_a) of 12°C. Females

Table 13.3 Studies that have measured the incubation metabolic rate (MR$_{inc}$) of birds incubating experimentally enlarged or reduced clutches. Regardless of incubation system, the female incubated alone in all cases, and all studies adopted a paired experimental design. Thermal neutral: Denotes whether measurements were performed within, i.e. 'Yes', or below, i.e. 'No', the birds' thermal neutral zones (TNZs). Clutch size: Denotes the original, i.e. unmanipulated clutch size of females; and 'N/A' means 'Not applicable'. ⁺mean clutch size during two subsequent years. * Clutch size did not affect MR$_{inc}$ when measurements were made within the TNZ. ** A reduction in clutch size of two or three eggs did not affect MR$_{inc}$.

Species	Setting	Thermal neutral	Clutch size	Manipulation (no. eggs)	Effect size per egg (% increase in MR$_{inc}$)	N	Reference
Common starling (*Sturnus vulgaris*)	Laboratory	Yes, No	N/A	+ 1 to + 8	4.0*	2, 8	Biebach (1981, 1984)
Atlantic canary (*Serinus canaria*)	Laboratory	Yes, No	4 or 5	−2 or −3	3.5*	3	Weathers (1985)
Blue tit (*Cyanistes caeruleus*)	Field	No	13	−5	3.6	1	Haftorn and Reinertsen (1985)
Great tit (*Parus major*)	Field	No	8⁺	+ 3	2.7**	29	de Heij et al. (2007)
Zebra finch (*Taeniopygia guttata*)	Laboratory	No	4	+ 2	2.8	12	Nord et al. (2010)

increased MR$_{inc}$ by some 3% for each egg added to the clutch. Similar results were later obtained for zebra finches (*Taeniopygia guttata*) incubating in the laboratory at a T_a of 10°C (Figure 13.2; Table 13.3; Nord et al. 2010). Clutch size does not appear to affect MR$_{inc}$ when birds are incubating within their thermal neutral zone (TNZ) (Biebach 1981, 1984; Weathers 1985). However, thermal neutral conditions are rarely met during the incubation stage, at least for free-living birds at temperate latitudes. We conclude that the incubation of enlarged clutches requires increased MR$_{inc}$ when birds are incubating below their lower critical temperature.

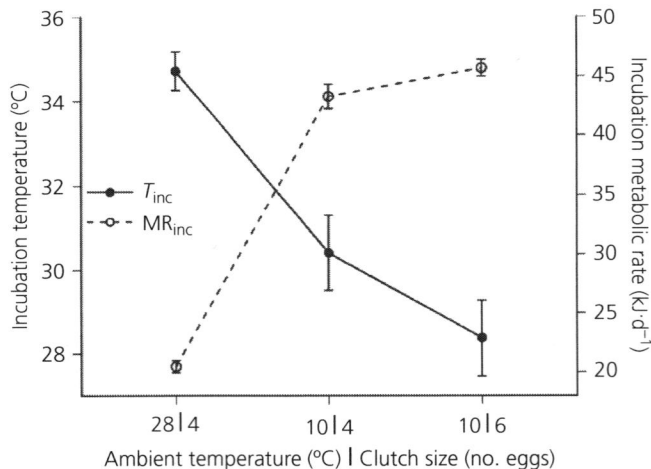

Figure 13.2 The relationship between the demands for, and energetic investment in, incubation by female zebra finches (*Taeniopygia guttata*) during steady-state incubation at night when females took the sole responsibility for incubation. The energy cost of incubation was manipulated either by reducing ambient temperature or by increasing clutch size. The investment in incubation was measured as variation in incubation temperature (T_{inc}; filled circles, solid line, left-hand axis) and oxygen consumption (incubation metabolic rate, MR$_{inc}$; open circles, dashed line, right-hand axis) during steady-state incubation at night. Accordingly, females invested more energy in incubation as demands increased but they did not fully compensate for higher costs because T_{inc} was lower in lower ambient temperature and during incubation of larger clutches. Points show means ± 1 SE. Redrawn from Nord et al. (2010) with permission from John Wiley & Sons.

13.3.1 Effects of clutch size on the energy budget of incubating birds

Although MR_{inc} increases as a function of clutch size, this does not provide unequivocal evidence in support of an energetic constraint on the evolution of clutch size, because higher MR_{inc} *per se* might be of little evolutionary importance unless it translates into a corresponding increase in FMR_{inc}, thereby constraining a bird's energy budget. However, there is limited evidence to suggest that experimental manipulation of clutch size affects FMR_{inc}. Moreno et al. (1991) reported higher FMR_{inc} of collared flycatchers (*Ficedula albicollis*) incubating enlarged clutches compared with those incubating (manipulated) smaller clutches (with neither category different from unmanipulated [control] clutches). In contrast, Moreno and Sanz (1994) reported lower FMR_{inc} of the Spanish subspecies of European pied flycatchers (*F. hypoleuca iberiae*) incubating reduced clutches, with no change in FMR_{inc} during incubation of enlarged clutches. In comparison, clutch-size manipulation did not affect FMR_{inc} in a Swedish population of European pied flycatchers (*F. h. hypoleuca*; Moreno and Carlson 1989). More recent estimates of FMR_{inc} in relation to clutch-size manipulation conform to the latter. Engstrand et al. (2002) found no evidence that an experimental reduction or increase in clutch size of white-throated dippers (*Cinclus cinclus*) affected FMR_{inc} (although its variation was larger in females incubating enlarged clutches). Similarly, female great tits did not expend more energy when incubating enlarged clutches (de Heij et al. 2008), despite birds in the same population having increased MR_{inc} whilst on the nest (de Heij et al. 2007). It should be noted, however, that clutch enlargement is not necessarily cost-free despite consistent values of FMR_{inc} across different manipulations. For instance, females with enlarged clutches in the study of Moreno and Carlson (1989) took longer to hatch their eggs, and great tits incubating enlarged clutches in the Netherlands hatched fewer eggs and survived less well between breeding seasons (de Heij et al. 2006). In summary, whereas the costs of incubation have been important in the evolution of avian clutch size, increased energy expenditure *per se* only sometimes prevents females from laying additional eggs. However, studies are scarce and this prevents unambiguous assessment of the role of energetic constraints in determining optimal clutch size.

13.3.2 The thermal consequences of clutch size

Manipulation of clutch size not only alters the instantaneous energetic investment, i.e. MR_{inc}, required for maintaining equilibrium egg temperature (T_{egg}) but also the physical properties of the clutch, as thermal mass of the clutch is directly proportional to the number of eggs in the nest cup. As a result, the rate of heat loss from eggs during incubation recesses is lower in larger clutches (Boulton and Cassey 2012). It follows that, for equal recess durations, females on enlarged clutches should on average return to eggs that are warmer than should females on unmanipulated clutches. In line with this, the cooling rate of unattended European pied flycatcher eggs varied inversely, and non-linearly, with manipulated clutch size (Nord and Nilsson 2012). Consequently, females on enlarged clutches, despite taking longer recesses from incubation, returned to eggs that were of similar temperature as those of females on unmanipulated clutches, and 1.1°C higher than those of females incubating reduced clutches (Nord and Nilsson 2012). Hence, females on enlarged clutches might accrue energy savings from a reduced need to rewarm the clutch at the beginning of a new incubation session, a task requiring 30% higher energy expenditure than steady-state incubation (Vleck 1981; Biebach 1986). This could attenuate any increase in FMR_{inc} pertaining to clutch enlargement, thus potentially explaining the discrepancy between studies that measured MR_{inc} and those that measured FMR_{inc}. To investigate this possibility, we estimated the cooling rate of great tit eggs in clutches of six, nine and 12 eggs (corresponding to the experimental clutch sizes in de Heij et al. 2007, 2008), using the equation for the cooling rate of great tit eggs from Boulton and Cassey (2012) and assuming a mean egg mass of 1.79 g (Boulton and Cassey 2012), a mean ambient temperature (T_a) of 9.3°C and that birds were 9.5 days into incubation (de Heij et al. 2008). We then used clutch-specific cooling rates to derive estimates of the energy costs of rewarming (EE_{warm}) differently sized clutches at the end of the recess, using the equation of Biebach (1986) and assuming a mean recess duration of 8.6 min (de Heij et al. 2008) and a T_{egg} of 35.4°C at the time females left the nest (cf. Haftorn 1983). Across clutch sizes, EE_{warm} contributed approximately 17% to total FMR_{inc}, but also decreased with increasing clutch size, attaining daily mean values of 14.3, 13.6 and 12.8 kJ·d^{-1} for clutches of six, nine and 12 eggs, respectively. Thus, as females on enlarged clutches in the study by de Heij et al. (2007) allocated 3.3 kJ·d^{-1} to keep additional eggs warm during steady-state incubation, the energy savings accrued from reduced EE_{warm} alone would have attenuated this increased energy cost by 23%.

In summary, there is evidence that the thermal benefits of larger clutches confer energy savings on incubating parents. As such, a shift in the requirements for

EE$_{warm}$ may partly explain why most studies that have asked if the energy costs of incubation constrain the evolution of clutch size have found no effects thereof, and why experimental manipulations of clutch size do not always follow predictions relating to reduced or increased parental fitness.

13.4 How are energy costs of incubation paid?

In intermittently incubating birds outside of thermal neutral temperatures, demands of incubation can result in significant energy shortage (e.g. Ardia et al 2010). In situations where the energy costs of incubation are sufficiently large to prevent parents from successfully meeting resource demands of other somatic functions, costs of incubation may ultimately even result in reduced fitness (Visser and Lessells 2001; de Heij et al. 2006). Attempts have been made to quantify the behavioural basis for any fitness costs associated with incubation (reviewed by Reid et al. 2002b). Such costs are typically characterised either by a temporal shift in the balance between incubation and other mutually exclusive activities such as foraging and display, i.e. incubation to self-maintenance trade-offs (Voss et al. 2006), or as carry-over effects from sub-optimal incubation conditions to subsequent parental performance during chick rearing manifested as, for example, reduced chick-provisioning rates (Heaney and Monaghan 1996). By contrast, few have attempted to understand the mechanistic underpinnings of the costs of incubation. This is unfortunate because behavioural observations alone can rarely explain how reproductive costs (such as reduced survival of hard-working parents) (e.g. de Heij et al. 2006) arise. Thus, although the evidence suggests that incubation can be energetically costly to parents, we currently have limited knowledge of *what* is causing these costs. To remedy this, there is now a need to switch focus from studies of consequences (that teach us about fitness costs) to studies of causes (that teach us about function) to understand further how costs of incubation arise and evolve. We will briefly introduce two possible pathways that could link the energy costs of incubation to fitness.

13.4.1 Oxidative stress as a proximate cost of incubation

One possibility is that oxidative stress might play a role in mediating reproductive costs during incubation (e.g. Selman et al. 2012; Metcalfe and Monaghan 2013).

Oxidative stress is an imbalance between the rate of production of reactive oxygen species (ROS) in the mitochondria (which occurs as a by-product of cellular metabolism) and the body's capacity to neutralise these via antioxidant defences and repair mechanisms (Murphy et al. 2011). Because ROS are highly reactive, and can damage lipids, proteins and DNA (Hulbert et al. 2007; Murphy et al. 2011), oxidative stress might invoke substantial somatic costs that potentially increase cellular senescence and ultimately even result in cell death (cf. discussion in Monaghan et al. 2009). Research to date provides mixed evidence for an increased susceptibility to oxidative stress during reproduction. Metcalfe and Monaghan (2013) reviewed 20 observational studies that measured markers of oxidative stress in breeding animals and found that three studies reported increased oxidative damage during reproduction, two reported a decrease, seven reported no change, and the remaining nine reported inconsistent results. By contrast, studies that experimentally increased reproductive effort at the chick-rearing stage provided stronger evidence for the idea that higher workloads might cause oxidative stress. Studies that elevated reproductive effort in zebra finches (Alonso-Alvarez et al. 2004; Wiersma et al. 2004) and great tits (Christe et al. 2012) by manipulating brood size found that parents provisioning enlarged broods were less able to resist an experimental ROS attack on red blood cells (Alonso-Alvarez et al. 2004; Christe et al. 2012) and pectoral muscle tissue (Wiersma et al. 2004). However, because these studies did not assess cellular ROS levels or resultant oxidative damage, it is not known if hard-working parents suffered increased oxidative stress. However, in support of this view Fletcher et al. (2013) quantified oxidative damage to blood plasma proteins, antioxidant capacity and FMR in American red squirrels (*Tamiasciurus hudsonicus*) at three discrete time-points across the annual cycle and found that oxidative damage was highest during lactation, i.e. the period associated with the highest energetic demands, and that females with higher FMR showed reduced antioxidant capacity and increased oxidative damage. By supplementing subjects with food during both winter and breeding, Fletcher et al. (2013) also found that alleviating constraints on energy acquisition resulted in increased antioxidant capacity and, consequently, reduced oxidative damage. To date, no studies have assessed the potential role of oxidative stress as a mediator of incubation costs. Nevertheless, as incubation and chick rearing can be equally costly in terms of energy, one would predict that increased incubation effort might sometimes invoke oxidative stress such as in species with a GII strategy in thermally challenging or resource-deficient environments. This, in turn, might provide a causal explanation of the association between demands of incubation and fitness. It should be kept in mind, however, that the relationship between energy expenditure and ROS production is not necessarily straightforward (Monaghan et al. 2009), and that experimentally increased reproductive effort does not always increase oxidative damage (at least in the laboratory under *ad libitum* food availability; cf. Xu et al. 2014). Future studies in this area have the potential to provide exciting new insights into the evolution of, and constraints on, avian incubation and clutch size.

13.4.2 Immunosuppression in incubating birds

Another potentially interesting aspect of incubation costs is the relationship between incubation effort and immune system functionality. Although maintenance of the immune system might be relatively inexpensive in energetic terms (Klasing 2004), up-regulation of immunity in response to disease typically requires augmentation of metabolic rate (Marais et al. 2011; Hawley et al. 2012) and increased use of energy-conserving behaviours (Råberg et al. 2000; Nord et al. 2014). Similarly, immunosuppression is sometimes observed in energetically demanding situations (e.g. Svensson et al. 1998; Merino et al. 2006). Accordingly, incubating birds might be required to trade-off energy allocation between egg warming and an immune response should infection occur during the incubation stage. For instance, reduced investment in immune function or increased susceptibility to disease was proposed as a potential mechanism to explain reduced survival of great tits incubating enlarged clutches (de Heij et al. 2006). Hanssen et al. (2005) increased clutch size of incubating common eiders (*Somateria mollissima*) and measured the corresponding effects on background immunity and the birds' capacity to respond to an experimental challenge to the antibody-mediated immune system. Females incubating enlarged clutches did not only show a decrease in white blood cells and a reduced propensity to respond to the immune challenge (measured as antibody production), but they also had a reduced fecundity in the year following infection (Hanssen et al. 2005). Subsequent work with common eiders by Hanssen and co-workers lends further support to the idea that the fitness costs of incubation might be mediated via constraints on immune function, in providing evidence for a direct link between the activation of the immune system during incubation and survival to the next breeding season (Hanssen

et al. 2004; Hanssen 2006). Such effects might arise from increased susceptibility to disease in incubating birds as was shown for chronic malaria infections (which are known to carry fitness costs; Asghar et al. 2011) in European pied flycatchers incubating experimentally enlarged clutches (Siikamäki et al. 1997).

Demands from the immune system can constrain parental effort also during current reproductive attempts. For instance, energy costs increased and the incubation period was prolonged in incubating common eiders subjected to an experimental immune challenge (Hanssen 2006), suggesting a trade-off between self-maintenance (to cover costs of immunity) and incubation. Moreover, experimental onset of an antibody-mediated immune response during chick rearing reduces parental feeding effort (Ilmonen et al. 2000; Råberg et al. 2000), thereby also affecting offspring quality and fledging success (Ilmonen et al. 2000). Whereas parental activity is relatively low during incubation, it is possible that the energy needed for an immune response could constrain heat supply to the clutch or carry over to reduced feeding effort during subsequent chick rearing.

Knowledge of immunosuppression during incubation is limited (Cichon 2000) but there is evidence to suggest that constraints on immune function might serve as a mediator of reproductive trade-offs during incubation. Better establishing the generality and functional characteristics of this link within and between species should prove an interesting and fruitful endeavour for future research efforts.

13.5 How are costs of incubation transferred from adults to their offspring?

Fitness costs of incubation are rarely restricted to the incubating adult but, instead, are frequently transferred to embryos and chicks. For example, suboptimal incubation conditions reduced growth rate of nestling zebra finches (Gorman et al. 2005) and blue tits (*Cyanistes caeruleus*; Nilsson et al. 2008) and led to reduced body condition and immune function in tree swallows (*Tachycineta bicolor*; Ardia et al. 2010). By analogy, improved conditions during incubation increased body mass and body condition in tree swallow nestlings (Pérez et al. 2008). Such trans-generational costs of incubation often occur independently of any simultaneous effects acting on parents, suggesting that fitness consequences on chicks manifest as early as the embryonic stage and not predominantly post-hatching (cf. Heaney and Monaghan 1996). However, few attempts have been made to explain how costs to parents are translated to young.

13.5.1 The relationship between incubation demands and egg temperature

Normal development of avian embryos depends on a relatively tightly controlled temperature during incubation (Chapter 14; Webb 1987; DuRant et al. 2013b), which with few exceptions (cf. Booth and Jones 2002; De Marchi et al. 2008), is supplied by active heat transfer from the incubating parent via the brood patch. Because the latter is energetically costly (Tøien et al. 1986; Haftorn and Reinertsen 1990), it is likely that increased demands of incubation will influence the thermal environment parents are able to provide for their clutch. In domestic fowl (*Gallus gallus*) temperature deviations during incubation can cause abnormal embryonic development and increase *in ovo* mortality (Lundy 1969). While this notion has been paid relatively little attention in wild species, natural and semi-natural experiments suggest that even subtle variation in developmental temperature can have far-reaching consequences for many aspects of growth and development in non-domesticated taxa. Specifically, an experimental 1.5°C reduction in average incubation temperature in blue tits significantly increased hatching failure, decreased nestling size and increased metabolic rate close to fledging (Nord and Nilsson 2011), while in wood ducks (*Aix sponsa*) a temperature reduction of less than 1°C affected multiple measures of hatchling growth and physiological maturation, including body condition (Chapter 14; Hepp et al. 2006; DuRant et al. 2010), locomotory development (Hopkins et al. 2011), immunocompetence (DuRant et al. 2012a), and thermoregulatory physiology (DuRant et al. 2012b; DuRant et al. 2013a). Given the close correspondence between results of studies directly manipulating incubation temperature and those manipulating the incubation environment of parents, it can be assumed that changes in chick phenotype following increased demands on incubating parents are proximately mediated by energetic constraints on egg temperature maintenance.

Table 13.4 summarises data from studies that experimentally changed the energy costs of incubation and subsequently measured incubation temperature. With one exception, results from these studies point consistently to a causal association between incubation demands and T_{egg}, regardless of taxonomy, incubation system, or type of manipulation. Accordingly, T_{egg} decreases when the energy cost of incubation increases, and increases when costs are reduced. The observation that parents invest more in warming eggs when demands on their energy budget are alleviated (Ardia et al. 2009; D'Alba et al. 2009) supports the idea that parents are energetically constrained from maintaining optimal T_{egg}. These findings fit observational data in the field: Haftorn (1978, 1979, 1983) found that T_{egg} during steady-state incubation at night by three species of boreal passerines typically reflected changes in T_a. Thus, variation in T_{egg} may serve as a functional link in explaining trans-generational costs of incubation. This notion is confirmed by two studies (Table 13.4) that extended beyond hatching and cross-fostered nestlings to separate effects acting on adults from those acting on eggs. One found that nestlings incubated in experimentally cooled nests (but sired by control females) suffered reduced body condition and immunocompetence (Ardia et al. 2010), while the other, i.e. Ardia et al. (2009), found that nestlings incubated in experimentally warmed nests (but, again, sired by controls) by analogy experienced increased body condition (Pérez et al. 2008).

13.5.2 Why might egg temperature be lower under more demanding incubation conditions?

If low T_{egg} is harmful to developing young, why do parents on the nest decrease heat transfer to the clutch when incubation conditions deteriorate, thereby passing costs of incubation to their offspring? If parents are energetically constrained from maintaining eggs at optimal developmental temperatures (see section 13.5.1), it is likely that reduced T_{egg} under more demanding incubation conditions serves to reduce the energy costs to the incubating adult, and in so doing perhaps also those to the assisting partner (in AGI systems) (cf. Nilsson and Smith 1988). However, a reduction in heat transfer to eggs is not likely to allow parents to avoid increased energy costs altogether when conditions deteriorate. For instance, when measuring metabolic rate and incubation temperature in captive zebra finches during nocturnal incubation in different T_as and with different clutch sizes, Nord et al. (2010) found that while parents increased MR_{inc} in response to increasing demands of incubation, incubation temperature (T_{inc}) still decreased predictably with increasing demands (Figure 13.2). Similar results were obtained by Haftorn and Reinertsen (1985) for an incubating blue tit subject to gradual changes in T_a. Thus, although parents do not escape metabolic costs by investing less in keeping eggs warm, passing costs to offspring likely alleviates the energy costs of incubation to adults. Accordingly, a voluntary

Table 13.4 Studies that have manipulated the demands of incubation in different species undergoing different incubation systems (BCI: shared incubation [bi-parental continuous incubation]; AGI: female-only incubation, with male incubation feeding [assisted gyneparental incubation—Williams 1996]; and GCI: female-only, continuous and unassisted incubation [gyneparental continuous incubation]), and subsequently recorded the effect on incubation temperature (T_{inc}): [a] Females incubated alone during all measurements. * Clutch sizes were increased (from the modal clutch of two) with either one or two eggs, with no temperature difference between modal (two) and three egg clutches. Accordingly, the effect size (defined below) denotes the temperature difference between the mean of two- and three-egg clutches versus four-egg clutches; ** Effect only for minimum on-bout temperature in first-year breeders with effect size (see below) denoting difference between second-year or older versus first-year breeders; and *** Effect only present between reduced (N−2 eggs) and enlarged (N + 2 eggs) during early incubation. Effect size: The difference between the least and the most challenging incubation regime (e.g. between warm and cold temperatures, control and enlarged clutch sizes etc.). N: Total sample size across experimental categories in the respective studies. [b] All birds were subjected to all experimental treatments.

Species	System	Manipulation	Effect on T_{inc}	Effect size (°C)	N	Reference
Black-tailed gull (*Larus crassirostris*)	BCI	Clutch enlargement	Decrease*	1.8	34	Niizuma et al. (2005)
Tree swallow (*Tachycineta bicolor*)	AGI	Handicapping	Decrease**	2.8	48	Ardia and Clotfelter (2007)
Tree swallow	AGI	Nest heating	Increase	3.3	30	Ardia et al. (2009)
Glaucous gull (*Larus hyperboreus*)	BCI	Clutch enlargement	Decrease	1.7	19[b]	Verboven et al. (2009)
Common eider (*Somateria mollissima*)	GCI	Nest shelter	Increase	1.0	26	D'Alba et al. (2009)
Tree swallow	AGI	Nest cooling	Decrease	1.4	36	Ardia et al. (2010)
Zebra finch (*Taeniopygia guttata*)	BCI[a]	Nest cooling	Decrease	3.9	12[b]	Nord et al. (2010)
Zebra finch	BCI[a]	Clutch enlargement	Decrease	2.4	12[b]	Nord et al. (2010)
European pied flycatcher (*Ficedula hypoleuca*)	AGI	Clutch size	Decrease***	0.5	81	Nord and Nilsson (2012)
Western jackdaw (*Corvus monedula*)	AGI	Clutch enlargement	No change	–	93	A. Nord and J.-Å. Nilsson (2014) (unpublished data)

reduction in T_{inc} could also explain why experimentally increased demands of incubation do not always translate into increased FMR_{inc} (e.g. Engstrand et al. 2002; de Heij et al. 2008). However, given the potentially severe consequences to chicks, it is unlikely that reduced T_{inc} is a ubiquitous feature of incubation when demands on parents increase. Rather, heat transfer should be constrained only as a 'last resort' when environmental refuelling options are limited or when parents risk a detrimentally high sustained metabolic rate whilst on the nest. Consistent with this, studies that have manipulated clutch size of European pied flycatchers suggest that females incubating enlarged clutches pass costs on to their offspring (in terms of reduced hatching success) only under relatively cold and wet breeding conditions (Siikamäki 1995; Nord and Nilsson 2012), which is in agreement with the observation that energetic constraints on incubation are not omnipresent but may arise only when environmental conditions deteriorate (Moreno 1989). To this end, incubation represents a trade-off between the energetic investment in egg warming and self-maintenance, the resolution of which is context-dependent as determined by the quality of

parents on the one hand and ambient conditions on the other. Future research should combine traditional measurements of energy expenditure by parents with simultaneous studies of variation in T_{inc} and/or T_{egg}. This will allow us to separate the amount of energy expended by incubating parents from that directly invested in keeping eggs warm, thereby allowing more detailed estimates of the true energetic investment in incubation.

13.6 Are energy costs of incubation sometimes prohibitively high?

Does the resolution of the trade-off between heat transfer to eggs by parents and self-maintenance ever reach the point when full priority is given to the latter and, if so, what happens when it does?

13.6.1 Temporary nest abandonment as a cost of incubation

One scenario that may indicate a complete switch to self-maintenance in incubating birds is when parents cease to incubate for several hours, resulting in T_{egg} dropping well below temperatures required for normal embryonic development, i.e. 25–27°C (Haftorn 1988), for a sufficiently long time to reduce mean T_{egg} significantly (Figure 13.3). Sometimes temporary egg abandonment is a part of the normal incubation routine (Lill 1979; Jia et al. 2010), but in most cases nest abandonment occurs during periods of inclement weather (Morton and Pereyra 1985; Haftorn 1988; Williams 1996; MacDonald et al. 2013). For instance, when monitoring incubation behaviour of alpine-breeding horned larks (*Eremophila alpestris*) at 86 nests during four years, MacDonald et al. (2013) found that extended incubation recesses, i.e. recesses ≥ 5.4 × mean recess duration, while infrequent (< 1% of total number of recesses), occurred more frequently in the two coldest years of the study (60% of all extended recesses), and often during spells of cold and/or wet weather (e.g. 39% of all extended recesses occurred during three storms). Despite similar observations in several other studies (e.g. Morton and Pereyra 1985; Haftorn 1988; Williams 1996 and references therein), prolonged egg neglect as a strategy of last resort by parents to cope with the demands of

incubation has been largely overlooked. To this end, we urge investigators not to regard T_{egg} records indicating anomalous parental absences from the nest as artefacts of disturbance or similar, but instead to evaluate such data properly in an ecological context.

13.6.2 Does prolonged egg neglect carry fitness costs?

It is currently poorly understood if prolonged periods of egg neglect (and consequently, prolonged periods of embryonic hypothermia) carry fitness costs to young. Contrary to expectation (e.g. Webb 1987), such behaviour often does not seem to impair hatching success (Morton and Pereyra 1985; Haftorn 1988; Wang and Beissinger 2009). However, there is evidence to suggest that egg neglect can carry fitness costs: horned larks that took extended recesses, i.e. recesses ≥ 5.4 × mean recess, from incubation hatched only 81% of their eggs whereas those that did not had a hatching success of 91% (MacDonald et al. 2013). Thus, unlike studies that experimentally introduced slight, but continuous, thermal stress (e.g. Hepp et al. 2006; DuRant et al. 2010;

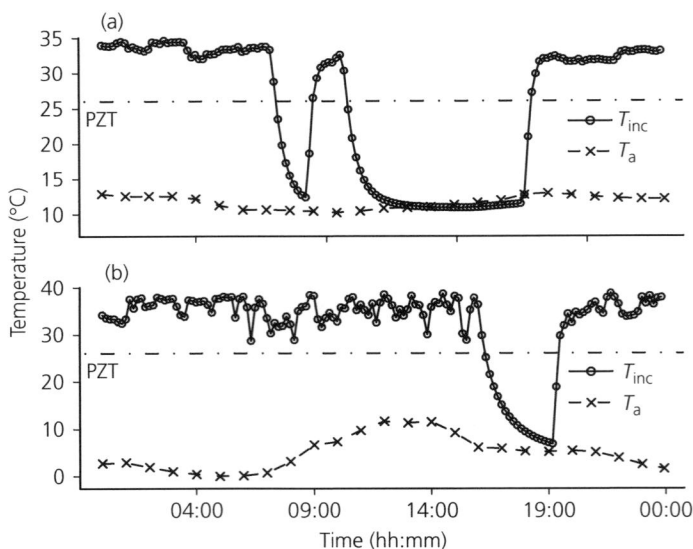

Figure 13.3 Twenty-four-hour temperature recordings of incubation temperature (T_{inc}) from a nest of (a) a European pied flycatcher (*Ficedula hypoleuca*), and (b) a blue tit (*Cyanistes caeruleus*) incubating in southern Sweden (A. Nord and J.-Å. Nilsson, 2014 unpublished data), showing representative examples of prolonged egg neglect where eggs are left unattended for several hours, thereby experiencing temperatures at or close to ambient (T_a) for prolonged periods. The horizontal dotted-dashed lines denote the physiological zero temperature (PZT; 25–27°C), below which no embryonic development occurs. Females in both nests hatched their eggs and subsequently fledged young. The incubation period of the flycatcher was two days longer, i.e. 15 days, than the annual mean of 13.1 days, whereas the blue tit clutch hatched after an incubation period of 13 days, which was equivalent to the annual mean of 13.1 days.

Nord and Nilsson 2011), it seems that periodic cooling of eggs only sometimes results in hatching failure.

Because egg development is thought to cease at temperatures below 25–27°C, i.e. the physiological zero (Webb 1987; Haftorn 1988), egg neglect could also increase the length of the incubation period. This, in turn, could increase total energy expenditure of embryos (Booth 1987; DuRant et al. 2011) such that body condition at hatching is reduced (Hepp et al. 2006), or increase the rate of nest predation provided the probability thereof is time-dependent (Tombre and Erikstad 1996; Remeš and Martin 2002). Surprisingly, correlative studies evaluating the effects of extended recesses on incubation period have not found support for this idea (Wang and Beissinger 2009; MacDonald et al. 2013; but see Miskelly 1989). However, by experimentally doubling recess duration in wood ducks such that average incubation temperature decreased by 0.4°C (but without inducing the detrimentally low temperatures characteristic of prolonged egg neglect), Carter et al. (2014) found that the incubation period increased by two days. Importantly, however, when the thermal manipulation compensated for increased recess duration by maintaining the same average incubation temperature, no effect on incubation duration was observed (Carter et al. 2014). Thus, we could expect to record obvious effects on embryonic growth and maturation only in situations where extended recesses are sufficiently frequent or long to reduce significantly the average temperature experienced by developing eggs during the incubation period.

In summary, there is currently no consistent evidence to suggest when prolonged egg neglect occurs and whether or not it is associated with fitness costs. This calls for systematic experimental studies to identify the ecological and evolutionary causes of this behaviour, as well as its consequences (however slight) for embryonic development, growth and post-hatching performance.

13.7 Water economy of desert birds during incubation

Desert regions present a unique challenge for incubating birds because the combination of intense solar radiation, frequent temperature extremes, low primary productivity, and scarcity of drinking water can mean that water and energy are in short supply (Louw and Seely 1982; Tieleman et al. 2008). Some species of desert passerines in which only the female incubates, such as the dune lark (*Certhilauda erythrochlamys*), which is endemic

to the Namib Desert, construct domed nests to protect eggs from direct sunlight (Williams 1996). During the breeding season in January to February, T_a rarely exceeds 40°C. Thus, females can leave nests during the day without their eggs reaching lethal temperatures (J.B. Williams, 2014 unpublished data). In other deserts, such as the Rub' al Khali in Saudi Arabia, species build an open cup nest that requires both parents to tend to eggs continuously during midday to prevent lethal temperatures from developing. Success of incubation under such circumstances is remarkable given that birds, in general, have relatively high rates of water loss, a physiological feature that accompanies high metabolic rates (Aschoff and Pohl 1970). Even in thermal neutral temperatures, high metabolic rates result in core body temperatures (T_bs) averaging approximately 41°C (Prinzinger et al. 1991). Thus, small birds routinely have T_bs close to the presumptive upper lethal limit of 46–47°C (Dawson and Schmidt-Nielsen 1964; Portner and Farrell 2008). As a result, water economy may be compromised during periods of heat stress, such as while incubating eggs on a nest exposed to the full sun, because birds prevent T_b from exceeding the lethal limit by evaporative cooling, a mechanism that mandates substantial water loss.

Tieleman and Williams studied reproductive ecology and physiology of greater hoopoe-larks (*Alaemon alaudipes*) in the desert of central Saudi Arabia for over a decade. The region receives on average 90 mm of rain per year, usually between February and May, and is composed of flat gravel plains sparsely populated with grasses (e.g. *Panicum, Stipagrostis* and *Lasiurus* spp.) and small bushes (e.g. *Salsola* and *Indigofera* spp.), and dissected by wadis with acacia trees (Williams and Tieleman 2008). In these Areas, greater hoopoe-larks breed from February to May coincident with rainfall when food is more abundant. Rather than placing their nests beneath acacia trees where shade would be abundant, greater hoopoe-larks build nests on the open gravel plain early in the season (February) when solar radiation is not so intense and T_a not as high. As the season progresses, they place their nests at the base of, or more often in the top of, small *Salsola* bushes where eggs are exposed to solar radiation when the parent is not attending the eggs (Tieleman et al. 2008).

The general environment of incubation in this species varies depending on the month of rainfall. During May and June, T_a can reach 45°C during midday and soil surface temperatures can reach 65°C (Tieleman and Williams 2005). Clearly, birds incubating under these regimes should have a difficult time maintaining T_b without experiencing a cost in terms of excess water loss for evaporative cooling. To estimate heat stress of birds directly, Tieleman et al. (2008) also monitored the temperature of copper models, to obtain 'operative temperature' (T_e), defined as the body temperature of an animal if it were in thermal equilibrium with the environment in the absence of metabolic heating or evaporative cooling (Bakken 1976; Dzialowski 2005). Birds that placed nests on the gravel plain experienced the hottest conditions (Figure 13.4a-c), and although maximum T_es in base nests and nests on top of bushes were similar, parents on base nests experienced up to 5°C higher T_es during a portion of the day compared with those on nests on top of *Salsola* bushes (Figure 13.4b and c).

13.7.1 Temperature of unattended eggs

Although greater hoopoe-larks normally cover eggs continuously, parents leave eggs unattended when attempting to lead approaching predators away from the nest. Tieleman and Williams were interested to know what temperature embryos would experience if the parents were to leave the eggs unattended (Tieleman et al. 2008). This would likely have substantial consequences for egg survival because the upper lethal

temperature limit for embryos is approximately 44°C (Webb 1987; Williams 1996), i.e. well below temperatures experienced at greater hoopoe-lark nests (Figure 13.4a-c). Between early April and late May, average daytime temperature of an unattended egg ($T_{egg-unatt}$) in an artificial nest increased from 30.0 to 46.4°C (gravel nests), 28.5 to 40.5°C (base nests), and 28.6 to 43.1°C (top of *Salsola* bush nests). Maximum values increased from 41.1 to 59.2°C (gravel nests), 35.3 to 49.3°C (base nests), and 37.3 to 54.8°C (bush nests). Early in the season, $T_{egg-unatt}$ was similar in base and bush nests and remained below 44°C throughout the day (Figure 13.4d–e). Later in the season, eggs in bush nests reached temperatures exceeding those in base nests by up to 9°C, thereby resembling gravel nests and surpassing the lethal limit for eggs for up to seven hours per day (Figure 13.4f). The number of hours that $T_{egg-unatt}$ exceeded 44°C increased over the course of the season at all three nest sites and was significantly higher in gravel nests compared with bush nests, but was not significantly different between bush and base nests (Figure 13.4d–f). So, later in the season parents must attend eggs continuously despite the risk of incurring substantial water costs from evaporative cooling of themselves and eggs.

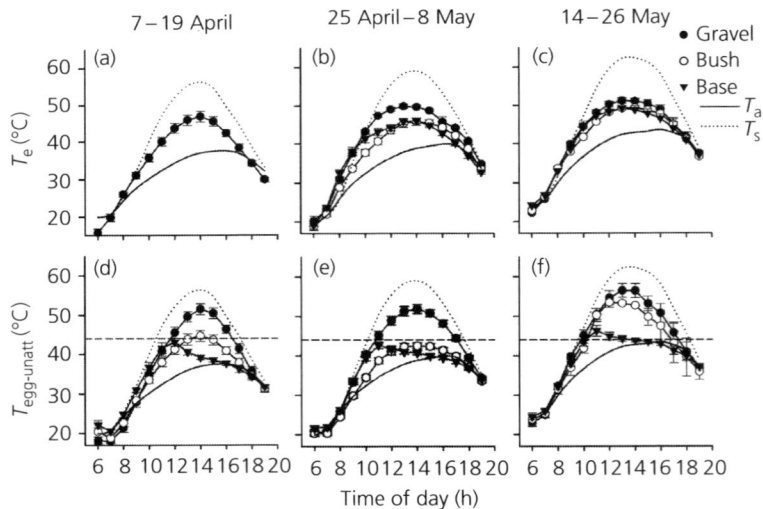

Figure 13.4 Mean (± 1 SE) operative temperature (T_e; [a]–[c]) and temperature of unattended eggs ($T_{egg-unatt}$; [d]–[f]) as a function of time of day, measured in artificial nests of greater hoopoe-larks (*Alaemon alaudipes*) in the desert of Saudi Arabia (GMT + 3 h) in mid-April (a and d), late April to early May (b and e), and mid-May (c and f). Nests were located either on the gravel plain (● 'Gravel'), at the base of Salsola bushes (▼ 'Base'), or in the top of Salsola bushes (○ 'Bush'). Solid lines show average ambient temperatures (T_a) and dotted lines show average soil temperatures (T_s). The horizontal dashed lines show the predicted lethal T_{egg}, i.e. 44°C. See main text for details. Reproduced with permission from Tieleman et al. (2008).

13.7.2 Body temperature of incubating birds

By implanting temperature-sensitive radio transmitters into the peritoneal cavity of incubating greater hoopoe-larks, Tieleman and Williams (Williams and Tieleman 2008; B.I. Tieleman and J.B. Williams, 2014 unpublished data) recorded T_bs of incubating parents during on- and off-bouts. In May, after about 9am (GMT + 3 h), parents on the nest began panting to control the rise in T_b caused by increasing T_a. Despite birds panting, T_b would rise to near and sometimes surpass 45°C, a high body temperature for birds. During the middle of the day, the same birds, when relieved of incubation duties by their partner, would often fly to a burrow of a spiny-tailed lizard (*Uromastyx* sp.), and press their chest to the cooler soil surface (which was measured to be 37°C) within the burrow. After they entered burrows, their T_bs declined by a rate of 0.22°C per minute (Williams and Tieleman 2008). Hence, a bird with a T_b of 45°C would reduce its T_b to 42°C in about 14 minutes by visiting a lizard burrow. Birds that flew from their nest to shade only reduced their T_bs by 0.03°C per minute. So, using the lizard burrows was a better option for losing body heat.

Use of hyperthermia by incubating greater hoopoe-larks undoubtedly saved considerable water. Heat storage (H_s) can be estimated from the equation of Bakken and Gates (1975):

$$H_s = C_p (T_{b\text{-}max} - T_{b\text{-}min})M_b \qquad \textbf{(equation 13.1)}$$

where C_p is specific heat of animal tissue, i.e. 3.48 J $g^{-1}.°C^{-1}$, $T_{b\text{-}max}$ and $T_{b\text{-}min}$ are the maximum and minimum core body temperatures, respectively, and M_b is body mass in grammes. For a greater hoopoe-lark weighing 45 g, H_s during incubation amounts to 470 J

when the bird's T_b rose from 42 to 45°C. Thus, if the adult had controlled its T_b at 42°C rather than using hyperthermia, it would have cost 0.2 mL water per incubation bout. Assuming approximately 12 bouts of incubation during the day, a greater hoopoe-lark would save 2.4 mL of water daily by using hyperthermia, which constitutes some 15–20% of total evaporative water loss (TEWL) at a T_a of 40°C (Tieleman et al. 2002).

13.7.3 Water costs during incubation

Desert birds often experience considerable heat stress during incubation and, as a result, operate at detrimentally high T_bs. Although the amount of water used for thermoregulation can be partly ameliorated by hyperthermia and replenished by selective foraging for insects with higher water content, the question remains as to what extent incubating desert birds incur water costs. Tieleman et al. (2008) estimated this for greater hoopoe-larks by extrapolating laboratory measurements of evaporative water loss to field conditions. They used TEWL data (Tieleman et al. 2002) to translate T_e (Figure 13.4a–c) to TEWL costs of incubation at the three different nest sites described in section 13.7 (Figure 13.5a–c). While TEWL at these sites largely mirrored site-specific differences in T_e, the most striking findings were the marked consequences for water loss with only small shifts in T_e. Moreover, the 10- to 15-fold increase in TEWL between the early morning hours and the middle part of the day highlighted the importance of the male's contribution to incubation. Specifically, whereas the paternal share of total nest attendance was small early and late in the day, during the water-demanding middle part

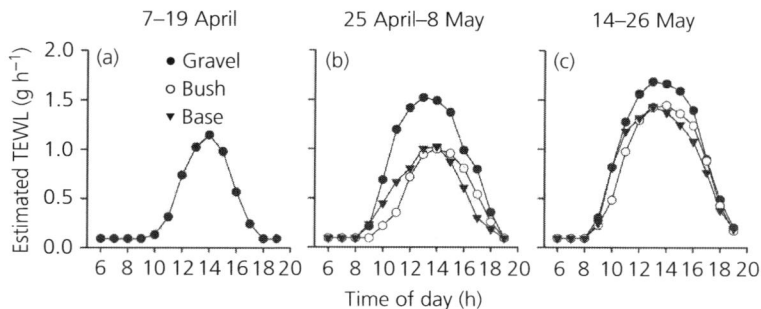

Figure 13.5 Water costs of incubation (expressed as estimated total evaporative water loss [TEWL]) in greater hoopoe-larks (*Alaemon alaudipes*) incubating at three different nest sites in the desert of Saudi Arabia (GMT + 3 h) in mid-April (a), late April to early May (b), and mid- to late May (c). The levels of sun exposure and heat stress differ between nest locations (as detailed in the main text). This is largely mirrored in the daily pattern of estimated TEWL, with birds nesting on the gravel plain incurring the highest water cost of incubation. Reproduced with permission from Tieleman et al. (2008).

of the day, male attendance was almost identical to that of females.

13.8 Conclusions

Our analyses support the view that incubation is a time of elevated parental energy expenditure (Williams 1996; Tinbergen and Williams 2002). For all species studied to date, the field metabolic rate during incubation (FMR_{inc}) is elevated an average of 3.4 times above basal metabolism (BMR). This estimate varies considerably with geography and taxonomy. Thus, terrestrial birds breeding at Arctic latitudes accrue the largest energy costs during incubation, i.e. 4.1 × BMR, whereas terrestrial birds at lower latitudes augment FMR_{inc} by 2.9 × BMR. Across all terrestrial birds for which FMR has been measured during both incubation and chick rearing, FMR_{inc} (2.9 × BMR) is lower than the field metabolic rate during chick rearing (FMR_{chick}; 3.4 × BMR). However, these estimates suffer from the lack of consistent measurements in the same individuals or populations during both the incubation and chick-rearing stages. Life-history strategies must also be taken into account when generalising across taxa: our analysis suggests that incubation and chick rearing are equally costly in species with single-sex incubation systems (cf. Figure 13.1c). Seabirds, whose incubation behaviour is typically characterised by fewer and longer on- and off-bouts than amongst terrestrial birds, display an intermediate increase in FMR during incubation, i.e. 3.3 × BMR. However, the actual cost pertaining to egg warming is modest in this group, i.e. 1.7 × BMR, and most of the increase in FMR_{inc} is accounted for by energetically expensive foraging bouts at sea (estimated to be 4.8 × BMR). Behaviours off the nest likely also contribute considerable variation to FMR_{inc} in terrestrial birds breeding in challenging environments, such as the Arctic (cf. Piersma et al. 2003). To understand such effects better, future studies should record birds' activity budgets when integrating measurements of energy expenditure over long time periods.

Many studies have assessed when and how energy costs of incubation arise, and equally many have manipulated the demands of incubation to record corresponding effects on parental fitness. However, few attempts have been made to understand the physiological pathways linking these. We highlighted two possible mechanisms—increased oxidative stress and reduced immune function—whereby fitness costs could arise through reduced energy allocation during incubation.

In some environments, the primary currency for incubation costs need not be energy. We exemplified this using data from desert-breeding larks where costs of incubation are paid largely by increased difficulties in maintaining water balance during heat stress, but there are likely many other currencies for costs of incubation yet to be described. To further our understanding of the evolution of avian incubation behaviour and physiology, future studies should therefore seek to identify new (and manipulate established) mechanisms, whether energy-based or not, by which costs of incubation are proximally manifested. Likewise, there is a need to understand better how incubation demands might constrain offspring phenotype (e.g. body condition and growth) and survival independently of effects acting on parents. Our review of the literature suggests that a reduction in egg temperature (T_{egg}) in challenging environments could be a prime candidate in explaining these trans-generational costs of incubation. In this context, it should be noted that reduced T_{egg} (together with a lower energy cost for rewarming eggs) could also explain why few have found increased FMR_{inc} in response to experimental clutch enlargement. Accordingly, T_{egg} should be regarded as an important response variable in studies manipulating reproductive effort during incubation.

Systematic studies in the above regards are highly warranted. These should involve not only manipulation of incubation effort to record subsequent effects on parental energy expenditure and nest environment but, more importantly, should also identify possible constraints on key physiological functions with known links to fitness.

Acknowledgements

AN was supported by the Swedish Research Council (grant no. 637-2013-7442) and JBW was funded by a grant from the National Science Foundation (grant no. IOS-1036914). The collection of previously unpublished data in Sweden was supported by the Helge Ax:son Johnson Foundation, the Lund Animal Protection Foundation (both to AN) and the Swedish Research Council (to Jan-Åke Nilsson, Lund University). Comments from Jim Reynolds, Simone Webber, Maria Caules and Elisabeth Calhoon improved a previous version of the chapter.

CHAPTER 14

Influence of incubation temperature on offspring phenotype and fitness in birds

G.R. Hepp, S.E. DuRant, and W.A. Hopkins

14.1 Introduction

Incubation is an important reproductive cost in birds, particularly in species that use contact incubation to warm eggs (Visser and Lessells 2001; Tinbergen and Williams 2002). Costs manifest as constraints on the allocation of energy, nutrients, and time to competing processes such as self-maintenance. Metabolic costs to parents increase during incubation, especially at ambient temperatures below thermoneutrality and in species where only one parent incubates (Williams 1996; Tinbergen and Williams 2002). However, incubating parents must spend time on the nest providing heat to eggs, so less time is available for foraging to meet their increased energy needs. Nutrient reserves can help to satisfy these energy demands, but the ability to use endogenous energy declines with body mass of species and individuals (Afton and Paulus 1992). In some species, feeding of incubating females by their male partner also can help females to balance daily energy costs (Lyon and Montgomerie 1985; Galván and Sanz 2011). However, it is clear that incubation costs are important and can influence both current and future reproductive success (Visser and Lessells 2001; Reid et al. 2000a, 2002b; Hanssen et al. 2005; Hepp and Kennamer 2012; DuRant et al. 2013a) and impact the evolution of life histories (Monaghan and Nager 1997; Martin 2008). Incubating individuals, therefore, are challenged to balance the competing demands of self-maintenance with the thermal requirements of developing embryos.

14.2 Trade-offs and incubation temperature

Optimal incubation temperature for avian embryos generally falls between 35.5°C and 38.5°C (Webb 1987).

Low incubation temperature slows development leading to longer incubation periods, reduced neonate 'quality', and increased risk of nest predation (Tombre and Erikstad 1996; Hepp et al. 2006; Martin et al. 2007, 2013a; DuRant et al. 2013a). Even small differences in incubation temperature can have significant impacts (reviewed in DuRant et al. 2013a). Therefore, it is important for incubating parents to provide a thermal environment that allows for normal embryo development although incubation temperatures can fall below this level especially during incubation recesses (Martin and Schwabl 2008).

Energy constraints and predation risk are two of the prominent factors that influence nest attentiveness and incubation temperatures. For example, females of several passerine species spent less time on the nest and were unable to maintain optimum incubation temperatures after incubation costs were experimentally increased (Ardia et al. 2010; Nord et al. 2010; Nord and Nilsson 2012). However, when incubation costs were reduced by heating nests, females increased nest attentiveness and incubation temperatures (Bryan and Bryant 1999; Ardia et al. 2009). These studies show that energy constraints during incubation are important and influence trade-offs between self-maintenance and embryo development. Predation risk can also influence parental investment decisions, independent of energy constraints. High risks of predation, for example, cause parents to reduce nest attentiveness resulting in lower incubation temperatures and longer incubation periods (Conway and Martin 2000b; Martin 2002; Martin et al. 2007).

Variation in parental quality certainly influences how parents respond to reproductive costs (Cam et al. 1998; Sanz-Aguilar et al. 2008), including those of

Hepp, G.R., DuRant, S.E., and Hopkins, W.A., *Influence of incubation temperature on offspring phenotype and fitness in birds*. In: *Nests, Eggs, and Incubation*. Edited by: D.C. Deeming and S.J. Reynolds, Oxford University Press (2015). © Oxford University Press.
DOI 10.1093/acprof:oso/9780198718666.003.00014

incubation. For example, as incubation costs increased with increasing clutch size in common eiders (*Somateria mollissima*), females in high body condition invested more in incubation and subsequent brood care than did females in low body condition (Hanssen et al. 2003). Ardia and Clotfelter (2007) altered the energy constraints of incubating female tree swallows (*Tachycineta bicolor*) by clipping their primary feathers forcing them to work harder during foraging. They found that early-nesting females in high body condition were affected less by the feather-clipping manipulation and spent more time incubating and produced nestlings in higher body condition than late-nesting females in low body condition. In wood ducks (*Aix sponsa*), females in low body condition nested later in the season than females in higher body condition (Hepp and Kennamer 1993). Late-nesting wood ducks reduced incubation costs by spending less time on the nest, but incubation temperatures were similar to those of early-nesting females because ambient temperatures were higher late in the season (Hepp and Kennamer 2011). These experimental studies make it clear that incubation costs are not homogeneous but depend on parental body condition.

14.3 Factors influencing incubation temperature

Parents use a variety of methods to achieve and maintain optimal egg temperatures during incubation (Ar and Sidis 2002; Deeming 2002e). Incubation behaviour is a key factor affecting incubation temperature, and variation in behaviours underlying the number and length of recesses and overall incubation constancy are influenced by factors such as ambient temperature, precipitation, predation risk, egg size, and body condition (Conway and Martin 2000b; Deeming 2002e; Tombre et al. 2012). Behavioural plasticity allows parents to modify their response to current environmental conditions. Female tree swallows responded to the experimental heating and cooling of nests by increasing and decreasing, respectively, time spent incubating (Ardia et al. 2009, 2010). Under natural conditions, female tree swallows modulate the amount of time they stay off the nest according to precipitation and temperature patterns (Coe et al. 2014). During cold and dry conditions, females stayed off their nests longer than under warm and dry conditions but this pattern was reversed under periods of heavy rainfall; females tended to take shorter off-bouts when it was rainy and cold compared to longer off-bouts during warmer rain events. As expected, variation in female tree swallow

behaviour was associated with differences in overall incubation temperature, such that females that took shorter, more frequent off-bouts maintained higher incubation temperatures than those that opted for longer, less frequent off-bouts (Coe et al. 2014). Similarly, female wood ducks successfully maintained optimal egg temperatures by adjusting daily incubation behaviours in response to variation in both ambient temperature and clutch cooling rates (McClintock et al. 2014). Martin et al. (2007) examined 41 passerine species on three continents and showed that reduced levels of parental nest attentiveness resulted in lower egg temperatures and longer incubation periods. Together, these studies clearly demonstrate that parental incubation behaviour is a key determinant of incubation temperature, but understanding the proximate causes of variable parental incubation behaviour remains an area ripe for future study.

Size, structure, and composition of nests influence their thermal properties and affect the energetic costs of incubation (Smith et al. 1974) as well as the parent's ability to provide an appropriate microclimate for the eggs (Skowron and Kern 1980; Lombardo et al. 1995; Hansell and Deeming 2002; Hilton et al. 2004; Heenan 2013; Mainwaring et al. 2014b; Chapter 4). Yellow warblers (*Dendroica petechia*) nesting at Churchill, Manitoba, for example, built larger, thicker nests that cooled more slowly than nests from a more southerly location in Ontario (Rohwer and Law 2010). Similarly, blue tits (*Cyanistes caeruleus*) and great tits (*Parus major*) consistently built better-insulated nests at higher latitudes in the UK where ambient temperatures were cooler (Mainwaring et al. 2012). Many bird species improve the insulatory quality of nests by adding feathers (see Chapter 3). McClintock et al. (2014) manipulated the amount of down insulation in wood duck nests and showed that nests with reduced down (0.5 g) cooled 20% faster than nests with normal amounts of down (4.0 g). Tree swallow eggs also cooled faster in nests without feathers than in nests with them (Windsor et al. 2013). Furthermore, tree swallow nests in which feathers were experimentally removed had longer incubation periods and produced fewer fledglings than nests with feathers (Lombardo et al. 1995).

Nest location affects nest microclimate by influencing exposure of nests to environmental conditions (Hartman and Oring 2003; Ardia et al. 2006b). Therefore, selection of nest sites is important because of the potential impact that nest microclimate has on reproductive success (Lloyd and Martin 2004; Salaberria et al. 2014). Tree pipits (*Anthus trivialis*), for example, preferred to build nests with an eastward orientation that protected eggs from direct sunlight in the middle of the day which improved hatching success of these nests compared to nests with different orientations (Burton 2006). Similarly, vegetative cover at nest sites selected by vesper sparrows (*Pooecetes gramineus*) and horned larks (*Eremophila alpestris*) protected nests from solar radiation and resulted in a more favourable nest microclimate (Nelson and Martin 1999).

14.4 Incubation temperature and avian phenotypes

Effect of incubation temperature on hatchling phenotypes has been widely studied in reptiles (e.g. Brown and Shine 2004; Deeming 2004b; Booth 2006; Mitchell et al. 2013; Du and Shine 2015) and domestic poultry (Deeming and Ferguson 1991b), but only recently has it been examined in wild birds (DuRant et al. 2013a).

Poultry scientists have been interested in incubation temperature because of its potential economic importance related to hatching success and early growth and performance of chicks (Lourens et al. 2005; Hulet et al. 2007). In domestic fowl (*Gallus gallus*), for example, chronic low incubation temperature (35°C) lengthens the incubation period, and reduces hatching success and the ability of neonates to thermoregulate (Black and Burggren 2004a, 2004b). Changes in temperature during just a portion of the incubation period also can impact embryo development. When temperatures were altered in late incubation (days 10–18), chicken eggs incubated at low temperatures (33.3–36.7°C) had longer incubation periods and produced chicks with reduced body mass at hatching and at 7 days of age compared to eggs incubated at mid (37.8–38.2°C) and high (38.9–40.0°C) temperatures (Ipek et al. 2014). Incubation temperatures early in the incubation period also influence embryo development. Eggs incubated at low temperatures (36.6°C) early in incubation (days 0–10) and then incubated at normal temperatures (37.8°C) for the remainder of incubation had reduced hatching success and chicks were smaller at 1, 3, and 6 weeks of age compared to those incubated constantly at 37.8°C (Joseph et al. 2006). Studies of incubation temperature in poultry may not be rooted in an ecological framework, but nevertheless they provide important insights into the potential importance of incubation temperature in wild birds.

Parental effects such as maternal transfer of hormones to eggs can help to prepare neonates for environmental conditions they will experience after hatching (Mousseau and Fox 1998b; Love and Williams 2008a; Bentz et al. 2013) but this is not always the case

(Monaghan 2008). Growth and development of nestling great tits, for example, were not enhanced when food resources were equivalent pre- and post-hatch (Giordano et al. 2014). Similarly, phenotypic changes to wild avian neonates caused by sub-optimal incubation temperatures do not appear to be adaptive (DuRant et al. 2013a). Recent studies, especially of wood ducks and tree swallows, have examined effects of incubation temperature on a suite of characters that are important to breeding success. Together, these studies suggest that sub-optimal temperatures produce phenotypes that would perform unfavourably under most environmental conditions. Here, we summarize those findings and discuss their potential effects on fitness.

14.4.1 Body size, body composition, and growth

Several studies have artificially incubated eggs at ecologically relevant temperatures, and measured effects of temperature on the mass, structural size, composition, and subsequent growth of recently hatched chicks. In a megapode, the Australian brush-turkey (*Alectura lathami*), eggs were incubated at 32°C, 34°C, and 36°C, and chicks from higher temperatures had shorter incubation periods and tended to be structurally smaller, have more residual yolk and reduced yolk-free body mass compared with chicks incubated at lower temperatures (Eiby and Booth 2009). Hepp and Kennamer (2012) incubated wood duck eggs at 35.0°C, 35.9°C, and 37.3°C, and found that body mass of recently hatched ducklings declined and tarsus length increased with lower incubation temperatures. Furthermore, ducklings incubated at the lowest temperature used more lipids during a longer incubation period and hatched with 20% fewer lipids than ducklings incubated at the higher temperatures (Hepp and Kennamer 2012). In other studies, the effect of incubation temperature (35.0°C, 35.9°C, and 37.0°C) on wood duck hatchling mass varied somewhat. Body mass and condition of hatchlings were greatest for hatchlings incubated at the highest temperature (DuRant et al. 2013b), but DuRant et al. (2010, 2012a) reported no effect of incubation temperature on hatchling mass. Tarsus length was always greater for ducklings incubated at the lowest temperature (DuRant et al. 2010, 2012a, 2013b). Taken together, these results suggest that prolonged development associated with low incubation temperature generates greater structural size in hatchlings at the expense of energy reserves.

Incubation temperature influences development of altricial species in a similar manner. Olson et al. (2006) incubated zebra finch (*Taeniopygia guttata*) eggs at constant temperature (37.5°C) or periodically cooled them to 20°C or 30°C. Periodic cooling of eggs reduced average incubation temperatures and caused eggs to develop more slowly and use more yolk. Body mass and size of yolk reserves of zebra finch embryos that were cooled to 20°C were 30% and 12% smaller, respectively, at day 12 of incubation than embryos from eggs incubated constantly at 37.5°C (Olson et al. 2006).

Effects of incubation temperature also carry over and impact the early growth and locomotor performance of hatchlings. Body mass and condition of wood ducks incubated at the coolest temperature (35°C) and fed an *ad libitum* diet were lower at 9, 12, and 20 days-post-hatch (dph) compared with ducklings that were incubated at higher temperatures (Figure 14.1; DuRant et al. 2010, 2012a, 2013b; but see Hopkins et al. 2011).

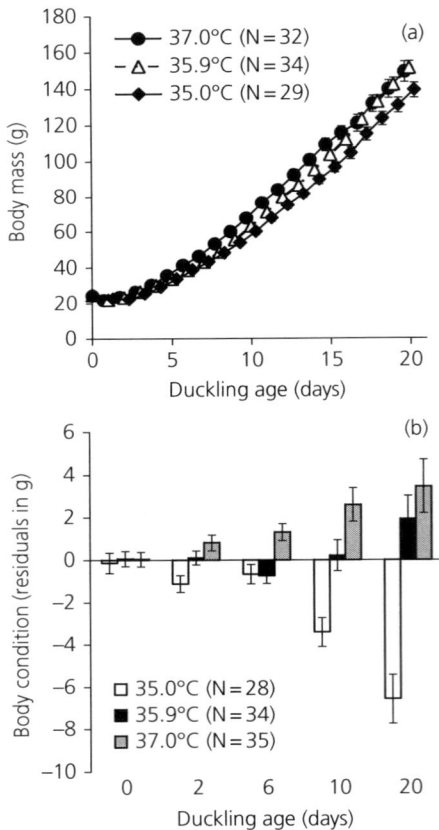

Figure 14.1 Mean (± 1 SE) body mass (a) and body condition (b) of wood duck (*Aix sponsa*) ducklings by age and incubation temperature. Body condition is the residual from the regression of body mass on tarsus length. Reproduced with permission from DuRant et al. (2012a).

Wood ducks that were incubated at 35°C also ran and swam more slowly at 15 dph than ducklings incubated at the higher temperatures, but locomotor performance at 20 dph was no longer affected by incubation temperature (Hopkins et al. 2011).

In tree swallows, cooling of nests lengthened the incubation period and resulted in lower body mass and condition of nestlings up to 13 dph compared to nestlings from nests that were not cooled (Ardia et al. 2010). When tree swallow nests were heated, the body mass and condition of nestlings at 4 and 7 dph were greater than nestlings from unheated nests, but these differences disappeared for older nestlings (Pérez et al. 2008). By contrast, incubation temperature (35.0°C, 36.5°C, and 38.0°C) in blue tits did not affect body mass or wing length of nestlings at 2, 6, and 14 dph, but nestlings incubated at the lowest temperature had shorter tarsi at 14 dph (Nord and Nilsson 2011).

14.4.2 Metabolism and thermoregulatory ability

Relatively few studies have examined the effects of incubation temperature on metabolism and energy expenditure of avian embryos. In two species of megapodes, energy used by embryos increased 56% in Australian brush-turkeys and 77% in malleefowl (*Leipoa ocellata*) when incubation temperatures were reduced by 4°C (Booth and Jones 2002). In both species, low incubation temperature resulted in a longer 'plateau phase' of oxygen consumption at the end of incubation, which caused increased energy expenditure (Booth and Jones 2002).

Similarly, DuRant et al. (2011) incubated wood ducks eggs at 35.0°C, 35.9°C, and 37.0°C, and measured respiration every 3 days during incubation. They found embryos incubated at the highest temperature generally had the greatest daily metabolic rates and developed faster than embryos at lower temperatures, but embryos incubated at the lowest temperature expended more energy (120 kJ) over the course of incubation than embryos incubated at 35.9°C (83 kJ) and 37.0°C (94 kJ; Figure 14.2). These differences in energy expenditure occurred during the hatching process. Wood duck embryos incubated at the lowest temperature took 1–2 days longer to hatch once the eggshell was pipped than embryos incubated at higher temperatures (DuRant et al. 2011). Olson et al. (2006) reported zebra finch embryos from eggs that had been periodically cooled to 20°C during incubation had 14% higher metabolic rates at day 12 of incubation than embryos incubated constantly at 37.5°C. In blue tits, incubation temperature significantly influenced metabolic rates of nestlings.

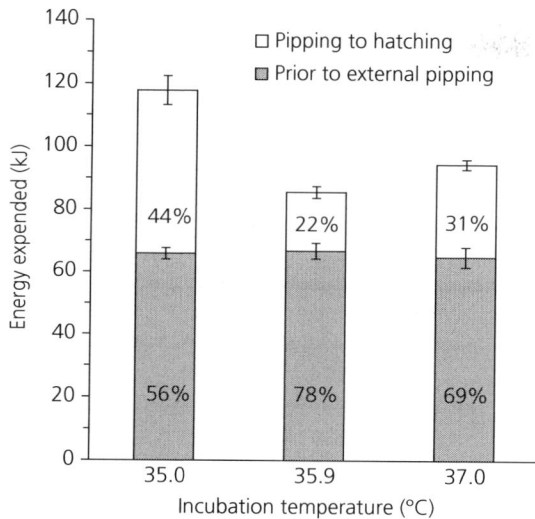

Figure 14.2 Mean (± 1 SE) total energy expended during incubation by wood duck (*Aix sponsa*) embryos incubated at three different temperatures. The grey portion of each bar represents energy used up to external pipping, and the white portion is the energy expended during the hatching process. Percentages represent the relative amounts of energy used by embryos up to pipping and from pipping to hatching. Reproduced with permission from DuRant et al. (2011).

The resting metabolic rate of nestling blue tits that had been incubated at 35.0°C was 7.5% and 8.1% higher at 14 dph than nestlings incubated at 36.5°C and 38.0°C, respectively (Nord and Nilsson 2011).

Experimental studies clearly show that incubation temperature has another important carry-over effect, that of influencing the development of thermoregulation (DuRant et al. 2011, 2012b). In one study, wood duck eggs were incubated at three temperatures (35.0°C, 35.9°C, and 37.0°C), and the thermoregulatory ability of recently hatched ducklings, i.e. 1 dph, was tested using a 1-hour thermal challenge at four temperatures (5°C, 10°C, 15°C, and 20°C; DuRant et al. 2011). Incubation temperature had no effect on a duckling's ability to maintain body temperature during these thermal challenges, but larger ducklings were more effective at maintaining body temperature than smaller ones. However, in a second experiment a thermal challenge at 15°C showed that ducklings from the lowest (35.0°C) incubation temperature were only able to maintain their body temperature by spending 27% and 40% more energy than ducklings that were incubated at 35.9°C and 37.0°C, respectively (Figure 14.3a; DuRant et al. 2011). In another study, using the same three incubation temperatures, wood duck

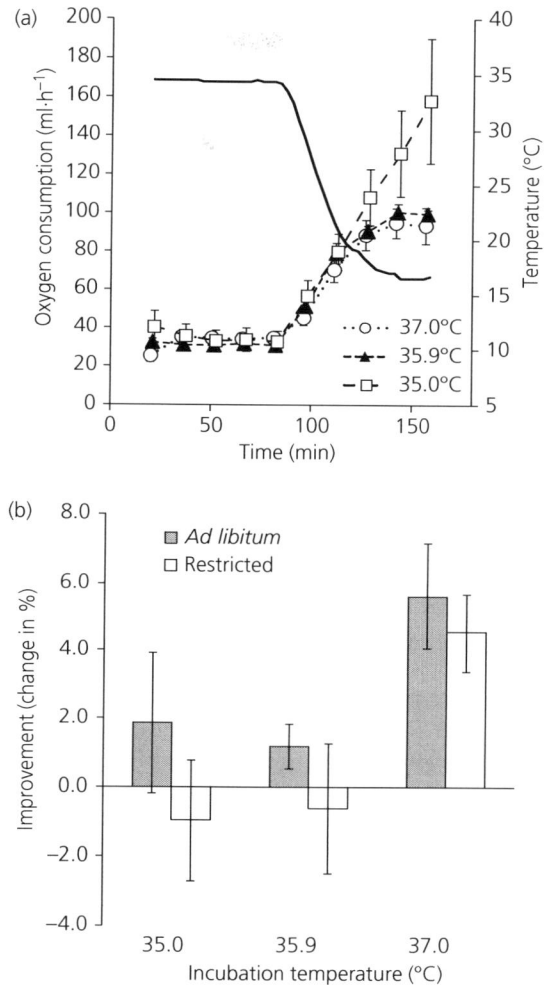

Figure 14.3 (a) Oxygen consumption of 1-day old wood duck (*Aix sponsa*) chicks at thermoneutrality (36°C) and during a 1-hour thermal challenge (15°C). Ducklings had been incubated at one of three temperatures. Oxygen consumption (least-squares mean ± 1 SE) was corrected for body mass. The solid line represents temperature of the respiration chamber. Reproduced with permission from DuRant et al. (2012b). (b) Improvement in the ability of wood duck chicks to maintain body temperature (mean ± 1 SE) from 1 to 4 days post-hatch when exposed to a 1-hour thermal challenge at 10°C. Ducklings were reared on an *ad libitum* diet or a restricted diet, and were incubated at one of three temperatures. Reproduced with permission from DuRant et al. (2013b).

chicks were thermally challenged (10°C) at 1 dph and again at 4 dph to test improvement in their ability to thermoregulate (DuRant et al. 2013b). These ducklings were also assigned a restricted or an *ad libitum* diet to test effects of both pre- and post-hatch conditions on the development of thermogenesis. Ducklings incubated at the highest temperature showed the greatest improvement in their ability to thermoregulate following the thermal challenge (Figure 14.3b). Importantly, improvement was influenced only by the pre-hatch conditions (incubation temperature) and not by post-hatch conditions (food availability; DuRant et al. 2013b).

14.4.3 Immune function

In several avian species, low incubation temperature slows development, and embryos use more energy and hatch with reduced body mass and nutrient reserves. Since immune activity is energetically expensive (Martin et al. 2003), low incubation temperature, because of its influence on the bioenergetics of development,

may reduce the immune response of young birds. To examine this idea, DuRant et al. (2012a) incubated wood ducks eggs at 35.0°C, 35.9°C, and 37.0°C, and tested immunocompetence of ducklings at 6 dph using two novel antigens: phytohaemagglutinin (PHA) and sheep red blood cells (SRBCs). They found that the response to PHA and SRBCs was lower for ducklings incubated at the lowest temperature than the two higher temperatures (Figure 14.4). Tree swallows from nests that were cooled during incubation showed reduced innate immunity at 13 dph compared to nestlings from nests that had not been cooled (Ardia et al. 2010). Similarly, in 22 passerine species from Venezuela and Arizona, there was a strong positive effect of egg temperature on levels of natural antibody titres at hatching (Arriero et al. 2013).

Interspecific variation in intrinsic developmental rates of birds did not explain differences in the immune function of young birds without first accounting for variation in incubation temperatures (Martin et al. 2011). Taken together, the few studies that have examined the effect of incubation temperature on immune responses in birds indicate a positive relationship.

14.5 Effect of temperature-induced phenotypes on fitness

The environmental conditions experienced by birds during early development can have long-term effects on survival and reproduction (Lindström 1999; Monaghan 2008). We think it is unlikely that phenotypic changes induced by low incubation temperatures are adaptive in the majority of environmental contexts. For example, wood ducks incubated at low temperatures used more energy and hatched with reduced body mass and residual lipid reserves, and required more energy to thermoregulate than ducklings incubated at higher temperatures (DuRant et al. 2011, 2013b; Hepp and Kennamer 2012). Residual nutrient reserves are especially important to precocial young that must leave the nest soon after hatching, travel from nest sites to brood-rearing habitats, feed independently, and meet thermoregulatory challenges. Not surprisingly, mortality of precocial young is greatest within the first two weeks of hatching, and there is often a positive relationship between body mass of recently hatched ducklings and survival (Pelayo and Clark 2003; Traylor and Alisauskas 2006; Amundson and Arnold 2011). Hatchling mass in birds generally has a positive effect on survival (Krist 2011).

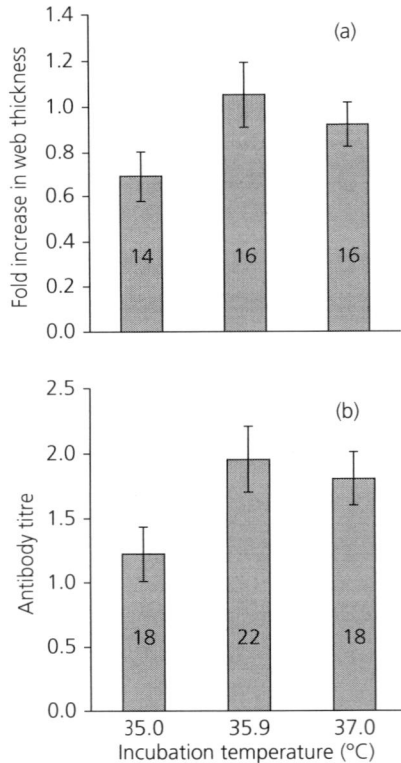

Figure 14.4 Immune responses (mean ± 1 SE) of wood ducks (*Aix sponsa*) that hatched from eggs incubated at different temperatures. (a) Fold increase in foot web thickness of ducklings 24 hours after exposure to phytohaemagglutinin. (b) Antibody titres of ducklings exposed to sheep red blood cells (SRBCs). Ns are shown within the bars. Reproduced with permission from DuRant et al. (2012a).

Low incubation temperature also influences growth rates of young birds, which can have long-lasting effects. However, dispersal of young often makes it difficult to follow individuals to adulthood and first reproduction. In some species, like those of the Anatidae, strong natal philopatry of females makes it easier to examine effects of conditions experienced during early development on fitness. In brant geese (*Branta bernicla*), for example, body size of goslings in their first summer was influenced by early environmental conditions, i.e. food (Sedinger and Flint 1991), and body size of female goslings had positive effects on first-year survival and future fecundity (Sedinger et al. 1995, 2004). Furthermore, brant geese showed no evidence of compensatory growth; larger goslings became larger adults (Sedinger et al. 1995). Similarly, there was a positive relationship between first-year survival and body mass at fledging in female snow geese (*Chen caerulescens*), but the relationship was complex and occurred only when environmental conditions were constrained, i.e. during late hatching (Cooch 2002).

Immune responses and thermoregulatory abilities can be affected by incubation temperature and are generally weaker for individuals incubated at low temperatures. Robust immune responses are potentially important to young birds. In a meta-analysis of several species of birds, survival was greater for individuals with strong immune responses (Møller and Saino 2004). Nestling barn swallows (*Hirundo rustica*) with more robust immune responses relative to their brood mates also had increased survival and longevity (Saino et al. 2012). Similarly, normal development of homeothermy is critical for avian neonates. Energy expenses are high after hatching, especially in precocial species (Krijgsveld et al. 2012). Elevated energy demands by neonates to maintain homeothermy may slow early growth rate and tissue maturation, and affect their thermoregulatory ability (e.g. Rhymer 1988; Visser and Ricklefs 1995). Mortality is often greatest soon after young birds leave the nest, and hypothermia can be an important source of early mortality (Talent et al. 1983; Korschgen et al. 1996).

Trade-offs that influence the parents' ability to provide an optimal thermal environment for developing embryos can compromise the phenotypic quality of their offspring in a number of important ways. It is likely that temperature-induced changes in offspring phenotype impact fitness of birds, much like they have been shown to influence that of reptiles (Warner and Shine 2008). However, studies linking incubation conditions of birds to fitness are still needed.

In the first study to test effects of incubation temperature experimentally on neonate survival and reproduction in a wild bird, Hepp and Kennamer (2012) artificially incubated wood duck eggs at three temperatures (35.0°C, 35.9°C, and 37.3°C) similar to those used by Hepp et al. (2006), DuRant et al. (2010, 2011, 2012a, 2012b, 2013b), and Hopkins et al. (2011). Recently hatched ducklings were individually marked, and broods were formed that contained ducklings from each of the incubation temperatures and placed in nestboxes of foster mothers. Since female wood ducks exhibit strong natal philopatry, recapture of females as breeding adults was used to test effects of incubation temperature on survival and recruitment to the breeding population. Reproductive success of these females also was monitored over four breeding seasons. Females incubated at 35°C were four times less likely to be recaptured than those incubated at higher temperatures. Females incubated at higher temperatures also had more than twice as many successful nests as those incubated at lower temperatures. This is the first study to show carry-over effects of low incubation temperature on survival, recruitment to the breeding population and subsequent reproductive success of birds.

14.6 Summary and future research perspectives

We have shown that incubation temperature influences the development of avian embryos and impacts an assortment of phenotypic traits of neonates that can carry over to later life stages. These effects are similar in many ways to those described for reptiles (Brown and Shine 2004; Deeming 2004b; Booth 2006; Mitchell et al. 2013; Du and Shine 2015). There are potential impacts on offspring survival and reproduction (Hepp and Kennamer 2012), so temperature-induced phenotypic effects in birds may play an important role in parental investment decisions, fitness, and the evolution of life histories (e.g. Monaghan and Nager 1997; Lyon and Eadie 2008; Martin 2008). However, further research is needed to test the strength, generality, and consequences of these relationships.

Experimental studies are needed to examine effects of incubation temperatures on hatchling phenotypes and the mechanisms responsible for these changes (e.g. DuRant et al. 2014). We have identified some of these effects, but others may also exist. For example, neural development and cognitive performance are especially sensitive to conditions experienced during early development (Nowicki et al. 2002) and may be influenced by incubation temperature (Amiel and Shine 2012). Furthermore, species may differ in how they respond to variation in incubation temperatures. We encourage studies that manipulate egg temperatures either by artificially incubating eggs for all or part of the incubation period or by manipulating egg temperatures in active nests (see DuRant et al. 2013a). These studies will be more meaningful if ecologically relevant egg temperatures are used.

Future research should also investigate whether effects of incubation temperature carry over to adulthood. We know that some temperature-induced effects can last for several days and weeks after hatching, but are these effects permanent or do they disappear? In some organisms, individuals that are stunted during early development can accelerate growth so that adult phenotypes are not compromised, but compensatory growth can be costly (Metcalfe and Monaghan 2001, 2003). Furthermore, some species show no evidence of compensatory growth (Sedinger et al. 1995). If phenotypic changes are relatively permanent, how do they affect fitness? Could conditions experienced during incubation help to alter an individual's adult phenotype and be partly responsible for the variation in individual 'quality' that is so common in avian populations?

Finally, it will be essential to evaluate the influence of early developmental effects induced by incubation temperature on components of fitness. This can best be accomplished using an experimental approach (Hepp and Kennamer 2012) that creates different hatchling phenotypes by manipulating incubation temperatures, then marks individuals, cross-fosters them to remove effects of temperature from those of parental quality, and examines the effects of incubation temperature on survival and reproduction through long-term capture-mark-recapture studies. In many species of migratory passerines, the young disperse and do not return to natal areas as adults; therefore, not all species will be amenable to this type of long-term experimental study. However, in many species, like waterfowl and non-migratory passerines, individuals show high levels of natal philopatry and breeding-site fidelity and can be monitored as breeding adults (Weatherhead and Forbes 1994; Hepp and Kennamer 2012). Future investigations such as these will help to clarify the importance of incubation temperature on development of avian neonates and the significance of investment decisions made by incubating parents on overall reproductive success.

Acknowledgements

We thank our field and lab crews, Bobby Kennamer, and the staff at the University of Georgia's Savannah River Ecology Laboratory. We also thank Charles Deeming for his editorial assistance. Our research was supported by National Science Foundation (NSF) grant IOB-0615361 to GRH and WAH, NSF Doctoral Dissertation Improvement Grant DEB-1110386 to SED, the Alabama Agricultural Experiment Station, Fralin Life Science Institute, and the U.S. Department of Energy under award DE-FC09-07SR22506 to the University of Georgia Research Foundation.

CHAPTER 15

Advances in techniques to study incubation

J.A. Smith, C.B. Cooper, and S.J. Reynolds

15.1 Introduction

Birds have long fascinated people (Birkhead 2008) and we have remained closely associated with them for centuries. They provide us with food in the form of meat and eggs, they serve as companion animals, and they are at the focus of pastimes such as pigeon racing and hunting. Yet, without an understanding of reproductive processes, specifically incubation, it would be challenging to produce birds and maintain these important relationships. Knowledge of incubation in the wild is also vital for conservation (Kuehler and Witman 1988; Elliott et al. 2001; Chapter 16), and for management of wild birds that provide recreational opportunities and that are integral for ecosystem function (McGowan and Simons 2006; Coates and Delehanty 2008). However, despite being a critical component of the avian life cycle, incubation remains one of the most problematic avian life-history stages to study because this is when many species are at their most secretive (see Reynolds and Schoech 2012 and references therein). In spite of these challenges, since the late 1700s direct observations of nests, often conducted from hides, have permitted ornithologists to document and quantify incubation traits such as: session and recess lengths, i.e. time spent on and off the eggs during incubation, respectively; nest attentiveness, i.e. time for which eggs are incubated over a given period; incubation periods, i.e. total time between onset of incubation and hatching; incubation rhythms, i.e. patterns of successional sessions and recesses; and egg turning, for a wide range of species (e.g. Jenner 1788; Gross 1912; Skutch 1962).

Direct observation allows the observer to detect subtle behaviours that may be missed through remote measurement or interpretation of data. As direct observation requires little or no expensive equipment, it is used extensively and especially in countries where modern technologies remain prohibitively expensive and thus inaccessible. However, direct observation has inherent disadvantages. Observers can disturb breeding adults and increase the likelihood of nest abandonment (Ellison and Cleary 1978; Henson and Grant 1991), decrease hatching success (Blackmer et al. 2004), disrupt reproductive behaviours (Verhulst et al. 2001; Holm and Laursen 2009), and increase predation risk (Bêty and Gauthier 2001; Verboven et al. 2001; but see Ibáñez-Álamo et al. 2012 and Reynolds and Schoech 2012). Since direct observation is time-consuming and labour-intensive, fatigue can result in observers collecting data of poor quality. Therefore, under most circumstances we must rely on methods other than, and sometimes in addition to, direct observation to generate data to test hypotheses about avian incubation.

Over the last century numerous methods have been developed to investigate incubation that automate direct observations. In this chapter we provide an account of various methods used to monitor incubation remotely. In particular, we critically appraise the advantages and disadvantages of a relatively recently developed automated device that is widely used in studies of captive and free-living species, namely the *iButton*® (Maxim Integrated Products 2014). In so doing we provide advice (including decision rules) that can be employed to adopt the most appropriate method to study incubation within various ornithological contexts.

15.2 Methods to investigate incubation time budgets

15.2.1 Perch-operated switches

Perch-operated switches work on the principle that a bird arriving at, or departing from, a nest lands on

Smith, J.A., Cooper, C.B., and Reynolds, S.J., *Advances in techniques to study incubation*. In: *Nests, Eggs, and Incubation*. Edited by: D.C. Deeming and S.J. Reynolds, Oxford University Press (2015). © Oxford University Press.
DOI 10.1093/acprof:oso/9780198718666.003.00015

Figure 15.1 Methods to investigate combined parental investment during incubation using: (a) a perch-operated switch used in a nestbox of tree swallows (*Tachycineta bicolor*; Photo: D.R. Ardia); (b) a load cell used in a nestbox of wood ducks (*Aix sponsa*; Photo: R.A. Kennamer); (c) a time-lapse camera system used with Mississippi kites (*Ictinia mississippiensis*; Photo: S. Chiavacci); and (d) a nestbox camera used with tree swallows (Photo: D.R. Ardia).

a perch which completes an electric circuit and activates a recording device (Figure 15.1a). One of the first perch-operated switches was used to determine incubation behaviour of house wrens (*Troglodytes aedon*; Kendeigh and Baldwin 1930). This device used two perches; one perch was placed at the entrance to the nestbox, and one just inside the entrance hole. This allowed researchers to establish both the time and the direction of each movement and, subsequently, assess incubation attentiveness. In addition to cavity nesters, perch-operated switches also allow open-cup nesting species to be studied. This requires careful positioning of the perch and branches around a nest site to ensure that the attending bird only alights on the perch switch

each visit (Marples and Gurr 1943). Alternatively, a bird can be forced to land on a perch switch through the use of a wire cage that it must pass through to access the nest (Kendeigh and Baldwin 1930). The use of perch-operated switches in studies of open-cup nesting species have permitted researchers to calculate average session and recess lengths and to assess temporal changes in incubation rhythms over multiple days (Marples and Gurr 1943).

With species such as burrow nesters that do not use perches, electrical switches can be incorporated into pressure plates at burrow entrances (e.g. petrels of the Procellariidae; Schramm 1983). When stepped on, an electric circuit is completed resulting in a pen making a time-specific mark on a rotating clockwork chart-drum. Despite their use in a range of species such as house wrens, common blackbirds (*Turdus merula*) and American robins (*Turdus migratorius*), perch-operated switches can sometimes cause nest abandonment (Kendeigh and Baldwin 1930; Marples and Gurr 1943). Therefore, we recommend that their use be limited to species that are tolerant of large foreign objects at the nest. In addition, perch-operated switches do not allow for individual recognition of parental birds unless combined with other techniques, such as colour ringing and video surveillance. Instead, data reflect combined parental investment.

15.2.2 Other mechanical devices

Nest switches employing a lever that extends into the nest to make contact with an incubating bird function similarly to perch-based switches and, similarly, can be used to assess combined parental investment. When depressed by a bird arriving at the nest, an electric circuit is completed and this activates a recording device. Nest switches have been used to determine temporal changes in incubation rhythms and nest attentiveness, and have provided insight into the effects of weather on incubation rhythms especially for gamebirds, including grey partridges (*Perdix perdix*; Fant 1953) and common pheasants (*Phasianus colchicus*; Klonglan et al. 1956). However, nest switches only allow measurement of session and recess lengths, but not activity on the nest. Power is supplied by batteries, so they also require researchers to visit nests repeatedly. As an alternative, Gurr (1955) constructed a simple device that did not require an external power source and that consisted of one copper tambour (a small, shallow circular drum connected to a sheet-rubber membrane) placed within a nest cup and another placed away from the nest. Movement of the nest tambour by the incubating bird was recorded through air pushed through connected rubber tubing that was recorded via the other tambour that excited a writing device. The tambour apparatus not only recorded continuous movements of birds on and off the nest, but it also provided some of the first insights into other subtle behaviours during incubation, such as fidgeting, by the incubating bird.

15.2.3 Photoelectric cells

A bird arriving at, or departing from, a nest can break a beam of light that is directed across the nest entrance and onto a photoelectric cell. This triggers a relay-operated pen that records arrival/departure times of the bird in attendance allowing incubation rhythms to be monitored. They are especially effective for cavity-nesting species where the light beam can be directed across the entrance hole, and have been used, for example, to assess the effects of weather on the incubation behaviour of purple martins (*Progne subis*; Finlay 1976). In addition, with careful placement they have been used with open-cup nesting species, such as common pheasants (Kessler 1962) and New Zealand ('Snares') snipe (*Coenocorypha aucklandica huegeli*; Miskelly 1989) to assess the relationship between stage of incubation and nest attentiveness. In the latter study, photoelectric cells were used in conjunction with direct observation of colour-ringed birds to assess incubation regimes of individuals. However, it should be stressed that when used alone, data collected using photoelectric cells reflect combined parental investment. Despite poor placement resulting in interference from sunlight, Finlay (1976) determined through verification using direct observations that well-placed devices can provide data of up to 99% accuracy.

15.2.4 Load cells

Load cells are less intrusive than other methods because they do not require the introduction of foreign objects into the nest cup. Load cells use a platform that can be inserted into a nestbox prior to a breeding attempt and on which birds subsequently construct their nests (Figure 15.1b). The presence/absence of a bird is monitored through changes in body mass detected by the load cell, which is attached to a data logger and power source located away from the nestbox in waterproof casing (Kennamer et al. 1990; Mallory and Weatherhead 1992, 1993). Load cells are effective in all weather conditions, can be installed quickly and provide an accurate measure of combined parental attentiveness over entire incubation periods. They have been used effectively in studies of nesting in common goldeneyes (*Bucephala clangula*; Mallory and Weatherhead 1993) and wood ducks (*Aix sponsa*; Kennamer et al. 1990). However, load cells are expensive and relatively power 'hungry' compared to alternative methods such as simple mechanical devices (Mallory and Weatherhead 1992).

15.2.5 Pressure-sensitive wireless devices

The presence/absence of a bird on eggs can also be detected using a pressure-sensitive switch inside a dummy egg that, when sat on, triggers a wireless transmitter to send an 'on' signal to a remote receiver. When a bird moves off the egg, the switch is de-activated and an 'off' signal is sent. Thus, movements on and off the dummy egg allow continuous recordings of session and recess lengths throughout the entire incubation period. Such data are more accurate than those collected from direct observation when short nest movements and nocturnal behaviours are hard or impossible to observe (Bottitta et al. 2002). Data may also reflect incubation behaviour more accurately than those collected using other automated devices because pressure-sensitive devices record the exact time that an incubating bird sits on its eggs as opposed to the time that an incubating bird returns to the nest.

Despite the advantages of using pressure-sensitive devices, their use is restricted to species that have eggs large enough, i.e. approximately 8 × 5.5 cm, to accommodate a wireless transmitting device that contains a pressure-sensitive switch and a battery (Bottitta et al. 2002). In addition, caution must be exercised because dummy eggs containing pressure-sensitive wireless devices can occasionally cause nest abandonment as was the case in 11% of common eider (*Somateria mollissima*) nests containing them (Bottitta et al. 2002). However, it was noted that this degree of abandonment was equivalent to that of another study where females were captured on the nest during incubation (Korschgen 1997).

15.2.6 Photography and time-lapse cameras

Cameras that are either linked to mechanical or infrared (IR) triggers, or that are pre-programmed to take pictures at defined intervals, are often used to study nest predation, but may also be used in studies of incubation (reviewed in Cox et al. 2012; Figure 15.1c). Early studies used time-lapse cameras to take frames at regular intervals allowing snapshots of activity and subsequent calculation of, for example, incubation attentiveness and incubation period (Temple 1972; Weller and Derksen 1972; Cadwell and Cornwell 1975), and more discrete behaviours such as egg turning (Derksen 1977). Such cameras typically used standard rolls of Super-8 film that contained approximately 3,600 frames; over a 2.5-day period exposure rates were limited to approximately 1 frame per minute introducing a ± 1 minute error rate into calculations of session and recess lengths. The introduction of affordable video cassette recorders (VCRs) in the mid-1980s and digital video recorders (DVRs) in the early 2000s has provided higher capacity storage options allowing higher exposure rates (e.g.

1–5 frames per second) and more accurate estimates of session and recess lengths (Coates and Delehanty 2008; Cooper et al. 2009; Bentzen et al. 2010).

Cameras triggered by the incubating bird can also be used. For example, Craig and Craig (1974) monitored the incubation behaviour of purple swamphens (*Porphyrio porphyrio*) that nest in tall vegetation in swamps. They used cameras that were triggered when a bird depressed a micro-switch incorporated into a footpad placed on a ramp leading to the nest. However, many of these early systems required frequent battery and film replacements. More recently, advances in digital technology have allowed researchers to take advantage of commercially available trail cameras that take time-stamped photographs to study incubation of previously poorly studied secretive species such as turkey vultures (*Cathartes aura*; Rollack et al. 2013). Trail cameras are often fitted with IR triggers and are affordable (approximately £200), energy efficient, and capable of storing large amounts of data on memory cards. Combining trail cameras with marking techniques such as patagial tags can provide insight into sex-specific incubation roles, such as both male and female turkey vultures incubating (Rollack et al. 2013). However, unless individuals are easily recognizable as in focal species that are sexually dichromatic or marked in some way, data collected using trail cameras or other photographic techniques reflect combined parental investment. Moreover, cameras with IR triggers need to be carefully placed; they can be sensitive to sunlight, wind, and moving vegetation, and may not always detect smaller birds (reviewed in Cox et al. 2012). Therefore, camera systems that integrate IR triggers are best-suited to large birds that nest in open spaces, such as grasslands and woodlands with a minimal understorey, or birds that breed in sheltered locations, such as nestboxes (Ospina et al. 2015).

15.2.7 Video cameras—real-time recording

Miniaturization and reduced costs of video technology over the past two decades have resulted in a wide range of video-based methods that allow continuous recording of incubation (reviewed in Ribic et al. 2012). For example, commercially available, portable camcorders can readily be mounted on tripods at nests to record arrival and departure times of birds (Smith and Montgomerie 1992; McGowan and Simons 2006). Although camcorders are user-friendly and portable, they are seldom weather-resistant and often have limited battery life and memory capacity.

As an alternative to camcorders, commercially available video surveillance systems that incorporate deep-cycle car batteries and portable DVRs are available for continuous nest monitoring over multiple weeks (Gula et al. 2010; Pietz et al. 2012). In secondary cavity-nesting species, such as western bluebirds (*Sialia mexicana*), birds will utilize nestboxes that contain custom-built video cameras mounted inside (Wang and Weathers 2009; Figure 15.1d). In primary cavity-nesters that use nestboxes or species that breed in natural cavities, such as the Eurasian three-toed woodpecker (*Picoides tridactylus*), custom-built cameras can be installed in the cavity's ceiling upon detection of the start of a breeding attempt (Pechacek 2005). Surveillance and nestbox cameras often possess IR lights, meaning that they can also be used to study nocturnal incubation (Wang and Weathers 2009). While many involve directly connected DVRs, data can be transmitted to a remote receiver using wireless technology over, for example, 2 km (Pechacek 2005). Despite requiring less nest visits compared with camcorders, nests still need to be visited during surveillance to replace batteries.

Some logistical challenges can be met using remote solar-powered video cameras that transmit data to receivers located up to 1 km from the nest. Remote video camera systems have allowed us to study incubation in species that are sensitive to human disturbance, and that nest in remote, inaccessible locations. Such studies include those on bearded vultures (*Gypaetus barbatus*; Margalida et al. 2006), ospreys (*Pandion haliaetus*; Kristan et al. 1996), and great northern loons (*Gavia immer*; Goodale et al. 2005). However, they may not be suitable for species that are hyper-sensitive to disturbance and/or that breed in inaccessible places because of associated bulky hardware, such as solar panels and waterproof containers.

Video-based methods have proliferated in recent years and they continue to increase our understanding of factors that influence incubation, such as weather (Smith and Montgomerie 1992), food availability (Barnett and Briskie 2010), ectoparasite burden (Heeb et al. 1998) or predation risk (Burns et al. 2013). Where sexes can be easily differentiated, such as for species that are sexually dichromatic, video-based methods can also allow assessment of single-parent investment. However, one drawback is the generation of large volumes of data from sometimes many hours of reviewing of video footage. Where possible, we recommend an instantaneous sampling method such as fixed-interval time point analysis (Martin and Bateson 1993), or reviewing footage in 1.5× or 2× speed. For a full account of the theoretical and applied aspects of video surveillance of incubation and other reproductive stages we refer the reader to Ribic et al. (2012).

15.3 Methods to investigate incubation time budgets: individual parental investment

15.3.1 Radio telemetry

While the methods described previously provide insight into attendance at the nest, they do not readily allow us to assess investment by individual birds in bi-parental incubators unless the species is sexually dichromatic or dimorphic in weight. For example, load cells that measure weight of nest, nest contents and incubator may be used to determine sex of the incubator where females and males show significant weight differences. Technological advances allow us to overcome this problem by attaching VHF transmitters ('radio tags') with unique radio frequencies to incubating birds and placing receivers close to their nests. Signal attenuation caused by their movements can then be used to assess attentiveness of species such as American black ducks (*Anas rubripes*; Ringelman et al. 1982) and American woodcocks (*Scolopax minor*; Licht et al. 1989). Alternatively, a custom-made identification device connected to the receiver can be used to register the presence and identity of an incubator at the nest at pre-defined time intervals (Bulla et al. 2014). The major drawback is that birds need to be captured to deploy tags and often this is during sensitive breeding stages when they are prone to abandonment (Graul 1975; Kania 1992). Furthermore, tagging efforts are time-consuming and often require numerous personnel and licences from federal agencies and/or non-governmental organizations. Therefore, we suggest that radio telemetry is most appropriate in studies of spatial ecology where it can be utilized to study movements across multiple life stages (e.g. incubation, brood-rearing, non-breeding) and when birds can be captured outside of sensitive phases of the annual cycle.

15.3.2 Passive integrated transponder (PIT) tags

PIT tags are small (< 2 mm in length) and durable microchips that can be mounted onto colour rings that are, subsequently, attached to a bird's leg (Fiedler 2009). When a PIT-tagged bird passes through an antenna which is placed around the entrance hole to a natural cavity or a nestbox (Figure 15.2), or around the perimeter of an open cup nest, the date, time, and unique identification code of the tag are recorded.

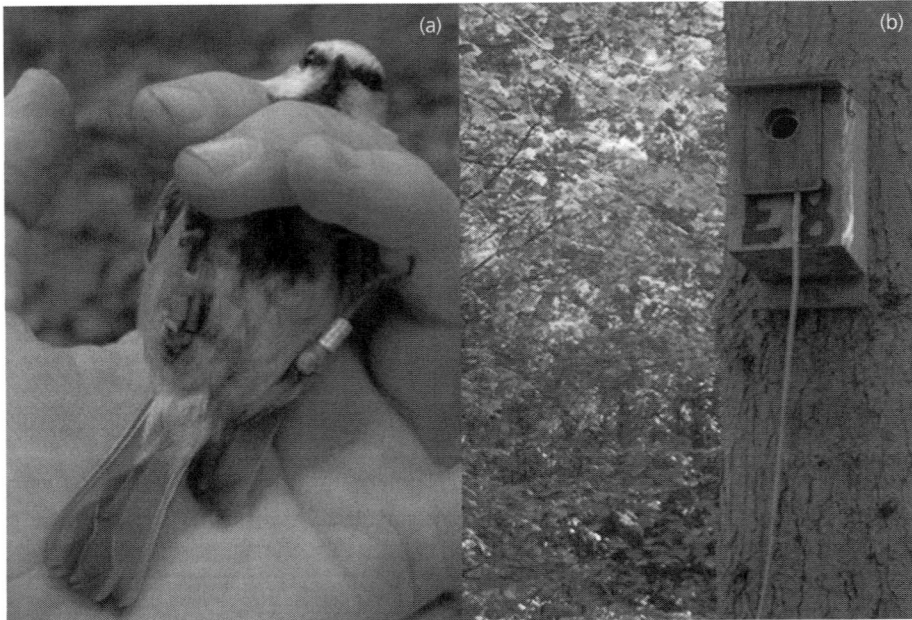

Figure 15.2 A passive integrated transponder (PIT) tag system used to study the incubation behaviour of blue tits (*Cyanistes caeruleus*) by recording the movements of (a) a bird with a PIT tag mounted on a leg ring (here shown on the left of the image - the bird's right leg) as it moves through (b) a receiver that is positioned around the nestbox entrance hole (Photos: JAS).

Although this allows the collection of time-specific presence data for individual birds, PIT tags do not allow differentiation of arrival and departure events. When combined with other approaches using, for example, temperature-sensitive devices such as *i*Buttons® (see section 15.4.4), recess and session lengths, and individual investment in incubation can be determined (Cresswell et al. 2003; Zangmeister et al. 2009). The deployment of PIT tags requires birds to be captured, which has to be conducted under licence issued by the appropriate authority.

15.3.3 Combining technologies with marking techniques

Technologies that allow assessment of arrival and departure times of birds at the nest can be combined with marking techniques to determine individual investment in incubation. For example, combining photography with marking techniques, such as plumage dyes, colour rings, patagial tags or flipper tags, allows the investigation of sex-specific roles during incubation (Weller and Derksen 1972; Craig and Craig 1974). However, uniquely marking birds is not always

straightforward and in most cases requires prior approval from the appropriate licencing authorities (Bub 1996; Balmer et al. 2008).

15.4 Methods to investigate egg temperature

15.4.1 Thermocouples and thermistors in nest cups

Advances in automated technology during the early 20th Century led to the development and use of temperature-sensitive thermocouples and thermistors (hereafter referred to as 'temperature probes') in nests to study incubation. Early devices used a potentiometer to record temperatures directly onto paper (Baldwin and Kendeigh 1927; Howell and Dawson 1954). Temperature probes allow us to detect the presence of birds by measurement of temperature through direct contact with incubating adults. They yield temperature traces that can provide high-resolution peaks and troughs reflecting patterns of nest attentiveness. Temperature probes record continuously, so they provided the first insights into temporal changes in incubation behaviour over multiple days or, in some cases, the entire incubation period. This highlighted that nest attentiveness is not consistent throughout the incubation period, and that clutches are incubated at night, as shown in house wrens (Baldwin and Kendeigh 1927). Potentiometers can be calibrated so temperature probes also allowed accurate estimation of nest temperature for the first time. Hence, this technique elucidated the energetics of incubation by showing that nest temperatures of Anna's hummingbirds (*Calypte anna*) are maintained above ambient temperatures even when the latter drop below optimum temperatures for incubation (Howell and Dawson 1954). However, despite such advances in knowledge, the use of temperature probes still requires bulky equipment, an electric current, and the introduction of foreign objects into focal nests that potentially disrupt natural incubation behaviours.

15.4.2 Thermocouples and thermistors within eggs

In the 1930s researchers inserted temperature probes into eggs of wild birds that allowed measurement of egg temperature for the first time, showing that it is not constant throughout incubation. Instead, it varies with behaviour of the incubator and environmental conditions such as wind and ambient temperature (Baldwin and Kendeigh 1932; Huggins 1941). Initially, small holes were drilled into eggs and probes were placed either near the embryo or cemented to the inner eggshell surface. Wires extending from probes were threaded through the nest structure or buried in the nest substrate and connected to a recording potentiometer (Huggins 1941). It is not clear how many eggs failed to hatch during such investigations but we suspect that it was many (Haftorn 1988).

The latest temperature probes have been developed for use with dummy eggs filled with silicon that have thermal properties similar to real eggs (Arnold et al. 2006; Cooper and Voss 2013). They are also used in conjunction with purpose-built commercially available data loggers, e.g. HOBO™—Onset Computer Corporation, Bourne, MA, USA (Ardia et al. 2009) instead of potentiometers. As well as allowing the assessment of incubation rhythms, within-egg temperature probes have increased understanding of subtle changes in egg temperature that underlie avian incubation strategies (Haftorn 1988; Cooper and Voss 2013). However, they result in clutch-size manipulation that may influence incubation behaviour (Smith 1989) and thus researchers should take this into account when interpreting data.

Furthermore, caution should be exercised when using dummy eggs with species that may be sensitive to 'foreign' eggs, particularly those species susceptible to brood parasitism. In addition, researchers using thermistors to record egg temperatures should be aware that they release heat while in use and therefore they may bias temperature recordings of small eggs (Cooper and Voss 2013). More work is needed to quantify the extent of bias during data collection using such devices.

15.4.3 Telemetric eggs

Miniature telemetering devices can be placed inside eggs that make them temperature- and/or light-sensitive (Figure 15.3). They transmit data to a receiver by means of a loop antenna placed above, or around the base of, the nest, or concealed below the nest material (Eklund and Charlton 1959; Varney and Ellis 1974; Fox et al. 1978). The receiver is concealed away from the nest thereby minimizing disturbance of incubating birds. Telemetric eggs have been used to study incubation rhythms and egg temperatures in many species, including the South Polar skua (*Stercorarius maccormicki*), the Adelie penguin (*Pygoscelis adeliae*; Eklund and Charlton 1959) and the golden eagle (*Aquila chrysaetos*; Varney and Ellis 1974), and abnormal incubation behaviour of herring gulls (*Larus argentatus*) exposed to pollutants: gulls exposed to pollutants were less attentive than control birds (Fox et al. 1978).

Telemetric eggs fitted with temperature and position sensors have also enabled the monitoring of egg positions within clutches and egg-turning rates (Howey et al. 1984; Boone and Mesecar 1989; Gee et al. 1995; Deeming 2002f). For example, we now know that mallards (*Anas platyrhynchos*) and domestic fowl (*Gallus gallus*) turn their eggs approximately 60° per hour (Caldwell and Cornwell 1975; Boone and Mesecar 1989). Telemetric eggs offer advantages over other approaches in assessing egg-turning rates, such as repeat visits to the nest to track the positions of marked eggs (Drent 1970), because they permit collection of continuous data without disrupting natural behaviour at the nest disturbed by frequent nest visits. Furthermore, telemetric eggs can be fitted with bi-axial accelerometers that allow egg positions to be recorded every 10 seconds (Beaulieu et al. 2010; Thierry et al. 2013b), permitting investigations of the effects of incubator stress (Thierry et al. 2013b) and stage of incubation (Beaulieu et al. 2010) on egg-turning rates. However, telemetric eggs fitted with bi-axial accelerometers by definition cannot provide

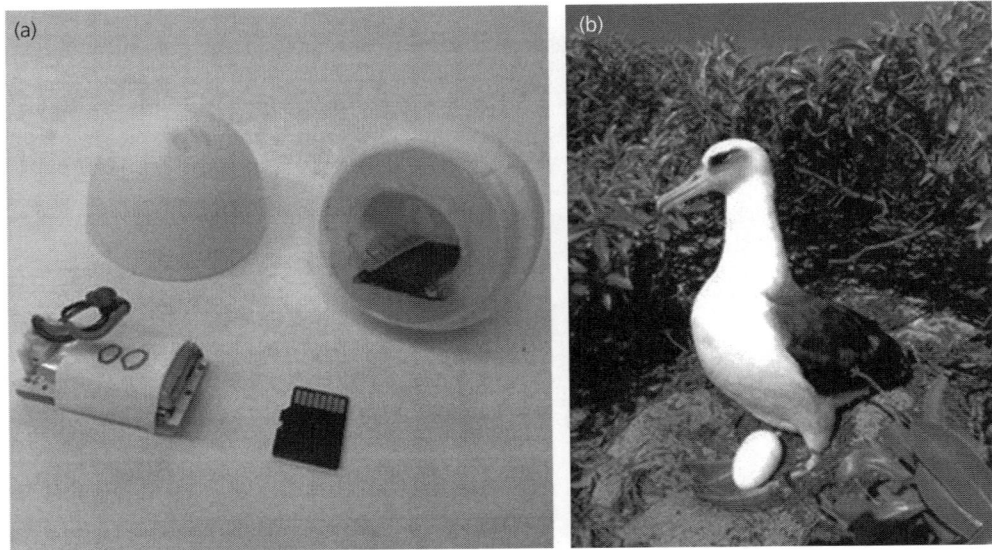

Figure 15.3 A telemetric egg (a) containing a tri-axial accelerometer, magnetometer, and a temperature-sensitive thermistor used to study egg-turning behaviour (Photo: S.A. Shaffer), and (b) placed in the nest of a Laysan albatross (*Phoebastria immutabilis*; Photo: L. Young).

insights into the three-dimensional movements of eggs during egg turning (Shaffer et al. 2014).

Recent developments have seen the emergence of telemetric eggs containing tri-axial accelerometers that also incorporate magnetometers that measure egg orientation in relation to the Earth's magnetic field. The latest telemetric eggs now provide a more accurate account of egg turning showing, for instance, that mean egg-turning rates are relatively conserved across seabird taxa, and that diurnal patterns in egg-turning rates vary between species; night- and day-time turning rates are consistent for western gulls (*Larus occidentalis*) and Laysan albatrosses (*Phoebastria immutabilis*), whereas turning rates are 44% higher at night compared to during the day for Cassin's auklets (*Ptychoramphus aleuticus*; Shaffer et al. 2014). Therefore, telemetric eggs allow continuous monitoring of *in situ* processes, such as egg turning, without the need for marking of eggs and regular flushing of birds from the nest (Caldwell and Cornwell 1975; but see Boulton and Cassey 2012).

15.4.4 *i*Buttons®

While automated technologies clearly allow for reduced disturbance of subjects and more streamlined collection of high-precision data relating to incubation, they still present some major challenges to researchers.

For example, their extensive use may be prohibitively expensive, may involve the carriage of bulky components, and may require an external power supply. The latter currently presents insurmountable problems at remote field locations but the situation should improve as power supplies become more lightweight and reliable. Technological advances have mitigated some of these challenges. For example, *i*Buttons® are self-contained temperature loggers (Figure 15.4) that are powered from within by batteries, contain computer chips that offer a capacity of up to 8,192 bit readings that can be taken at intervals from 1 second to 273 hours and have an accuracy of ± 0.5°C (Maxim Integrated Products 2014). This provides the user with an inexpensive (e.g. a hi-accuracy *i*Button® costs approximately £30) means of recording and collating high resolution temperature data and eliminates the need for bulky equipment. Moreover, an *i*Button® is 16 mm in diameter and 6 mm in thickness, weighs 3.24 g, and has a stainless steel casing and a shelf life of up to 10 years depending upon sampling rate (Maxim Integrated Products 2014), making it attractive for studying incubation of a wide range of avian species.

The use of *i*Buttons® to study incubation has escalated throughout the past decade (Table 15.1). Not only can they provide high-resolution peaks and troughs on temperature traces reflecting patterns of attentiveness (Figure 15.5), they potentially may allow study of

Figure 15.4 An *i*Button® logger used to study avian incubation behaviour. It is placed next to a 10 mm scale (Photo: N. Day).

more subtle behaviours. For example, during egg turning, an *i*Button® placed in a nest cup will be exposed to warm egg surfaces that have been in contact with the brood patch and, subsequently, colder egg surfaces that have been in contact with the nest substrate. Therefore, thermal data collected over bouts of egg turning may provide insights into egg-turning rates, but further investigations are needed to confirm this.

*i*Buttons® have also been used to determine characteristics of nests *ex situ*. Insulatory properties of nests of a variety of passerine species revealed using *i*Buttons® have demonstrated the influence of temperature and latitude on nest construction (Mainwaring et al. 2012, 2014b; L. Deacon et al., 2014 unpublished data; see Chapter 4).

15.5 Using *i*Buttons® to study incubation

Although *i*Buttons® are widely used and appear to have become the 'default' device for many studying incubation, they may not always provide the most suitable method for data collection. Their deployment requires researchers to consider their suitability for their study system and, in some cases, to overcome

considerable methodological and analytical challenges. We now consider the challenges that may arise when using *i*Buttons®, and present possible solutions to overcome them. Our aim here is to enable the reader to make informed decisions during the planning of research that will ultimately yield high-quality data. We focus on *i*Buttons® because of their widespread adoption by field ornithologists (Table 15.1) and many of the other available technologies (e.g. thermocouples, telemetric eggs) require similar resolutions to shared problems in the planning and execution phases of their deployment.

15.5.1 *A priori* considerations

The suitability of *i*Buttons® should be assessed within the context of the biology of the focal species (e.g. whether it is commonly parasitized, average clutch size), its environment (e.g. ambient temperature) and the nature of the data to be collected (e.g. sampling period and interval, temperature range). The sampling interval must be shorter than the minimum bout/recess length to ensure that incubation rhythms are resolved over the entire incubation period. Therefore, it is advisable to carry out pilot work to generate temperature data from *i*Buttons® and Figure 15.6 provides a step-by-step guide to carrying out such pilot work.

A number of modes of deployment are available to prevent the removal of *i*Buttons® from, or their burial in, nests. For species that build nests that consist of a solid structural layer, such as the common chaffinch (*Fringilla coelebs*) and the European greenfinch (*Carduelis chloris*), or in cavities, such as great tits (*Parus major*), blue tits (*Cyanistes caeruleus*), or European pied flycatchers (*Ficedula hypoleuca*; Ferguson-Lees et al. 2011), devices can be wired into the nest by attaching them first to a shirt button using Velcro® (or double-sided adhesive), which is then attached to garden wire. The wire is then wound around the nesting material or pushed through the nest and secured on the outside. Using Velcro® allows the *i*Button® to be easily retrieved without the need to remove the wire, a process which could cause damage to the nest's structure. Alternatively, a device can be secured to the nest by wrapping an elastic band around it through which a hair grip is threaded which can then be put through the bottom of the nest (from above) keeping the *i*Button® flush with the bottom of the nest (Bayard and Elphick 2011). For larger nests like those of common blackbirds (Snow 1958), it is possible to secure the device within a small wire cage that is then attached to the nest with wire (D. Leech, 2008 personal communication).

Table 15.1 Focal species in a sample of incubation studies in which small temperature loggers have been deployed. Common and scientific names of species and taxonomic order are in accordance with Gill and Wright (2006).

Common name	Scientific name	Reference
Common pochard	*Aythya ferina*	Legagneux et al. (2011)
Magellanic penguin	*Spheniscus magellanicus*	Rebstock and Boersma (2011)
Osprey	*Pandion haliaetus*	M. Voss (2008, personal communication)
Northern lapwing	*Vanellus vanellus*	T. Bodey (2009, personal communication)
Piping plover	*Charadrius melodus*	Schneider and McWilliams (2007)
Long-billed curlew	*Numenius americanus*	Hartman and Oring (2006)
Common redshank	*Tringa totanus*	Cervencl et al. (2011)
Least tern	*Sternula antillarum*	M. Voss (2008, personal communication)
Black-bellied sandgrouse	*Pterocles orientalis*	Mougeot et al. (2014)
Chuck-will's-widow	*Caprimulgus carolinensis*	O'Connor and Ritchison (2013)
Superb fairywren	*Malurus cyaneus*	N. Margraf (2008, personal communication)
Stitchbird	*Notiomystis cincta*	R. Thorogood (2008, personal communication)
Black-capped chickadee	*Poecile atricapillus*	M. Voss (2008, personal communication)
Great tit	*Parus major*	Smith (2011)
Blue tit	*Cyanistes caeruleus*	Smith (2011)
Tree swallow	*Tachycineta bicolor*	Ardia and Clotfelter (2007)
Chilean swallow	*Tachycineta meyeni*	M. Liljesthrom (2008, personal communication)
House wren	*Troglodytes aedon*	E. Clotfelter (2008, personal communication)
Eastern bluebird	*Sialia sialis*	Cooper and Mills (2005)
Western bluebird	*Sialia mexicana*	Duckworth (2006)
Mountain bluebird	*Sialia currucoides*	Johnson et al. (2013)
Common blackbird	*Turdus merula*	D. Leech (2008, personal communication)
Sociable weaver	*Philetairus socius*	Spottiswoode (2009)
House finch	*Carpodacus mexicanus*	Badyaev et al. (2003)
Red-winged blackbird	*Agelaius phoeniceus*	T. Rusch (2008, personal communication)
White-throated sparrow	*Zonotrichia albicollis*	J. Bodwell (2008, personal communication)
Savannah sparrow	*Passerculus sandwichensis*	J. Johnston (2008, personal communication)
Seaside sparrow	*Ammodramus maritimus*	R. Boulton (2011, personal communication)
Saltmarsh sparrow	*Ammodramus caudacutus*	Gjerdrum et al. (2008)

In waders that lay eggs in scrapes a method described by Hartman and Oring (2006) can be adapted depending upon the species. They studied incubation of long-billed curlews (*Numenius americanus*) using *i*Buttons® that were glued to the top of 250 mm-long aluminium stakes, covered with black polyester and secured with nylon thread. The stake was then pushed into the centre of the nest until the *i*Button® was flush with the bottom of the nest. Similar approaches have been used with common redshanks (*Tringa totanus*), northern lapwings (*Vanellus vanellus*) (T. Bodey, 2009 personal communication), and common eiders (M. Parsons, 2009 personal communication). In a study of little penguins (*Eudyptula minor*) *i*Buttons® were attached using Araldite® (Huntsman Advanced Materials, Salt Lake City, UT, USA) to plastic spikes which were then inserted into the ground in burrows so that the *i*Button® was positioned over the nest cup (L. Horne, 2009

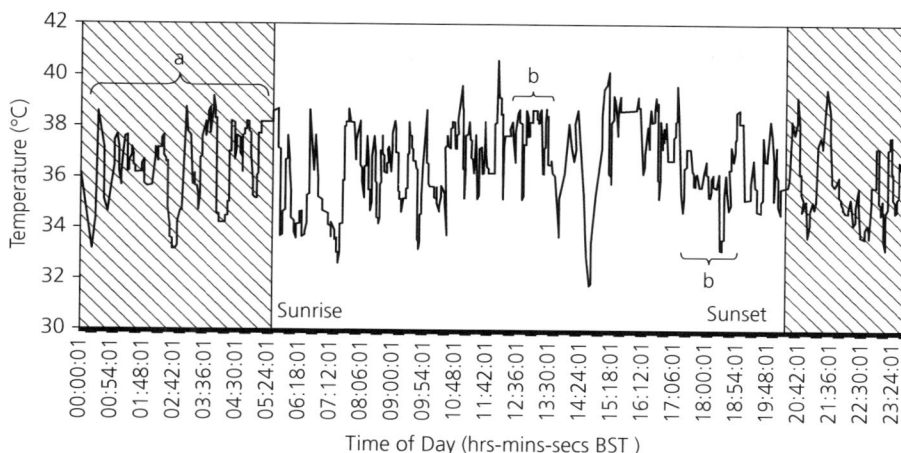

Figure 15.5 A temperature trace from the nest of an incubating blue tit (*Cyanistes caeruleus*) in Chaddesley Woods National Nature Reserve in Worcestershire, UK in 2007. The trace shows high resolution of nest temperature over a 24-hour period on day 7 of the incubation period with marked temperature fluctuations (a) throughout the night, and more subtle temperature fluctuations (b) punctuating the trace throughout the day (Source: Smith 2011).

personal communication). It should be noted, however, that during burrow excavation little penguins dislodged some of them resulting in their loss (L. Horne, 2010 personal communication). In wandering albatrosses (*Diomedea exulans*), it is possible to mount

*i*Buttons® on 300 mm-long aluminium strips which can be inserted into pre-cut horizontal slots at mid-height in nests (Sinclair and Chown 2006).

Instead of physically attaching an *i*Button® in nest substrate, or to a stake to prevent its removal, it can be

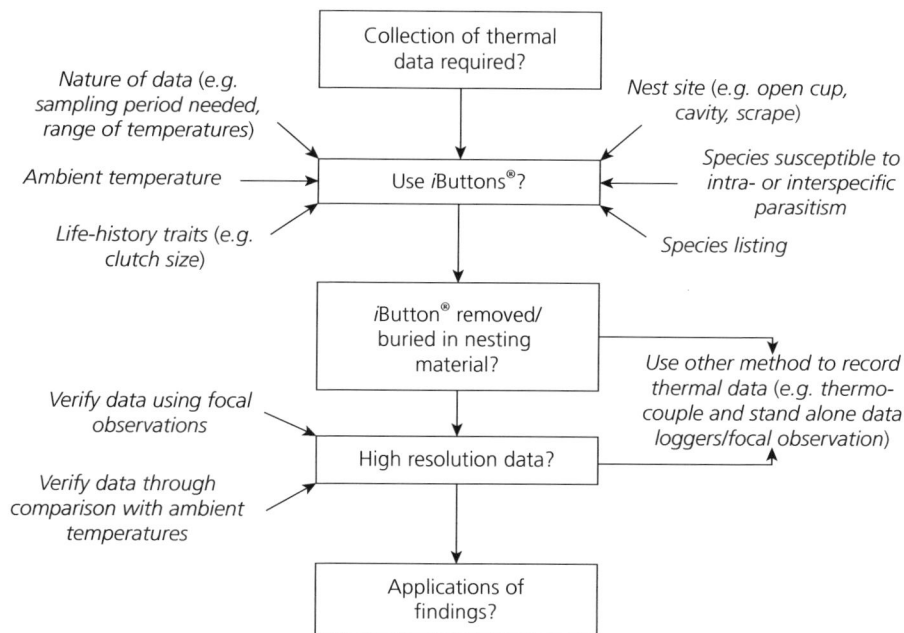

Figure 15.6 Steps that should be undertaken in carrying out pilot work to test the suitability of *i*Buttons® for studying incubation of birds. Textboxes pose questions that should be asked, and italicized statements are important considerations before and during their deployment.

sealed in Parafilm® (Pechiney Plastic Packaging, Chicago, IL, USA) and embedded within a wax gel-filled dummy egg that mimics the thermal properties of eggs of the study species. This is then placed within a clutch of eggs that may have been manipulated, i.e. an egg is removed, to compensate for its addition (M. Voss, 2008 personal communication). Rebstock and Boersma (2011) encapsulated *i*Buttons® inside hollow plastic eggs filled with sand to measure egg temperature of Magellanic penguins (*Spheniscus magellanicus*). Obviously, this method is restricted to species with eggs larger than *i*Buttons®.

In species with smaller eggs that may detect *i*Buttons® visually (e.g. open-cup nesters), devices can still be deployed successfully in nests by reducing their visibility by colouring them with matt paints or permanent markers, wrapping them in pieces of sheer stocking that reduce their lustre and increase their resemblance to the nest substrate, and by covering them with nest-lining material. The coloured *i*Buttons® can then be either left loosely within the clutch or secured to the nest substrate (J. Bodwell, 2008 personal communication; T. Rusch, 2008 personal communication; T. Schaming, 2008 personal communication). Furthermore, such devices can be made less conspicuous by wrapping them in small pieces of nest substrate-coloured fabric before securing them to the nest material (N. Margraf, 2008 personal communication; S. Wilson, 2014 personal communication). By careful consideration of the nesting conditions and of the mechanisms of *i*Button® detection by birds (e.g. through visual, tactile, auditory cues), it is possible to incorporate *i*Buttons® into natural nest structures without the breeding bird(s) removing them.

15.5.2 Challenges of using *i*Buttons®: deployment

*i*Buttons® need to be placed where they can detect shifts in temperature resulting from the arrival or departure of the incubating bird. Ideally, the device is placed loosely within a clutch where it is in close contact with the adult (Figure 15.7). However, some species, such as long-billed curlews (Hartman and Oring 2006), salt-marsh sparrows (*Ammodramus caudacutus*; C. Elphick, 2008 personal communication), superb fairywrens (*Malurus cyaneus*; N. Margraf, 2008 personal communication), and great tits (Smith 2011), will remove an *i*Button® from their nests. If not removed, some species such as stitchbirds (*Notiomystis cincta*), will bury the *i*Button® in the nest-lining material (R. Thorogood, 2008 personal communication), while others such as blue tits and great tits (Smith 2011) will bury them deep in nest material. This markedly reduces the resolution of data because of thermal impedance (compare Figure 15.8a with 15.8b). In addition, some species may move *i*Buttons® within the nest reducing their ability to detect diagnostic temperature fluctuations (Bayard

Figure 15.7 An *i*Button® within a clutch of a blue tit (*Cyanistes caeruleus*) where it was deployed to record incubation behaviour in Chaddesley Woods National Nature Reserve in Worcestershire, UK in 2008. The *i*Button® was secured to the nest using garden wire that was threaded through a shirt button attached to the back of the *i*Button® (Photo: JAS).

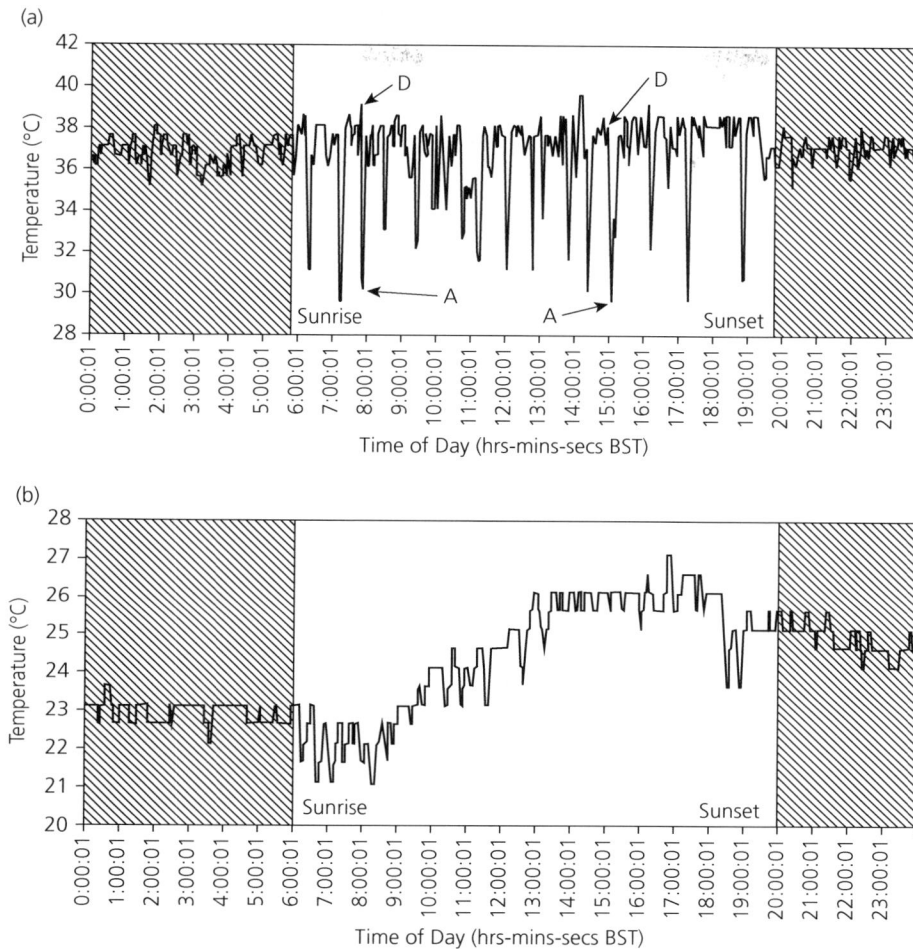

Figure 15.8 Temperature traces from *i*Buttons® (a) placed in close proximity to an incubating blue tit (*Cyanistes caeruleus*) showing high resolution of nest temperature over 24 hours of recording on day 7 of the incubation period, and (b) buried beneath nest material in a different nest showing low resolution of nest temperature over a similar period. Changes in the temperature trace denote the incubation behaviour of the bird with a decline due to her departure (D) and an increase due to her arrival (A) at the nest. Vertical lines indicate sunrise and sunset. *i*Buttons® were used in Chaddesley Woods National Nature Reserve in Worcestershire, UK in 2007 (Source: Smith 2011).

and Elphick 2011; Smith 2011). Consequently, calculation of session and recess lengths during incubation is often problematic and sometimes impossible. On rare occasions *i*Buttons® can result in nest desertion by incubating birds, such as in blue and great tits (Smith 2011) and seaside sparrows (*Ammodramus maritimus*; R. Boulton, 2011 personal communication) especially when placed in nests during egg laying (Smith 2011). Nevertheless, some species including common starlings (*Sturnus vulgaris*; G. Martin, 2010 personal communication), eastern bluebirds (*Sialia sialis*; CBC, 2008 personal observation), and tree swallows (*Tachycineta*

bicolor; E. Clotfelter, 2008 personal communication) are tolerant of such devices in the nest and may even appear to ignore them.

Interspecific differences in tolerance to such devices may partly be explained by life-history traits. Species that are susceptible to either intra- or interspecific brood parasitism have an innate or learned response (Rothstein 1974) which results in removal of foreign objects such as a parasite's eggs. Their response may even be explained in terms of a more fundamental physiological reaction to ultrasound emitted by *i*Buttons® (Willis et al. 2009). For example, several bat species avoided *iButtons*® when they were being used to monitor microclimates of their roosting sites. Although birds appear to demonstrate a low sensitivity to ultrasonic frequencies (Pytte et al. 2004), few species have been studied and detection of *iButtons*® by birds through auditory cues remains to be investigated in detail.

Detection and subsequent removal or burial of *iButtons*® in nest cups may occur during nest sanitation behaviours that combat nest parasitism (Moskát et al. 2003; see Chapter 8). *iButtons*® are likely to be detected, then removed or buried when an adult detects it against its brood patch or other body part during incubation or during egg turning (Deeming 2002f). Their detection may also disrupt tremble-thrusting (Haftorn 1994) which is thought to assist in egg turning (Hartshorne 1962) and to maintain the insulatory properties of the nest by creating small air pockets (Haftorn 1994; see Chapter 4).

The likelihood of removal of *iButtons*® is also dependent on the physical characteristics of the incubating species with, for example, size and shape of bill limiting a bird's grasping abilities (Underwood and Sealy 2006). Furthermore, for species that are intolerant to foreign objects, the removal of *iButtons*® is likely to disrupt normal incubation and their use would be entirely inappropriate in such species that are likely to expend substantial time and energy in effecting their removal. Therefore, it is essential that detailed knowledge of the reproductive behaviour of the focal species at the nest (including tolerance to foreign objects) exists prior to deployment of *iButtons*®.

15.5.3 Challenges of using *iButtons*®: technical issues

iButtons® were not originally designed to be self-contained waterproof devices and sometimes they malfunction as a result of water entering the steel casing, preventing data retrieval (Wolaver and Sharp 2007). This can be a problem when they become submerged for extended periods, such as when Gjerdrum et al. (2008) used them to investigate nest attendance during nest flooding events. Therefore, in studies where the nest environment is subject to inundation sufficient waterproofing measures should be considered. For example, Maxim Integrated Products have produced the *iButton*® logger capsule mechanical enclosure to protect their devices from moisture, solvents and pressure. However, the capsule is resistant to sunlight, ozone, oxygen and UV light (Maxim Integrated

Products 2014). Alternative waterproof casing is available in iBCod (submersible *iButton*® loggers) which can withstand pressure to a depth of up to 900 m (Alpha Mach Inc. 2014). It should be noted, however, that some waterproofing can disrupt accuracy of temperature measurements (Roznik and Alford 2012). Therefore, we suggest pilot work to validate temperatures recorded from waterproofed devices before they are deployed in full studies.

15.5.4 Challenges of using *iButtons*®: analytical issues

Their use can also be analytically challenging because they can accumulate large data sets, e.g. up to 8.2 kB (Maxim Integrated Products 2014). While scrutiny of temperature traces from *iButtons*® has the potential to identify relationships between environmental variables (and therefore climate change) and breeding biology of birds, data can be open to misinterpretation. For example, Figure 15.5 illustrates a temperature trace from an incubating blue tit from which it appears possible to determine incubation bout and recess lengths. However, on closer inspection, marked temperature fluctuations throughout the night cast doubt over the commonly held belief that the incubator remains in the nestbox (de Heij et al. 2007) where she sits 'tightly' on the eggs (Haftorn 1981). Throughout the day the temperature trace can be punctuated with numerous more subtle fluctuations that reflect poorly understood behaviours (Haftorn 1994; Deeming 2002e, 2002f; G.A. Rebstock, 2008 personal communication). Such 'noisy' data can also result from the position of an *iButton*® in the nest cup such as if it is under the eggs or buried within nest material, which may impede its ability to monitor brood patch temperature (Legagneux et al. 2011; Smith 2011). To date, such difficulties with their deployment and resultant performance are poorly understood but they are likely to vary with nest material type. Finally, 'noisy' data may occur when the bout and recess lengths are shorter than the sampling interval of the *iButton*®.

Figure 15.6 raises concerns over the reliability of temperature traces in reflecting specific incubation events. The extent of misinterpretation of data from devices is likely to vary with life history within and between species. For example, Schneider and McWilliams (2007) found that *iButtons*® did not accurately determine patterns of nest attendance of piping plovers (*Charadrius melodus*) that lay small clutches of four eggs on average. However, plovers frequently covered *i*Buttons® in sand and repositioned them, resulting in them not being in close contact with incubating birds. The resolution from a trace of temperature fluctuations may also be affected by ambient temperature and *i*Buttons® may only be effective where significant thermal gradients exist, but even then incubation rhythms can be difficult to resolve. For example, *i*Buttons® were used to investigate incubation patterns of white-tailed ptarmigans (*Lagopus leucura*) in the Yukon, Canada, where ambient temperatures during breeding are approximately 15–20°C. Video footage indicated that devices that were programmed to record temperatures every 30 seconds did not accurately reflect arrival and departure times of incubating birds such that 20-minute recess lengths could not be determined (S. Wilson, 2014 personal communication). It is possible that the mass of the clutch of white-tailed ptarmigans, which contains, on average, 7.1 eggs of 18.8 g mass each (S. Wilson, 2014 personal communication), retained enough heat during recesses when ambient temperatures exceeded 15°C to mask the effect of the incubator leaving the nest.

To overcome analytical challenges resulting from bi-parental incubation, *i*Buttons® should be combined with other technologies (e.g. PIT tags) that allow the temporal matching of temperature traces with the bird's identity. Another approach that is especially effective in open-nesting species is offered by video monitoring (Ribic et al. 2012) used in conjunction with colour ringing of birds.

To simplify analysis of *i*Button® data, software called 'Rhythm' (Cooper and Mills 2005) can be used in conjunction with the 'Raven Pro' bioacoustical analysis software (Bioacoustics Research Program 2014). Rhythm software can convert temperature files for viewing, tagging, and extracting data in Raven Pro. Rhythm automatically detects incubation recesses in temperature time series while Raven Pro displays the associated data (e.g. duration of recess, maximum temperature during recess) in tabular form. If two temperature series were recorded in tandem (e.g. within-nest and ambient), Raven Pro can display and extract data from both at once. Rhythm can also be used to insert sunrise and sunset times (or civil twilight) in order to facilitate computation of latency between sunrise and the first recess of the focal incubating bird that day, for example. By using Rhythm and Raven Pro in concert, incubation patterns can be quantified in a few minutes from temperature time series data collected by *i*Buttons® as opposed to lengthy periods to achieve similar data extraction through visual inspection (Cooper and Mills 2005). Despite the relative ease of extraction of

temperature data using this software, their interpretation may be challenging especially when data are of low resolution and/or when temperature fluctuations are subtle (Figures 15.5b and 15.8b). We suggest that *i*Button® data that cannot be analyzed using automated software should be randomly sub-sampled to generate a smaller data set which then can be analyzed using a series of carefully selected decision rules (Figure 15.9).

15.5.5 Challenges of using *i*Buttons®: other considerations

Direct observations of nests containing *i*Buttons® should routinely be conducted in parallel with their deployment. For example, observations of arrival and departure times should be compared with nest events as inferred from *i*Button® data to determine their accuracy such as in the cases of house finches (*Carpodacus mexicanus*; Badyaev et al. 2003) and tree swallows (D. Ardia, 2007 personal communication). In addition, *i*Buttons® used in conjunction with high-precision nestbox cameras allow verification of subtle nocturnal behaviours of the bird at the nest (Pendlebury and Bryant 2005). It is useful to use at least one additional *i*Button® to measure ambient temperature; it should be placed close enough to study nest microclimate but not so close so as to be directly influenced by the bird's thermoregulation.

By using multiple technologies in conjunction with *i*Buttons®, it is possible to calculate accurately arrival or departure times of birds at the nest and, thus, to derive session and recess lengths of incubation, and timing of predation and nest abandonment events. They also permit quantification of more subtle incubation activities either independently from, or in combination with, bout and recess lengths. In the latter, temperature shifts in an *i*Button® trace reflecting a 24-hour period are counted. The resulting metric, 'daily incubation activity', encapsulates incubation activities as well as more subtle discrete behaviours such as tremble-thrusting, bill pressing and egg turning (Smith 2011).

15.6 What about the next 20 years?

There is no doubt that the development of methods over the last century has increased our understanding of incubation of a wide range of avian taxa. However, how will technological advances in the future help in this regard? We suggest that continued miniaturization of solar-powered tags that use the Global Positioning System (GPS) and that transmit data to an internet-based

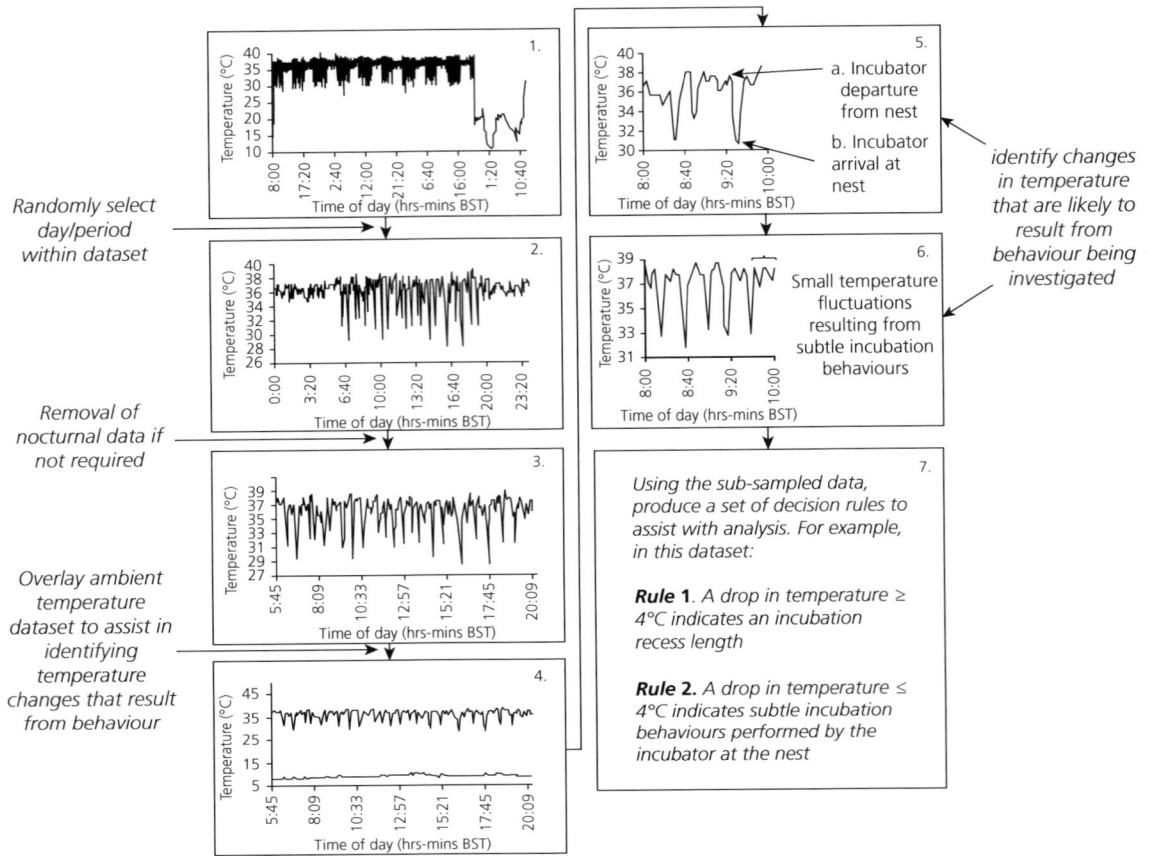

Figure 15.9 A flow diagram outlining the processing of raw data from an *i*Button® containing (1) the full data set to (4) a reduced, sub-sampled data set for analysis. After sub-sampling, a set of decision rules can be used to assist with analysis (5 and 6), and then implemented (7). Italicized text describes procedures during the sub-sampling and analysis of *i*Button® data.

account using the Global System for Mobile Communications (GPS/GSM tags), that currently weigh 20–100 g and that allow GPS fixes every 30 seconds (Cellular Tracking Technologies 2014), will permit researchers to collect fine-scale spatial data that will provide insights not only into activity at the nest but also away from it (e.g. foraging distances, habitat use). As portable thermal imaging cameras become more affordable, further investigations of the role of egg temperatures in mediating egg-turning rates and incubation rhythms will be possible (Boulton and Cassey 2012; see Chapter 4 for further details).

In this chapter we have highlighted how modern techniques such as video monitoring, temperature-sensing devices, and PIT-tagging of birds can be allied to more traditional approaches such as direct nest observations and ringing to study incubation behaviour.

We suggest that researchers should consider integrating technologies further to assess incubation from both the adult and the embryo perspective. We conclude by suggesting that our understanding of incubation will surely accelerate if ornithologists are creative in working with engineers, statisticians and other biologists in harnessing the potential of the myriad new technologies and methods.

Acknowledgements

We are grateful to all who provided information concerning various techniques and problems encountered during deployment of technologies. These include Dan Ardia, Emily Argyle, Trinia Bayard, Thomas Bodey, Jenny Bodwell, Rebecca Boulton, Ethan Clotfelter, Paul Doherty, Chris Elphick, Lyndal Horne, Martin

Jacobson, William Jensen, Jason Clarence Johnston, Dave Leech, Marcela Liljesthrom, Nicolas Margraf, Graham Martin, Scott McWilliams, Monika Parsons, Ginger Rebstock, Travis Rusch, Taza Schaming, Brent Sinclair, Rose Thorogood, Nigel Titley, Margaret Voss, and Scott Wilson. For help and support during field-work we thank Rebecca Boulton, Tim Harrison, and Simone Webber. The Worcestershire Wildlife Trust, es-pecially Mervyn Needham, kindly granted permission for us to work at Chaddesley Woods National Nature Reserve, Worcestershire, UK to study blue and great tits. For comments on drafts of this chapter we are grateful to Charles Deeming, Paul Donald, Brian Hind-man, Darryl Jones, Irene Tieleman, and Jeremy Wilson. We would like to thank the Natural Environment Re-search Council for funding JAS through a studentship (NER/S/A/2006/14054), the British Trust for Orni-thology for acting as a CASE partner on the student-ship and the School of Biosciences at the University of Birmingham for funding some equipment.

CHAPTER 16

Applications of incubation science to aviculture and conservation

D.C. Deeming and N.S. Jarrett

16.1 Introduction

The study and application of avian incubation have long been driven by the needs of the commercial poultry industry. Pre-19th Century farmyard poultry were often unassisted during breeding with eggs harvested or chicks slaughtered once they were sufficiently large for the table. Although production of domestic fowl (*Gallus gallus*) eggs has been undertaken by people for millennia, the development of new technologies from the 19th Century onwards translated into larger numbers of birds producing more hatchlings (Etches 1996). The industrialization of the poultry industry during the second half of the 20th Century was accompanied by rapid development of artificial incubation techniques and thus increased hatchability. Modern poultry producers now have hatching successes of approximately 80–85% of eggs incubated and often over 95% of fertile eggs will produce chicks (DCD, 2014 personal observations). Commercial pressures drove development of modern machines capable of incubating tens of thousands of eggs simultaneously. Thus, modern technology allows the commercial production of domestic fowl for eggs and meat, as well as domestic turkeys (*Meleagris gallopavo*) and domestic ducks (*Anas platyrhynchos*), and even less common species, such as common pheasants (*Phasianus colchicus*), partridges (*Alectoris* and *Perdix* spp.), common pigeons (*Columba livia*), and ratites (e.g. the common ostrich [*Struthio camelus*]), at a scale that could have hardly been imagined only 50–100 years ago.

The artificial incubation of bird eggs has not been restricted to domesticated poultry. Eggs of birds of prey, especially falcons (Falconiformes; Fox 1995), and parrots (Psittaciformes; Jordan 1989) are often incubated in small numbers to produce birds for recreational or conservation purposes. Again, commercial pressures have produced incubators to accommodate these species although there is often a reliance on the parent birds to incubate under captive conditions.

Collar and Butchart (2014) describe conservation breeding as when captive breeding is undertaken for conservation purposes ranging from reducing or preventing extinction risk for some species, such as the Mauritius kestrel (*Falco punctatus*; Jones et al. 1995), to raising awareness of conservation issues for others like the greater flamingo (*Phoenicopterus roseus*; Batty et al. 2006). Many bird conservation breeding programmes have been established from founder stock hatched from eggs arriving from wild populations (e.g. Lubbock 1978, 1980a, 1980b, 1981a, 1981b; Bennet 1986; Saint Jalme 1999) or from extant captive populations (Richardson 1997) and sometimes to initiate *in situ* reintroduction programmes (e.g. Ounsted 1983, 1988). The entire population of wildfowl at the Wildfowl and Wetlands Trust (WWT) centre at Llanelli in Wales, UK, was founded from 600 birds of 81 species arriving as eggs (or occasionally hatchlings) from seven other WWT centres in England and Northern Ireland, UK (Richardson 1997). Significant experience with wildfowl has taught us that birds which hatch from eggs imported to a captive setting establish and breed more readily than birds imported to the same place as fully grown adults. This is exemplified by pink-eared ducks (*Malacorhynchus membranaceus*) hatched from wild birds' eggs that were translocated in a portable incubator from Australia to Slimbridge (Lubbock 1980a) and bred when aged ten months (Lubbock 1981a), whereas freckled ducks (*Stictonetta naevosa*) imported as adults to Slimbridge from Australia in 1988 (Hall 1986) did not breed until six years later (Jarrett 1992). This is possibly because the processes of translocation and acclimatization of full-grown birds from wild to captive conditions disrupt temporal patterns of non-breeding and breeding.

Deeming, D.C. and Jarrett, N.S., *Applications of incubation science to aviculture and conservation*. In: *Nests, Eggs, and Incubation*. Edited by: D.C. Deeming and S.J. Reynolds, Oxford University Press (2015). © Oxford University Press. DOI 10.1093/acprof:oso/9780198718666.003.00016

In this chapter we explore how our understanding of incubation in both incubators and nests is being used to good effect in aviculture. This can often apply to commercially orientated operations but increasingly conservation programmes are drawing on scientific knowledge and experience (Saint Jalme and van Heezik 1996). Captive birds are often left to breed under 'natural' conditions with the adults incubating eggs and rearing chicks. Under many circumstances this approach is preferred but it often limits the potential for high productivity within a season. Egg-pulling, i.e. removal of a clutch prior to incubation, may induce birds to re-lay and so increase reproductive output as is the case with Californian condors (*Gymnogyps californianus*; Snyder and Snyder 2000) and houbara bustards (*Chlamydotis undulata*; Saint Jalme and Heezik 1996). This could be crucial for conservation programmes but it means that the aviculturalist is reliant on artificial methods for incubation and chick-rearing. Although previous publications (Deeming and Ferguson 1991a; Tullett 1991; Deeming 2002a) have shown our scientific understanding of various aspects of incubation is detailed, it is likely that information is not being effectively transferred to aviculturalists. Thus, while we acknowledge that information needs to be freely available and communicated, there have been few attempts to date to develop incubation 'manuals' for aviculturalists. For example, Deeming (2003) effectively summarized Deeming (2002a) in layman's terms, and Deeming (1997, 2000, 2002h) provided practical incubation advice for ratite, game, and poultry farmers, respectively.

Here, we will initially briefly consider small-scale incubation equipment before describing how incubation conditions for the vast majority of bird species may differ from those for domestic poultry. We will demonstrate the fundamental importance of effective interaction between professional ornithologists and applied avian conservationists. As part of this we will provide the case study of artificial incubation of the Madagascar pochard (*Aythya innotata*), one of the world's rarest birds, under extreme field conditions at the start of a conservation breeding programme to save the species from extinction.

16.2 Incubation equipment

Artificial incubation has a long history largely concerned with poultry eggs (Sykes 1991). Non-commercial aviculture invariably uses small-scale incubators with capacities of typically fewer than 200 eggs and often < 30. They comprise two basic types: still air and fan-assisted. Still air incubators have a heater in the space (often within the lid of the cabinet) above the eggs which warms the eggs from above. Humidity is often provided by evaporation from a water pan in the bottom of the cabinet. The cabinet is aerated, i.e. fresh air is introduced, and ventilated by convection of warm air through holes in the top of the cabinet. Cooler air then enters the cabinet from the room through holes in the base of the cabinet. Turning of the single layer of eggs can be automatic or by hand depending upon the incubator's complexity.

Fan-assisted incubators have a heater to warm the air and a fan to circulate air within the cabinet. More technologically advanced incubators may have an automatic spray or external water pump system to regulate humidity but most use a water pan for the same purpose. Air is often 'introduced' via a single vent over the fan but in many incubators there are no other holes in the cabinet to allow exchange of air (DCD, 2014 personal observations). Poor hatching in such machines can often be attributed to lack of oxygen in the later stages of incubation which is exacerbated when the machines are full of eggs (DCD, 2014 personal observations). The more efficient incubators use a fan that circulates air in the cabinet, and pushes air out of, and draws replacement air into, the cabinet through holes in the floor or sides. There can be multiple layers of eggs and these are often fixed in position. Egg turning is typically automatic with eggs moving through ± 45° every hour.

Hatchability of these incubator types varies between operators and species but there has been no systematic study to explain variability. Claims by manufacturers should always be treated with caution and it is better to take advice from those with operational experience of the incubators. For more information about incubator design see Owen (1991).

One novel incubation system that was trialled in the early 2000s was a contact incubator manufactured by Brinsea Products® in the UK. This machine attempted to replicate the top-down incubation of a still air incubator by mimicking natural incubation in a nest. Eggs were heated by contact between the upper shell of the eggs and a plastic bubble, which was full of air heated to the incubation temperature. Using eggs of red-legged partridges (*Alectoris rufa*), Deeming and Riches (2005) used three machines to incubate eggs, each with a different temperature within the plastic bubble: 38.5°C, 39.5°C, and 40.5°C; relative humidity (RH) was 55%. Eggs were set between plastic rollers and were turned through 120° every hour when the bag was automatically lifted off them. The control incubator

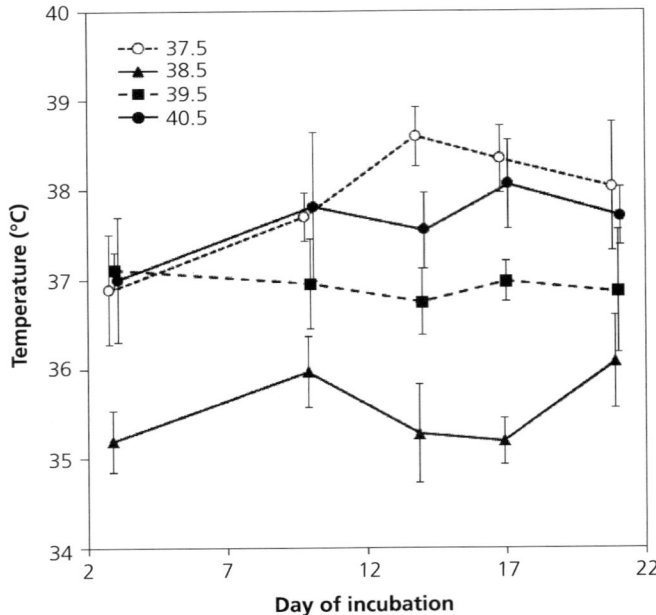

Figure 16.1 Variation in mean (± 1 SD) temperature of the upper surface of eggshells of red-legged partridges (*Alectoris rufa*) with day of incubation. Eggs were incubated in a force-draught incubator (37.5°C) or in a contact-incubator with the air bubble temperature set at 38.5°C, 39.5°C or 40.5°C. Reproduced with permission from Deeming and Riches (2005).

was a fourth set of partridge eggs incubated in a force-draught machine ('Octagon 100'; Brinsea Products Ltd) set at 37.5°C, 55% RH, and with egg turning through 90° every 30 minutes.

The temperature of the upper surface of the eggshell was ~3°C lower than the air temperature in the bubble displayed by the machine (Figure 16.1). The control eggs had the shortest incubation period in both trials followed by eggs from contact incubators set at 40.5°C and hatchability of these eggs was not significantly different from the control machine. However, as bubble temperature decreased then the incubation was progressively prolonged and hatchability was reduced. Eggs under a bubble temperature of 38.5°C contained a large number of small, underdeveloped embryos that were alive upon opening the hatch debris but would never have hatched 'normally'. Deeming and Riches (2005) concluded that contact incubation was an effective alternative to conventional artificial incubators used for small-scale incubation if the bubble temperature was at least 40.5°C but more research was needed to determine the appropriate temperatures for other species. At the time of writing Brinsea Products® continue to market a 'Contaq' model but we do not know of the extent of take-up of such models by aviculturalists or enthusiasts.

16.3 Incubation conditions

16.3.1 Temperature

Incubation temperature in incubators is often set between 37.0°C and 38.0°C, a thermal range determined from many years of empirical studies and practical experience. This is largely a consequence of incubators being used with galliform eggs in domesticated situations. However, operating incubation temperatures are determined by egg size with those of common ostriches in multi-stage incubators (which include eggs of varying ages) needing to be a maximum of 36.5°C (Jarvis et al. 1985; Deeming 1997). Common ostrich embryos are tolerant of higher temperatures (e.g. 37.0°C) early in development but the metabolic heat generated by older embryos means that lower incubator set points are required to promote heat loss from eggs to prevent embryos from overheating towards the end of incubation (Deeming 1995).

Developmental temperature of all bird species has effectively been assumed to be the same as that for galliforms, particularly the domestic fowl. However, in nests of free-living species eggs experience a wide range of temperatures (Webb 1987) and some embryos can tolerate temperatures that are deleterious to those of domestic fowl embryos. Being ectotherms, avian embryos are sensitive to changes in temperature, which can have short-term and long-term physiological and developmental effects (Deeming and Ferguson 1991b). This is despite the fact that birds that exhibit intermittent incubation leave their eggs to cool to temperatures close to ambient and yet hatchability is rarely affected.

The incubation temperature adopted by aviculturalists for non-galliform species is often dependent on the incubation equipment, which is often designed for poultry eggs and has limited scope for using a temperature outside of the range used for poultry. This is despite the fact that mean egg temperature is often much lower than 37.0°C. For example, eggs of penguins (Sphenisciformes) are often recorded in the field at mean temperatures below 35°C (Webb 1987). However, caution should be exercised when thinking of deviating from a poultry-based incubation temperature. Egg temperature is often recorded using dummy or infertile eggs and Rahn (1991) reported a mean egg temperature across a range of 80 species from 15 orders of 35.7°C. Unfortunately, these do not represent the actual temperatures that embryonated eggs experience, particularly during the latter stages of incubation. This means that egg temperatures apparently experienced in nests can significantly underestimate embryo temperature (Deeming 2008) and adopting them in an incubator can have adverse effects on development.

This does not mean that an embryo's temperature should lie between 37.0°C and 38.0°C across all species. Avian incubation is complicated by the thermal dynamics of heat transfer from a brood patch (Turner 1991, 2002) to the surface of eggs of different size (Deeming 2002c). Rahn (1991) reported an average brood patch temperature of 40.7°C, some 5°C warmer than the eggs. Deeming (2008) investigated the interrelationships between core body and brood patch temperatures, and found that larger species had lower brood patch temperatures and this correlated with incubation period. More crucially, the difference between body temperature and brood patch temperature was ~2°C irrespective of body mass. So, whilst an embryo of the domestic fowl was shown empirically to develop optimally at a temperature of approximately 37.8°C (Meijerhof and van Beek 1993), the brood patch temperature was 40.4–40.7°C and body temperature was approximately 1°C higher. This implies that if brood patch temperature is as low as 37.5°C, the optimal temperature for development might also be lower. This has not been widely appreciated but embryos of Adelie penguins (*Pygoscelis adeliae*) were reported to develop optimally at 34.7°C (Eklund and Charlton 1959) despite average brood patch temperature of penguins being 37.4°C (Deeming 2008). Artificial incubation of eggs of

Humboldt penguins (*Spheniscus humboldti*) at 36.5°C resulted in a hatchability in excess of 90% (Deeming et al. 1991).

One of the key issues is that eggs are rarely artificially incubated at lower than 'normal' temperatures because of the inherent risks that development will not be optimal and chicks will fail to hatch. However, the data suggest that for some orders such as Sphenisciformes, Procellariiformes and Pelecaniformes, brood patch temperature is relatively low, i.e. < 37.5°C (Figure 16.2), and that perhaps optimal developmental temperature is lower than that currently used in poultry-based incubators. Similarly, Passeriformes have a high average brood patch temperature, i.e. 41.2°C, perhaps indicating that they need a higher temperature during artificial incubation. What is clear is that under natural incubation conditions birds are unable to heat their eggs such that they attain either the parent's brood patch or body temperature (Deeming 2008).

Temperatures used in artificial incubation vary. Galliform gamebirds are generally incubated at temperatures typical for female domestic fowl, i.e. 37.5–37.7°C (Game Conservancy Advisory Service 1993). However, Kuehler et al. (2000) successfully incubated eggs of omaos (*Myadestes obscurus*) weighing ~5 g at 37.2–38.1°C with hatchability ranging between 40% and 80%. The hatching success of eggs of parrots (Psittaciformes) was higher at incubation temperatures above 37.2°C (Jordan 1989). Californian condor eggs that weigh between 210 g and 290 g have been incubated successfully at 36.4–36.7°C (Snyder and Snyder 2000). Kuehler and Good (1990) found that smaller birds, which lay smaller eggs and have higher brood patch temperatures, develop well at relatively high incubation temperatures, resulting in high hatching successes (Table 16.1).

Typically, these results have been produced rather by trial and error and there is scope for a more systematic approach to understanding an optimal temperature for development in more species. As aviculture becomes increasingly important in species conservation there will be a requirement to use artificial means to improve productivity during incubation and hence increase population sizes. Reliance on birds to breed naturally within captivity may hamper efforts to increase population sizes rapidly, despite the perception that 'birds do it [incubation] best'. Hence, species like the houbara bustard, which are the focus of considerable conservation effort, relies on artificial insemination and artificial incubation of eggs taken from nests, partly to stimulate further egg production (Saint Jalme and van Heezik 1996; Baggott et al. 2003). The time has certainly come

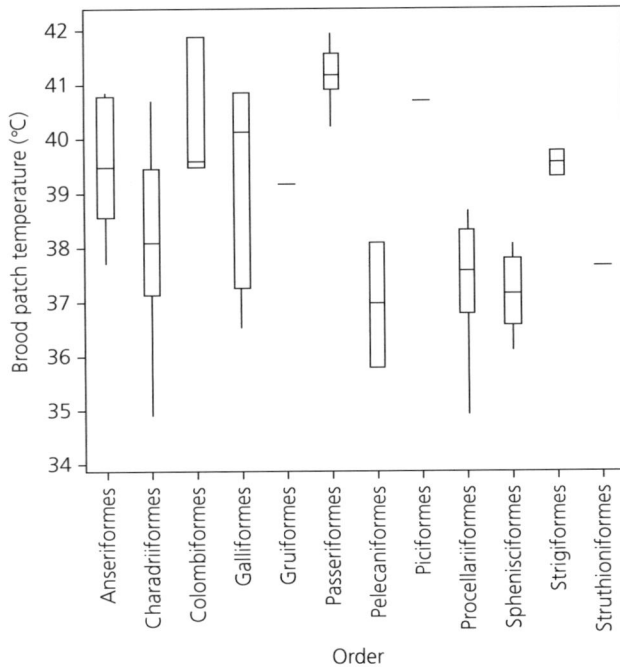

Figure 16.2 Box and whisker plots of brood patch temperatures (°C) for various avian orders. The line within each box indicates the median, the upper and lower peripheries of each box indicate the 75% and 25% quartiles, respectively, and the whiskers indicate the range of temperatures. Data derived from table published by Deeming (2008).

Table 16.1 Ranges of incubation temperatures and median (and ranges of) hatching successes reported by Kuehler and Good (1990) for aviculture of various orders.

Taxon	Number of species represented	Incubation temperature (°C)	Median % hatching success (range)
Large ratites	4	35.8–36.4	51.5 (46–58)
Tinamiformes	1	37.2	80
Pelecaniformes	1	37.2	100
Anseriformes	18	36.9–37.2	62.5 (19–100)
Falconiformes	6	36.1–37.2	84.5 (67–100)
Galliformes	37	37.2–37.5	75 (15–100)
Gruiformes	12	37.2–37.5	82 (50–100)
Charadriiformes	3	37.2–37.5	100 (87–100)
Columbiformes	4	37.5–38.1	91.5 (8–100)
Psittaciformes	34	36.9–37.5	72.5 (27–100)
Cuculiformes	5	37.2–37.5	50 (40–100)
Strigiformes	2	37.2	69 (67–71)
Caprimulgiformes	1	37.2	70
Coraciiformes	3	37.2	94 (80–100)
Piciformes	1	38.1	100
Passeriformes	9	37.2–38.1	100 (70–100)

for research to be carried out on non-poultry species to examine their thermal optima during artificial incubation to help us to understand better the artificial conditions required for individual taxa.

16.3.2 Gaseous environment

Bird eggs incubate in warm humid environments and exchange gases through their eggshell pores (Rahn and Paganelli 1990; Paganelli 1991; Rahn 1991; Ar and Sidis 2002). Humidity control is crucial for normal development because bird eggs in general have evolved to lose an average of 15% of their mass between oviposition and external pipping, i.e. the first breakage of the shell at the start of hatching (Ar and Rahn 1980). There is a range of optimal rates of mass loss between 10% and 20% for different species (Ar and Rahn 1980). However, a good goal for a novel species undergoing artificial incubation for the first time would be 15% mass loss. This loss of mass is solely down to water vapour because the mass of oxygen diffusing into the egg is balanced by that of carbon dioxide (CO_2) leaving the egg (Tullett 1984). The volume of air that replaces the water vapour, and which accumulates in the air cell, is equivalent to the volume of air of the respiratory system of the chick at full-term development (Ar 1991b).

It is known that nest humidity can impact on the water vapour conductance of the eggshells within the clutch. Well-insulated nests (Deeming 2011b) or 'wet' nests (Portugal et al. 2014a) are associated with high conductance eggshells. Again, there is little understanding of the natural nesting environments for most bird species and so future aviculture would benefit from more field-based studies of a wider variety of species to document nest humidity and gaseous environments.

Although mass loss of eggs varies between species (Ar and Rahn 1980), it is accepted that if there is significant departure from the optimum rate of mass loss then there are deleterious effects (Ar and Deeming 2009). A low humidity results in reduced hatchability because the embryo, or in the worst cases the whole egg, becomes dehydrated (Tullett 1984). Regular weighing of eggs and modification of the humidity by altering the humidity within the incubator, or moving the egg to a different incubator with the same air temperature but different humidity (French and Board 1983), to accommodate irregular mass loss can generally allow most embryos to survive and hatch (Tullett 1984). If humidity is higher than normal then such conditions are also associated with reduced hatchability due to embryonic mortality. Whilst this is often associated with

water retention by the embryo, which can restrict the hatching process (e.g. in the common ostrich; Deeming 1995), the key problem is lack of oxygen because of low eggshell porosity.

Provision of sufficient oxygen to developing embryos is a key aspect of artificial incubation. Hypoxia is a major threat to embryos because the eggshell is an effective mediating barrier for gas exchange (Paganelli 1991). Any reduction in the oxygen content of the air within a nest or an incubator will directly affect the oxygen consumption of the embryos, particularly as they approach peak rates close to internal pipping. Assuming incubators have holes to allow air in and out of the cabinet (see section 16.2), maintaining a good rate of aeration of the incubator cabinet and the containing room can be crucial to successful outcomes of artificial incubation.

Moreover, if the oxygen conductance of the eggshell is naturally low (e.g. Tullett and Deeming 1982), embryos can experience hypoxia that may prove fatal. In domesticated ducks the presence of the cuticle creates such a problem that it has to be washed off with a hypochlorite solution prior to incubation (Cherry and Morris 2008). In wild waterfowl this problem does not occur because of bacterial digestion of the cuticle during incubation (Baggott and Graeme-Cook 2002). There is also evidence that the cuticle on eggshells of Macqueen's bustards (*Chlamydotis macqueenii*) may adversely impact upon hatchability if it is not abraded during incubation (Baggott et al. 2003). There is certainly a need to learn more about the general role of cuticular integrity during incubation in birds.

Despite the key role that oxygen plays in development it is CO_2 levels that are often considered to be critical. This emphasis reflects the ease by which CO_2 can be measured rather than its potential deleterious effects. However, embryos of domestic fowl are known to tolerate high concentrations of CO_2. For the first three days of incubation concentrations of ~1% have no discernible impact on their survival (Taylor et al. 1956) and the same is true during the last three days of incubation when they can tolerate 7.5% CO_2 (Taylor et al. 1971).

Our knowledge of the concentrations of respiratory gas levels and how they change during incubation in nests is extremely poor (Rahn 1991). This is a reflection of the methodological difficulties of measuring such entities *in situ*. Nests in cavities are relatively well understood in this regard (Birchard et al. 1984; Lill and Fell 2007; Mersten-Katz et al. 2013; see Chapter 4) but our knowledge for more open nests is practically non-existent. It is possible that this may not be crucial for developing artificial incubation conditions for non-poultry species but it would be useful to improve the body of knowledge for this critical aspect of incubation.

16.3.3 Egg turning

Another critical variable in incubation is the rate at which eggs are turned (Deeming 2002f). Failure to turn eggs has deleterious effects on hatchability, rates of embryonic growth and use of albumen (Deeming 1991). Artificial incubators are set up for poultry eggs which have been shown empirically to be turned approximately once per hour through approximately ± 45° (Deeming 1991). However, Deeming (2002f) showed that rates of egg turning differed between bird orders perhaps reflecting different relative albumen fractions of eggs. Data collected subsequently continue to support the hypothesis that egg-turning rates differ across orders (Figure 16.3). This means that rates of egg turning during artificial incubation may need to be supplemented over and above rates at which incubators turn eggs. As with developmental temperature, different species may require different rates of egg turning. Albumen-rich eggs of altricial species that are incubated over relatively short periods require higher rates of turning than do 'standard' poultry eggs. Most

commercially available incubators turn eggs every 30 or 60 minutes and so supplementary turning by hand may be necessary for many species (Kasielke 2007). Deeming et al. (1991) hand-turned Humboldt penguin eggs seven times a day in addition to the hourly turning provided by the incubator. There is no evidence that egg turning at a rate higher than needed has any deleterious effect on embryonic development but there is a real need for more research to document rates of egg turning in a wider range of species.

16.4 Incubation in conservation aviculture

Collar and Butchart (2014) justified conservation breeding action for 257 bird species suggesting it as necessary to prevent extinction of 13 species and an integral part of conservation of another 32 species for which population reinforcement through releases was recommended or ongoing. If conservation breeding programmes are to succeed, knowledge about species' incubation will be critical. For many years artificial rearing systems have relied on use of surrogate 'broody' domestic fowls to incubate eggs of other species (e.g. Jordan 1989) because of lack of success and/or aviculturalists' lack of confidence using artificial incubators. That the 'bird does it best' limits potential productivity

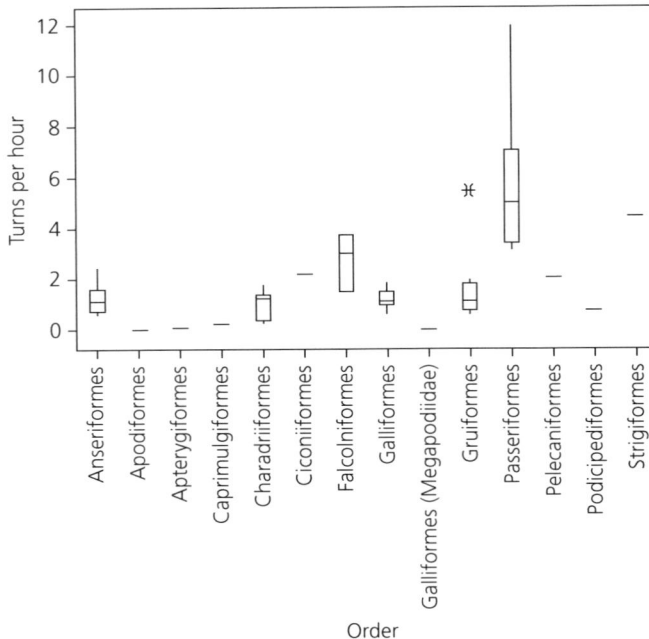

Figure 16.3 Box and whisker plots of egg-turning rates for eggs of various avian orders. The line within each box indicates the median, the upper and lower peripheries of each box indicate the 75% and 25% quartiles, respectively, and the whiskers indicate the range of turning rates. * = an outlier. Data derived from those reported by Deeming (2002f).

of captive populations as egg-pulling for artificial incubation may result in two or more replacement clutches being produced, which may increase productivity of a pair during a breeding season; for example, in Californian condors laying in captivity during 1998 seventeen females laid one or two more eggs per year if the first single-egg clutch was removed (Snyder and Snyder 2000). There is a potential increase in hatching success of eggs because risks associated with bird incubation, in particular predation of the nest contents and the attending adult(s), are removed. However, egg-pulling does not necessarily induce a replacement clutch; in 1998 five captive Californian condors only laid one single-egg clutch (Snyder and Snyder 2000).

Egg collection and artificial incubation can be extremely effective in producing chicks but this can depend on the species involved; in the black-bellied sandgrouse (*Pterocles orientalis*) production of viable hatchlings was better from wild-laid eggs than captive eggs (Aourir et al. 2013). By contrast, removal of kakapo (*Strigops habroptilus*) eggs from wild nests was not successful (Elliott et al. 2001). Reintroductions of a species into areas where it has previously become extinct can also rely on use of eggs collected from the wild and artificially incubated. The programme for the reintroduction of great bustard (*Otis tarda*) into the UK during the mid-2000s involved artificial incubation of 232 eggs collected and incubated in Russia (Burnside et al. 2012). Hatching success averaged 66% and 102 chicks were translocated to the UK, where 86 were released at an average age of 100 days (Burnside et al. 2012).

As a consequence there is a need to apply our understanding of artificial incubation in more common species to those where relatively little is known. Understanding how incubation temperature influences avian development and how it is influenced by taxonomic order or egg size can be crucial in the success of any artificial incubation programme. Sub-optimal temperatures may reduce embryonic development and thereby adversely impact on the embryo's survival through to hatching. Similarly, rates of egg turning appear to be crucial in promoting a typical pattern of development; in this case an understanding of the egg composition may allow prediction of the rate of egg turning as there is a positive correlation between percentage albumen content and rate of turning (Deeming 2002f). Regulation of humidity during incubation is relatively easy to accommodate given that mass loss from eggs can be fine-tuned through control of water loss. Ensuring that there is sufficient exchange of respiratory gases is crucial but can be most easily achieved by using well-designed equipment. All of these aspects of artificial incubation rely on engagement between aviculturalists and the scientific community. Moreover, scientists need to be more engaged with aviculturalists directly and via targeted publications to ensure that improvements in scientific understanding are disseminated widely.

16.4.1 Interaction between incubation and egg science with aviculture

The Wildfowl and Wetlands Trust (WWT) has assisted the re-establishment of the locally extinct common crane (*Grus grus*) in South-west England by raising and releasing 96 juveniles hatched and raised from 110 of 115 eggs collected from wild nests and translocated approximately 1,000 km from Germany. They have also established a captive population of the Critically Endangered spoon-billed sandpiper (*Calidris pygmaea*) at the WWT Slimbridge centre by translocating eggs from North-east Russia, more than 9,000 km away (with 95% egg hatchability; NSJ, 2014 personal observation). The development of portable incubators has enabled aviculturalists to provide the appropriate incubation requirements for eggs during translocation and so the potential stress to the embryo in transit is reduced. Moreover, animal translocations bring the risk of novel pathogens being introduced into an existing population, particularly with full-grown birds. The risk of pathogenic transfer is likely to be greatly reduced with eggs, the shells of which can be cleaned and disinfected prior to (and post-) importation, without harm to the developing embryo.

There is an increasing tendency to establish new populations using eggs rather than full-grown birds. If eggs are collected from birds' nests before, or just after, incubation has started the parents are likely to lay repeat fertile clutches. This means that collecting eggs for conservation purposes is unlikely to reduce to a great extent the source population's contribution to annual productivity, particularly in those species which are naturally double-brooded each year and lay repeat clutches even when first broods are hatched and raised or when clutches are lost at the hatching stage (e.g. through actions of hungry predators or well-intentioned conservationists).

To ensure successful egg translocations, knowledge of the species' biology is required, not just to assist nest finding at the outset but to inform all management practices to achieve captive propagation. Importantly, a fuller understanding of interactions between adults and their eggs during incubation, both under natural and artificial conditions, and technological advances in incubators are paramount. In combination this information enables a thorough risk assessment to identify hazards and to devise counter-controls. Below, we outline how the practical application of experience and knowledge of egg incubation processes can be used in the field to establish a conservation breeding programme for the critically endangered Madagascar pochard.

16.4.2 A case study taking artificial incubation into the field

The Madagascar pochard, a medium-sized diving duck endemic to Madagascar, was believed extinct in 1991 (Young and Kear 2006). The species was rediscovered in 2006 when a small population was found breeding on a remote lake in Northern Madagascar (Réné de Roland et al. 2007; Figure 16.4a). In 2008, 25 mature individuals were counted at the site but these failed to raise any ducklings despite repeated breeding attempts (Bamford et al. 2015). In 2009, WWT, Durrell Wildlife Conservation Trust (Durrell), and The Peregrine Fund joined with the Government of Madagascar's Ministère de L'Environnement des Eaux et Forêts to develop a plan to increase the pochard's population size by developing a conservation breeding programme in the country.

In July 2009 project partners, including aviculturalists from WWT and Durrell, visited the pochard's breeding site and nearby towns to assess the feasibility of beginning a breeding programme for the species. When they visited the breeding lake, the team counted 19 pochards, only six of which were females. The low number of females, coupled with the failure of the previous breeding season prompted partners to devise an emergency rescue plan: to collect the next laid clutches in 2009 and to translocate them to a hotel with the nearest reliable electricity supply in the town of Antsohihy, some 160 km from the lake. Any hatched ducklings would be raised and held in temporary facilities until a propagation centre could be built in the following years.

The main logistical problem in using founder stock in such a programme was not the lack of a propagation centre but access to the lake. A 160 km pot-holed road and dirt track connected the hotel and the lake (Figure 16.4b). The last 40 km of this route were barely passable on foot and impassable for long periods during the rainy season, the beginning of which was thought to mark the start of the pochard's breeding season. Although in dry weather a 4×4 vehicle could negotiate the route in approximately 6 hours, the extremely rutted and uneven road conditions precluded safe egg translocation. Thus, a plan was developed to incubate and hatch eggs in battery-powered incubators at the lakeside so that ducklings, instead of eggs, could be transported safely in a well-padded container to the hotel set-up for rearing.

Madagascar has 220–240V domestic electricity supply, which is at best unreliable outside large towns and non-existent in remote locations. Therefore, aviculturalists decided to use Brinsea® 12V still air incubators to incubate collected eggs. Brinsea machines were favoured as the team were familiar with their operation and they could be carried into the field easily as they weighed relatively little. Two Brinsea 'Hatchmaker' incubators (weighing 3.5 kg each) were carried to the lakeside field camp (Figures 16.4c and d). These 2.5 A machines were known to function for 36 hours at room temperature when powered by a 90 A·hour^{-1} lead acid battery. One 'Polyhatch' incubator (5.5 kg) was set up at the hotel. This would be used to incubate any eggs that had to be evacuated because rain threatened to make roads impassable, or if a clutch had an asynchronous hatch-off due to it containing 'dumped' eggs. At room temperature, this 4.0 A machine was known to operate for 22.5 hours when powered by a 90 A·hour^{-1} lead acid battery. Both Brinsea machines would operate from a mains supply if required via a transformer. A 12V (4.6 A) AB Portable Houbara Incubator® (weighing 9 kg) was also carried to the lake. If required, this machine would be used to translocate any unhatched eggs from the lake to the town and would be powered from the 4×4 car battery.

As intra- and interspecific brood 'parasitism' is a known reproductive strategy for *Aythya* spp. (Sayler 1992; Baldassarre 2014; Hannu et al. 2014), aviculturalists carried three incubators into the field in case eggs in the same clutch were at different incubation stages and therefore required different egg-turning rates, and temperature and humidity settings. One of these incubators was a specially designed portable incubator to translocate any unhatched eggs alongside 'day-old' ducklings from the 'same' clutch. At WWT, 11 of the 12 extant *Aythya* spp. are managed in captivity and all of these are known to defecate on their eggs if suddenly disturbed from the nest (NSJ, 2014 personal observation) in the same way as common eiders (*Somateria mollissima*; McDougall and Milne 1978). Therefore, aviculturalists took egg disinfectant into the field so any soiled eggs, collected from nests where females defecated when flushed, could be cleaned prior to transfer to the Brinsea or AB incubators for artificial incubation.

Figure 16.4 (a) A pair of the Critically Endangered Madagascar pochard (*Aythya innotata*). This diving duck species was thought extinct until a small population of approximately 20 birds was discovered on a remote lake in North Madagascar in 2006. Ornithologists recorded many failed breeding attempts at the lake and this prompted conservationists to begin a conservation breeding programme to safeguard birds against extinction. (Photo: S. Dench/WWT.) (b) The poorly maintained 160 km track to the lake precluded egg translocation from wild nests to a rearing facility in a hotel. (Photo: P. Cranswick/WWT.) (c) The incubation tent set-up located under the canopy of the lakeside forest. (Photo: M. Roberts/WWT.) (d) A Brinsea 'Hatchmaker' with accessory incubation equipment and materials. (Photo: M. Roberts/WWT.) (e) Madagascar pochards nest in dense papyrus (*Cyperus* spp.) marsh at the lakeside. (Photo: M. Brown/WWT and NSJ/WWT.) (f) A nine-egg clutch of a Madagascar pochard. (Photo: M. Brown/WWT and NSJ/WWT.) (g) Ducklings and pipped eggs in the 'Hatchmaker'. (Photo: G. Young/Durrell.) (h) Adult drake Madagascar pochard: at the beginning of 2015 the conservation breeding programme comprised 54 ducks. (Photo: P. Cranswick/WWT.) (See Plate 15.)

Aviculturalists decided to collect eggs close to hatching to reduce the length of time required to provide ideal incubation conditions using artificial incubators powered by rechargeable batteries in a field camp scenario. They expected this would not restrict females' opportunities to lay replacement clutches and raise their own broods as observations since 2006 had revealed pairs were capable of multiple nesting attempts each year (T. Seing Sam, 2009 personal communication).

Egg laying was suspected when females were absent from the lake for short intervals on seven or more consecutive mornings. Incubation was confirmed when females were absent from the lake during most of the daytime except for approximately one hour each morning and late afternoon when they joined their mate on open water to feed by diving almost continuously. Nest searches were made from water (by boat) rather than from land (by foot; Figure 16.4e) as it was anticipated this might lead to abandonment or predation by black rats (*Rattus rattus*). In this way three nests were found in papyrus (*Cyperus* spp.) marsh extending approximately 500 m along one side of the lake (Figure 16.4f).

At the time of collection all equipment (e.g. carrying box and foam, callipers and scales) was wiped with a disinfectant solution before and after use. Eggs were handled wearing nitrile gloves, examined at collection for signs of pipping and candled, i.e. shining a bright light through the eggshell to visualize the contents, to reveal cracks and embryonic development, and clearly marked with clutch number using a soft-lead pencil. Linear measurements (length and breadth) of eggs were recorded with dial callipers to the nearest 0.1 mm and egg mass recorded to the nearest 0.1 g with electronic scales. As eggs were not badly soiled at collection, they were not disinfected in case this destroyed any advantageous microbes on the eggshell (see Chapter 7). Eggs were arranged together as a batch in the centre of the 'Hatchmaker' incubator to allow embryos to communicate (see Chapters 11 and 12). A calibrated digital Brinsea 'Spot Check' thermometer was positioned between eggs and another mercury thermometer was placed so it was just above the eggs in the centre of the clutch.

Fresh egg mass (FEM; in g) was estimated for each egg using the equation reported by Hoyt (1979):

$$FEM = 0.548LB^2 \qquad \textbf{(equation 16.1)}$$

where L is egg length (cm), B is egg breadth (cm) and 0.548 is the average mass coefficient (Hoyt 1979). It was assumed that egg mass loss during incubation would average 15% (Ar and Rahn 1980; French and Board 1983). With egg mass known at collection and the incubation time to pipping assumed to be 24 days as the incubation period to hatch had been reported as 25–27 days (T. Seing Sam, 2009 personal communication), it was possible to estimate the incubation period (I_p) that had elapsed:

$$\text{Days of incubation} = (24*\text{egg mass at collection})/ FEM*0.15 \quad \textbf{(equation 16.2)}$$

The 'Hatchmaker' incubation temperature was set to record between 38.5°C and 39.0°C when the mercury thermometer was positioned just above the eggs. This was decreased to 37.5–38.0°C for hatching eggs. Relative humidity was raised from 55% to 80% during hatching by filling water pans in bases of incubators with disinfectant solution at 37.0°C and spraying eggs with boiled and cooled water at 36.0–37.0°C at two–three hourly intervals. Any hatching eggs were placed pip upwards.

When ducklings hatched, the incubator temperature at duckling head height was lowered to 36.0°C. Hatching eggs were transferred to a plastic box within the incubator to prevent ducklings from catching their toes in the metal frame floor of the incubator. This box was lined with paper or cotton towel to prevent ducklings from slipping (Figure 16.4g). Temperature in the box was monitored closely.

Three clutches were collected before their estimated hatching dates. Clutch 1 comprised nine eggs collected on estimated day 24 of incubation. At collection one egg had pipped at the equator and was 'dead-in-shell'. Visual examination and field candling showed the other eight eggs to be hatching: three were externally pipped at the low edge of the airspace which is the normal position for duck eggs, and five were internally pipped. Transfer time from nest to incubator was 25 minutes at an ambient temperature of 20°C. All eight viable eggs hatched successfully on what was estimated to be day 25 of incubation. The day-old ducklings were successfully translocated to the hotel rearing set-up and all were successfully raised to fledging.

The second clutch also comprised nine eggs and was collected on estimated day 21 of incubation, which was supported by candling. Three and six eggs pipped on days 23 and 24 of incubation, respectively. Seven eggs hatched on day 25 and two hatched on day 26 of incubation (the last egg hatched in the AB portable incubator 3 hours into the 10-hour journey to the hotel). Eight of nine ducklings were raised to fledging with one duckling dying at 43 days of age of unknown causes.

Seven eggs were in the third clutch collected at approximately day 19 of incubation. As heavy rain was forecast this clutch was translocated in the AB portable incubator to the hotel. The portable incubator operated at 36.0°C for the entire journey. At the hotel, the eggs were checked, marked, candled and their linear dimensions measured. Ambient temperature was 12°C at the lake, 25–35°C in the vehicle and 32°C at the hotel in the town. Eggs were incubated in the 'Polyhatch' incubator at the hotel at 38.5–39.0°C at the top of the egg position and 55% RH until internally pipped when temperature was lowered and RH increased as before. Eggs were turned hourly through 180° until internally pipped. All eggs pipped on day 25 or 26 of incubation, and hatched by the end of day 26. All seven ducklings were raised to fledging (Figure 16.4h).

16.5 Conclusions

Artificial incubation of bird eggs can be easily achieved as long as the appropriate conditions are known. There is an increasing realization that poultry eggs may not be a suitable model for the incubation parameters of many species of birds. Temperature and humidity conditions will vary between species and egg-turning rates may be crucial in determining their hatching success. Improving our understanding of the incubation conditions in a wider range of species could be crucial in the future success of *ex situ* conservation programmes. There is an urgent need to collect data on natural incubation and for empirical experimentation on incubation conditions in species other than commercially important galliforms.

The case of the Madagascar pochard is only one of many that could have been described here but it illustrates well how the application of the biology of a species together with knowledge of artificial incubation can be used in conservation efforts for a targeted threatened species. Familiarity with incubation of waterfowl eggs particularly helped in this situation (resulting in a captive population of 54 pochards by the beginning of 2015), but such experience can be applied to birds from other orders. The future survival of many bird species depends on us and an improved knowledge of incubation for eggs *ex situ* will help to ensure that our conservation efforts yield the highest rewards. The need for incubation scientists and aviculturalists to interact more freely and to share their knowledge and experience has never been so important.

Acknowledgements

We thank Jim Reynolds, Peter Cranswick, Baz Hughes, and Glyn Young for comments on earlier drafts of this chapter. The Mission Madagascar project is a collaborative project of Wildfowl & Wetlands Trust, Durrell Wildlife Conservation Trust, The Peregrine Fund, Asity Madagascar, and the Government of Madagascar. It has been generously supported by: the Darwin Initiative, Mitsubishi Corporation Fund for Europe and Africa, Fota Wildlife Park, BBC Wildlife Fund, a private donor, the Mohamed bin Zayed Species Conservation Fund, US Fish & Wildlife Service, Aviornis UK, Synchronicity Earth, WWT and Durrell members, and many generous individuals.

CHAPTER 17

The role of citizen science in studies of avian reproduction

C.B. Cooper, R.L. Bailey, and D.I. Leech

17.1 Introduction

In 2014, one of the new additions to the Oxford English Dictionary was the phrase 'citizen science'. The earliest documented use of the phrase, in 1989, was to describe the efforts of volunteers with the National Audubon Society who were collecting data on acid rain. In 1995, the term citizen science started to be used heavily within the context of ornithology, referring to styles of public participation in genuine research (reviewed in Bonney et al. 2009). Amateur ornithologists (bird watchers, nest monitors, ringers, and banders, and naturalists with varied interests) have been, and will continue to be, key resources for ornithological research. Citizen science and professional ornithology differ in many ways, but their roots have a common origin with amateurs. Historically, amateurs made observations and applied their observations to research and also made contributions to museum collections for posterity.

As the term citizen science slowly gained acceptance, non-professional contributors to ornithological research were called a variety of terms, including volunteers, amateur ornithologists, birders, and bird watchers. Like many amateurs in history, those engaged in citizen science possess an enormous amount of expertise. While 'amateur' has positive connotations in citizen science in some disciplines, such as astronomy, it can also have negative connotations associated with being a 'novice'. We feel that ornithological expertise straddles any demarcation to distinguish professional and volunteer. Volunteers not only participate in ornithology as independent researchers but can be part of large crowd-sourcing projects.

Even though non-professionals have contributed to ornithology in many ways for centuries, their efforts are rarely explicitly acknowledged. Research papers that rely on citizen science data have not typically used the phrase 'citizen science'. For example, of 171 original research papers that were used by Knudsen et al. (2011) to formulate claims about migratory birds and climate change, 82 relied at least in part on citizen science data, but none used the term 'citizen science' (Cooper et al. 2014).

Given that citizen science is an important but underappreciated approach, in this chapter we will review studies and discoveries related to avian reproduction (anything prior to chick rearing) that resulted from the efforts of citizen scientists. We also explore issues related to pushing new frontiers at the intersection of citizen science and avian reproduction.

17.2 Contributions of citizen science

17.2.1 Historic collections

Birds' nests and eggs have captured the interest and imagination of professional and amateur ornithologists alike for centuries. Part of the fascination may stem from the fact that this mode of reproduction differs so greatly, at least superficially, from that of mammals. However, the fact that these organisms, so unrestricted in movement due to the power of flight, are at their most accessible during the nesting cycle must also be significant. Prior to the widespread use of high-powered optical equipment, alternative opportunities to study bird behaviour and ecology must have been few and far between. It is therefore unsurprising that the earliest examples of citizen science within the field of ornithology relate to the study of the reproductive cycle.

Nest monitoring originated from the scientific discipline of oology, which blossomed during the 19th Century. While the collection of eggs had the obvious

Cooper, C.B., Bailey, R.L., and Leech, D.I., *The role of citizen science in studies of avian reproduction.* In: *Nests, Eggs, and Incubation.* Edited by: D.C. Deeming and S.J. Reynolds, Oxford University Press (2015). © Oxford University Press.
DOI 10.1093/acprof:oso/9780198718666.003.00017

potential to impact negatively on species' population trajectories, it was clear that many practitioners were genuinely motivated by the desire to broaden their knowledge of avian ecology. In his book 'The Science of Birdnesting' Gosnell (1947) advised that 'an egg should never be taken unless something can be learned from it' and espoused the virtues of clearly labelling each egg or clutch in a standardized format. The study and collection of birds' nests and eggs began before ornithology was a profession, and continued as a vocation until very recent times. Many of the nests and eggs in museum collections were gathered by amateur enthusiasts. The advent of photography during the 20th Century countered many of the scientific arguments for destructive sampling, including the equally widespread practice of shooting birds to further knowledge of identification and physiology. As public interest in conservation grew, so the pressure on politicians and law enforcement agencies to protect wildlife increased. As a result, the Migratory Bird Treaty Act of 1918 in the USA and the Protection of Birds Act of 1954 in the UK banned the practice of egg collecting.

17.2.2 Nests

Even though most museum collections include birds' nests, these are generally underused as a resource in research (Russell et al. 2013). Nests are highly variable in structure, shape, and size, and in their composition, which includes fur, feathers, arthropod silk, mud, and plant materials (Hansell 2000). Advances in DNA extraction and analytical techniques are rapidly increasing the accuracy with which floral and faunal materials can be assigned to species. Nests also provide a substrate for eggs and chicks and protection from weather and predators, while influencing microclimatic conditions (Hansell 2000; Chapter 4). Heenan and Seymour (2011) used over 200 cup-shaped nests from museums in Australia to study the functional drivers of nest design. Nests in museums could be used to study the materials used by different species, construction methods, the structural changes that occur during nesting, the presence of ectoparasites in nest material, and even studies of avian cognition (Russell et al. 2013). Aubrecht et al. (2013) investigated whether the horse hair fungus (Marasmius spp.) functioned as a possible antibacterial agent in nests. In evolutionary studies, nest structure can be a taxonomic trait used to examine behavioural radiation and speciation (Russell et al. 2013).

One of the potential limitations when using museum collections to investigate nest structure is that studies frequently necessitate dismantling the specimen. New

technologies, including X-ray computed tomography (CT scan) and digitization, provide less destructive alternatives in some situations, but often there is no substitute for dissection. Increasingly, academics are asking volunteers involved in national nest monitoring surveys to send them used nests. Mainwaring et al. (2014b) used common blackbird (Turdus merula) nests collected by a mixture of academics and volunteers to demonstrate a negative relationship between insulatory properties and ambient temperature. Whitworth (2003a, 2003b) crowd-sourced nests from volunteer collectors around the USA to devise a key to the puparia of 27 species of North American bird blowfly.

17.2.3 Eggs

Egg collections can be used to detect key environmental changes and analysis of specimens can also have important consequences for public health, epidemiology, and national security (Suarez and Tsutsui 2004). Ratcliffe (1967) used egg specimens from British museums to demonstrate a reduction in the thickness of peregrine falcon (Falco peregrinus) and Eurasian sparrowhawk (Accipiter nisus) eggshells over time, corresponding to the initiation of dichlorodiphenyl-trichloroethane (DDT) application to crops in 1945–6. Similar patterns for raptors and herring gulls (Larus argentatus) in the USA were reported the following year, also revealing declines beginning in 1947 (Hickey and Anderson 1968). Subsequent studies linked eggshell thinning to increased failure rates of British raptors (Newton 1973; Newton and Bogan 1974) and comparisons of museum specimens and contemporary samples continue to reveal the historical impacts of DDT and polychlorinated biphenyls (PCBs, e.g. Bignert et al. 2005). Researchers have noted the potential use of eggs from widespread species such as common starlings (Sturnus vulgaris) and house sparrows (Passer domesticus) as a bio-monitoring tool with which to study organohalogenated contaminants (Eens et al. 2013; Eng et al. 2014) and heavy metals (Swaileh and Sansur 2005). Given that these species are non-native in many parts of the world, their eggs are not subject to legal protection in all countries and therefore provide opportunities for reviving citizen oology in this limited context.

Egg collections have also been used to advance research regarding the evolution of pigmentation and colouration (Maurer et al. 2011b; Chapter 11). Maurer et al. (2012) sampled eggs of 230 European bird species to investigate patterns of variation in shell structure and thickness. Cassey et al. (2012) quantified the amounts of protoporphyrin IX and biliverdin occurring in the eggs of a diverse range of Neoaves breeding in Britain and related structural components and pigmentation to life-history traits. Egg collections can also act as genetic archives, enabling whole genome amplification using DNA extracted from specimens (Lee and Prys-Jones 2008).

Data associated with egg collections, such as clutch size and dates, can potentially be used to examine temporal trends from historical archived data to those generated by current nest monitoring schemes. Scharlemann (2001) constructed a 150-year laying-date trend for white-throated dippers (*Cinclus cinclus*) and song thrushes (*Turdus philomelos*), demonstrating a significant negative relationship with March temperatures. McNair (1985, 1987) argued that the standardized museum protocol used for archiving eggs produced more reliable data sets than those generated by volunteer surveys. However, such data should be interpreted with care as large or unusual clutches can be more attractive to collectors (Lack 1946), leading to unrepresentative sampling.

17.3 Nest monitoring schemes

Some amateur ornithologists have established nest monitoring schemes of sufficient size and duration to enable the investigation of long-term trends and spatial variation in breeding success. One such project is located in Treswell Wood, Nottinghamshire, UK where data on tit species (Paridae) breeding in nestboxes have been collected since the 1970s. This information has been used to explore aspects of breeding ecology as diverse as selection of lining material (Surgey et al. 2012), behavioural strategies for minimizing nest-parasite loads (du Feu 1992) and interspecific variation in egg and nestling failure rates (Deeming and du Feu 2011). Cowley (1983) studied multi-brooding in a UK sand martin (*Riparia riparia*) population, while Slater (2001) investigated common woodpigeon (*Columba palumbus*) breeding ecology in the suburbs of North-west England.

While there are doubtless other examples of publications arising from independent amateur nest monitoring projects, it can be difficult to distinguish these from outputs of studentships or professional studies when searching the literature and the authors acknowledge that they may be under-represented in this chapter. However, in the countries with which we are most familiar, the major sources of citizen science publications are national nest monitoring schemes, run by conservation and research institutions, which impose standardized monitoring methodologies and collate data submitted by large numbers of individual volunteers on an annual basis. The larger schemes include a series of projects administered by the Cornell Lab of Ornithology: the historic North American Nest Record Card Program, which began in 1965, and the extant NestWatch project (which began as the Cornell Nest Box Network, and then was called the Birdhouse Network). Hereafter, these are jointly referred to as 'Cornell nest records' and combined they hold over half a million records. Another significant programme is the Nest Record Scheme initiated in 1939 and administered by the British Trust for Ornithology (hereafter, 'BTO NRS'), which currently holds 1.7 million nest histories (Crick et al. 2003a). The Nest Record Schemes run by Bird Studies Canada (hereafter, 'BSC nest records') engage many professionals as well as volunteers. The Australian Nest Record Scheme (hereafter 'RAOU NRS') has been operated by BirdLife Australia since 1963 (Marchant 1974) and the New Zealand Nest Record Scheme was initiated by Birds New Zealand in 1959 (Heather and Robertson 1996). Many countries in Europe (Belgium, Czechoslovakia, Denmark, Estonia, Hungary, Finland, Germany, Iceland, Italy, the Netherlands, Norway, Poland, Spain, Sweden, and Switzerland) and Africa (South Africa, Zambia, and Zimbabwe) also established nest recording schemes. However, some have struggled in recent decades due to a lack of sufficiently skilled volunteers and/or adequate funding and have either stopped collecting data or reduced coverage to a restricted range of target species.

The exact methodology employed differs between schemes, but all record information about nest contents, location, and surrounding habitat (Crick 1992). Most collect observations on a series of dated visits made throughout the nesting cycle. Counts of eggs and chicks provide data to estimate clutch and brood sizes, respectively. Estimates of whole-nest failure rates can be derived for attempts visited on multiple occasions using the analyses developed by Mayfield (1961, 1975). Observations of changes in egg numbers across visits and the stage of chick development on a given date can also be used to back-calculate laying dates, either on the basis of observer estimates, e.g. some of the Cornell nest records (Cooper et al. 2014), or using parameter files generated from the published literature (BTO NRS; Crick et al. 2003a).

17.3.1 Reproductive biology and natural history

As the quality of field guides increased dramatically during the 20th Century in response to a growing demand from the birding community, so did advances in our understanding of basic ecology. This was particularly true for breeding biology, in part thanks to volunteer observations. Baicich (1986) highlighted important knowledge gaps in the first edition of *A Field Guide to Nests, Eggs, and Nestlings of North American Birds* (Harrison 1978), criticizing the author for not including data from Cornell nest records. Of the 249 nesting species for which Harrison (1978) claimed to have little or no data, 82% had at least one submitted nest record card that could have been referenced. These data were included in the second edition of *A Guide to the Nests, Eggs, and Nestlings of North American Birds* (Baicich and Harrison 1997). More recently, Ferguson-Lees et al. (2011) produced *A Field Guide to Monitoring Nests*, which used BTO NRS data as the basis for the majority of species' accounts. The text details breeding locations, phenology, nest structure, appearance of eggs, and typical clutch and brood sizes for the majority of UK bird species.

Many of the initial papers produced using data from the BTO NRS were summaries of basic breeding ecology, including phenology, clutch and brood sizes, and nest locations. One of the earliest examples (Silva 1949) focussed on the 173 records available for song thrush at that time. Data from the BTO NRS were used in a number of publications authored by Lack (1946, 1948a, 1948b, 1955a, 1955b, 1958b), exploring the breeding ecology of European robins (*Erithacus rubecula*), common starlings, and a variety of tit (Paridae) species. Snow (1955a, 1955b, 1966, 1969a, 1969b, 1969c) also analyzed several BTO NRS thrush (*Turdus* spp.) data sets. By the 1970s, participation had grown to the extent that BTO staff members Rob Morgan and David Glue were able to publish a steady stream of papers investigating nesting behaviour of species for which annual sample sizes were relatively small, either because they were scarce or required a considerable effort to monitor (Flegg and Glue 1973, 1975; Glue and Morgan 1974; Morgan and Shorten 1974; Fuller and Glue 1977; Glue 1977, 1990; Morgan and Glue 1977, 1981; Morgan 1982a, 1982b; Glue and Boswell 1994). To date, BTO NRS data have contributed to over 100 publications summarizing the breeding behaviour of individual species.

Use of citizen science data sets to detail basic nesting ecology is by no means unique to the UK. According to Phillips and Dickinson (2009), Cornell nest records have contributed to approximately 130 papers, including peer-reviewed articles and species accounts appearing in *The Birds of North America Online* (Poole 2005). Mills (1986) combined Cornell data holdings

for whip-poor-will (*Caprimulgus vociferus*) with professional data sets to demonstrate the influence of the lunar cycle on laying dates, a phenomenon also researched in Britain by Perrins and Crick (1996) with respect to European nightjars (*Caprimulgus europaeus*). Apfelbaum and Seelbach (1983) used Cornell nest records to quantify habitat use (nest height, tree species used, and habitat selection) of seven North American raptors. McNair (2000) documented the breeding season of lark sparrows (*Chondestes grammacus*) in Tennessee. Similar examples utilizing RAOU NRS data include studies of the willie wagtail (*Rhipidura leucophrys*; Marchant 1974), the welcome swallow (*Hirundo neoxena*; Marchant and Fullagar 1983), and the eastern yellow robin (*Eopsaltria australis*; Marchant 1984).

The size of citizen science data sets and the range of sites over which information is collected can be incredibly valuable when providing practical management guidance. Many agricultural and forestry operations have the potential to damage or destroy birds' nests, so robust knowledge of breeding locations and representative estimates of laying initiation and season duration are vital tools for minimizing impacts. Characterization of white-headed woodpecker (*Picoides albolarvatus*) nest sites using Cornell nest records (Milne and Hejl 1989) shaped US Forest Service snag retention protocols for the Pacific Southwest.

17.3.2 Brood parasites

In North America, Cornell nest record cards have been used in studies of brood parasitism. The brown-headed cowbird (*Molothrus ater*) is not host-specific and occurs at high densities along edges of forest patches in agricultural landscapes. This species has been implicated as a factor in the decline of many forest-breeding passerines (Robinson et al. 1995). Cornell nest records specifically request data on cowbird presence, enabling researchers to investigate both inter- and intraspecific variation in the probability of parasitism. Coker and Confer (1990) reported no difference in cowbird parasitism rates of golden-winged warblers (*Vermivora chrysoptera*) and blue-winged warblers (*Vermivora pinus*), despite the parasite occurring sympatrically with the latter host over a longer period. Wootton (1986) found that parasitism rates of house finch (*Carpodacus mexicanus*) in the introduced population in the eastern USA were higher than in the native western population. This is possibly because cowbirds parasitizing the former were naive to the unsuitability of this host's diet (Kozlovic et al. 1996). Hoover and Brittingham (1993) identified a positive relationship between cowbird density and the proportion of parasitized wood thrush (*Hylocichla mustelina*) nests, but parasitism declined with decreasing wood thrush density. In a test of the parasite-predator hypothesis (Arcese et al. 1996), Hauber (2000) used Cornell nest records to demonstrate that song sparrow (*Melospiza melodia*) nests parasitized by brown-headed cowbirds were less likely to be predated prior to hatching, suggesting that cowbirds were removing clutches to stimulate re-laying.

BTO NRS data have been used to study nest parasitism by the common cuckoo (*Cuculus canorus*), the UK's only brood parasite. Populations of this species have declined by 73% over the past 50 years (Baillie et al. 2014). Glue and Morgan (1972) identified 26 species in the NRS data set recorded as hosting cuckoo eggs, with parasitism most prevalent in nests of dunnocks (*Prunella modularis*), Eurasian reed warblers (*Acrocephalus scirpaceus*), and meadow pipits (*Anthus pratensis*). Brooke and Davies (1987) used over 73,000 records of the six main hosts, collected over 40 years, to demonstrate that parasitism of reed warblers had increased, while that of all other species had declined. Douglas et al. (2010) suggested that this trend may be linked to climate-induced shifts in the timing of breeding, but that host availability was unlikely to have played a significant role in the population decline of the parasite.

17.3.3 Phenology

Volunteer nest observations are useful to document nesting phenology, which can inform land management practices. For example, Ells (1995) used the Cornell nest records and other sources of data to compare bobolink (*Dolichonyx oryzivorus*) egg-laying dates reported in other areas with observations from the field in Massachusetts. The aim was to determine dates beyond which local hay could safely be harvested without destroying the majority of bobolink nests. In the UK uplands, BTO NRS have been used to adjust the timing of moorland burning, which promotes heather (*Erica* spp.) regeneration, in response to climate-mediated shifts in laying dates (Joys and Crick 2004; Moss et al. 2005; Newson et al. 2007).

Many early studies explored the relationship between weather conditions and the timing of breeding. Erskine (1971) used data from BSC nest records to study the breeding ecology of common grackles (*Quiscalus quiscula*) in Canada. It was found that the start of egg laying was well correlated with temperature, and that clutch size was typically correlated with laying date. Erskine (1971) also suggested that nesting attempts in dense conifers may have been initiated earlier and at

lower temperatures than those in less protected sites. BTO NRS have been used to examine the impacts of weather on the laying dates of species such as the spotted flycatcher (*Muscicapa striata*; O'Connor and Morgan 1982), a late-arriving migrant, and to quantify phenological responses to extreme events (Mayer-Gross 1964).

Volunteers also provide important data for long-term and large-scale ecological questions, such as range limitation and climate change impacts. James and Shugart (1974) used Cornell nest records to study the nesting phenology of American robins (*Turdus migratorius*). A combination of April dry-bulb and wet-bulb temperatures was the best predictor of the beginning of the nestling period. The large sample size of cards meant that they were able to model the climate space of acceptable limits of temperature and humidity graphically across the contiguous United States.

Seminal papers by Crick et al. (1997) and Crick and Sparks (1999) utilized BTO NRS data to demonstrate that the majority of British species across a broad taxonomic range had progressively advanced their laying dates over a 40-year period and that this phenological shift was linked to increasing spring temperatures. Temperature-mediated advances in clutch initiation have also been demonstrated using Cornell and BSC nest records (Dunn and Winkler 1999; Torti and Dunn 2005). Dunn and Winkler (1999) found that tree swallows (*Tachycineta bicolor*) started breeding nine days earlier between 1959 and 1991 due primarily to increasing surface air temperatures over the same period. They were able to study the species across the majority of its breeding range due to the large volume of nest record cards submitted to the Cornell and BSC nest record schemes. Using the same data set, Winkler et al. (2002) also determined that for every day later that a swallow began laying, its final clutch size averaged 0.03 fewer eggs. Although spring temperatures did not have a direct effect on clutch size, there was a strong indirect effect on the number of eggs produced in response to laying dates. Conversely, volunteer data from the RAOU nest records indicate that timing of breeding for 16 Australian species was driven primarily by rainfall (Gibbs et al. 2011).

If the rate of response to warming is faster at lower trophic levels, advancement of laying may not be sufficient to prevent mismatches between the timing of peak food availability and the energetic demands of their nestlings, which can have significant implications for reproductive output and subsequent population trends (van Noordwijk et al. 1995; Visser et al. 1998; Both et al. 2006; but see Salido et al. 2012; Reed et al.

2013). The potential for mismatch has been demonstrated in the UK using BTO NRS data in conjunction with a range of professional and volunteer data sets relating to primary producers and consumers (Thackeray et al. 2010). BTO NRS data have also been used to demonstrate that climatic conditions on the wintering grounds (Ockendon et al. 2013) and along migratory routes (Finch et al. 2014) can also influence migrant breeding success.

Torti and Dunn (2005) found that four of six North American species initiated laying earlier when spring temperatures were warmer, and two species showed advancement in lay date over the 50-year period of the study. Clutch size did not show a significant relationship with temperature, but it was correlated with laying date and latitude. This study was conducted using nest record cards from the Cornell and BSC nest record schemes.

Heterogeneity in temporal coverage may be an issue in citizen science surveys where effort is not standardized across observers. The BTO NRS asks participants to allocate a score to each month of the season, indicating whether they spent none, a few, or many days monitoring nests, giving an indication of variation in seasonal effort. Figure 17.1 shows the mean proportion of volunteers who were active in each month over the period 1970–2013 (inclusive). Unsurprisingly, the peak period of activity occurs between April and July, the core breeding season in the UK. However, > 0.50 of volunteers were active in March and August, and this proportion has remained relatively constant over time (1970s: March—0.61; August—0.48; 2000s: March—0.56; August—0.54). As there is individual variation in the species targeted by recorders, the small reduction in early-season effort may reflect an increased focus on species utilizing nestboxes, which tend to breed slightly later than open-nesting passerines. Rodrigues and Crick (1997) directly compared data on common chiffchaff (*Phylloscopus collybita*) breeding performance collected by an intensive academic study with those gathered by the BTO NRS and found no significant difference in the resultant estimates of phenology or productivity. However, Weidinger (2001) did demonstrate seasonal biases in sampling in the Czech nest record scheme data set across a range of passerine species through comparison with more focussed local studies; the extent of such biases was in part dependent on the degree to which detectability varied through the season, but it is likely that variation in observer effort was also implicated in the differences identified between data sets.

17.3.4 Difficult-to-detect phenomena

Citizen science puts many eyes on alert, making it more likely that uncommon or difficult-to-observe events will be detected, providing a mechanism by which rare (or rarely-detected) phenomena can be shared with the scientific community. The adoption of novel breeding locations is well-documented by citizen science nest monitoring schemes. For example, Brooks (1993) used Cornell nest record cards to verify a previously

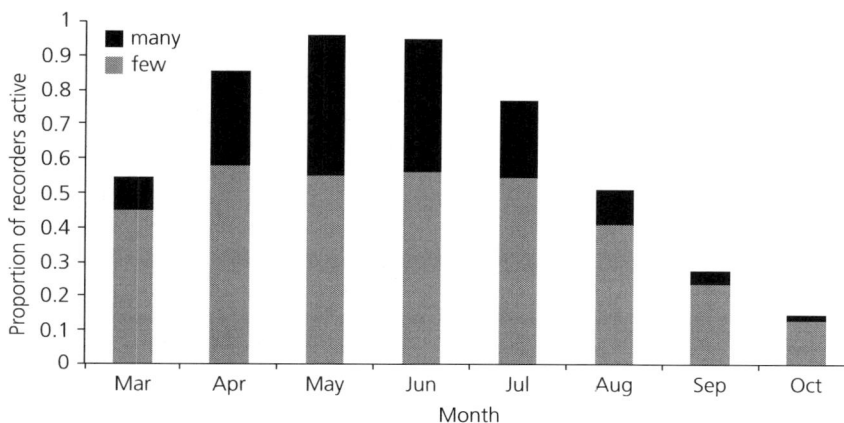

Figure 17.1 Seasonal variation in monitoring effort of the British Trust for Ornithology's (BTO's) Nest Record Scheme volunteers between 1970 and 2013 (inclusive). Columns show proportion of participants actively recording nests for 'a few' (grey bar) and 'many' (black bar) days in each month.

undocumented nesting substrate, coniferous trees, for American yellow warblers (*Dendroica petechia*). Hudson (1994) described the use of reeds (*Phragmites australis*) by the American robin, and Yunick (1990) documented the first case of winter wren (*Troglodytes troglodytes*) breeding in a North American nestbox. Bailey and Clark (2014) recorded the first known occurrence of twin eastern bluebirds (*Sialia sialis*) based on reports from a volunteer to the Cornell nest records; twinning had only been previously described in 13 wild bird species.

In the UK, BTO NRS volunteers highlighted the use of cultivated rape (*Brassica* sp.) as breeding habitat of sedge warblers (*Acrocephalus schoenobaenus*; Bowey 1999) and the adoption of artificial baskets by long-eared owls (*Asio otus*; Garner and Milne 1998). The occurrence of 'runt' eggs, i.e. those considerably smaller than the others in the clutch, was studied by Crick (1995), again using data from the BTO NRS. A number of participants have published observations of unusual behaviour, such as nest reuse (McNair 1984a), or unusually high productivity (Radford 1980).

17.3.5 Complex patterns in clutch size

Soon after ornithology emerged as a discipline, key figures began to use avian information to formulate life-history theory (reviewed by Ricklefs 2000). Several studies relying on volunteer observations have identified intraspecific latitudinal trends in a range of breeding parameters. Peakall (1970) conducted the first major analysis of the Cornell nest records and was able to utilize more data than any comparable single-species study in North America at that time. Breeding season length and clutch size of the eastern bluebird throughout its range appeared to be greater in the centre of the range, decreasing towards the northern and southern peripheries (Peakall 1970). Cooper et al. (2005a) used more Cornell nest records on eastern bluebirds to demonstrate latitudinal trends in breeding season length and number of potential nest attempts.

Several explanations for latitudinal trends in clutch size have been proposed (Ricklefs 2000), and citizen science nest monitoring schemes present an opportunity to test the various hypotheses using standardized data sets. Ashmole (1961, 1963) hypothesized that relative, rather than absolute, levels of resource availability during the breeding season influence clutch size, and consequently, clutch sizes should be larger at higher latitudes. Dunn et al. (2000) were able to test Ashmole's hypothesis across the majority of the tree swallow's breeding range due to the large volume of Cornell and BSC nest records. They found that mean clutch size of tree swallows increased with latitude and relative resource abundance, supporting Ashmole's hypothesis. Engstrom (1986) also demonstrated that the complex pattern of spatial variation involving both latitudinal and longitudinal gradients in red-winged blackbird (*Agelaius phoeniceus*) clutch sizes could be explained by heterogeneity in resource availability. Dunn et al. (2000) found that mean clutch size of tree swallows increased with latitude and relative resource abundance, as predicted by Ashmole's hypothesis. They were able to study the species across the majority of its breeding range due to the large volume of Cornell and BSC nest records. Young (1994b) used Cornell nest records and other data to model variation in house wren (*Troglodytes aedon*) clutch size. He found that the number of eggs laid was positively correlated with latitude but, interestingly, not with a proxy measure of primary productivity in either North American (migratory) or South American (resident) populations. Citizen science data have also been used to examine other hypotheses to account for latitudinal trends in clutch size. For example, a study by Cooper et al. (2006) using eastern bluebird data provided support for the egg viability hypothesis (Stoleson and Beissinger 1999), where higher temperatures experienced at lower latitudes result in increased failure rates of untended eggs, thereby limiting clutch size.

Attempts to identify latitudinal trends within Britain using BTO NRS data (Evans et al. 2009) for single-brooded species failed to find strong relationships in the seven species studied. At 8°, the latitudinal range involved was relatively small but abiotic factors proposed as drivers did vary spatially and significant relationships between clutch size and day length were identified. These relationships were generally positive for diurnal species and negative for tawny owl (*Strix aluco*), the one nocturnal species included. Evans et al. (2005) analyzed data for 11 passerine species introduced to New Zealand from the UK using data from the respective national nest record schemes. Their results and those from a more intensive study of common blackbird data (Samaš et al. 2013) provide some support for Ashmole's hypothesis (1961, 1963), whereby smaller clutches are related to reduced seasonality, but suggest that this is not necessarily the general driver of all latitudinal trends.

Seasonal variation in clutch sizes has also been well-documented using citizen science data. Crick et al. (1993) used BTO NRS data to explore trends across a range of British passerines, demonstrating that clutch sizes of single-brooded residents declined as the season progressed but clutch sizes of multi-brooded residents displayed a mid-season peak. This resulted from a trade-off between the probability of producing repeat or replacement broods and resource availability. Interestingly, migrants displayed a seasonal decline whether single or multi-brooded, suggesting that early breeding is constrained by arrival dates, a conclusion supported by a comparison of trends in migratory and resident eastern bluebird populations (Dhondt et al. 2002). Silva (1949) used BTO NRS to show non-linear changes in song thrush clutch size over the season. The average clutch size increased from early spring to a peak in May, followed by a decline. Howard (1967) analyzed American robin nesting records from Cornell and BSC schemes covering New York, Massachusetts, and the maritime provinces of Canada. It was found that in all areas but Massachusetts, clutch size peaked in June, although most robins were double-brooded. Clutch size was also significantly lower in the maritime provinces than in the New York–Massachusetts area, where the breeding season was approximately two weeks earlier. Increase in clutch size with latitude was not indicated in this study, possibly because maritime/island environments tend to produce lower clutch sizes (Lack 1948c).

Dhondt et al. (2002) demonstrated a similar seasonal pattern in clutch size that varied with latitude, based on Cornell nest records for eastern bluebirds. These patterns were revisited by Cooper et al. (2005b) in relation to constraints during the egg-laying and incubation stages, rather than selection during the chick-rearing stage.

Using a combination of museum records, Cornell nest records, and data from the literature and professionals, McNair (1984b) found that brown-headed nuthatches (*Sitta pusilla*) showed a significant positive correlation between clutch size and latitude, but this was primarily due to the influence of Florida clutches, which were significantly smaller. By contrast, Leck and Cantor (1979) found no geographic variation in clutch size for cedar waxwings (*Bombycilla cedrorum*) using Cornell nest records. Both cedar waxwings (Leck and Cantor 1979) and brown-headed nuthatches (McNair 1984b) exhibited a significant negative correlation between clutch size and date of clutch initiation, consistent with a late-season reduction in clutch sizes of other passerine species (Lack 1947, 1948a).

Wootton (1986) identified geographic variation in house finch clutch sizes, which were significantly higher in the introduced eastern population in the USA than in the native western population. The latter was not, however, explained by latitude or altitude and may be the result of higher predation rates or population density in the west.

17.3.6 Behavioural observations

Traditionally, nest record schemes have collected data during a series of short visits made by observers in person, but technological improvements in video equipment and a subsequent reduction in the price of nest cameras (see Chapter 15) have resulted in a rapid uptake by volunteers. Extensive filming of nests by citizen scientists unlocks the potential for spatial and temporal variation in nesting behaviour to be explored. Cooper et al. (2009) found that in eastern bluebirds, ultimate eggs are laid later in the day and after a longer interval than preceding eggs. This discovery was made possible by volunteers who placed video cameras in nestboxes and archived images for the Cornell Lab of Ornithology's NestCams project. This approach has the potential to greatly increase our understanding of bird behaviour during laying, incubation and chick rearing, as well as providing reliable identification of nest predators.

17.3.7 Breeding distributions

Although atlas programmes involving systematic collection of data by volunteers across assigned grid quadrats are best suited for modelling breeding distributions, data from nest monitoring schemes have supplemented our understanding, particularly for species that occur at low densities or are difficult to detect using standard survey protocols. The Cornell nest records data set helped to determine the breeding distributions of northern mockingbirds (*Mimus polyglottos*; Arnold 1980) and Bendire's thrashers (*Toxostoma bendirei*; England and Laudenslayer 1989) in California. McNair (2000) used records in conjunction with other information to describe the historical breeding range of the lark sparrow in Tennessee. Information from the BTO NRS significantly increased the number of sites at which breeding could be confirmed for a range of species, particularly nocturnal birds such as tawny owls and barn owls (*Tyto alba*), during production of the latest British and Irish bird atlas (Balmer et al. 2013). However, care must be taken when interpreting data sets where sampling is unstructured, as a lack of records may be indicative of poor coverage rather than true absence of breeding individuals.

17.3.8 Spatial variation in productivity

Land-use changes can have significant impacts on breeding success. Studies utilizing BTO NRS data have primarily focussed on agricultural habitats, exploring the relationship between productivity and the post-war intensification of farming experienced in Britain and Ireland. Siriwardena et al. (2000a) analyzed nest records for 10 farmland passerines and determined that breeding success was generally poorest in arable habitats. Shrubb (1990) found that northern lapwing (*Vanellus vanellus*) breeding success was highest on tilled land, but that sowing of winter cereals was reducing the availability of this habitat. Leech et al. (2009) demonstrated increased barn owl breeding success in rough pasture relative to areas of more intensive cultivation. Spatial variation in the quality of urban habitats may also influence breeding success of house sparrows (Shaw et al. 2008).

British Trust for Ornithology NRS data have also been used to investigate the impacts of nest predators on trends in breeding success. Paradis et al. (2000) reported that nest failure rates of common blackbirds and song thrushes during incubation were positively related to corvid abundance. This may explain some spatial variation in population size but is unlikely to be implicated in national declines (Gooch et al. 1991). Newson et al. (2010) identified a similar relationship between grey squirrel (*Sciurus carolinensis*) abundance and nest failure rates of common blackbirds and Eurasian collared doves (*Streptopelia decaocto*), although no such correlation was apparent for other woodland bird species studied.

17.3.9 Population modelling

Understanding the factors driving variation in breeding success is the first step in addressing species' declines. In order to determine the role productivity has played in driving trends in abundance and to predict future impacts, it is necessary to incorporate data on nesting success into population models. This process would ideally integrate information on long-term changes in population size, gathered by census surveys, with the results of schemes simultaneously monitoring productivity and survival rates, the latter provided by bird ringing studies (Baillie 1990).

Estimates of breeding success generated from citizen science surveys can help to assess the viability of populations at a given point in time when used in conjunction with survival data (Conway et al. 1994). Initial attempts to explore the relationship between annual variation in productivity and bird numbers employed key factor analysis (Baillie and Peach 1992), or correlated individual demographic parameters with population trends in isolation (Baillie 1990), often taking advantage of spatial variation in trajectories (Peach et al. 1995). The efficacy of the former technique has since been questioned by some ecologists (Royama 1996), while the latter approach is hampered by the fact that estimates of the impact of breeding success do not take into account simultaneous changes in survival. In an attempt to address this latter issue, researchers developed models that incorporated data on both survival and productivity simultaneously, holding one constant while allowing the other to vary annually as observed in the field and comparing the modelled results to recorded trends in abundance. Siriwardena et al. (2000b) fitted such models to BTO NRS data to assess the processes implicated in the widespread fall in farmland bird numbers across the UK during the 20th Century. The study identified reduced overwinter survival as the primary determinant of population size for the majority of the 12 species studied, although reduced reproductive output does appear to have underpinned common linnet (*Carduelis cannabina*) declines. Declining survival was also found to be responsible for the drop in British common reed bunting (*Emberiza schoeniclus*; Peach et al. 1999) and spotted flycatcher (Freeman and Crick 2003) abundance, although the driver of Eurasian bullfinch (*Pyrrhula pyrrhula*) population trends is less clear (Siriwardena et al. 2001; Profitt et al. 2004).

Freeman et al. (2007) refined the analytical models to explore common starling declines in the UK. These identified decreased juvenile survival as the driver, and recent advances in computing power have enabled the development of fully integrated modelling using a Bayesian framework. Robinson et al. (2012) pioneered this approach with respect to BTO NRS data, demonstrating that common blackbird breeding numbers were determined by adult survival rates but that the relative contribution of demographic parameters displayed significant regional variation. Robinson et al. (2014a) extended this approach to 17 British bird species, reporting that recruitment tended to limit the population size of declining species while adult survival limited that of increasing species. The next step in the development of these models is to incorporate the effects of density dependence (Paradis et al. 2002), whereby a decline in abundance can lead to increased survival or productivity due to relaxation of intraspecific competition.

17.3.10 Numbers of breeding attempts

Several studies have highlighted the impact of climate change on the length of the breeding season. Multi-brooded species display a positive relationship with temperature, indicative of an increased capacity to produce repeat or replacement broods at higher ambient temperatures (Møller et al. 2010; Dunn and Møller 2014). Breeding season length may also be influenced by land use. Exploring the BTO NRS data set, Chamberlain and Crick (1999) suggested that a reduction in the incidence of double-brooding resulting from changes to cropping regimes may have led to the decline of Eurasian skylarks (*Alauda arvensis*) in Britain, although this relationship was inferred rather than quantified, because estimating the number of attempts initiated per individual using citizen science data sets can be extremely challenging. Cooper et al. (2005b) compared *relative* breeding season length across latitude, assuming any biases among participants did not vary by location.

Assuming unbiased seasonal coverage (see discussion in Section 17.3.3 above), continued development of the Bayesian modelling approach adopted by Cornulier et al. (2009) to analyze the BTO NRS yellowhammer (*Emberiza citrinella*) data set, could be used to identify multiple peaks in the annual laying date distribution of a wider range of species. This study identified well-defined, habitat-mediated differences in numbers of attempts, but results are likely to be less clear-cut where breeding is less synchronous, for example in species where failure rates are high and relaying is therefore frequent.

An alternative approach involves attribution of successive nesting attempts to individual birds. Ideally, this could involve observation of uniquely marked parents, although territory mapping may be sufficient for species breeding at relatively low densities (Brickle and Harper 2002). Increasing fieldwork intensity is potentially problematic for citizen science surveys, but development of new technologies may help to reduce the effort required to collect this type of data. In particular, in a limited context where volunteers can obtain necessary training and permits, passive integrated transponder (PIT) tags provide a relatively accessible and efficient means of recording the identity of free-flying birds, particularly in situations where they return regularly to a predictable location, e.g. a nest or food source (Fiedler 2009; Chapter 15).

17.3.11 Post-fledging survival

Streby et al. (2014) identified the constraints of using per-nesting breeding success as a measure of productivity in population models, terming this the 'nest success paradigm'. One potential shortcoming is the failure to detect annual variation in numbers of breeding attempts. Such models also fail to account for variation in post-fledging mortality, which can have a significant influence on population size (Cox et al. 2014).

A composite measure of season-long productivity can be provided for some species using juvenile:adult ratios generated by standardized mist-netting activities (Desante et al. 1995; Peach et al. 1996). Data from the BTO Constant Effort Sites scheme in Britain and Ireland have contributed to several population modelling analyses (Robinson et al. 2007; Cave et al. 2010). Alternative methodologies have been proposed, including a count data approach utilizing observations of the size of post-fledging family parties (Sage et al. 2011). Regardless of the efficiency of these schemes, if variation in annual productivity is identified then stage-specific monitoring will still be required to determine the mechanism, be it nest monitoring or nestling ringing (banding). Each year, bird ringers in Britain and Ireland mark 150,000–200,000 nestlings of over 150 species (Robinson and Clark 2014). This data set has the potential to generate huge amounts of information on post-fledging survival and dispersal, which are two major knowledge gaps in our understanding of population limitation. However, while the extensive network of volunteers provides some information about annual variation in recruitment probabilities for frequently encountered species and those exhibiting high site fidelity, data are currently collected in an unstructured manner. Development of standardized citizen science protocols, particularly those that maximize capture during the period immediately following fledging, could greatly increase our understanding of the role early stage mortality plays in determining population size.

17.4 The science of citizen science

Citizen science projects must be designed carefully in order to achieve their research goals, matching aims and protocols with participants' skills and interests. For example, in the UK, where there are significant numbers of trained and licenced volunteer ringers, collection of data that requires ringing or attachment of transponders is feasible within a citizen science framework; in the US, such projects would not be feasible at a large scale.

The two nest schemes primarily mentioned in this chapter have some similarities and differences. Over 650 volunteer nest recorders currently contribute to the

BTO NRS, the highest number in the scheme's 75-year history, compiling a data set of over 40,000 individual nest histories each year involving an estimated 55,000 hours of fieldwork per annum, not including travel (BTO, 2014 unpublished data). Almost all data are submitted by amateurs, although occasional submissions are received from intensive professional studies and students. Historically there has been no formal training programme, although field courses were established in 2009 and introduce an estimated 30–40 participants a year to nest monitoring. In 2014, a mentoring network was launched with the aim of encouraging experienced volunteers to provide more intensive field training for new recruits. While some investment is still required to manage the volunteer base, provide support for participants and feed results back to maintain participation, it is a fraction of that which would be required should all data collection be undertaken by paid fieldworkers.

There are 1,500 participants currently enrolled in Cornell nest records, many of whom coordinate small groups of unregistered volunteers, and much training is distributed through state-based organizations. For example, each US state operates a 'Bluebird Society' and many counties have Audubon clubs. All participants retain access to their own submitted data and the collective project-wide data, via online interactive mapping and reporting tools.

One of the greatest advantages of data collection via a network of citizen scientists is the scale at which information can be obtained. This is particularly important for nest monitoring schemes, where the time required to locate individual nests and follow their progress may be substantial. However, the focal period of activity, the breeding season, is relatively short-lived, with the majority of nesting attempts across species occurring over a 3–4 month period in temperate regions.

While recording can be spatially extensive, a potential disadvantage of data collection via a volunteer network is the difficulty in directing effort to achieve representative coverage. For example, analysis of species composition in data submissions to the BTO NRS in the mid-2000s demonstrated a gradual decline in the monitoring of open-nesting passerines over the preceding three decades (Figure 17.2). The situation has since begun to improve as a result of targeted promotion and training but this reversal has required significant investment by BTO and volunteer trainers. From 1996 to 2006, online data entry for the Cornell nest records was only possible for observations of cavity-nesting species. Although contributions of data on cup-nesting species have slowly been increasing (Figure 17.3), participants are receptive to calls for increased efforts on particular species.

Another form of potential bias is geographic. Ideally, collection of nest monitoring data would occur systematically with standardized coverage across a regular grid, as in state-level or national atlas programmes (e.g. Balmer et al. 2013), or, failing that, via a network of stratified random sites, as per the BTO/JNCC/RSPB Breeding Bird Survey in the UK (Newson et al. 2005; Harris et al. 2014). However, due to the paucity of potential volunteers, citizen science nest monitoring programmes generally permit participants to select their own study sites. Resulting biases in coverage can limit studies of reproduction if the research questions require data from underpopulated regions

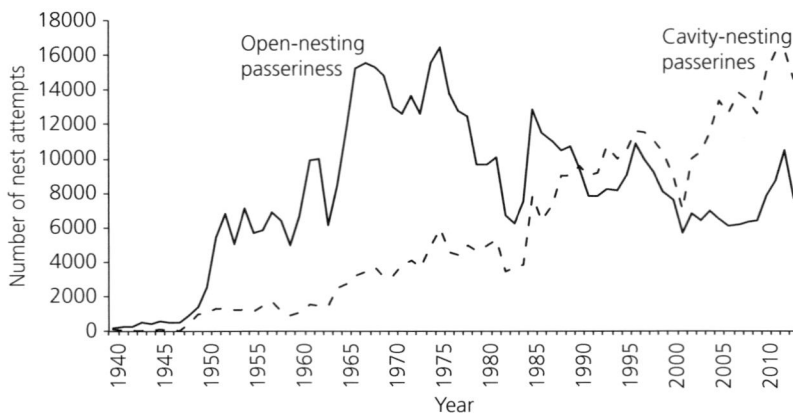

Figure 17.2 Number of individual breeding attempts of cavity-nesting passerines (dashed line) and open-nesting passerines (solid line) recorded per annum for the British Trust for Ornithology's (BTO's) Nest Record Scheme between 1940 and 2013 (inclusive).

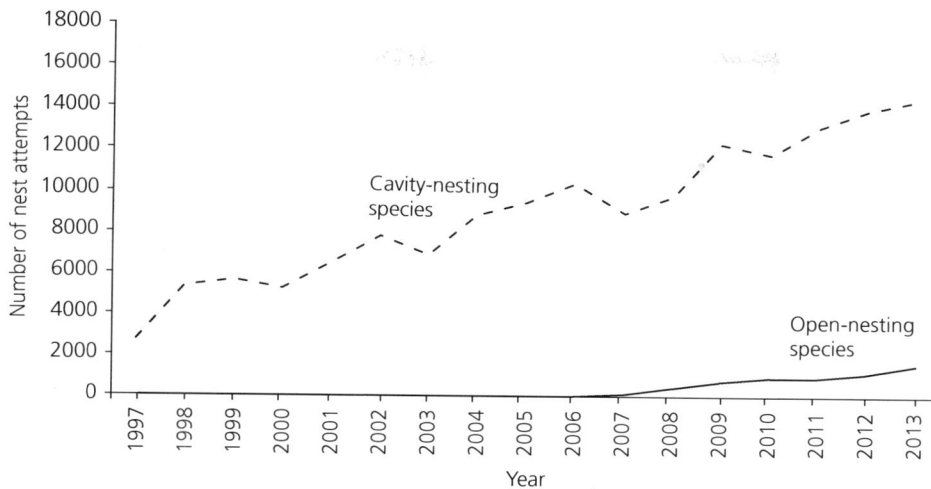

Figure 17.3 Number of individual breeding attempts of cavity-nesting species (dashed line) and open-nesting species (solid line) recorded per annum for the Cornell Lab of Ornithology's online nest record programs between 1997 and 2013 (inclusive).

and/or inaccessible habitats (e.g. uplands, seabird islands). For scarce species, observer site selection can also lead to over-representation of 'honeypot' populations in the data set. Volunteers gravitate towards those sites where birds are more abundant and the probability of locating a nest is therefore greater, which can create bias for conservation research. Given the types of potential bias that can occur when relying on volunteer observations, citizen science is well-suited to research questions in residential areas and along urban gradients.

There is a perception that the quality of data collected by 'amateurs' is likely to be inferior compared to those collected by professionals, but volunteers may actually be significantly more experienced fieldworkers. Data from the BTO NRS show that the average span of submissions (1970–2013) is six years, with almost 20% contributing data for 10 or more seasons, much longer than the typical duration of a research grant (see Chapter 18). In addition, the protocols established by most citizen science nest monitoring schemes aim to minimize the need for interpretation by the observer, with participants asked to record only what they see. For example, BTO nest recorders note the stage of feather development of any nestlings encountered rather than being asked to assess age directly, with hatching dates back-calculated by BTO staff on receipt of data using standardized parameter files derived from published literature (Crick et al. 2003a).

Careful design of data collection and incorporation of methods for assessing the skill and/or experience levels of volunteers can improve analyses. Bird et al. (2014) summarized common statistical techniques useful for the analysis of large-scale data, which is relevant to citizen science. Other studies have specifically addressed issues of data quality in ornithological citizen science, including verifying rare observations (Bonter and Cooper 2012) and adjusting for weekend bias (Sparks et al. 2008; Courter et al. 2012; Cooper 2013). The use of novel online data filters and workflows for detecting erroneous submissions (Sullivan et al. 2014), combined with modified project design and analysis (Cooper 2013), represent promising advances in the vetting and dissemination of citizen science data sets.

17.4.1 Bird welfare

Bird welfare is of utmost concern, both from an ethical perspective and in terms of data integrity. The impact of visitation has been studied using information from both the BTO (Mayer-Gross et al. 1997) and Cornell schemes (Bart 1977). Analysis of data from both academic research projects and citizen science schemes has demonstrated that nest visitation need not impact negatively on the outcome of the attempt (reviewed in Ibáñez-Álamo et al. 2012; Chapter 1). All nest monitoring schemes provide clear guidance as to best practices when collecting data, e.g. *BTO NRS Code of Conduct* (Crick et al. 2003b) and the *NestWatch Nest Monitoring Manual* (Martin et al. 2013b). Again, many volunteers will have as much, if not more, experience in nest monitoring than professional researchers. For instance, a comparison of the outcome of nests monitored by these two groups using Neighborhood Nestwatch data from North America showed no difference in failure rates (Ryder et al. 2010).

17.5 The future

Some common challenges faced by ornithologists running citizen science programmes relate to managing human capital. Newly organized societies such as the Citizen Science Association (citizenscienceassociation. org), the European Citizen Science Association (ecsa. biodiv.naturkundemuseum-berlin.de), and the Citizen Science Network Australia (csna.gaiaresources.com.au) are providing opportunities for professional exchanges of knowledge and experience with the practice of citizen science. The complexity of engaging people in research raises legal, ethical, logistical, statistical, and practical issues that traditional training in ornithology does not provide. As a result, citizen science has become the subject of study in a variety of fields including: informal science education, informatics, human–computer interactions, communications, evaluation, and science, technology, and society studies. Projects frequently require multidisciplinary teams with the expertise needed to manage people and manage large data sets, the analysis of which may require development of new statistical methods (Kelling et al. 2009; Hochachka et al. 2012).

Acknowledgements

Foremost, the authors wish to thank the legions of volunteers who collected nesting data for the various nest record schemes; their cumulative efforts represent a truly astonishing contribution to the field of ornithology. The British Trust for Ornithology's (BTO's) Nest Record Scheme is funded by the BTO/Joint Nature Conservation Committee (JNCC) partnership that the JNCC undertakes on behalf of the statutory nature conservation bodies: Council for Nature Conservation and the Countryside, Natural England, Natural Resources Wales and Scottish Natural Heritage. NestWatch is supported by the Cornell Lab of Ornithology and was developed under the National Science Foundation grant No. 0540185. Melcolm Crutchfield helped with preparation of materials for the manuscript. Diane Tessaglia-Hymes and Tony Leech helped prepare some of the figures.

CHAPTER 18

Perspectives on avian nests and eggs

D.C. Deeming and S.J. Reynolds

18.1 The scientific study of nests, eggs, and incubation

The original premise of *Avian Incubation* (Deeming 2002a) was that it would bring together a wide-ranging group of topics based around incubation in birds in one volume. It was hoped that this book would serve as an important reference work for existing and new researchers interested in studying nests, eggs, and incubation. It seems that it has served its purpose and continues to do so for major areas of avian reproductive biology. However, our understanding of nests, eggs, and incubation has progressed and new areas of research have come to light as a result of emerging new technologies and refinements of established technological and analytical approaches. Although nests, eggs, and incubation may be relatively poorly understood compared with other aspects of avian biology (see Chapter 1 for further context), they remain of great interest to ornithologists and offer a multitude of areas for further research as outlined in previous chapters. This broad diversity of research opportunities is evidenced by the extensive reference list at the end of this book with over 50% of the citations having been published after Deeming (2002a). The term 'avian incubation' encompasses many diverse subject areas, such as the biology of nests and eggs, and the energetic and temporal investments of incubation as a process, and continues to yield areas for novel research. However, neither Deeming (2002a), nor this book, has dealt with the rearing of the offspring post-hatching, which is especially difficult to study in precocial and semi-precocial species. We make this point not to highlight shortcomings of the two books but to emphasize that answers to questions posed in these two books stimulate further lines of exciting enquiry.

In his concluding chapter of *Avian Incubation* Deeming (2002g) revisited preceding chapters in order to assess the key areas of interest from his perspective.

It was clear at that time that while much was already known about the processes and associated behaviours of incubation in nests, much still remained to be discovered. During the intervening years research in this subject area has continued in a variety of pre-existing and novel directions (Chapter 1). Most areas of research employ modern techniques that provide a better understanding of how birds meet the challenges of incubation. Here, we follow the lead of Deeming (2002g) and briefly highlight some of the key areas of research covered in the preceding chapters in the present book before we conclude this latest review of avian incubation by considering what the future holds for research in this fundamental area of avian biology.

18.2 Advances in avian reproduction

The fossil record of avian reproduction remains extremely limited (Chapter 2) and it can be strongly argued that it still tells us little about the evolution of this fundamental aspect of avian life history. However, since 2002 there has been a plethora of new finds and an increasing number of extinct species is becoming available to study. One key aspect of the fossil record that should be remembered during future studies is that new finds should be treated with caution and novel specimens should not be over-interpreted (Deeming 2002b; Chapter 2).

The remainder of this book documents and reviews four main areas of research: nests, eggs, incubation, and the broader aspects of the study of incubation. Hansell (2000) and Hansell and Deeming (2002) reviewed variation in nests but it is clear that our appreciation of the biology of nests has significantly improved since that time. There has been a renewed interest in the behaviour of nest construction particularly from a cognitive perspective (Chapter 3). However, it is clear that we know little about this key process in the behavioural repertoire of birds in readiness for incubation and we

Deeming, D.C. and Reynolds, S.J., *Perspectives on avian nests and eggs*. In: *Nests, Eggs, and Incubation*. Edited by: D.C. Deeming and S.J. Reynolds, Oxford University Press (2015). © Oxford University Press.
DOI 10.1093/acprof:oso/9780198718666.003.00018

need to focus on documenting the diversity and details of nest-building processes undertaken by birds. Such a basic understanding would allow significantly greater insights into the decisions made by individuals and into the diversity of nest-building strategies explained from a life-history perspective (*sensu* Bennett and Owens 2002).

A key change from Deeming (2002a) is that interest has been sparked in how nests actually function as a location for incubation. The concept of the 'bird-nest-incubation unit' introduced by Deeming (2002b) has been superseded by a greater appreciation of the effect that adult body size plays on nests as functional units (Chapter 4). Larger birds are using nests simply as a location for incubation when they act merely as platforms with no real thermodynamic role—hence, the concept of 'bird-as-incubator' applies. However, as adult body size decreases, its interaction with increasing relative clutch size translates into nests playing an ever more important role in incubation—it is now realized that the 'bird-nest-incubation unit' is a fundamental consideration in effective incubation by species of smaller body sizes (Chapter 4).

For the first time, there has been a synthesis of the role that ecology, particularly the effects of predation, plays in nest location and function (Chapter 5). The decisions taken by birds when placing their nests in particular locations appear to be driven by minimizing the risk of predation but our increasing understanding of nest construction behaviour and the various functions of nests may provide critical insights into other factors, such as microclimate, that may influence nest-site location. This may be crucial in understanding the impact of climate change on avian reproduction (Chapter 6). To date, the foci have remained fixed on the timing of breeding (Deeming and du Feu 2011) and on the chick-rearing stage, when for some small passerines phenological mismatches occur between peak nutritional demands of nestlings and the availability of their invertebrate prey (e.g. Visser et al. 1998; Webber 2012). However, we understand little about how birds 'track' changing environmental conditions and respond to them behaviourally, physiologically, ecologically etc. However, climate change could have direct impacts on the process of incubation and perhaps more crucially the nesting biology of species. For example, a warming environment, or one with more frequent (and less predictable) precipitation events, might change the availability of nest materials and force birds to be more versatile in their nest-building behaviour. Moreover, it might result in species markedly changing their nesting biology as a result of expansion or restriction of breeding ranges (Huntley et al. 2007). This could result in exciting new insights into how individual species meet such challenges and thereby markedly change how we assess the vulnerability of avian taxa in an ever-changing world.

Nests are not simply receptacles for eggs, nestlings or incubating birds and we are improving our understanding of how other inhabitants of nests interact with the anticipated occupants. Incubation microbiology has been previously considered (Baggott and Graeme-Cook 2002) but has now been updated to consider the role that nest materials play in controlling the microbial environment to ensure embryonic survival (Chapter 7). Invertebrate occupants of nests play a key role in the reproductive biology of birds (Chapter 8) with ectoparasites not only residing within the nest but also visiting nests to discover avian hosts. So, birds have had to develop defences in order to minimize the impacts of these invertebrates on breeding efforts and ultimately on their fitness. In some cases such 'defences' can result in the wholesale movement of entire breeding assemblages of birds, such as seabird colonies, to areas where encounters with such ectoparasites are significantly less likely.

Previous studies have extensively focussed on egg characteristics but our improving understanding of this aspect of avian reproduction has been covered in this book (Chapter 9). It has long been recognized that phylogeny must be controlled for in interspecific allometric analyses of the inter-relationships between various egg characteristics, but here its importance has been further explored and clarified. Phylogenetically corrected regression has little impact on allometric relationships but the effects of higher taxonomic levels such as order can provide a critical level for analysis, especially when framing our understanding of egg biology within the altricial–precocial framework extolled by Starck and Ricklefs (1998). Our understanding of the influence of egg components on embryonic development and post-hatching effects continues to improve (Chapter 10). An integrated approach has investigated the effects of overall egg quality on embryonic development and the post-hatching phenotype. Since Deeming (2002a), the concept that eggs, have roles in reproduction other than simply providing passive receptacles for the embryo has revolutionized how we consider these critical units for successful reproduction. Chapter 11 describes how eggs, can be used as signals in various aspects of reproduction, and how there may be a greater degree of communication between the embryo and the incubating adult. In particular, it highlighted the need to 'look' beyond conventional visual systems of signallers and receivers, and instead to explore other sensory modalities for a fuller appreciation of eggs, as signals.

Incubation behaviour continues to be an important area of study, particularly when considering the behaviour of embryos, which may not be quite so passive as was once thought (Brua 2002) in their response to the incubation environment (Chapter 12). The energetics of incubation remains a key theme of the study of avian reproduction and evidence continues to point to the high energetic cost of incubation (Chapter 13). The energy required to produce an egg may be relatively easy to define through careful investigation but how birds respond to energetic demands of incubation remains a critical aspect of study, particularly as our understanding of the thermal characteristics of nests improves. We expect this area to attract considerably more research attention in the next few years as climate change and an urbanizing world influence all aspects of avian breeding biology (Marzluff et al. 2001; Gaston 2010).

The impact of the incubation environment on post-hatching development of chicks was first recognized in

poultry but similar effects on nestlings of wild species have been appreciated over the past decade; it is now realized that variation in incubation conditions, such as temperature, in many nests can have a marked impact on post-hatching phenotypes, such as body condition at hatching, and hence fitness (Chapter 14). To date, this aspect of incubation has been studied in a very limited number of species and there is real scope for further study through expansion of the species base.

Finally, this book has sections on new areas of incubation science previously not covered. Technology has changed the way that incubation can be studied over the past decade (Chapter 15). Although the use of self-contained, remote, miniature, and archival devices, such as *i*Buttons®, has proliferated in recent years and allowed relatively simple study at nests of free-living species, their use has in turn illuminated many issues that must be considered *before* (planning), *during* (reactive modification), and *after* (analysis and interpretation) their deployment. The conservation crisis caused by human activity on a global scale has meant that increasing numbers of avian species face the real threat of

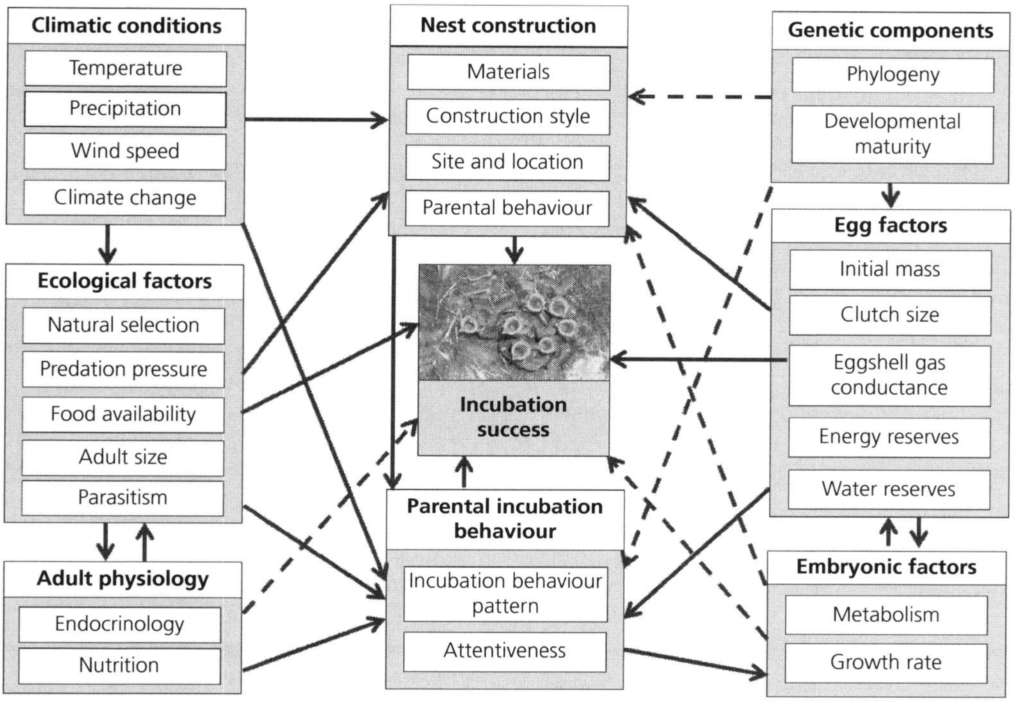

Figure 18.1 Patterns of interaction of factors affecting incubation temperature during avian reproduction. Solid arrows indicate interactions where evidence exists to support the interaction, whereas dashed arrows are postulated interactions that have yet to be investigated. Image of the great tit (*Parus major*) brood is by DCD.

extinction (Collar and Butchart 2014). One solution to this crisis is the more effective use of artificial incubation to aid conservation efforts. We have come to realize that simply trusting to incubation parameters used for domesticated poultry is not applicable to most bird species and to enhance the success of artificial incubation there needs to be better exchange of ideas and knowledge between poultry and non-poultry aviculturalists (Chapter 16). Information exchange is also *the* key theme of citizen science programmes that have been highlighted as a key contributor to advance our understanding of the nesting biology of birds for the first time (Chapter 17). A good example is provided by nest monitoring that, if well organized, can use the 'power' of citizen science over broad spatial and long temporal scales to facilitate exchange of information between amateur and professional ornithological communities and to bring scientific outputs to wider audiences.

A major feature of the study of avian incubation is its multidisciplinary nature, which is necessary to interpret the complex interactions between the various factors that could affect incubation success. We attempt to summarize this in Figure 18.1 (which is significantly expanded from a similar attempt by Deeming et al. 2006). It highlights the critical aspects of avian life history and how they impact on incubation success but also it attempts to illustrate how they can interact. For example, some factors can directly impact on incubation success. Food availability can determine whether adult birds have sufficient energy reserves to maintain incubation. Other factors, such as climatic conditions, do not directly impact on incubation success but can affect bird behaviour through changes in patterns of nest construction and/or incubation behaviour, both of which can indirectly affect the number of eggs, hatched. Figure 18.1 also postulates some possible interactions, such as whether phylogeny is critical in determining nest construction behaviour, that have yet to be explored fully.

We hope that the topics covered within this book will serve as much of a reference as *Avian Incubation* has proved to be. We feel that this new book complements, rather than supersedes, the first book because it builds on a body of work that largely remains relevant. Our hope is that *Nests, Eggs, and Incubation* will become a key reference text for the future.

18.3 The future of avian incubation research?

It is rather paradoxical that research on avian reproduction seems to be simultaneously becoming harder *and* easier. Funding to support research efforts (including that for equipment, analyses, and staff salaries) has been difficult to obtain for many years, a situation exacerbated by economic problems facing many governments. With mainstream research funding provided in discrete and relatively short-lived 'packages' (typically over 3–5 years), long-term ecological research programmes have been especially hard hit (Birkhead 2014), despite their outputs continuing to shape the policies of many governments around the world. Interest in studying how birds nest may not seem very relevant in such times of austerity. Yet, ornithologists readily embrace and drive forward the development of novel technologies that make field research easier and more time- and cost-effective. There is a growing appetite outside of the established scientific community for involvement in practical scientific research that is fuelled by experience of the natural world that cannot be 'bought off the shelf'. Engagement of the public in monitoring bird nesting habits seems to be growing ever more popular, perhaps because of relatively recent demographic changes characterized by a healthier, well-educated, older population who by definition have more to give to research efforts. Channelling available resources, be they financial, temporal or energetic, is a positive challenge that could be crucial to the future of not only research into bird nests, eggs, and incubation, but also to other areas of bird biology.

We believe that this book has demonstrated our view that the study of avian incubation is an exciting and innovative area of research. There is still so much more to learn. The distribution of bird species globally means that there are likely to be a wide range of solutions to similar problems that emerge from local breeding conditions. For instance, there is increasing evidence that the nests that birds construct reflect the evolved response of a species to a local environment; the same problems can be solved in a variety of ways. Switch environments and the problems, and the solutions, change. We have only touched upon the way(s) that nest construction can impact on reproductive behaviour, physiology, and success.

Given that we have recognized that the environment is rapidly changing, we need to consider novel ways of monitoring the impact that these changes will have. Studying nesting biology may provide an insight into how individual species respond to change. Unfortunately, in many respects we know very little about the reproductive biology of the vast majority of bird species and this could directly impact on their chances of survival. We need to address this now to help us to predict and therefore to understand the nature of future changes.

What is clear from this book is that we have learned a huge amount since *Avian Incubation* was published in 2002 but there is still so much more to learn about bird nests, eggs, and incubation. They have the potential to provide an exciting and vibrant area of research for many years to come.

Acknowledgements

This book would not have been possible without the enthusiasm and commitment of the contributing authors and we take a final opportunity to thank them again for their great efforts in contributing to what has been a really worthwhile project.

References

Abad-Gómez, J.M., Masero, J.A., Villegas, A., Albano, N., Gutiérrez, J.S. and Sánchez-Guzmán, J.M. (2012). Sex-specific deposition and survival effects of maternal antibodies: a case study with the gull-billed tern *Gelochelidon nilotica*. *Journal of Avian Biology*, **43**, 491–5.

Abeyrathne, E.D.N.S., Lee, H.Y. and Ahn, D.U. (2013). Egg white proteins and their potential use in food processing or as nutraceutical and pharmaceutical agents—a review. *Poultry Science*, **92**, 3292–9.

Abraham, C.L. and Evans, R.M. (1999). Metabolic costs of heat solicitation calls in relation to thermal need in embryos of American white pelicans. *Animal Behaviour*, **57**, 967–75.

Adamík, P. and Král, M. (2008a). Climate- and resource-driven long-term changes in dormice populations negatively affect hole-nesting songbirds. *Journal of Zoology*, **275**, 209–15.

Adamík, P. and Král, M. (2008b). Nest losses of cavity nesting birds caused by dormice (Gliridae, Rodentia). *Acta Theriologica*, **53**, 185–92.

Addison, B., Ricklefs, R.E. and Klasing, K.C. (2010). Do maternally derived antibodies and early immune experience shape the adult immune response? *Functional Ecology*, **24**, 824–9.

Adkins-Regan, E. (2005). *Hormones and animal social behavior*. Princeton University Press, Princeton and Oxford.

Aebischer, A. and Meyer, D. (1998). Brutbiologie des Rohrschwirls *Locustella luscinioides* am Neuenburgersee [Breeding biology of Savi's warbler *Locustella luscinioides* at Lake Neuchatel]. *Der Ornithologische Beobachter*, **95**, 177–202.

Afonso, S., Vanore, G. and Batlle, A. (1999). Protoporphyrin IX and oxidative stress. *Free Radical Research*, **31**, 161–70.

Afton, A.D. (1980). Factors affecting incubation rhythms of northern shovelers. *The Condor*, **82**, 132–7.

Afton, A.D. and Paulus, S.L. (1992). Incubation and brood care. In BDJ Batt, AD Afton, MG Anderson, CD Ankney, DH Johnson, JA Kadlec, GL Krapu, eds. *Ecology and management of breeding waterfowl*, pp. 62–108. University of Minnesota Press, Minneapolis.

Ahlborn, G.J., Clare, D.A., Sheldon, B.W. and Kelly, R.W. (2006). Identification of eggshell membrane proteins and purification of ovotransferrin and beta-NAGase from hen egg white. *The Protein Journal*, **25**, 71–81.

Alabrudzińska, J., Kaliński, A., Slomczyński, R., Wawrzyniak, J., Zieliński, P. and Bańbura, J. (2003). Effects of nest characteristics on breeding success of great tits *Parus major*. *Acta Ornithologica*, **38**, 151–4.

Albrecht, T., Hořák, D., Kreisinger, J., Weidinger, K., Klvaňa, P. and Michot, T.C. (2006). Factors determining pochard nest predation along a wetland gradient. *Journal of Wildlife Management*, **70**, 784–91.

Aldrich, T.W. and Raveling, D.G. (1983). Effects of experience and body-weight on incubation behavior of Canada geese. *The Auk*, **100**, 670–9.

Allen, J.A. (1878). An inadequate theory of birds nests. *Bulletin of the Nuttall Ornithological Club*, **3**, 23–32.

Alonso-Alvarez, C., Bertrand, S., Devevey, G., Prost, J., Faivre, B. and Sorci, G. (2004). Increased susceptibility to oxidative stress as a proximate cost of reproduction. *Ecology Letters*, **7**, 363–8.

Alpha Mach Inc. (2014). http://www.alphamach.com/temperature/ibcod-2/. [Accessed: 14 July 2014]

Al-Safadi, M. and Al-Aqas, J. (2006). Observations on the breeding biology of the Swallow *Hirundo rustica transitiva* in Gaza Strip, Palestine. *Journal Al-Azhar University*, **10**, 13–19.

Álvarez, E. and Barba, E. (2008). Nest quality in relation to adult bird condition and its impact on reproduction in Great Tits *Parus major*. *Acta Ornithologica*, **43**, 3–9.

Álvarez, E. and Barba, E. (2011). Nest characteristics and reproductive performance in Great Tits *Parus major*. *Ardeola*, **58**, 125–36.

Álvarez, E., Belda, E.J., Verdejo, J. and Barba, E. (2013). Variation in Great Tit nest mass and composition and its breeding consequences: a comparative study in four Mediterranean habitats. *Avian Biology Research*, **6**, 39–46.

Amat, J.A. and Masero, J.A. (2004). Predation risk on incubating adults constrains the choice of thermally favourable nest sites in a plover. *Animal Behaviour*, **67**, 293–300.

Amat, J.A., Visser, G.H., Pérez–Hurtado, A. and Arroyo, G.M. (2000). Brood desertion by female shorebirds: a test of the differential parental capacity hypothesis on Kentish plovers. *Proceedings of the Royal Society of London Series B: Biological Sciences*, **267**, 2171–6.

Ambros, A. (1989). Technique of nest-building of the snowfinch *Montifringilla nivalis nivalis* L. *Egretta*, **32**, 58–71.

Amiel, J.J. and Shine, R. (2012). Hotter nests produce smarter young lizards. *Biology Letters*, **8**, 372–4.

Amo, L., Galvan, I., Tomás, G. and Sanz, J.J. (2008). Predator odour recognition and avoidance in a songbird. *Functional Ecology*, **22**, 289–93.

Amundsen, T. (2000). Why are female birds ornamented? *Trends in Ecology and Evolution*, **15**, 149–55.

Amundson, C.L. and Arnold, T.W. (2011). The role of predator removal, density-dependence, and environmental factors on Mallard duckling survival in North Dakota. *Journal of Wildlife Management*, **75**, 1330–9.

Anderson, A. (2003). *Prodigious birds: moas and moa-hunting in prehistoric New Zealand.* Cambridge University Press, Cambridge.

Anderson, D., Reeve, J. and Bird, D. (1997). Sexually dimorphic eggs, nestling growth and sibling competition in American kestrels *Falco sparverius. Functional Ecology*, **11**, 331–5.

Anderson, D.W. and Keith, J.O. (1980). The human influence on seabird nesting success: conservation implications. *Biological Conservation*, **18**, 65–80.

Anderson, R.C. (1956). The life cycle and seasonal transmissions of *Ornithofilaria falliesensis* Anderson: a parasite of domestic and wild ducks. *Canadian Journal of Zoology*, **34**, 485–525.

Andersson, M. (1994). *Sexual selection.* Princeton University Press, Princeton.

Angelier, F., Barbraud, C., Lormée, H., Prud'homme, F. and Chastel, O. (2006). Kidnapping of chicks in emperor penguins: a hormonal by-product? *The Journal of Experimental Biology*, **209**, 1413–20.

Angelier, F. and Chastel, O. (2009). Stress, prolactin and parental investment in birds: A review. *General and Comparative Endocrinology*, **163**, 142–8.

Angst, D., Buffetaut, E., Lécuyer, C., Amiot, R., Smektala, F., Giner, S., Méchin, A., Méchin, P., Amoros, A. Leroy, L., Guiomar, M., Tong, H. and Martinez, A. (2015). Fossil avian eggs from the Palaeogene of southern France: new size estimates and a possible taxonomic identification of the egg layer. *Geological Magazine*, **152**, 70–9.

Antczak, M., Hromada, M. and Tryjanowski, P. (2005). Research activity induces change in nest position of the Great Grey Shrike *Lanius excubitor. Ornis Fennica*, **82**, 20–5.

Antczak, M., Hromada, M., Czechowski, P., Tabor, J., Zabłocki, P., Grzybek, J. and Tryjanowski, P. (2010). A new material for old solutions – the case of plastic string used in Great Grey Shrike nests. *Acta Ethologica*, **13**, 87–91.

Antonov, A. (2004). Smaller Eastern Olivaceous Warbler *Hippolais pallida elaeica* nests suffer less predation than larger ones. *Acta Ornithologica*, **39**, 87–92.

Antonov, A. and Atanasova, D. (2003). Re-use of old nests versus the construction of new ones in the Magpie *Pica pica* in the city of Sofia (Bulgaria). *Acta Ornithologica*, **38**, 1–4.

Aourir, M., Znari, M., Radi, M. and Melin, J.-M. (2013). Wild-laid versus captive-laid eggs in the black-bellied sandgrouse: Is there any effect on chick productivity? *Zoo Biology*, **32**, 592–9.

Apanius, V. (1998). The ontogeny of immune function. In JM Starck, RE Ricklefs, eds. *Avian growth and development: evolution within the altricial–precocial spectrum*, pp. 203–22. Oxford University Press, Oxford.

Apfelbaum, S.I. and Seelbach, P. (1983). Nest tree, habitat selection and productivity of seven North American raptor species based on the Cornell University Nest Record Card Program. *Journal of Raptor Research*, **17**, 97–113.

Ar, A. (1991a). Roles of water in avian eggs. In DC Deeming, MWJ Ferguson, eds. *Egg incubation: its effects on embryonic development in birds and reptiles*, pp. 229–43. Oxford University Press, Oxford.

Ar, A. (1991b). Egg water movements during incubation. In SG Tullett, ed. *Avian incubation*, pp. 157–73. Butterworth-Heinemann, London.

Ar, A., Arieli, R., Belinsky, A. and Yom-Tov, Y. (1987). Energy in avian eggs and hatchlings: utilization and transfer. *Journal of Experimental Zoology*, Supplement **1**, 151–64.

Ar, A. and Deeming, D.C. (2009). Water and gas exchange in determining hatchability success. *Avian Biology Research*, **2**, 61–6.

Ar, A., Ifergan, O., Feldman, A., Zelik, L. and Reizis, A. (2004). Possible role of nitric oxide emission from bird embryos. *Avian and Poultry Biology Reviews*, **15**, 105–6.

Ar, A., Ifergan, O., Reizis, A., Zelik, L. and Feldman, A. (2000). Does nitric oxide (NO) play a role in embryo-bird communication during incubation? *Avian and Poultry Biology Reviews*, **11**, 284.

Ar, A., Paganelli, C.V., Reeves, R.B., Greene, D.G. and Rahn, H. (1974). The avian egg: water vapor conductance, shell thickness, and functional pore area. *The Condor*, **76**, 153–8.

Ar, A. and Rahn, H. (1980). Water in the avian egg: overall budget of incubation. *American Zoologist*, **20**, 373–84.

Ar, A. and Rahn, H. (1985). Pores in avian eggshells: Gas conductance, gas exchange and embryonic growth rate. *Respiration Physiology*, **61**, 1–20.

Ar, A., Rahn, H. and Paganelli, C.V. (1979). The avian egg: mass and strength. *The Condor*, **81**, 331–7.

Ar, A. and Sidis, Y. (2002). Nest microclimate during incubation. In DC Deeming, ed. *Avian incubation: behaviour, environment, and evolution*, pp. 143–60. Oxford University Press, Oxford.

Arcese, P., Smith, J.N.M. and Hatch, M.I. (1996). Nest predation by cowbirds and its consequences for passerine demography. *Proceedings of the National Academy of Sciences of the USA*, **93**, 4608–11.

Ardia, D.R. and Clotfelter, E.D. (2007). Individual quality and age affect responses to an energetic constraint in a cavity-nesting bird. *Behavioral Ecology*, **18**, 259–66.

Ardia, D.R., Cooper, C.B. and Dhondt, A.A. (2006a). Warm temperatures lead to early onset of incubation, shorter incubation periods and greater hatching asynchrony in tree swallows *Tachycineta bicolor* at the extremes of their range. *Journal of Avian Biology*, **37**, 137–42.

Ardia D.R., Pérez, J.H., Chad, E.K., Voss M.A. and Clotfelter, E.D. (2009). Temperature and life history: experimental heating leads female Tree Swallows to modulate egg temperature and incubation behaviour. *Journal of Animal Ecology*, **78**, 4–13.

Ardia, D.R., Pérez, J.H. and Clotfelter, E.D. (2006b). Nest box orientation affects internal temperature and nest site selection by Tree Swallows. *Journal of Field Ornithology*, **77**, 339–44.

Ardia, D.R., Pérez, J.H. and Clotfelter, E.D. (2010). Experimental cooling during incubation leads to reduced innate immunity and body condition in nestling Tree Swallows. *Proceedings of the Royal Society B: Biological Sciences*, **277**, 1881–8.

Ardia, D.R. and Schat, K.A. (2008). Ecoimmunology. In F Davison, B Kaspers, KA Schat, eds. *Avian immunology*, pp. 421–41. Elsevier, Amsterdam.

Arendt, W.J. (1985). *Philornis* ectoparasitism of pearly-eyed thrashers. II. Effects on adults and reproduction. *The Auk*, **102**, 281–92.

Arnold, J.M., Oswald, S.A., Voigt, C.C., Palme, R., Braasch, A., Bauch, C. and Becker, P.H. (2008). Taking the stress out of blood collection: comparison of field blood-sampling techniques for analysis of baseline corticosterone. *Journal of Avian Biology*, **39**, 588–92.

Arnold, J.M., Sabom, D., Nisbet, I.C.T. and Hatch, J.J. (2006). Use of temperature sensors to monitor patterns of nocturnal desertion by incubating Common Terns. *Journal of Field Ornithology*, **77**, 384–91.

Arnold, J.R. (1980). Distribution of the mockingbird in California. *Western Birds*, **11**, 97–102.

Arriero, E., Majewska, A. and Martin, T.E. (2013). Ontogeny of constitutive immunity: maternal vs. endogenous influences. *Functional Ecology*, **27**, 472–8.

Aschoff, J. and Pohl, H. (1970). Rhythmic variations in energy metabolism. *Federation Proceedings*, **29**, 1541–52.

Asghar, M., Hasselquist, D. and Bensch, S. (2011). Are chronic avian haemosporidian infections costly in wild birds? *Journal of Avian Biology*, **42**, 530–7.

Ashmole, N.P. (1961). *The biology of certain terns*. PhD dissertation, Oxford University, Oxford.

Ashmole, N.P. (1963). The regulation of numbers of tropical oceanic birds. *Ibis*, **103**, 458–73.

Atkinson, C.T., Woods, K.L., Dusek, R.J., Sileo, L.S. and Iko, W.M. (1995). Wildlife disease and conservation in Hawaii: Pathogenicity of avian malaria (*Plasmodium relictum*) in experimentally infected Iiwi (*Vestiaria coccinea*). *Parasitology*, **111**, S59–S69.

Aubrecht, G., Huber, W. and Weissenhofer, A. (2013). Coincidence or benefit? The use of *Marasmius* (horse-hair fungus) filaments in bird nests. *Avian Biology Research*, **6**, 26–30.

Augustine, D.J. and Skagen, S.K. (2014). Mountain plover nest survival in relation to prairie dog and fire dynamics in shortgrass steppe. *Journal of Wildlife Management*, **78**, 595–602.

Avilés, J.M., Pérez-Contreras, T., Navarro, C. and Soler, J.J. (2008). Dark nests and conspicuousness in color patterns of nestlings of altricial birds. *The American Naturalist*, **171**, 327–38.

Avilés, J.M., Pérez-Contreras, T., Navarro, C. and Soler, J.J. (2009). Male spotless starlings adjust feeding effort based on egg spots revealing ectoparasite load. *Animal Behaviour*, **78**, 993–9.

Avilés, J.M., Soler, J.J. and Pérez-Contreras, T. (2006). Dark nests and egg colour in birds: a possible functional role of ultraviolet reflectance in egg detectability. *Proceedings of the Royal Society of London Series B: Biological Sciences*, **273**, 2821–9.

Badyaev, A.V., Hill, G.E. and Beck, M.L. (2003). Interaction between maternal effects: onset of incubation and offspring sex in two populations of a passerine bird. *Oecologia*, **135**, 386–90.

Badyaev, A.V. and Uller, T. (2009). Parental effects in ecology and evolution: mechanisms, processes and implications. *Philosophical Transactions of the Royal Society B*, **364**, 1169–77.

Baggott, G.K., Deeming, D.C., Hémon, S. and Paillat, P. (2003). Relationships between eggshell pigmentation, ultrastructure and water vapour conductance in the Houbara bustard (*Chlamydotis undulata macqueeni*). In FT Scullion, TA Bailey, eds. *Proceedings of the World Association of Wildlife Veterinarians Wildlife Sessions at the 27th World Veterinary Congress, Tunisia, 26th September 2002*, pp. 85–8. World Association of Wildlife Veterinarians, Northern Ireland.

Baggott, G.K. and Graeme-Cook, K. (2002). Microbiology of natural incubation. In DC Deeming, ed. *Avian incubation: behaviour, environment, and evolution*, pp. 179–91. Oxford University Press, Oxford.

Baicich, P.J. (1986). Probing the unknown: incubation, nestling, and nestling period among North American birds. *Birding*, **18**, 254–9.

Baicich, P.J. and Harrison, C.J.O. (1997). A guide to the nests, eggs, and nestlings of North American birds, 2nd edition. Academic Press, San Diego.

Bailey, I.E., Morgan, K.V., Bertin, M., Meddle, S.L. and Healy, S.D. (2014). Physical cognition: birds learn the structural efficacy of nest material. *Proceedings of the Royal Society B*, **281**, e20133225.

Bailey, I.E., Muth, F., Morgan, K.V., Meddle, S.L. and Healy, S.D. (2015). Birds build camouflaged nests. *The Auk*, **132**, 11–15.

Bailey, R.L. and Clark, G.E. (2014). Occurrence of twin embryos in the Eastern Bluebird. *PeerJ*, **2**, e273.

Baillie, S.R. (1990). Integrated population monitoring of breeding birds in Britain and Ireland. *Ibis*, **132**, 151–66.

Baillie, S.R., Marchant, J.H., Leech, D.I., Massimino, D., Eglington, S.M., Johnston, A., Noble, D.G., Barimore, C., Kew, A.J., Downie, I.S., Risely, K. and Robinson, R.A. (2014). *BirdTrends 2013: trends in numbers, breeding success and survival for UK breeding birds*. BTO Research Report No. 652. British Trust for Ornithology, Thetford. http://www.bto.org/birdtrends.

Baillie, S.R. and Peach, W.J. (1992). Population limitation in Palaearctic-African migrant passerines. *Ibis*, **134**, (Supplement 1), 120–32.

Bain, M.M. (1991). A reinterpretation of eggshell strength. In SE Solomon, ed. *Egg and eggshell quality*, pp. 131–42. Wolfe Publishing Limited, Aylesbury.

Baird, T., Solomon, S.E. and Tedstone, D.R. (1975). Localisation and characterisation of egg shell porphyrins in several avian species. *British Poultry Science*, **16**, 201–8.

Baker, E.C.S. (1913). The evolution of adaptation in parasitic cuckoos' eggs. *Ibis*, **13**, 384–98.

Baker, J.R. (1967). A review of the role played by the Hippoboscidae (Diptera) as vectors of endoparasites. *Journal of Parasitology*, **53**, 412–8.

Bakken, G.S. (1976). A heat transfer analysis of animals: unifying concepts and the application of metabolism chamber data to field ecology. *Journal of Theoretical Biology*, **60**, 337–84.

Bakken, G.S. and Gates, D.M. (1975). Heat-transfer analysis of animals: some implications for field ecology, physiology, and evolution. In GS Bakken, DM Gates, eds. *Perspectives of biophysical ecology*, pp. 225–90, Springer Verlag, New York.

Bakken, G.S., Vanderbilt, V.C., Buttemer, W.A. and Dawson, W.R. (1978). Avian eggs: thermoregulatory value of very high near-infrared reflectance. *Science*, **200**, 321–3.

Balaban, E., Desco, M. and Vaquero, J.-J. (2012). Waking-like brain function in embryos. *Current Biology*, **22**, 852–61.

Balanoff, A.M. and Rowe, T. (2007). Osteological description of an embryonic skeleton of the extinct elephant bird, *Aepyornis* (Palaeognathae: Ratitae). *Journal of Vertebrate Paleontology*, **27**, (Supplement 4), 1–53.

Baldassarre, G. (2014). *Ducks, geese, and swans of North America, Volume 2*. Johns Hopkins University Press, Baltimore.

Baldwin, S.P. and Kendeigh, S.C. (1927). Attentiveness and inattentiveness in the nesting behavior of the House Wren. *The Auk*, **44**, 206–16.

Baldwin, S.P. and Kendeigh, S.C. (1932). Physiology of the temperature of birds. *Scientific Publications of the Cleveland Museum of Natural History*, **3**, 1–196.

Balmer, D., Coiffait, L., Clark, J. and Robinson, R. (2008). *Bird ringing: A concise guide*. British Trust for Ornithology, Thetford.

Balmer, D., Gillings, S., Caffrey, B., Swann, R., Downie, I. and Fuller, R. (2013). *Bird Atlas 2007–11*. British Trust for Ornithology, Thetford.

Balthazart, J. and Adkins-Regan, E. (2002). Sexual differentiation of brain and behavior in birds. In DW Pfaff, AP Arnold, AM Etgen, SE Fahrbach, RT Rubin, eds. *Hormones, brain and behavior*, pp. 223–301. Academic Press, Amsterdam.

Balthazart, J. and Taziaux, M. (2009). The underestimated role of olfaction in avian reproduction? *Behavioural Brain Research*, **200**, 248–59.

Bamford, A.J., Seing Sam, T., Razafindrajao, F., Robson, H., Woolaver, L.G. and René de Roland, A.L. (2015). The status and ecology of the last wild population of Madagascar Pochard *Aythya innotata*. *Bird Conservation International*, **25**, 1–14.

Barbosa, A., Merino, S., de Lope, F. and Møller, A.P. (2002). Effects of feather lice on flight behavior of male barn swallows (*Hirundo rustica*). *The Auk* **119**, 213–6.

Bargar, T.A., Scott, G.I. and Cobb, G.P. (2001). Maternal transfer of contaminants: Case study of the excretion of three polychlorinated biphenyl congeners and technical-grade endosulfan into eggs by white leghorn chickens (*Gallus domesticus*). *Environmental Toxicology and Chemistry*, **20**, 61–7.

Barnes, D.M. (2001). Expression of P-glycoprotein in the chicken. *Comparative Biochemistry and Physiology A: Molecular and Integrative Physiology*, **130**, 301–10.

Barnett, C.A. and Briskie, J.V. (2010). Silvereyes *Zosterops lateralis* increase incubation attentiveness in response to increased food availability. *Ibis*, **152**, 169–72.

Barnett, C.A., Clairardin, S.G., Thompson, C.F. and Sakaluk, S.K. (2011). Turning a deaf ear: a test of the manipulating androgens hypothesis in house wrens. *Animal Behaviour*, **81**, 113–20.

Baron, F., Jan, S., Gonnet, F., Pasco, M., Jardin, J., Giudici, B., Gautier, M., Guérin-Dubiard, C. and Nau, F. (2014). Ovotransferrin plays a major role in the strong bactericidal effect of egg white against the *Bacillus cereus* group. *Journal of Food Protection*, **77**, 955–62.

Barrow, P.A. and Lovell, M.A. (1991). Experimental infection of egg-laying hens with *Salmonella enteritidis* phage type 4. *Avian Pathology*, **20**, 335–48.

Bar-Shira, E., Cohen, I., Elad, O. and Friedman, A. (2014). Role of goblet cells and mucin layer in protecting maternal IgA in precocious birds. *Developmental and Comparative Immunology*, **44**, 186–94.

Bart, J. (1977). Impact of human visitations on avian nesting success. *Living Bird*, **16**, 187–92.

Barta, Z. and Székely, T. (1997). The optimal shape of avian eggs. *Functional Ecology*, **11**, 656–62.

Bartholomew, G.A. and Howell, T.R. (1964). Experiments on nesting behaviour of Laysan and black-footed albatrosses. *Animal Behaviour*, **12**, 549–59.

Batty, M., Jarrett, N.S., Forbes, N., Brown, M.J., Standley, S., Richardson, T., Oliver, S., Ireland, B., Chalmers, K.P. and Fraser, I. (2006). Hand-rearing greater flamingos *Phoenicopterus ruber roseus* for translocation from WWT Slimbridge to Auckland Zoo. *International Zoo Yearbook*, **40**, 261–70.

Bayard, T.S. and Elphick, C.S. (2011). Planning for sea level rise: quantifying patterns of Saltmarsh Sparrow (*Ammodramus caudacutus*) nest flooding under current sea level conditions. *The Auk*, **128**, 393–403.

Beaulieu, M., Thierry, A.-M., Handrich, Y., Massemin, S., Le Maho, Y. and Ancel, A. (2010). Adverse effects of instrumentation in incubating Adélie penguins. *Polar Biology*, **33**, 485–92.

Beckett, M., Baggott, G.K. and Graeme-Cook, K. (2003). Bacterial degradation of eggshell cuticle of the Mandarin Duck (*Aix galericulata*). *Avian and Poultry Biology Reviews*, **14**, 196–7.

Bedrani, L., Helloin, E., Guyot, N. and Nys, Y. (2013b). Systemic administration of lipopolysaccharide in laying hens stimulates antimicrobial properties of egg white against *Staphylococcus aureus*. *Veterinary Immunology and Immunopathology*, **152**, 225–36.

Bedrani, L., Helloin, E., Guyot, N., Réhault-Godbert, S. and Nys, Y. (2013a). Passive maternal exposure to environmental microbes selectively modulates the innate defences of chicken egg white by increasing some of its antibacterial activities. *BMC Microbiology*, **13**, 128.

Behl, C., Widmann, M., Trapp, T. and Holsboer, F. (1995). 17-β estradiol protects neurons from oxidative stress-induced cell death *in vitro*. *Biochemical and Biophysical Research Communication*, **216**, 473–82.

Beissinger, S.R., Cook, M.I. and Arendt, W.J. (2005). The shelf life of bird eggs: testing egg viability using a tropical climate gradient. *Ecology*, **86**, 2164–75.

Bennet, R. (1986). Around the trust—special projects. *Wildfowl World, The Wildfowl Trust Magazine*, **95**, 12–13.

Bennett, A.F. and Dawson, W.R. (1979). Physiological responses of embryonic Heermann's Gulls to temperature. *Physiological Zoology*, **52**, 413–21.

Bennett, G.F. (1960). On some ornithophilic blood-sucking diptera in Algonquin Park, Ontario, Canada. *Canadian Journal of Zoology*, **38**, 377–89.

Bennett, G.F. and Whitworth, T.L. (1991). Studies on the life histories of some species of *Protocalliphora* (Diptera: Calliphoridae). *Canadian Journal of Zoology*, **69**, 2048–58.

Bennett, P.M. and Owens, I.P.F. (2002). *Evolutionary ecology of birds: life histories, mating systems and extinction*. Oxford University Press, Oxford.

Bentz, A.B., Navara, K.J. and Siefferman, L. (2013). Phenotypic plasticity in response to breeding density in Tree Swallows: An adaptive maternal effect? *Hormones and Behavior*, **64**, 729–36.

Bentzen, R.L., Powell, A.N., Phillips, L.M. and Suydam, R.S. (2010). Incubation behavior of King Eiders on the coastal plain of northern Alaska. *Polar Biology*, **33**, 1075–82.

Berg, M.L., Beintema, N.H., Welbergen, J.A. and Komdeur, J. (2006). The functional significance of multiple nest-building in the Australian Reed Warbler (*Acrocephalus australis*). *Ibis*, **148**, 395–404.

Berghof, T.V.L., Parmentier, H.K. and Lammers, A. (2013). Transgenerational epigenetic effects on innate immunity in broilers: An underestimated field to be explored? *Poultry Science*, **92**, 2904–13.

Berlin, K.E. and Clark, A.B. (1998). Embryonic calls as care-soliciting signals in budgerigars, *Melopsittacus undulatus*. *Ethology*, **104**, 531–44.

Bernardo, J. (1996). The particular maternal effect of propagule size, especially egg size: patterns, models, quality of evidence and interpretations. *American Zoologist*, **36**, 216–36.

Bertin, A., Richard-Yri, M.-A., Möst, E. and Lickliter, R. (2009). Increased yolk testosterone facilitates prenatal perceptual learning in northern bobwhite quail (*Colinus virginianus*). *Hormones and Behavior*, **56**, 416–22.

Bertram, B.C.R. (1979). Ostriches recognise their own eggs and discard others. *Nature*, **279**, 233–4.

Bertram, B.C.R. and Burger, A.E. (1981). Are ostrich, *Struthio camelus*, eggs the wrong colour? *Ibis*, **123**, 207–10.

Bêty, J. and Gauthier, G. (2001). Effects of nest visits on predator activity and predation rate in a Greater Snow Goose colony. *Journal of Field Ornithology*, **72**, 573–86.

Biancucci, L. and Martin, T.E. (2010). Can selection on nest size from nest predation explain the latitudinal gradient in clutch size? *Journal of Animal Ecology*, **79**, 1086–92.

Biddle, L., Deeming, D.C. and Goodman, A. (2015). Morphology and biomechanics of the nests of the Common Blackbird *Turdus merula*. *Bird Study*, **65**, 87–95.

Biebach, H. (1981). Energetic costs of incubation on different clutch sizes in starlings (*Sturnus vulgaris*). *Ardea*, **69**, 141–2.

Biebach, H. (1984). Effect of clutch size and time of day on the energy expenditure of incubating starlings (*Sturnus vulgaris*). *Physiological Zoology*, **57**, 26–31.

Biebach, H. (1986). Energetics of rewarming a clutch in starlings (*Sturnus vulgaris*). *Physiological Zoology*, **59**, 69–75.

Bignert, A., Litzen, K., Odsjo, T., Olsson, M., Persson, W. and Reutergardh, L. (2005). Time-related factors influence the concentration of sDDT, PCBs and shell parameters in eggs of Baltic Guillemot (*Uria aalge*), 1861–1989. *Environmental Pollution*, **89**, 27–36.

Bijlsma, R. (1984). On the breeding association between woodpigeons *Columba palumbus* and hobbies *Falco subbuteo*. *Limosa*, **57**, 133–9.

Bioacoustics Research Program (2014). http://www.birds.cornell.edu/brp/raven/RavenOverview.html. [Accessed: 1 August 2014]

Birchard, G.F. and Deeming, D.C. (2009). Scaling of avian eggshell thickness: implications for maximum body mass in birds. *Journal of Zoology*, **279**, 95–101.

Birchard, G.F., Kilgore, D.L. Jr and Boggs, D.F. (1984). Respiratory gas concentrations and temperatures within the burrows of three species of burrow-nesting birds. *The Wilson Bulletin*, **96**, 451–6.

Birchard, G.F., Ruta, M. and Deeming, D.C. (2013). Evolution of parental incubation behaviour in dinosaurs cannot be inferred from clutch mass in birds. *Biology Letters*, **9**, 20130036.

Bird, T.J., Bates, A.E., Lefcheck, J.S., Hill, N.A., Thomson, R.J., Edgar, G.J., Stuart-Smith, R.D., Wotherspoon, S., Krkosek, M., Stuart-Smith, J.F., Pecl, G.T., Barrett, N. and Frusher, S. (2014). Statistical solutions for error and bias in global citizen science datasets. *Biological Conservation*, **173**, 144–54.

BirdLife International. (2014). www.birdlife.org/datazone/species/factsheet/22705629. [Accessed: 29 July 2014]

Birkhead, T. (2008). *The wisdom of birds*. Bloomsbury, London.

Birkhead, T., Wimpenny, J. and Montgomerie, B. (2014). *Ten thousand birds: Ornithology since Darwin*. Princeton University Press, Princeton.

Birkhead, T.R. (2014). Stormy outlook for long-term ecology studies. *Nature*, **514**, 405.

Bize, P., Devevey, G., Monaghan, P., Doligez, B. and Christe, P. (2008). Fecundity and survival in relation to resistance to oxidative stress in a free-living bird. *Ecology*, **89**, 2584–93.

Bize, P., Roulin, A., Tella, J.L., Bersier, L.F. and Richner, H. (2004). Additive effects of ectoparasites over reproductive attempts in the long-lived alpine swift. *Journal of Animal Ecology*, **73**, 1080–8.

Black, J.L. and Burggren, W.W. (2004a). Acclimation to hypothermic incubation in developing chicken embryos (*Gallus domesticus*) I. Developmental effects and chronic and acute metabolic adjustments. *The Journal of Experimental Biology*, **207**, 1543–52.

Black, J.L. and Burggren, W.W. (2004b). Acclimation to hypothermic incubation in developing chicken embryos (*Gallus domesticus*) II. Hematology and blood O_2 transport. *The Journal of Experimental Biology*, **207**, 1553–61.

Blackmer, A.L., Ackerman, J.T. and Nevitt, G.A. (2004). Effects of investigator disturbance on hatching success and nest-site fidelity in a long-lived seabird, Leach's Storm-petrel. *Biological Conservation*, **116**, 141–8.

Blanco, G. and Bertellotti, M. (2002). Differential predation by mammals and birds: implications for egg-colour polymorphism in a nomadic breeding seabird. *Biological Journal of The Linnean Society*, **75**, 137–46.

Blanco, G., Martínez-Padilla, J., Serrano, D., Dávila, J.A. and Viñuela, J. (2003). Mass provisioning to different-sex eggs within the laying sequence: consequences for adjustment of reproductive effort in a sexually dimorphic bird. *Journal of Animal Ecology*, **72**, 831–8.

Blanco, G., Seoane, J. and Martínez-de la Puente, J. (1999). Showiness, non-parasitic symbionts, and nutritional condition in a passerine bird. *Annales Zoologici Fennici*, **36**, 83–91.

Blanco, G., Tella, J.L. and Potti, J. (1997). Feather mites on group living Red-billed Choughs: a non-parasitic interaction? *Journal of Avian Biology*, **28**, 197–206.

Blanco, G., Tella, J.L., Potti, J. and Baz, A. (2001). Feather mites on birds: costs of parasitism or conditional outcomes? *Journal of Avian Biology*, **32**, 271–4.

Bled, F., Royle, J.A. and Cam, E. (2011). Assessing hypotheses about nesting site occupancy dynamics. *Ecology*, **92**, 938–51.

Blem, C.R. and Blem, L.B. (1993). Do swallows sunbathe to control ectoparasites? An experimental test. *The Condor*, **95**, 728–30.

Blem, C.R. and Blem, L.B. (1994). Composition and microclimate of Prothonotary Warbler nests. *The Auk*, **111**, 197–200.

Blount, J.D., Houston, D.C. and Møller, A.P. (2000). Why egg yolk is yellow. *Trends in Ecology and Evolution*, **15**, 47–9.

Blount, J.D., Houston, D.C., Surai, P.F. and Møller, A.P. (2004). Egg-laying capacity is limited by carotenoid pigment availability in wild gulls *Larus fuscus*. *Proceedings of the Royal Society of London B: Biological Sciences*, **271**, S79–81.

Blount, J.D., Surai, P.F., Nager, R.G., Houston, D.C., Møller, A.P., Trewby, M.L. and Kennedy, M.W. (2002). Carotenoids

and egg quality in the lesser black-backed gull *Larus fuscus*: a supplemental feeding study of maternal effects. *Proceedings of the Royal Society of London B: Biological Sciences*, **269**, 29–36.

Blueweiss, L., Fox, H., Kudzma, V., Nakashima, D., Peters, R.H. and Sams, S. (1978). Relationships between body size and some life history parameters. *Oecologia*, **37**, 257–72.

Board, R.G. (1982). Properties of avian egg shells and their adaptive value. *Biological Reviews of the Cambridge Philosophical Society*, **57**, 1–28.

Board, R.G. and Fuller, R. (1974). Non-specific antimicrobial defences of the avian egg, embryo and neonate. *Biological Reviews of the Cambridge Philosophical Society*, **49**, 15–49.

Board, R.G. and Halls, N.A. (1973). The cuticle: A barrier to liquid and particle penetration of the shell of the hen's egg. *British Poultry Science*, **14**, 69–97.

Board, R.G., Loseby, S. and Miles, V.R. (1979). A note on microbial growth on hen egg-shells. *British Poultry Science*, **20**, 413–20.

Board, R.G. and Sparks, N.H.C. (1991). Shell structure and formation in avian eggs. In DC Deeming and MWJ Ferguson, eds. *Egg incubation: its effects on embryonic development in birds and reptiles*, pp. 71–86. Cambridge University Press, Cambridge.

Bohm, C. and Landmann, A. (1995). Nest-site selection and nest construction in the water pipit (*Anthus spinoletta*). *Journal of Ornithology*, **136**, 1–16.

Bolhuis, J.J. and van Kampen, H.S. (1992). An evaluation of auditory learning in filial imprinting. *Behaviour*, **122**, 195–230.

Boncoraglio, G., Rubolini, D., Romano, M., Martinelli, R. and Saino, N. (2006). Effects of elevated yolk androgens on perinatal begging behavior in yellow-legged gull (*Larus michahellis*) chicks. *Hormones and Behavior*, **50**, 442–7.

Bonisoli-Alquati, A., Rubolini, D., Romano, M., Cucco, M., Fasola, M., Caprioli, M. and Saino, N. (2010). Egg antimicrobials, embryo sex and chick phenotype in the yellow-legged gull. *Behavioral Ecology and Sociobiology*, **64**, 845–55.

Bonney, R., Cooper, C.B., Dickinson, J., Kelling, S., Phillips, T., Rosenberg, K.V. and Shirk, J. (2009). Citizen science: A developing tool for expanding science knowledge and scientific literacy. *BioScience*, **59**, 977–84.

Bonter, D.N. and Cooper, C.B. (2012). Data validation in citizen science: a case study from Project FeederWatch. *Frontiers in Ecology and the Environment*, **10**, 305–7.

Bonter, D.N., MacLean, S.A., Shah, S.S. and Moglia, M.C. (2014). Storm-induced shifts in optimal nesting sites: a potential effect of climate change. *Journal of Ornithology*, **155**, 631–8.

Book, C.M., Millam, J.R., Guinan, M.J. and Kitchell, R.L. (1991). Brood patch innervation and its role in the onset of incubation in the turkey hen. *Physiology and Behavior*, **50**, 281–5.

Boone, R.B. and Mesecar, R.S. III. (1989). Telemetric egg for use in egg-turning studies. *Journal of Field Ornithology*, **60**, 315–22.

Boonstra, R., Eadie, J.M., Krebs, C.J. and Boutin, S. (1995). Limitations of far infrared thermal imaging in locating birds. *Journal of Field Ornithology*, **66**, 192–8.

Boos, M., Zimmer, C., Carriere, A., Robin, J.P. and Petit, O. (2007). Post-hatching parental care behaviour and hormonal status in a precocial bird. *Behavioural Processes*, **76**, 206–14.

Booth, D.T. (1987). Effect of temperature on development of mallee fowl *Leipoa ocellata* eggs. *Physiological Zoology*, **60**, 437–45.

Booth, D.T. (2006). Influence of incubation temperature on hatchling phenotype in reptiles. *Physiological and Biochemical Zoology*, **79**, 274–81.

Booth, D.T., Clayton, D.H. and Block, B.A. (1993). Experimental demonstration of the energetic cost of parasitism in free-ranging hosts. *Proceedings of the Royal Society of London B: Biological Sciences*, **253**, 125–9.

Booth, D.T. and Jones, D.N. (2002). Underground nesting in the megapodes. In DC Deeming, ed. *Avian incubation: behaviour, environment, and evolution*, pp. 192–206. Oxford University Press, New York.

Borchelt, P.L. and Duncan, L. (1974). Dustbathing and feather lipid in bobwhite (*Colinus virginianus*). *The Condor*, **76**, 471–2.

Borgmann, K.L., Conway, C. and Morrison, M.L. (2013). Breeding phenology of birds: mechanisms underlying seasonal declines in the risk of nest predation. *PLoS ONE*, **8**, e65909.

Borgmann, K.L. and Rodewald, A.D. (2004). Nest predation in an urbanizing landscape: the role of exotic shrubs. *Ecological Applications*, **14**, 1757–65.

Borgo, J.S., Conner, L.M. and Conover, M.R. (2006). Role of predator odor in roost site selection of southern flying squirrels. *Wildlife Society Bulletin*, **34**, 144–9.

Bortolotti, G.R., Negro, J.J., Surai, P.F. and Prieto, P. (2003). Carotenoids in eggs and plasma of red-legged partridges: Effects of diet and reproductive output. *Physiological and Biochemical Zoology*, **76**, 367–74.

Both, C., Bouwhuis, S., Lessells, C.M. and Visser, M.E. (2006). Climate change and population declines in a long-distance migratory bird. *Nature*, **441**, 81–3.

Both, C. and Visser, M.E. (2005). The effect of climate change on the correlation between avian life-history traits. *Global Change Biology*, **11**, 1606–13.

Bottitta, G.E., Gilchrist, H.G., Kift, A. and Meredith, M.G. (2002). A pressure-sensitive wireless device for continuously monitoring avian nest attendance. *Wildlife Society Bulletin*, **30**, 1033–8.

Boulinier, T. and Staszewski, V. (2008). Maternal transfer of antibodies: raising immuno-ecology issues. *Trends in Ecology and Evolution*, **23**, 282–8.

Boulton, R.L. and Cassey, P. (2012). How avian incubation behaviour influences egg surface temperatures: relationships with egg position, development and clutch size. *Journal of Avian Biology*, **43**, 289–96.

Boulton, R.L., Cassey, P., Schipper, C. and Clarke, M.F. (2003). Nest site selection by yellow-faced honeyeaters *Lichenostomus chrysops*. *Journal of Avian Biology*, **34**, 267–74.

Bourin, M., Gautron, J., Berges, M., Attucci, S., Le Blay, G., Labas, V., Nys, Y. and Réhault-Godbert, S. (2011). Antimicrobial potential of egg yolk ovoinhibitor, a multidomain Kazal-like inhibitor of chicken egg. *Journal of Agricultural and Food Chemistry*, **59**, 12368–74.

Bowey, K. (1999). Sedge warblers nesting in rape crops. *British Birds*, **92**, 365–72.

Bradley, N.S. and Bekoff, A. (1990). Development of coordinated movement in chicks: I. Temporal analysis of hindlimb muscle synergies at embryonic days 9 and 10. *Developmental Psychobiology*, **23**, 763–82.

Breil, D.A. and Moyle, S.M. (1976). Bryophytes used in construction of bird nests. *The Bryologist*, **79**, 95–8.

Breuner, C.W. (2011). Stress and reproduction in birds. In DO Norris, KH Lopez, eds. *Hormones and reproduction of vertebrates—Birds*, pp. 129–51. Elsevier, Amsterdam.

Brewer, T.M. (1878). Variations in the nests of the same species of birds. *The American Naturalist*, **12**, 35–40.

Brickle, N.W. and Harper, D.G. (2002). Agricultural intensification and the timing of breeding of Corn Buntings *Miliaria calandra*. *Bird Study*, **49**, 219–28.

Briskie, J.V. (1995). Nesting biology of the Yellow Warbler at the northern limit of its range. *Journal of Field Ornithology*, **66**, 531–43.

Britt, J. and Deeming, D.C. (2011). First egg date and air temperature affect nest construction in Blue Tits *Cyanistes caeruleus* but not in Great Tits *Parus major*. *Bird Study*, **58**, 78–89.

Broggi, J. and Senar, J.C. (2009). Brighter Great Tit parents build bigger nests. *Ibis*, **151**, 588–91.

Brooke, M. de L. (1985). The effect of allopreening on tick burdens of molting eudyptid penguins. *The Auk*, **102**, 893–5.

Brooke, M.de L. and Davies, N.B. (1987). Recent changes in host usage by Cuckoos *Cuculus canorus* in Britain. *Journal of Animal Ecology*, **56**, 873–83.

Brooks, E.W. (1993). Use of conifers as nest sites by Yellow Warbler in New York. *The Kingbird*, **43**, 26.

Brouwer, L. and Komdeur, J. (2004). Green nesting material has a function in mate attraction in the European starling. *Animal Behaviour*, **67**, 539–48.

Brown, C.R. and Brown, M.B. (1986). Ectoparasitism as a cost of coloniality in cliff swallows (*Hirundo pyrrhonota*). *Ecology*, **67**, 1206–18.

Brown, C.R. and Brown, M.B. (2004). Empirical measurement of parasite transmission between groups in a colonial bird. *Ecology*, **85**, 1619–26.

Brown, C.R., Brown, M.B. and Rannala, B. (1995). Ectoparasites reduce long-term survival of their avian host. *Proceedings of the Royal Society of London B: Biological Sciences*, **262**, 313–9.

Brown, G.P. and Shine, R. (2004). Maternal nest-site choice and offspring fitness in a tropical snake (*Tropidonophis mairii*, Colubridae). *Ecology*, **85**, 1627–34.

Brown, J.H., Marquet, P.A. and Taper, M.L. (1993). Evolution of body size: consequences of an energetic definition of fitness. *The American Naturalist*, **142**, 573–84.

Brown, M.B. and Brown, C.R. (2009). Blood sampling reduces annual survival in cliff swallows (*Pterochelidon pyrrhonota*). *The Auk*, **126**, 853–61.

Brua, R.B. (2002). Parent-offspring interactions. In DC Deeming, ed. *Avian incubation: behaviour, environment, and evolution*. pp. 88–99, Oxford University Press, Oxford.

Brua, R.B., Nuechterlein, G.L. and Buitron, D. (1996). Vocal response of eared grebe embryos to egg cooling and egg turning. *The Auk*, **113**, 525–33.

Brulez, K., Cassey, P., Meeson, A., Mikšík, I., Webber, S.L., Gosler, A.G. and Reynolds, S.J. (2014a). Eggshell spot scoring methods cannot be used as a reliable proxy to determine pigment quantity. *Journal of Avian Biology*, **45**, 94–102.

Brulez, K., Choudhary, P.K., Maurer, G., Portugal, S.J., Boulton, R.L., Webber, S.L. and Cassey, P. (2014b). Visual scoring of eggshell patterns has poor repeatability. *Journal of Ornithology*, **155**, 701–6.

Bryan, S.M. and Bryant, D.M. (1999). Heating nest-boxes reveals an energetic constraint on incubation behaviour in Great Tits, *Parus major*. *Proceedings of the Royal Society of London Series B: Biological Sciences*, **266**, 157–62.

Bryant, D.M., Hails, C.J. and Tatner, P. (1984). Reproductive energetics of two tropical bird species. *The Auk*, **101**, 25–37.

Bub, H. (1996). *Bird trapping and bird banding: A handbook for trapping methods all over the world*. Cornell University Press, Ithaca.

Buckley, P.A. and Buckley, F.G. (1972). Individual egg and chick recognition by adult royal terns (*Sterna maxima maxima*). *Animal Behaviour*, **20**, 457–62.

Buechler, K., Fitze, P.S., Gottstein, B., Jacot, A. and Richner, H. (2002). Parasite-induced maternal response in a natural bird population. *Journal of Animal Ecology*, **71**, 247–52.

Bugden, S.C. and Evans, R.M. (1999). The development of a vocal thermoregulatory response to temperature in embryos of the domestic chicken. *The Wilson Bulletin*, **111**, 188–94.

Bukaciński, D. and Bukacińska, B. (2000). The impact of mass outbreaks of black flies (Simuliidae) in the parental behaviour and breeding output of colonial common gulls (*Larus canus*). *Annales Zoologici Fennici*, **37**, 43–9.

Bulla, M., Šálek, M. and Gosler, A.G. (2012). Eggshell spotting does not predict male incubation but marks thinner areas of a shorebird's shells. *The Auk*, **129**, 26–35.

Bulla, M., Valcu, M., Rutten, A.L. and Kempenaers, B. (2014). Biparental incubation patterns in a high-Arctic breeding shorebird: how do pairs divide their duties? *Behavioral Ecology*, **25**, 152–64.

Buntin, J.D. (1996). Neural and hormonal control of parental behavior in birds. In JS Rosenblatt, CT Snowdon, eds. *Advances in the study of behavior*, Vol. 25. pp. 161–213. Academic Press, New York.

Buntin, J.D., El Halawani, M.E., Ottinger, M.A., Fan, Y. and Fivizzani, A.J. (1998). An analysis of sex and breeding stage differences in prolactin binding activity in brain and hypothalamic GnRH concentration in Wilson's phalarope, a sex role-reversed species. *General and Comparative Endocrinology*, **109**, 119–32.

Burgess, S.C. and Marshall, D.J. (2014). Adaptive parental effects: the importance of estimating environmental predictability and offspring fitness appropriately. *Oikos*, **123**, 769–76.

Burhans, D.E., Dearborn, D., Thompson, F.R. and Faaborg, J. (2002). Factors affecting predation at songbird nests in old fields. *Journal of Wildlife Management*, **66**, 240–9.

Burley, R.W. and Vadhera, D.V. (1989). *The avian egg*. John Wiley, New York.

Burnham, K.P. and Anderson, D.R. (2004). *Model selection and multimodel inference: a practical information-theoretic approach*, 2nd edition. Springer, New York.

Burns, F., McCulloch, N., Székely, T. and Bolton, M. (2013). No overall benefit of predator exclusion cages for the endangered St. Helena Plover *Charadrius sanctaehelenae*. *Ibis*, **155**, 397–401.

Burnside, R.J., Carter, I., Dawes, A., Waters, D., Lock, L., Goriup, P. and Székely, T. (2012). The UK great bustard *Otis tarda* reintroduction trial: a 5-year progress report. *Oryx*, **46**, 112–21.

Burton, N.H.K. (2006). Nest orientation and hatching success in the Tree Pipit *Anthus trivialis*. *Journal of Avian Biology*, **37**, 312–7.

Burton, N.H.K. (2007). Intraspecific latitudinal variation in nest orientation among ground-nesting passerines: A study using published data. *The Condor*, **109**, 441–6.

Burtt, E.H. (1986). An analysis of physical, physiological, and optical aspects of avian coloration with emphasis on wood-warblers. *Ornithological Monographs*, **38**, 1–125.

Bush, S.E., Villa, S.M., Boves, T.H., Brewer, D. and Belthoff, J.R. (2012). Influence of bill and foot morphology on the ectoparasites of barn owls. *Journal of Parasitology*, **98**, 256–61.

Butterfield, P.A. and Crook, J.H. (1968). Annual cycle of nest building and agonistic behaviour in captive *Quelea quelea* with reference to endocrine factors. *Animal Behaviour*, **16**, 308–17.

Cabana, G., Frewin, A., Peters, R.H. and Randall, L. (1982). The effect of sexual size dimorphism on variations in reproductive effort of birds and mammals. *The American Naturalist*, **120**, 17–25.

Calder, W.A. III (1979). The kiwi and egg design: evolution as a package deal. *BioScience*, **29**, 461–7.

Calder, W.A. (1984). *Size, function, and life history*. Harvard University Press, Cambridge.

Caldwell, P.J. and Cornwell, G.W. (1975). Incubation behavior and temperatures of the Mallard Duck. *The Auk*, **92**, 706–31.

Calvelo, S., Trejo, A. and Ojeda, V. (2006). Botanical composition and structure of hummingbird nests in different habitats from northwestern Patagonia (Argentina). *Journal of Natural History*, **40**, 589–603.

Cam, E., Hines, J.E., Monnat, J.-Y., Nichols, J.D. and Danchin, E. (1998). Are adult nonbreeders prudent parents? The Kittiwake model. *Ecology*, **79**, 1917–30.

Campagna, S., Mardon, J., Celerier, A. and Bonadonna, F. (2012). Potential semiochemical molecules from birds: A practical and comprehensive compilation of the last 20 years' studies. *Chemical Senses*, **37**, 3–25.

Cancellieri, S. and Murphy, M.T. (2013). Experimental examination of nest reuse by an open-cup-nesting passerine: time/energy savings or nest site shortage? *Animal Behaviour*, **85**, 1287–94.

Canestrari, D., Bolopo, D., Turlings, T.C., Röder, G., Marcos, J.M. and Baglione, V. (2014). From parasitism to mutualism: unexpected interactions between a cuckoo and its host. *Science*, **343**, 1350–2.

Cantarero, A., López-Arrabé, J., Redondo, A.J. and Moreno, J. (2013). Behavioural responses to ectoparasites in pied flycatchers *Ficedula hypoleuca*: an experimental study. *Journal of Avian Biology*, **44**, 1–9.

Carere, C. and Balthazart, J. (2007). Sexual versus individual differentiation: the controversial role of avian maternal hormones. *Trends in Endocrinology and Metabolism*, **18**, 73–80.

Carey, C. (1996). Female reproductive energetics. In C Carey, ed. *Avian energetics and nutritional ecology*, pp. 325–74. Chapman and Hall, New York.

Carey, C., Rahn, H. and Parisi, P. (1980). Calories, water, lipid and yolk in avian eggs. *The Condor*, **82**, 335–43.

Cariello, M.O., Lima, M.R., Schwabl, H.G. and Macedo, R.H. (2004). Egg characteristics are unreliable in determining maternity in communal clutches of guira cuckoos *Guira guira*. *Journal of Avian Biology*, **35**, 117–24.

Caro, T.M. (2005). *Antipredator defences in birds and mammals*. Chicago University Press, Chicago.

Carpenter, K., Hirsch, K.F. and Horner, J.R. (1994). *Dinosaur eggs and babies*. Cambridge University Press, Cambridge.

Carter, A.W., Hopkins, W.A., Moore, I.T. and DuRant, S.E. (2014). Influence of incubation recess patterns on incubation period and hatchling traits in wood ducks *Aix sponsa*. *Journal of Avian Biology*, **45**, 273–9.

Casini, S., Fossi, M.C., Leonzio, C. and Renzoni, A. (2003). Porphyrins as biomarkers for hazard assessment of bird populations: destructive and non-destructive use. *Ecotoxicology*, **12**, 297–305.

Cassey, P. (2009). Biological optics: seeing colours in the dark. *Current Biology*, **19**, R1083–4.

Cassey, P., Ewen, J., Blackburn, T., Hauber, M., Vorobyev, M. and Marshall, N.J. (2008). Eggshell colour does not predict measures of maternal investment in eggs of *Turdus* thrushes. *Naturwissenschaften*, **95**, 713–21.

Cassey, P., Ewen, J.G., Marshall, N.J., Vorobyev, M., Blackburn, T.M. and Hauber, M.E. (2009). Are avian eggshell colours effective intraspecific communication signals in the Muscicapoidea? A perceptual modelling approach. *Ibis*, **151**, 689–98.

Cassey, P., Thomas, G.H., Portugal, S.J., Mauer, G., Hauber, M.E., Grim, T., Lovell, P.G. and Mikšík, I. (2012). Why are birds' eggs colourful? Eggshell pigments co-vary with life-history and nesting ecology among British breeding non-passerine birds. *Biological Journal of the Linnean Society*, **106**, 657–72.

Cave, V.M., King, R. and Freeman, S.N. (2010). An integrated population model from constant effort bird-ringing data. *Journal of Agricultural, Biological, and Environmental Statistics*, **15**, 119–37.

Cavitt, J.F., Pearse, A.T. and Miller, T.A. (1999). Brown Thrasher nest reuse: A time saving resource, protection from search-strategy predators, or cues for nest-site selection? *The Condor*, **101**, 859–62.

Cellular Tracking Technologies (2014). http://celltracktech. com/index.php/products/wildlife-telemetry-solutions/ ctt-1000-series/. [Accessed: 28 July 2014]

Cervencl, A., Esser, W., Maier, M., Oderdiek, N., Thyen, S., Wellbrock, A. and Exo, K.-M. (2011). Can differences in incubation patterns of Common Redshanks *Tringa totanus* be explained by variations in predation risk? *Journal of Ornithology*, **152**, 1033–43.

Chalfoun, A.D. and Martin, T.E. (2009). Habitat structure mediates predation risk for sedentary prey: experimental tests of alternative hypotheses. *Journal of Animal Ecology*, **78**, 497–503.

Chalfoun, A.D. and Martin, T.E. (2010a). Parental investment decisions in response to ambient nest-predation risk versus actual predation on the prior nest. *The Condor*, **112**, 701–10.

Chalfoun, A.D. and Martin, T.E. (2010b). Facultative nest patch shifts in response to nest predation risk in the Brewer's sparrow: a 'win-stay, lose-switch' strategy? *Oecologia*, **163**, 885–92.

Chalfoun, A.D. and Schmidt, K.A. (2012). Adaptive breeding-habitat selection: is it for the birds? *The Auk*, **129**, 589–99.

Chamberlain, D.E. and Crick, H.Q.P. (1999). Population declines and reproductive performance of Skylarks *Alauda arvensis* in different regions and habitats of the United Kingdom. *Ibis*, **141**, 38–51.

Chandler, R.M. and Wall, W.P. (2001). The first record of bird eggs from the early Oligocene of North America. National Park Service/Natural Resources Geologic Resources Division/Geologic Resources Division Technical Report 01/01. http://www.nature.nps.gov/geology/ paleontology/Publications/research_volumes/fossil_ conf6_2001.pdf .

Chaplin, S.B., Cervenka, M.L. and Mickleson, A.C. (2002). Thermal environment of the nest during development of Tree Swallow (*Tachycineta bicolor*) chicks. *The Auk*, **119**, 845–51.

Chase, I.D. (1980). Cooperative and noncooperative behavior in animals. *The American Naturalist*, **115**, 827–57.

Chastel, O., Lacroix, A., Weimerskirch, H. and Gabrielsen, G.W. (2005). Modulation of prolactin but not corticosterone responses to stress in relation to parental effort in a long-lived bird. *Hormones and Behavior*, **47**, 459–66.

Chatterjee, S. (1997). *The rise of birds: 225 million years of evolution*. Johns Hopkins University Press, Baltimore.

Chen, J.N., Liu, N.F., Yan, C. and An, B. (2011). Plasticity in nest site selection of Black Redstart (*Phoenicurus ochruros*): a response to human disturbance. *Journal of Ornithology*, **152**, 603–8.

Chen, P.J., Dong, Z.M. and Zhen, S.N. (1998). An exceptionally well-preserved theropod dinosaur from the Yixian Formation of China. *Nature*, **391**, 147–52.

Cheng, M.F. (1973). Effect of estrogen on behavior of ovariectomized ring doves (*Streptopelia risoria*). *Journal of Comparative and Physiological Psychology*, **83**, 234–9.

Cherel, Y., Mauget, R., Lacroix, A. and Gilles, J. (1994). Seasonal and fasting-related changes in circulating gonadal-steroids and prolactin in king penguins, *Aptenodytes patagonicus*. *Physiological Zoology*, **67**, 1154–73.

Cherry, M.I. and Gosler, A.G. (2010). Avian eggshell coloration: new perspectives on adaptive explanations. *Biological Journal of the Linnean Society*, **100**, 753–62.

Cherry, P. and Morris, T. (2008). *Domestic duck production: science and practice*. CAB International, Wallingford.

Chiappe, L.M. (2007). *Glorified dinosaurs: the origin and early evolution of birds*. University of California Press, Berkeley.

Chiappe, L.M., Marugán-Lobón, J., Ji, S. and Zhou, Z. (2008). Life history of a basal bird: morphometrics of the Early Cretaceous *Confuciusornis*. *Biology Letters*, **4**, 719–23.

Chiappe, L.M. and Witmer, L.M. (2002). *Mesozoic birds: above the heads of dinosaurs*. Wiley-Blackwell, New York.

Chinsamy, A., Chiappe, L.M., Marugán-Lobón, J., Chunling, G. and Fengjiao, Z. (2013). Gender identification of the Mesozoic bird *Confuciusornis sanctus*. *Nature Communications*, **4**, article 1381. doi: 10.1038/ncomms2377.

Christe, P., Glaizot, O., Strepparava, N., Devevey, G. and Fumagalli, L. (2012). Twofold cost of reproduction: an increase in parental effort leads to higher malarial parasitaemia and to a decrease in resistance to oxidative stress. *Proceedings of the Royal Society of London Series B: Biological Sciences*, **279**, 1142–9.

Christie, P., Møller, A.P., González, G. and de Lope, F. (2002). Intraseasonal variation in immune defense, body mass and hematocrit in adult house martins *Delichon urbica*. *Journal of Avian Biology*, **33**, 321–5.

Christie, P., Richner, H. and Oppliger, A. (1996). Of Great Tits and fleas: sleep baby sleep. *Animal Behaviour*, **52**, 1087–92.

Cichon, M. (2000). Costs of incubation and immunocompetence in the collared flycatcher. *Oecologia*, **125**, 453–7.

Clark, A.B. and Wilson, D.S. (1981). Avian breeding adaptations: Hatching asynchrony, brood reduction, and nest failure. *Quarterly Review of Biology*, **56**, 253–77.

Clark, J.M., Norell, M.A. and Chiappe, L.M. (1999). An oviraptorid skeleton from the Late Cretaceous of Ukhaa Tolgod, Mongolia, preserved in an avianlike brooding position over an oviraptorid nest. *American Museum Novitates*, **3265**, 1–36.

Clark, L. (1991). The nest protection hypothesis: the adaptive use of plant secondary compounds by European starlings. In JE Loye, M Zuk, eds. *Bird-parasite interactions: ecology, evolution and behaviour*, pp. 204–21. Oxford University Press, Oxford.

Clark, L. and Mason, J.R. (1985). Use of nest material as insecticidal and anti-pathogenic agents by the European Starling. *Oecologia*, **67**, 169–76.

Clark, L. and Mason, J.R. (1987). Olfactory discrimination of plant volatiles by the European starling. *Animal Behaviour*, **35**, 227–35.

Clark, L. and Mason, J.R. (1988). Effect of biologically active plants used as nest material and the derived benefit to starling nestlings. *Oecologia*, **77**, 174–80.

Clayton, D.H. (1990). Mate choice in experimentally parasitized rock dove: lousy males lose. *American Zoologist*, **30**, 251–62.

Clayton, D.H. (1991). Coevolution of avian grooming and ectoparasite avoidance. In JE Loye, M Zuk, eds. *Bird-parasite interactions: ecology, evolution and behaviour*, pp. 258–89. Oxford University Press, Oxford.

Clayton, D.H., Adams, R.J. and Bush, S.E. (2008). Phthiraptera, the chewing lice. In TC Atkinson, NJ Thomas, DB Hunter, eds. *Parasitic diseases of wild birds*, pp. 515–26. Wiley-Blackwell, Ames.

Clayton, D.H., Koop, J.A.H., Harbison, C.W., Moyer, B.R. and Bush, S.E. (2010). How birds combat ectoparasites. *The Open Ornithology Journal*, **3**, 41–71.

Clayton, D.H., Lee, P.L.M., Tompkins, D.M. and Brodie, E.D. III. (1999). Reciprocal natural selection on host-parasite phenotypes. *The American Naturalist*, **154**, 261–70.

Clayton, D.H. and Moore, J. (1997). *Host-parasite evolution: general principles and avian models*. Oxford University Press, Oxford.

Clayton, D.H., Moyer, B.R., Bush, S.E., Jones, T.G., Gardiner, D.W., Rhodes, B.B. and Goller, F. (2005). Adaptive significance of avian beak morphology for ectoparasite control. *Proceedings of the Royal Society B*, **272**, 811–7.

Clayton, D.H. and Vernon, J.G. (1993). Common grackles anting with lime fruit and its effect on ectoparasites. *The Auk*, **110**, 951–2.

Clayton, D.H. and Wolfe, N.D. (1993). The adaptive significance of self-medication. *Trends in Ecology and Evolution*, **8**, 60–3.

Clucas, B., Rowe, M.P., Owings, D.H. and Arrowood, P.C. (2008). Snake scent application in ground squirrels, *Spermophilus* spp.: A novel form of antipredator behaviour? *Animal Behaviour*, **75**, 299–307.

Clutton-Brock, T.H. (1991). *The evolution of parental care*. Princeton University Press, Princeton.

Coates, P.S. and Delehanty, D.J. (2008). Effects of environmental factors on incubation patterns of Greater Sage-Grouse. *The Condor*, **110**, 627–38.

Coatney, G. (1931). On the biology of the pigeon fly, *Pseudolynchia maura* Bigot (Diptera, Hippoboscidae). *Parasitology*, **23**, 525–32.

Cockle, K.L., Martin, K. and Wesołowski, T. (2011). Woodpeckers, decay, and the future of cavity-nesting vertebrate communities worldwide. *Frontiers in Ecology and the Environment*, **9**, 377–82.

Cody, M.L. (1966). A general theory of clutch size. *Evolution*, **20**, 174–84.

Cohen, J.B., Wunker, E.H. and Fraser, J.D. (2008). Substrate and vegetation selection by nesting Piping Plovers. *The Wilson Journal of Ornithology*, **120**, 404–7.

Coker, D.R. and Confer, J.L. (1990). Brown-headed Cowbird parasitism on Golden-winged and Blue-winged Warblers. *The Wilson Bulletin*, **102**, 550–2.

Cole, E.F. and Quinn, J.L. (2014). Shy birds play it safe: personality in captivity predicts risk responsiveness during reproduction in the wild. *Biology Letters*, **10**, 20140178.

Coleman, M.A. and McNabb, R.A. (1975). Photo acceleration of development in depigmented Japanese quail eggs. *Poultry Science*, **54**, 1849–55.

Coleman, R.M. and Whittall, R.D. (1988). Clutch size and the cost of incubation in the Bengalese finch (*Lonchura striata* var. *domestica*). *Behavioral Ecology and Sociobiology*, **23**, 367–72.

Collar, N., Fisher, C. and Feare, C. (2003). Why museums matter: avian archives in an age of extinction. *Bulletin of the British Ornithologists' Club Supplement*, **123A**, 1–360.

Collar, N.J. and Butchart, S.H.M. (2014). Conservation breeding and avian diversity: chances and challenges. *International Zoo Yearbook*, **48**, 7–28.

Collias, E.C. and Collias, N.E. (1962a). Development of nest-building in the Village Weaverbird (*Textor cucullatus* Muller). *American Zoologist*, **4**, 399.

Collias, E.C. and Collias, N.E. (1964). The development of nest-building behavior in a weaverbird. *The Auk*, **81**, 42–52.

Collias, N.E. (1997). On the origin and evolution of nest building by passerine birds. *The Condor*, **99**, 253–70.

Collias, N.E. and Collias E.C. (1962b). An experimental study of the mechanisms of nest building in a weaverbird. *The Auk*, **79**, 568–95.

Collias, N.E. and Collias, E.C. (1984). *Nest building and bird behavior*. Princeton University Press, Princeton.

Colombelli-Négrel, D., Hauber, M.E., Robertson, J., Sulloway, F.J., Hoi, H., Griggio, M. and Kleindorfer, S. (2012). Embryonic learning of vocal passwords in superb fairy-wrens reveals intruder cuckoo nestlings. *Current Biology*, **22**, 2155–60.

Colwell, M.A., Meyer, J.J., Hardy, M.A., McAllister, S.E., Transou, A.N., Levalley, R.R. and Dinsmore, S.J. (2011). Western Snowy Plovers *Charadrius alexandrinus nivosus* select nesting substrates that enhance egg crypsis and improve nest survival. *Ibis*, **153**, 303–11.

Conkling, T.J., Pope, T.L., Smith, K.N., Mathewson, H.A., Morrison, M.L., Wilkins, R.N. and Cain, J.W. (2012). Black-capped vireo nest predator assemblage and predictors for nest predation. *Journal of Wildlife Management*, **76**, 1401–11.

Conway, C.J., Eddleman, W.R. and Anderson, S.H. (1994). Nesting success and survival of Virginia Rails and Soras. *The Wilson Bulletin*, **106**, 466–73.

Conway, C.J. and Martin, T.E. (2000a). Effects of ambient temperature on avian incubation behavior. *Behavioral Ecology*, **11**, 178–88.

Conway, C.J. and Martin, T.E. (2000b). Evolution of passerine incubation behavior: influence of food, temperature, and nest predation. *Evolution*, **54**, 670–85.

Conover, M.R. (2007). *Predator-prey dynamics: The role of olfaction*. CRC Press, Boca Raton.

Conover, M.R., Borgo, J.S., Dritz, R.E., Dinkins, J.B. and Dahlgren, D.K. (2010). Greater Sage-grouse select nest sites to avoid visual predators but not olfactory predators. *The Condor*, **112**, 331–6.

Cooch, E.G. (2002). Fledging size and survival of Snow Geese: timing is everything (or is it?). *Journal of Applied Statistics*, **29**, 143–62.

Cook, M.I. (2005). Microbial infection affects egg viability and incubation behavior in a tropical passerine. *Behavioral Ecology*, **16**, 30–6.

Cook, M.I., Beissinger, S.R., Toranzos, G. and Arendt, W.J. (2005). Incubation reduces microbial growth on eggshells and the opportunity for trans-shell infection. *Ecology Letters*, **8**, 532–7.

Cook, M.I., Beissinger, S.R., Toranzos, G., Rodriguez, R. and Arendt, W.J. (2003). Trans-shell infection by pathogenic micro-organisms reduces the shelf life of non-incubated bird's eggs: a constraint on the onset of incubation? *Proceedings of the Royal Society of London Series B: Biological Sciences*, **270**, 2233–40.

Cooney, S.J. and Watson, D.M. (2008). An experimental approach to understanding the use of mistletoe as a nest substrate for birds: nest predation. *Wildlife Research*, **35**, 65–71.

Cooper, C.B. (2013). Is there weekend bias in clutch-initiation dates from citizen science: Implications for studies of avian breeding phenology. *International Journal of Biometeorology*, **58**, 1415–9.

Cooper, C.B., Hochachka, W., Butcher G. and Dhondt, A.A. (2005b). Egg viability as a constraint on seasonal and latitudinal trends in clutch size. *Ecology*, **86**, 2018–31.

Cooper, C.B., Hochachka, W.M. and Dhondt, A.A. (2005a). Latitudinal trends in within-year reoccupation of nest boxes and their implications. *Journal of Avian Biology*, **36**, 31–9.

Cooper, C.B., Hochachka, W., Phillips, T.B. and Dhondt, A.A. (2006). Geographic and seasonal gradients in hatching failure in eastern bluebirds reinforce clutch size trends. *Ibis*, **148**, 221–30.

Cooper, C.B. and Mills, H. (2005). New software for quantifying incubation behavior from time-series recordings. *Journal of Field Ornithology*, **76**, 352–6.

Cooper, C.B., Shirk, J. and Zuckerberg, B. (2014). The invisible prevalence of citizen science in global research: migratory birds and climate change. *PLoS ONE*, **9**(9), e106508.

Cooper, C.B. and Voss, M.A. (2013). Avian incubation patterns reflect temporal changes in developing clutches. *PLoS ONE*, **8**(6), e65521.

Cooper, C.B., Voss, M.A. and Zivkovic, B. (2009). Extended laying interval of ultimate eggs of the Eastern Bluebird. *The Condor*, **111**, 752–5.

Cornulier, T., Elston, D.A., Arcese, P., Benton, T.G., Douglas, D.J.T., Lambin, X., Reid, J., Robinson, R.A. and Sutherland, W.J. (2009). Estimating the annual number of breeding attempts from breeding dates using mixture models. *Ecology Letters*, **12**, 1184–93.

Costantini, D. (2014). The role of oxidative stress and hormesis in shaping reproductive strategies. In D Costantini, ed. *Oxidative stress and hormesis in evolutionary ecology and physiology*, pp. 205–39. Springer Heidelberg, New York.

Costantini, D. and Bonadonna, F. (2010). Patterns of variation of serum oxidative stress markers in two seabird species. *Polar Research*, **29**, 30–5.

Cotgreave, P. and Clayton, D.H. (1994). Comparative analyses of time spent grooming by birds in relation to parasite load. *Behaviour*, **131**, 3–4.

Coulson, J.C. (1963). Egg size and shape in the Kittiwake (*Rissa tridactyla*) and their use in estimating age composition of populations. *Proceedings of the Zoological Society of London*, **140**, 211–26.

Courter, J.R., Johnson, R.J., Stuyck, C.M., Lang, B.A. and Kaiser, E.W. (2012). Weekend bias in citizen science data reporting: implications for phenology studies. *International Journal of Biometeorology*, **57**, 715–20.

Cowley, E. (1983). Multi-brooding and mate infidelity in the sand martin. *Bird Study*, **30**, 1–7.

Cox, R.T. (1996). *Birder's dictionary*. Falcon Press, Helena.

Cox, W.A., Pruett, M.S., Benson, T.J., Chiavacci, S.J. and Thompson, F.R. III. (2012). Development of camera technology for monitoring nests. In CA Ribic, FR Thompson III, PF Pietz, eds. *Video surveillance of nesting birds*. Studies in Avian Biology, Vol. 43, pp. 185–210. University of California Press, Berkeley.

Cox, W.A., Thompson, F.R., Cox, A.S. and Faaborg, J. (2014). Post-fledging survival in passerine birds and the value of post-fledging studies to conservation. *Journal of Wildlife Management*, **78**, 183–93.

Craig, G.H. and Craig, J.L. (1974). An automatic nest recorder. *Ibis*, **116**, 557–61.

Cresswell, W. (1997). Nest predation: the relative effects of nest characteristics, clutch size and parental behaviour. *Animal Behaviour*, **53**, 93–103.

Cresswell, W. (2008).The non-lethal effects of predation in birds. *Ibis*, **150**, 3–17.

Cresswell, W., Holt, S., Reid, J.M., Whitfield, D.P. and Mellanby, R.J. (2003). Do energetic demands constrain incubation scheduling in a biparental species? *Behavioral Ecology*, **14**, 97–102.

Cresswell, W. and McCleery, R. (2003). How great tits maintain synchronization of their hatch date with food supply in response to long-term variability in temperature. *Journal of Animal Ecology*, **72**, 356–66.

Crick, H.Q.P. (1992). A bird-habitat coding system for use in Britain and Ireland incorporating aspects of land-management and human activity. *Bird Study*, **39**, 1–12.

Crick, H.Q.P. (1995). The strange case of the whistling oofoo. What are runt eggs? *British Birds*, **88**, 169–80.

Crick, H.Q.P., Baillie, S.R. and Leech, D.I. (2003a). The UK Nest Record Scheme: its value for science and conservation. *Bird Study*, **50**, 254–70.

Crick, H.Q.P., Dudley, C., Glue, D., Beaven, P. and Leech, D. (2003b). *The nest record scheme handbook*. British Trust for Ornithology, Thetford. www.bto.org/volunteer-surveys/nrs/coc.

Crick, H.Q.P., Dudley, C., Glue, D.E. and Thomson, D.L. (1997). UK birds are laying eggs earlier. *Nature*, **388**, 526.

Crick, H.Q.P., Gibbons, D.W. and Magrath, R.D. (1993). Seasonal changes in clutch size in British birds. *Journal of Animal Ecology*, **62**, 263–73.

Crick, H.Q.P. and Sparks, T. (1999). Climate change related to egg-laying trends. *Nature*, **399**, 423–4.

Crino, O.L., Driscoll, S.C. and Breuner, C.W. (2014). Corticosterone exposure during development has sustained but not lifelong effects on body size and total and free corticosterone responses in the zebra finch. *General and Comparative Endocrinology*, **196**, 123–9.

Criscuolo, F., Chastel, O., Gabrielsen, G.W., Lacroix, A. and Le Maho, Y. (2002). Factors affecting plasma concentrations of prolactin in the common eider *Somateria mollissima*. *General and Comparative Endocrinology*, **125**, 399–409.

Crook, J.H. (1960). Nest form and construction in certain West African weaver-birds. *Ibis*, **103**, 1–12.

Crook, J.H. (1963). A comparative analysis of nest structure in the weaver birds (Ploceinae). *Ibis*, **105**, 238–62.

Crossman, C.A., Rohwer, V.G. and Martin, P.R. (2011). Variation in the structure of bird nests between northern Manitoba and southeastern Ontario. *PLoS ONE*, **6**(4), e19086.

Crudo, A., Suderman, M., Moisiadis, V.G., Petropoulos, S., Kostaki, A., Hallett, M., Szyf, M. and Matthews, S.G. (2013). Glucocorticoid programming of the fetal male hippocampal epigenome. *Endocrinology*, **154**, 1168–80.

Cucco, M. and Malacarne, G. (1997). The effect of supplemental food on time budget and body condition in the Black Redstart *Phoenicurus ochruros*. *Ardea*, **85**, 211–21.

Curio, E. (1983). Why do young birds reproduce less well? *Ibis*, **125**, 400–4.

Curno, O., Behnke, J.M., McElligott, A.G., Reader, T. and Barnard, C.J. (2009). Mothers produce less aggressive sons with altered immunity when there is a threat of disease during pregnancy. *Proceedings of the Royal Society B*, **276**, 1047–54.

Currie, D.C. and Hunter, D.B. (2008). Black Flies (Diptera: Simuliidae). In TC Atkinson, NJ Thomas, DB Hunter, eds. *Parasitic diseases of wild birds*, pp. 537–45. Wiley-Blackwell, Ames.

Currie, P.J., Koppelhus, E.B., Shugar, M.A. and Wright, J.L. (2004). *Feathered dragons: studies on the transition from dinosaurs to birds*. Indiana University Press, Bloomington.

Czerkas, S.A. and Feduccia, A. (2014). Jurassic archosaur is a non-dinosaurian bird. *Journal of Ornithology*, **155**, 841–51.

Czeszczewik, D., Walankiewicz, W. and Stańska, M. (2008). Small mammals in nests of cavity-nesting birds: Why should ornithologists study rodents? *Canadian Journal of Zoology*, **86**, 286–93.

Czirják, G.A., Pap, P.L., Vágási, C.I., Giraudeau, M., Mureşan, C., Mirleau, P. and Heeb, P. (2013). Preen gland removal increases plumage bacterial load but not that of feather-degrading bacteria. *Naturwissenschaften*, **100**, 145–51.

Dacke, C.G. (2000). *The parathyroids, calcitonin and vitamin D.* Academic Press, Orlando.

Dacke, C.G., Aarkle, S., Cook, D.J., Wormstone, I.M., Jones, S., Zaidi, M. and Bascal, Z.A. (1993). Medullary bone and avian calcium regulation. *The Journal of Experimental Biology*, **184**, 63–88.

D'Alba, L., Jones, D.N., Badawy, H.T., Eliason, C.M. and Shawkey, M.D. (2014). Antimicrobial properties of a nanostructured eggshell from a compost-nesting bird. *The Journal of Experimental Biology*, **217**, 1116–21.

D'Alba, L., Monaghan, P. and Nager, R.G. (2009). Thermal benefits of nest shelter for incubating female eiders. *Journal of Thermal Biology*, **34**, 93–9.

D'Alba, L., Oborn, A. and Shawkey, M.D. (2010a). Experimental evidence that keeping eggs dry is a mechanism for the antimicrobial effects of avian incubation. *Naturwissenschaften*, **97**, 1089–95.

D'Alba, L., Shawkey, M.D., Korsten, P., Vedder, O., Kingma, S.A., Komdeur, J. and Beissinger, S.R. (2010b). Differential deposition of antimicrobial proteins in blue tit (*Cyanistes caeruleus*) clutches by laying order and male attractiveness. *Behavioral Ecology and Sociobiology*, **64**, 1037–45.

Danchin, E., Boulinier, T. and Massot, M. (1998). Breeding habitat selection based on conspecific reproductive success: implications for the evolution of coloniality. *Ecology*, **79**, 2415–28.

Darwin, C.R. (1859). *On the origin of species by means of natural selection, or the preservation of favoured races in the struggle for life*. John Murray, London.

Dauwe, T., Jaspers, V.L.B., Covaci, A. and Eens, M. (2006). Accumulation of organochlorines and brominated flame retardants in the eggs and nestlings of great tits, *Parus major*. *Environmental Science and Technology*, **40**, 5297–303.

Davies, N.B. (2000). *Cuckoos, cowbirds and other cheats*. T. and A.D. Poyser, London.

Davies, N.B. and Brooke, M.de L. (1988). Cuckoos versus reed warblers: adaptations and counteradaptations. *Animal Behaviour*, **36**, 262–84.

Davies, N.B. and Brooke, M. de L. (1989a). An experimental study of co-evolution between the cuckoo, *Cuculus canorus*, and its hosts. 1. Host egg discrimination. *Journal of Animal Ecology*, **58**, 207–24.

Davies, N.B. and Brooke, M.de L. (1989b). An experimental study of co-evolution between the cuckoo, *Cuculus canorus*, and its hosts. 2. Host egg markings, chick discrimination and general discussion. *Journal of Animal Ecology*, **58**, 225–36.

Davies, S.J.J.F. (1961). The nest-building behaviour of the magpie goose *Anseranas semipalmata*. *Ibis*, **104**, 147–57.

Dawson, R.D., Hillen, K.K. and Whitworth, T.L. (2005b). Effects of experimental variation in temperature on larval densities of parasitic *Protocalliphora* (Diptera: Calliphoridae) in nests of tree swallows (Passeriformes: Hirundinidae). *Environmental Entomology*, **34**, 563–8.

Dawson, R.D., Lawrie, C.C. and O'Brien, E.L. (2005a). The importance of microclimate variation in determining size, growth and survival of avian offspring: experimental evidence from a cavity nesting passerine. *Oecologia*, **144**, 499–507.

Dawson, R.D., O'Brien, E.L. and Mlynowski, T. J. (2011). The price of insulation: costs and benefits of feather delivery to nests for male tree swallows *Tachycineta bicolor*. *Journal of Avian Biology*, **42**, 93–102.

Dawson, W.R. and Schmidt-Nielsen, K. (1964). Terrestrial animals in dry heat: desert birds. In DB Dill, ed. *Handbook of Physiology. Part 4—Environment*, pp. 481–92. American Physiological Society, Washington D.C.

Dean, M. and Annilo, T. (2005). Evolution of the ATP-binding cassette (ABC) transporter superfamily in vertebrates. *Annual Review of Genomics and Human Genetics*, **6**, 123–42.

De Coster, G., de Neve, L. and Lens, L. (2012). Intraclutch variation in avian eggshell pigmentation: the anaemia hypothesis. *Oecologia*, **170**, 297–304.

Deeming, D.C. (1991). Reasons for the dichotomy in the need for egg turning during incubation in birds and reptiles. In DC Deeming, MWJ Ferguson, eds. *Egg incubation: its effects on embryonic development in birds and reptiles*, pp. 307–23. Cambridge University Press, Cambridge.

Deeming, D.C. (1995). Factors affecting hatchability during commercial incubation of ostrich (*Struthio camelus*) eggs. *British Poultry Science*, **36**, 51–65.

Deeming, D.C. (1997). *Ratite egg incubation—a practical guide*. Ratite Conference, High Wycombe.

Deeming, D.C. (2000). *Principles of artificial incubation for game birds—a practical guide*. Ratite Conference, Wallingford.

Deeming, D.C. (2002a). *Avian incubation: behaviour, environment, and evolution*. Oxford University Press, Oxford.

Deeming, D.C. (2002b). Importance and evolution of incubation in avian reproduction. In DC Deeming, ed. *Avian incubation: behaviour, environment, and evolution*, pp. 1–7. Oxford University Press, Oxford.

Deeming, D.C. (2002c). Functional characteristics of eggs. In DC Deeming, ed. *Avian incubation: behaviour, environment, and evolution*, pp. 28–42. Oxford University Press, Oxford.

Deeming, D.C. (2002d). Embryonic development and utilisation of egg components. In DC Deeming, ed. *Avian incubation: behaviour, environment, and evolution*, pp. 43–53. Oxford University Press, Oxford.

Deeming, D.C. (2002e). Behaviour patterns during incubation. In DC Deeming, ed. *Avian incubation: behavior, environment, and evolution*. pp. 63–87. Oxford University Press, New York.

Deeming, D.C. (2002f). Patterns and significance of egg turning. In DC Deeming, ed. *Avian incubation: behaviour, environment, and evolution*, pp. 161–78. Oxford University Press, Oxford.

Deeming, D.C. (2002g). Perspectives in avian incubation. In DC Deeming, ed. *Avian incubation: behaviour, environment, and evolution*, pp. 326–9. Oxford University Press, Oxford.

Deeming, D.C. (2002h). *Practical aspects of commercial incubation in poultry.* Ratite Conference Books, Lincoln.

Deeming, D.C. (2003). *Nest, birds and incubators: New insights into natural and artificial incubation.* Brinsea Products Ltd, Sandford.

Deeming, D.C. (2004a). *Reptilian incubation: environment, evolution and behaviour.* Nottingham University Press, Nottingham.

Deeming, D.C. (2004b). Post-hatching phenotypic effects of incubation in reptiles. In DC Deeming, ed. *Reptilian incubation: environment, evolution and behaviour,* pp. 211–28. Nottingham University Press, Nottingham.

Deeming, D.C. (2006). Ultrastructural and functional morphology of eggshells supports idea that dinosaur eggs were incubated buried in a substrate. *Paleontology,* **49,** 171–85.

Deeming, D.C. (2007a). Effects of phylogeny and hatchling maturity on allometric relationships between female body mass and the mass and composition of bird eggs. *Avian and Poultry Biology Reviews,* **18,** 21–37.

Deeming, D.C. (2007b). Allometry of mass and composition in bird eggs: Effects of phylogeny and hatchling maturity. *Avian and Poultry Biology Reviews,* **18,** 71–86.

Deeming, D.C. (2008). Avian brood patch temperature: Relationships with female body size, incubation period, developmental maturity and phylogeny. *Journal of Thermal Biology,* **33,** 345–54.

Deeming, D.C. (2011a). Laws governing research on bird and reptile embryos. *Lab Animal Europe,* **11**(2), 3.

Deeming, D.C. (2011b). Importance of nest type on the regulation of humidity in bird nests. *Avian Biology Research,* **4,** 23–31.

Deeming, D.C. (2011c). A review of the relationship between eggshell colour and water vapour conductance. *Avian Biology Research,* **4,** 224–30.

Deeming, D.C. (2013). Effects of female body size and phylogeny on avian nest dimensions. *Avian Biology Research,* **6,** 1–11.

Deeming, D.C. and Biddle, L.E. (2015). Thermal properties of bird nests depend on air-gaps between the materials. *Acta Ornithologica,* **50,** 121–5.

Deeming, D.C. and Birchard, G.F. (2007). Allometry of egg and hatchling mass in birds and reptiles: roles of developmental maturity, eggshell structure and phylogeny. *Journal of Zoology,* **271,** 78–87.

Deeming, D.C. and Birchard, G.F. (2008). Why were extinct gigantic birds so small? *Avian Biology Research,* **1,** 189–96.

Deeming, D.C., Birchard, G.F., Crafer, R. and Eady, P.E. (2006). Egg mass and incubation period allometry in birds and reptiles: the effects of phylogeny. *Journal of Zoology,* **270,** 209–18.

Deeming, D.C. and du Feu, C.R. (2011). Long-term patterns in egg mortality during incubation and chick mortality during rearing in three species of tits in an English woodland. *Bird Study,* **58,** 278–90.

Deeming, D.C. and Ferguson, M.W.J. (1991a). *Egg incubation: its effects on embryonic development in birds and reptiles.* Cambridge University Press, Cambridge.

Deeming, D.C. and Ferguson, M.W.J. (1991b). Physiological effects of incubation temperature on embryonic development in reptiles and birds. In DC Deeming, MWJ Ferguson, eds. *Egg incubation: its effects on embryonic development in birds and reptiles,* pp. 147–71, Cambridge University Press, Cambridge.

Deeming, D.C., Harvey, R.L., Harvey, L., Carey, S. and Leuchars, D. (1991). Artificial incubation and hand-rearing of Humboldt penguins *Spheniscus humboldti. International Zoo Yearbook,* **30,** 165–73.

Deeming, D.C., Mainwaring, M.C., Hartley, I.R. and Reynolds, S.J. (2012). Local temperature and not latitude determines the design of Blue Tit and Great Tit nests. *Avian Biology Research,* **5,** 203–8.

Deeming, D.C. and Pike, T.W. (2013). Embryonic growth and antioxidant provision in avian eggs. *Biology Letters,* **9,** 20130757.

Deeming, D.C. and Pike, T.W. (2015) Nest surface temperature measured using infrared thermography predicts fledging success. *Acta Ornithologica,* **50,** 247–51.

Deeming, D.C. and Riches, C. (2005). Importance of temperature and egg position for contact incubation of eggs of the red-legged partridge (*Alectoris rufa*). *Avian and Poultry Biology Reviews,* **15,** 51–2.

Deeming, D.C. and Ruta, M. (2014). Amniote egg morphology and the evolution of avian eggs. *Royal Society Open Science,* **1,** 140311.

de Heij, M.E., Ubels, R., Visser, G.H. and Tinbergen, J.M. (2008). Female great tits *Parus major* do not increase their daily energy expenditure when incubating enlarged clutches. *Journal of Avian Biology,* **39,** 121–6.

de Heij, M.E., van den Hout, P.J. and Tinbergen, J.M. (2006). Fitness cost of incubation in great tits (*Parus major*) is related to clutch size. *Proceedings of the Royal Society of London Series B: Biological Sciences,* **273,** 2353–61.

de Heij, M.E., van der Graaf, A.J., Hafner, D. and Tinbergen, J.M. (2007). Metabolic rate of nocturnal incubation in female great tits, *Parus major,* in relation to clutch size measured in a natural environment. *The Journal of Experimental Biology,* **210,** 2006–12.

del Hoyo, J., Elliott, A. and Christie, D.A. (2006). *Handbook of the Birds of the World. Volume 11. Old World flycatchers to Old World warblers.* Lynx Edicions, Barcelona.

de Lope, F., González, G., Pérez, J.J. and Møller, A.P. (1993). Increased detrimental effects of ectoparasites on their bird hosts during adverse environmental conditions. *Oecologia,* **95,** 234–40.

de Lope, F., Møller, A.P. and de la Cruz, C. (1998). Parasitism, immune response and reproductive success in the house martin *Delichon urbica. Oecologia,* **114,** 188–93.

Delotelle, R.S. and Epting, R.J. (2007). Variation in resin-well construction as an indicator of predation load and breeder experience in red-cockaded woodpecker populations. *Environmental Bioindicators,* **2,** 71–88.

De Marchi, G., Chiozzi, G. and Fasola, M. (2008). Solar incubation cuts down parental care in a burrow nesting tropical shorebird, the crab plover *Dromas ardeola. Journal of Avian Biology,* **39,** 484–6.

de Matos, R. (2008). Calcium metabolism in birds. *Veterinary Clinics of North America: Exotic Animal Practice*, **11**, 59–82.

de Neve, L. and Soler, J.J. (2002). Nest-building activity and laying date influence female reproductive investment in magpies: an experimental study. *Animal Behaviour*, **63**, 975–80.

de Neve, L., Soler, J.J., Soler, M., Pérez-Contreras, T., Martín-Vivaldi, M. and Martínez, J.G. (2004). Effects of a food supplementation experiment on reproductive investment and a post-mating sexually selected trait in magpies *Pica pica*. *Journal of Avian Biology*, **35**, 246–51.

Derksen, D.V. (1977). A quantitative analysis of the incubation behavior of the Adélie Penguin. *The Auk*, **94**, 552–66.

De Reu, K., Grijspeerdt, K., Messens, W., Heyndrickx, M., Uyttendaele, M., Debevere, J. and Herman, L. (2006). Eggshell factors influencing eggshell penetration and whole egg contamination by different bacteria, including *Salmonella enteritidis*. *International Journal of Food Microbiology*, **112**, 253–60.

Desante, D.F., Burton, K.M., Saracco, J.F. and Walker, B.L. (1995). Productivity indices and survival rate estimates from MAPS, a continent-wide programme of constant-effort mist-netting in North America. *Journal of Applied Statistics*, **22**, 935–48.

Dhondt, A.A., Kast, T.L. and Allen, P.E. (2002). Geographical differences in seasonal clutch size variation in multi-brooded bird species. *Ibis*, **144**, 646–51.

Dickinson, J.L. and Bonney, R. (2012). *Citizen science: public participation in environmental research*. Cornell University Press, Ithaca.

Dietrich, M., Gómez-Díaz, E. and McCoy, K.D. (2011). Worldwide distribution and diversity of seabird ticks: implications for the ecology and epidemiology of tick-borne pathogens. *Vector-Borne and Zoonotic Diseases*, **5**, 453–60.

Di Iorio, O., Turienzo, P., Bragagnolo, L., Santillán, M.A. and Grande, J.M. (2013). New family host and records of *Acanthocrios furnarii* (Cordero and Vogelsang, 1928) (Hemiptera: Cimicidae) from Argentina, and implications in the transmission mechanism of cimicid bugs among birds' nests. *Zootaxa*, **3630**, 582–90.

Dik, B. (2006). Erosive stomatitis in a White Pelican (*Pelecanus onocrotalus*) caused by *Piagetiella titan* (Mallophaga: Menoponidae). *Journal of Veterinary Medicine Series B*, **53**, 153–4.

Dingus, L. and Rowe, T. (1998). *The mistaken extinction: dinosaur evolution and the origin of birds*. W. H. Freeman and Co., New York.

Dinsmore, S.J., White, G.C. and Knopf, F.L. (2002). Advanced techniques for modeling avian nest survival. *Ecology*, **83**, 3476–88.

Dixon, C. (1902). *Birds' nests: An introduction to the science of caliology*. Frederick A. Stokes, New York.

Doligez, B., Danchin, E., Clobert, J. and Gustaffson, L. (1999). The use of conspecific reproductive success for breeding habitat selection in a noncolonial hole-nesting species, the collared flycatcher. *Journal of Animal Ecology*, **68**, 1193–206.

Dolmans, D., Fukumura, D. and Jain, R. (2003). Photodynamic therapy for cancer. *Nature Reviews Cancer*, **3**, 375–80.

Dol'nik, T.V. and Dol'nik, V.R. (1982). Production and productive energy during egg laying in birds. *Trudy Zoologicheskogo Instituta*, **113**, 124–43.

Donaire, M. and López-Martínez, N. (2009). Porosity of Late Paleocene *Ornitholithus* eggshells (Tremp Fm, south-central Pyrenees, Spain): Palaeoclimatic implications. *Palaeogeography, Palaeoclimatology, Palaeoecology*, **279**, 147–59.

Dong, Z.-M. and Currie, P.J. (1996). On the discovery of an oviraptorid skeleton on a nest of eggs at Bayan Mandahu, Inner Mongolia, People's Republic of China. *Canadian Journal of Earth Sciences*, **33**, 631–6.

Douglas, D.J.T., Newson, S.E., Leech, D.I., Noble, D.G. and Robinson, R.A. (2010). How important are climate-induced changes in host availability for population processes in an obligate brood parasite, the European cuckoo? *Oikos*, **119**, 1834–40.

Douglas, H.D., Co, J.E., Jones, T.H. and Conner, W.E. (2004). Interspecific differences in *Aethia* spp. auklet odorants and evidence for chemical defense against ectoparasites. *Journal of Chemical Ecology*, **30**, 1921–35.

Douglas, H.D., Co, J.E., Jones, T.H., Conner, W.E. and Day, J.F. (2005). Chemical odorant of colonial seabird repels mosquitoes. *Journal of Medical Entomology*, **42**, 647–51.

Dow, H. and Fredga, S. (1983). Breeding and natal dispersal of the goldeneye, *Bucephala clangula*. *Journal of Animal Ecology*, **52**, 681–95.

Dowling, D.K., Richardson, D.S. and Komdeur, J. (2001). No effects of a feather mite on body condition, survivorship, or grooming behavior in the Seychelles warbler, *Acrocephalus sechellensis*. *Behavioral Ecology and Sociobiology*, **50**, 257–62.

Dreitz, V.J., Conrey, R.Y. and Skagen, S.K. (2012). Drought and cooler temperatures are associated with higher nest survival in Mountain Plovers. *Avian Conservation and Ecology*, **7**, 6.

Drent, R.H. (1970). Functional aspects of incubation in the herring gull. *Behaviour*, Supplement **17**, 1–132.

Drent, R.H. (1975). Incubation. In DS Farner, JR King, eds. *Avian biology*, Vol. 5, pp. 333–420. Academic Press, New York.

Drent, R.H. and Piersma, T. (1990). An exploration of the energetics of leap-frog migration in Arctic-breeding waders. In E Gwinner, ed. *Bird migration: Physiology and ecophysiology*, pp. 399–412. Springer Verlag, Berlin.

Drouillard, K.G. and Norstrom, R.J. (2001). Quantifying maternal and dietary sources of 2,20,4,40,5,50-hexachlorobiphenyl deposited in eggs of the ring dove (*Streptopelia risoria*). *Environmental Toxicology and Chemistry*, **20**, 561–7.

Drury, W.H. (1962). Breeding activities, especially nest-building, of the yellowtail (*Ostinops decumanus*) in Trinidad, West Indies. *Zoologiea (New York)*, **47**, 39–58.

Du, W.G. and Shine, R. (2015). The behavioural and physiological strategies of bird and reptile embryos in response to unpredictable variation in nest temperature. *Biological Reviews of the Cambridge Philosophical Society*, **89**, 19–30.

Dubiec, A., Góźdź, I. and Mazgajski, T.D. (2013). Green plant material in avian nests. *Avian Biology Research*, **6**, 133–46.

Dubiec, A. and Mazgajski, T.D. (2013). Nest mass variation over the nesting cycle in the Pied Flycatcher (*Ficedula hypoleuca*). *Avian Biology Research*, **6**, 127–32.

Dubois, N.S. and Getty, T. (2003). Empty nests do not affect female mate choice or maternal investment in House Wrens. *The Condor*, **105**, 382–7.

Duckworth, R.A. (2006). Behavioral correlations across breeding contexts provide a mechanism for a cost of aggression. *Behavioral Ecology*, **17**, 1011–9.

du Feu, C. (2003). *The BTO nestbox guide*. British Trust for Ornithology, Thetford.

du Feu, C.R. (1992). How tits avoid flea infestation at nest sites. *Ringing & Migration*, **13**, 120–1.

Duffy, D.C. (1983). The ecology of tick parasitism on densely nesting Peruvian seabirds. *Ecology*, **64**, 110–9.

Dumbacher, J.P. (1999). Evolution of toxicity in pitohuis: I. Effects of homobatrachotoxin on chewing lice (order Phthiraptera). *The Auk*, **116**, 957–63.

Dumbacher, J.P., Beehler, B.M., Spandem, T.F., Garraffo, H.M. and Daly, J.W. (1992). Homobatrachotoxin in the genus *Pitohui*: chemical defense in birds? *Science*, **258**, 799–801.

Dunn, P.O. and Møller, A.P. (2014). Changes in breeding phenology and population size of birds. *Journal of Animal Ecology*, **83**, 729–39.

Dunn, P.O., Thusius, K.J., Kimber, K. and Winkler, D.W. (2000). Geographic and ecological variation in clutch size of Tree Swallows. *The Auk*, **117**, 215–21.

Dunn, P.O. and Winkler, D.W. (1999). Climate change has affected breeding date of Tree Swallows throughout North America. *Proceedings of the Royal Society of London Series B: Biological Sciences*, **266**, 2487–90.

Dunn, P.O. and Winkler, D.W. (2010). Effects of climate change on timing of breeding and reproductive success in birds. In AP Møller, W Fiedler, P Berthold, eds. *Effects of climate change on birds*, pp. 113–28. Oxford University Press, Oxford.

DuRant, S.E., Carter, A.W., Denver, R.J., Hepp, G.R. and Hopkins, W.A. (2014). Are thyroid hormones mediators of incubation temperature-induced phenotypes in birds? *Biology Letters*, **10**, 20130950.

DuRant, S.E., Hepp, G.R., Moore, I.T., Hopkins, B.C. and Hopkins, W.A. (2010). Slight differences in incubation temperature affect early growth and stress endocrinology of Wood Duck (*Aix sponsa*) ducklings. *The Journal of Experimental Biology*, **213**, 45–51.

DuRant, S.E., Hopkins, W.A., Carter, A.W., Stachowiak, C.M. and Hepp, G.R. (2013b). Incubation conditions are more important in determining early thermoregulatory ability than posthatch resource conditions in a precocial bird. *Physiological and Biochemical Zoology*, **86**, 410–20.

DuRant, S.E., Hopkins, W.A., Hawley, D.M. and Hepp, G.R. (2012a). Incubation temperature affects multiple measures of immunocompetence in young Wood Ducks (*Aix sponsa*). *Biology Letters*, **8**, 108–11.

DuRant, S.E., Hopkins, W.A. and Hepp, G.R. (2011). Embryonic developmental patterns and energy expenditure are affected by incubation temperature in Wood Ducks (*Aix sponsa*). *Physiological and Biochemical Zoology*, **84**, 451–7.

DuRant, S.E., Hopkins, W.A., Hepp, G.R. and Walters, J.R. (2013a). Ecological, evolutionary, and conservation implications of incubation temperature-dependent phenotypes in birds. *Biological Reviews of the Cambridge Philosophical Society*, **88**, 499–509.

DuRant, S.E., Hopkins, W.A., Wilson, A.F. and Hepp, G.R. (2012b). Incubation temperature affects the metabolic cost of thermoregulation in a young precocial bird. *Functional Ecology*, **26**, 416–22.

Duval, C., Cassey, P., Mikšík, I., Reynolds, S.J. and Spencer, K.A. (2013). Condition-dependent strategies of eggshell pigmentation: an experimental study of Japanese quail (*Coturnix coturnix japonica*). *The Journal of Experimental Biology*, **216**, 700–8.

Dyke, G., Vremir, M., Kaiser, G. and Naish, D. (2012). A drowned Mesozoic bird breeding colony from the Late Cretaceous of Transylvania. *Naturwissenschaften*, **99**, 435–42.

Dykstra, R.C., Hays, L.J. and Simon, M.M. (2009). Selection of fresh vegetation for nest lining by red-shouldered hawks. *The Wilson Journal of Ornithology*, **121**, 207–10.

Dzialowski, E.M. (2005). Use of operative and standard operative temperature models in thermal biology. *Journal of Thermal Biology*, **30**, 317–34.

Easterling, D.R., Meehl, G.A., Parmesan, C., Changnon, S.A. Lark, T.R. and Mearns, L.O. (2000). Climate extremes: observations, modelling, and impacts. *Science*, **289**, 2068–74.

Edelmann, H.M.L., Duchek, P., Rosenthal, F.E., Föger, N., Glackin, C., Kane, S.E. and Kuchler, K. (1999). Cmdr1, a chicken P-glycoprotein, confers multidrug resistance and interacts with estradiol. *Biological Chemistry*, **380**, 231–41.

Eens, M., Jaspers, V.L.B., van den Steen, E., Bateson, M., Carere, C., Clergeau, P., Costantini, D., Dolenec, Z., Elliott, J.E., Flux, J., Gwinner, H., Halbrook, R.S., Heeb, P., Masgajski, T.D., Moksnes, A., Polo, V., Soler, J.J., Sinclair, R., Veiga, J.P., Williams, T.D., Covaci, A. and Pinxten, R. (2013). Can starling eggs be useful as a biomonitoring tool to study organohalogenated contaminants on a worldwide scale? *Environment International*, **51**, 141–9.

Eeva, T. and Lehikoinen, E. (1995). Egg shell quality, clutch size and hatching success of the great tit (*Parus major*) and the pied flycatcher (*Ficedula hypoleuca*) in an air pollution gradient. *Oecologia*, **102**, 312–23.

Eeva, T., Lehikoinen, E. and Nurmi, J. (1994). Effects of ectoparasites on breeding success of great tits (*Parus major*) and pied flycatchers (*Ficedula hypoleuca*) in an air pollution gradient. *Canadian Journal of Zoology*, **72**, 624–35.

Eeva, T., Ruuskanen, S., Salminen, J.-P., Belskii, E., Järvinen, A., Kerimov, A., Korpimäki, E., Krams, I., Moreno, J., Morosinotto, C., Mänd, R., Orell, M., Qvarnström, A., Siitari, H., Slater, F., Tilgar, V., Visser, M., Winkel, W., Zang, H. and Laaksonen, T. (2011). Geographical trends in the yolk carotenoid composition of the pied flycatcher (*Ficedula hypoleuca*). *Oecologia*, **165**, 277–87.

Eggers, S., Griesser, M., Nystrand, M. and Ekman, J. (2006). Predation risk induces changes in nest-site selection and clutch size in the Siberian jay. *Proceedings of the Royal Society B*, **273**, 701–6.

Eiby, Y.A. and Booth, D.T. (2009). The effects of incubation temperature on the morphology and composition of Australian Brush-turkey (*Alectura lathami*) chicks. *Journal of Comparative Physiology B*, **179**, 875–82.

Eichholz, M.W., Dassow, J.A., Stafford, J.D. and Weatherhead, P.J. (2012). Experimental evidence that nesting ducks use mammalian urine to assess predator abundance. *The Auk*, **129**, 638–44.

Eising, C.M. and Groothuis, T.G.G. (2003). Yolk androgens and begging behaviour in black-headed gull chicks: an experimental field study. *Animal Behaviour*, **66**, 1027–34.

Eising, C.M., Muller, W., Dijkstra, C. and Groothuis, T.G.G. (2003). Maternal androgens in egg yolks: relation with sex, incubation time and embryonic growth. *General and Comparative Endocrinology*, **132**, 241–7.

Eising, C.M., Muller, W. and Groothuis, T.G.G. (2006). Avian mothers create different phenotypes by hormone deposition in their eggs. *Biology Letters*, **2**, 20–2.

Eklund, C.R. and Charlton, F.E. (1959). Measuring the temperatures of incubating penguin eggs. *American Scientist*, **47**, 80–6.

Elf, P.K. and Fivizzani, A.J. (2002). Changes in sex steroid levels in yolks of the leghorn chicken, *Gallus domesticus*, during embryonic development. *Journal of Experimental Zoology*, **293**, 594–600.

El Halawani, M.E., Burke, W.H. and Dennison, P.T. (1980). Effect of nest-deprivation on serum prolactin level in nesting female turkeys. *Biology of Reproduction*, **23**, 118–23.

El Halawani, M.E., Silsby, J.L., Behnke, E.J. and Fehrer, S.C. (1986). Hormonal induction of incubation behavior in ovariectomized female turkeys (*Meleagris gallopavo*). *Biology of Reproduction*, **35**, 59–67.

Elliott, G.P., Merton, D.V. and Jansen, P.W. (2001). Intensive management of a critically endangered species: the kakapo. *Biological Conservation*, **99**, 121–33.

Ellison, L.N. and Cleary, L. (1978). Effects of human disturbance on breeding of Double-crested Cormorants. *The Auk*, **95**, 510–7.

Ells, S.F. (1995). Bobolink protection and mortality on suburban conservation lands. *Bird Observer*, **23**, 98–112.

Elmberg, J. and Pöysä, H. (2011). Is the risk of nest predation heterospecifically density-dependent in precocial species belonging to different nesting guilds? *Canadian Journal of Zoology*, **89**, 1164–71.

Elts, J. (2005). On the nest material of the willow warbler: a quantitative analysis. *Hirundo*, **18**, 31–3.

Elzanowski, A. (1981). Embryonic bird skeletons from the Late Cretaceous of Mongolia. *Palaeontologia Polonica*, **42**, 147–79.

Encabo, S.I., Barba, E., Gil-Delgado, J. and Monrós, J.S. (2001). Fitness consequences of egg shape variation: a study on two passerines and comments on the optimal egg shape model. *Ornis Fennica*, **78**, 83–92.

Eng, M.L., Elliott, J.E., Letcher, R.J. and Williams, T.D. (2013). Individual variation in body burden, lipid status and reproductive investment affects maternal transfer of a brominated diphenyl ether (BDE-99) to eggs in the zebra finch. *Environmental Toxicology and Chemistry*, **32**, 345–52.

Eng, M.L., Williams, T.D., Letcher, R.J. and Elliott, J.E. (2014). Assessment of concentrations and effects of organohalogen contaminants in a terrestrial passerine, the European starling. *Science of The Total Environment*, **473–4**, 589–96.

England, A.S. and Laudenslayer, W.F. Jr. (1989). Distribution and seasonal movements of Bendire's Thrasher in California. *Western Birds*, **20**, 97–124.

Engstrand, S.M. and Bryant, D.M. (2002). A trade-off between clutch size and incubation efficiency in the barn swallow *Hirundo rustica*. *Functional Ecology*, **16**, 782–91.

Engstrand, S.M., Ward, S. and Bryant, D.M. (2002). Variable energetic responses to clutch size manipulations in white-throated dippers *Cinclus cinclus*. *Journal of Avian Biology*, **33**, 371–9.

Engstrom, R.T. (1986). *Geographic variation in clutch size and nesting phenology of the red-winged blackbird* (Agelaius phoeniceus L.). Ph.D. dissertation, Florida State University, Tallahassee.

Erckmann, W.J., Beletsky, L.D., Orians, G.H., Johnsen, T., Sharbaugh, S. and D'Antonio, C. (1990). Old nests as cues for nest-site selection: an experimental test with Red-winged Blackbirds. *The Condor*, **92**, 113–7.

Erskine, A.J. (1971). Some new perspectives on the breeding ecology of common grackles. *The Wilson Bulletin*, **83**, 352–70.

Etches, R.J. (1996). *Reproduction in poultry*. CAB International, Wallingford.

Evans, K.L., Duncan, R.P., Blackburn, T.M. and Crick, H.Q.P. (2005). Investigating geographical variation in clutch size using a natural experiment. *Functional Ecology*, **19**, 616–24.

Evans, K.L., Leech, D.I., Crick, H.Q.P., Greenwood, J.J. and Gaston, K.J. (2009). Latitudinal and seasonal patterns in clutch size of some single-brooded British birds. *Bird Study*, **56**, 75–85.

Evans, M.R. (1997). Nest building signals male condition rather than age in wrens. *Animal Behaviour*, **53**, 749–55.

Evans, M.R. and Burn, J.L. (1996). An experimental analysis of mate choice in the wren: a monomorphic, polygynous passerine. *Behavioral Ecology*, **7**, 101–8.

Evans, R.M. (1988). Embryonic vocalisations as care soliciting signals, with particular reference to the American white pelican. In H Ouellet, ed. *Acta XIX Congressus Internationalis Ornithologica*, pp. 1467–75. University of Ottawa Press, Ottawa.

Evans, R.M. (1992). Embryonic and neonatal vocal elicitation of parental brooding and feeding response in American white pelicans. *Animal Behaviour*, **44**, 667–75.

Evans, R.M., Whitaker, A. and Wiebe, M.O. (1994). Development of vocal regulation of temperature by embryos in pipped eggs of ring-billed gulls. *The Auk*, **111**, 596–604.

EWA. (2014). www.zoo.ox.ac.uk/egi/research-at-the-egi/ethno-ornithology-world-archive-ewa/. [Accessed: 01 August 2014]

Fallis, A.M. and Bennett, G.F. (1961). Sporogony of *Leucocytozoon* and *Haemoproteus* in simuliids and ceratopogonids and a revised classification of the Haemosporidiida. *Canadian Journal of Zoology*, **39**, 215–28.

Fallis, A.M. and Wood, D.M. (1957). Biting midges (Diptera: Ceratopogonidae) as intermediate hosts for *Haemoproteus* of ducks. *Canadian Journal of Zoology*, **35**, 425–35.

Fant, R.J. (1953). A nest-recording device. *Journal of Animal Ecology*, **22**, 323–7.

Fanti, F., Currie, P.J. and Badamgarav, D. (2012). New specimens of *Nemegtomaia* from the Baruungoyot and Nemegt formations (Late Cretaceous) of Mongolia. *PLoS ONE*, **7**(2), e31330.

Farajollahi, A., Crans, W.J., Nickerson, D., Bryant, P., Wolf, B., Glaser, A. and Andreadis, T.G. (2005). Detection of West Nile virus RNA from the louse fly *Icosta americana* (Diptera: Hippoboscidae). *Journal of the American Mosquito Control Association*, **21**, 474–6.

Fargallo, J.A., Blanco, G., Potti, J. and Viñuela, J. (2001). Nest-box provisioning in a rural population of Eurasian kestrels: breeding performance, nest predation and nest parasitism. *Bird Study*, **48**, 236–44.

Fargallo, J.A., de Leon, A. and Potti, J. (2001). Nest-maintenance effort and health status in chinstrap penguins, *Pygoscelis antarctica*: the functional significance of stone-provisioning behaviour. *Behavioral Ecology and Sociobiology*, **50**, 141–50.

Fargallo, J.A., López-Rull, I., Mikšík, I., Eckhardt, A. and Peralta-Sánchez, J.M. (2014). Eggshell pigmentation has no evident effects on offspring viability in common kestrels. *Evolutionary Ecology*, **28**, 627–37.

Fast, P.L., Gilchrist, H. and Clark, R.G. (2010). Nest-site materials affect nest-bowl use by Common Eiders (*Somateria mollissima*). *Canadian Journal of Zoology*, **88**, 214–8.

Fauth, T.P., Krementz, G.D. and Hines, E.J. (1991). Ectoparasitism and the role of green nesting material in the European starling. *Oecologia*, **88**, 22–9.

Feare, C.J. (1976). Desertion and abnormal development in a colony of Sooty Terns (*Sterna fuscata*) infested by virus-infected ticks. *Ibis*, **118**, 112–5.

Feduccia, A. (1996). *The origin and evolution of birds*, 2nd edition. Yale University Press, New Haven.

Feenders, G., Liedvogel, M., Rivas, M., Zapka, M., Horita, H., Hara, E., Wada, K., Mouritsen, H. and Jarvis, E.D. (2008). Molecular mapping of movement-associated areas in the avian brain: a motor theory for vocal learning origin. *PLoS ONE*, **3**(3), e1768.

Ferguson-Lees, J., Castell, R. and Leech, D. (2011). *A field guide to monitoring nests*. British Trust for Ornithology, Thetford.

Fernández, M.S., García, R.A., Fiorelli, L., Scolaro, A., Salvador, R.B., Cotaro, C.N., Kaiser, G.W. and Dyke, G.J. (2013). A large accumulation of avian eggs from the Late Cretaceous of Patagonia (Argentina) reveals a novel nesting strategy in Mesozoic birds. *PLoS ONE*, **8**(4), e61030.

Fessl, B., Sinclair, J.B. and Kleindorfer, B. (2006). The life-cycle of *Philornis downsi* (Diptera: Muscidae) parasitizing Darwin's finches and its impacts on nestling survival. *Parasitology*, **133**, 739–47.

Fessl, B. and Tebbich, S. (2002). *Philornis downsi*—a recently discovered parasite on the Galapagos Archipelago—a threat for Darwin's finches? *Ibis*, **144**, 445–51.

Festing, M.F.W., Overend, P., Gaines Das, R., Cortina Borja, M. and Berdoy, M. (2002). *The design of animal experiments: reducing the use of animals in research through better experimental design*. Laboratory Animal Handbooks No. 14. The Royal Society of Medicine Press, London.

Ficken, M.S. (1965). Mouth color of nestling passerines and its use in taxonomy. *The Wilson Bulletin*, **77**, 71–5.

Fiedler, W. (2009). New technologies for monitoring bird migration and behaviour. *Ringing & Migration*, **24**, 175–9.

Finch, T., Pearce-Higgins, J.W., Leech, D.I. and Evans, K.L. (2014). Carry-over effects from passage regions are more important than breeding climate in determining the breeding phenology and performance of three avian migrants of conservation concern. *Biodiversity and Conservation*, **23**, 2427–44.

Finlay, J.C. (1976). Some effects of weather on purple martin activity. *The Auk*, **93**, 231–44.

Fischer, O.A., Matlova, L., Dvorska, L., Svastova, P., Bartl, J., Weston, R.T. and Pavlik I. (2004). Blowflies *Calliphora vicina* and *Lucilia sericata* as passive vectors of *Mycobacterium avium* subsp. *avium*, *M. a. paratuberculosis* and *M. a. hominissuis*. *Medical and Veterinary Entomology*, **18**, 116–22.

Fitze, P.S., Tschirren, B. and Richner, H. (2004). Life history and fitness consequences of ectoparasites. *Journal of Animal Ecology*, **73**, 216–26.

Fjeldså, J. (2013). The discovery of new bird species. In J del Hoyo, A Elliott, J Sargatal, DA Christie, eds. *Handbook of the birds of the world. Special volume: new species and global index*, pp. 147–85. Lynx Edicions, Barcelona.

Flegg, J.J.M. and Glue, D.E. (1973). A water rail study. *Bird Study*, **20**, 69–70.

Flegg, J.J.M. and Glue, D.E. (1975). The nesting of the ring ouzel. *Bird Study*, **22**, 1–8.

Fletcher, Q.E., Selman, C., Boutin, S., McAdam, A.G., Woods, S.B., Seo, A.Y., Leeuwenburgh, C., Speakman, J.R. and Humphries, M.H. (2013). Oxidative damage increases with reproductive energy expenditure and is reduced by food-supplementation. *Evolution*, **67**, 1527–36.

Flint, P.L., Lindberg, M.S., MacCluskie, M.S. and Sedinger, J.S. (1994). The adaptive significance of hatching synchrony of waterfowl eggs. *Wildfowl*, **45**, 248–54.

Folstad, I. and Karter, A.J. (1992). Parasites, bright males, and the immunocompetence handicap. *The American Naturalist*, **139**, 603–22.

Forbes, S. and Wiebe, M. (2010). Egg size and asymmetric sibling rivalry in red-winged blackbirds. *Oecologia*, **163**, 361–72.

Forsman, J.T., Mönkkönen, M., Korpimäki, E. and Thomson, R.L. (2013). Mammalian nest predator feces as a cue in avian habitat selection decisions. *Behavioral Ecology*, **24**, 262–6.

Forstmeier, W. and Weiss, I. (2004). Adaptive plasticity in nest-site selection in response to changing predation risk. *Oikos*, **104**, 487–99.

Forsythe, D.M. (1971). Clicking in the egg-young of the long-billed curlew. *The Wilson Bulletin*, **83**, 441–2.

Fox, G.A., Gilman, A.P., Peakall, D.B. and Anderka, F.W. (1978). Behavioral abnormalities of nesting Lake Ontario Herring Gulls. *Journal of Wildlife Management*, **42**, 477–83.

Fox, N. (1995). *Understanding the bird of prey*. Hancock House, Surrey.

Fraga, R.M. (1984). Bay-winged Cowbirds (*Molothrus badius*) remove ectoparasites from their brood parasites, the screaming cowbirds (*M. rufoaxillaris*). *Biotropica*, **16**, 223–6.

Freeman, S.N. and Crick, H.Q.P. (2003). The decline of the Spotted Flycatcher *Muscicapa striata* in the UK: an integrated population model. *Ibis*, **145**, 400–12.

Freeman, S.N., Robinson, R.A., Clark, J.A., Griffin, B.M. and Adams, S.Y. (2007). Changing demography and population decline in the Common Starling *Sturnus vulgaris*: a multisite approach to integrated population monitoring. *Ibis*, **149**, 587–96.

French, N.A. and Board, R.G. (1983). Water vapour conductance of wildfowl eggs and incubator humidity. *Wildfowl*, **34**, 144–52.

Friedl, T.W.P. and Klump, G.M. (2000). Nest and mate choice in the Red Bishop (*Euplectes orix*): female settlement rules. *Behavioral Ecology*, **11**, 378–86.

Fry, D.M. and Toone, C.K. (1981). DDT-induced feminization of gull embryos. *Science*, **213**, 922–4.

Fuller, R.J. and Glue, D.E. (1977). The breeding biology of the stonechat and whinchat. *Bird Study*, **24**, 215–28.

Furrer, R.K. (1975). Breeding success and nest site stereotypy in a population of Brewer's Blackbirds (*Euphagus cyanocephalus*). *Oecologia*, **20**, 339–50.

Galligan, E.W., Bakken, G.S. and Lima, S.L. (2003). Using a thermographic imager to find nests of grassland birds. *The Wildlife Society Bulletin*, **31**, 865–9.

Galligan, T.H. and Kleindorfer, S. (2008). Support for the nest mimicry hypothesis in Yellow-rumped Thornbills *Acanthiza chrysorrhoa*. *Ibis*, **150**, 550–7.

Gallizzi, K., Gern, L. and Richner, H. (2008). A flea-induced pre-hatching maternal effect modulates tick feeding behaviour on great tit nestlings. *Functional Ecology*, **22**, 94–9.

Galván, I., Barba, E., Piculo, R., Cantó, J.L., Cortés, V., Monrós, J.S., Atiénzar, F. and Proctor, H. (2008). Feather mites and birds: an interaction mediated by uropygial gland size? *Journal of Evolutionary Biology*, **21**, 133–44.

Galván, I. and Sanz, J.J. (2006). Feather mite abundance increases with uropygial gland size and plumage yellowness in Great Tits *Parus major*. *Ibis*, **148**, 687–97.

Galván, I. and Sanz, J.J. (2011). Mate-feeding has evolved as a compensatory energetic strategy that affects breeding success in birds. *Behavioral Ecology*, **22**, 1088–95.

Game Conservancy Advisory Service. (1993). *Egg production and incubation*. The Game Conservancy, Fordingbridge.

García-Tarrasón, M., Sanpera, C., Jover, L. and Costantini, D. (2014). Levels of antioxidants in breeding female Audouin's gulls and their deposition in eggs across different environments. *Journal of Experimental Marine Biology and Ecology*, **453**, 116–22.

Garner, D.J. and Milne, B.S. (1998). A study of the Long-eared Owl *Asio otus* using wicker nesting baskets. *Bird Study*, **45**, 62–7.

Garvin, J.C., Reynolds, S.J. and Schoech, S.J. (2002). Conspecific egg predation by Florida Scrub-Jays. *The Wilson Bulletin*, **114**, 136–9.

Gasparini, J., Piault, R., Bize, P. and Roulin, A. (2009). Prehatching maternal effects inhibit nestling humoral immune response in the tawny owl *Strix aluco*. *Journal of Avian Biology*, **40**, 271–8.

Gaston, A.J. and Elliot, R.D. (1996). Predation by Ravens *Corvus corax* on Brunnich's Guillemot *Uria lomvia* eggs and chicks and its possible impact on breeding site selection. *Ibis*, **138**, 742–8.

Gaston, K.J. (2010). *Urban ecology*. Cambridge University Press, Cambridge.

Gautron, J., Murayama, E., Vignal, A., Morisson, M., McKee, M.D., Réhault, S., Labas, V., Belghazi, M., Vidal, M.-L., Nys, Y. and Hincke, M.T. (2007). Cloning of ovocalyxin-36, a novel chicken eggshell protein related to lipopolysaccharide-binding proteins, bactericidal permeability-increasing proteins, and plunc family proteins. *The Journal of Biological Chemistry*, **282**, 5273–86.

Gautron, J., Réhault-Godbert, S., Pascal, G., Nys, Y. and Hincke, M.T. (2011). Ovocalyxin-36 and other LBP/BPI/PLUNC-like proteins as molecular actors of the mechanisms of the avian egg natural defences. *Biochemical Society Transactions*, **39**, 971–6.

Gedeon, C.I., Markó, G., Mémeth, I., Nyitral, V. and Altbäcker, V. (2010). Nest material selection affects nest insulation quality for the European ground squirrels (*Spermophilus citellus*). *Journal of Mammalogy*, **91**, 636–41.

Gee, G.F., Hatfield, J.S. and Howey, P.W. (1995). Remote monitoring of parental incubation conditions in the Greater Sandhill Crane. *Zoo Biology*, **172**, 159–72.

Gehlbach, F.R. and Baldridge, R.S. (1987). Live blind snakes (*Leptotyphlops dulcis*) in eastern screech owl (*Otus asio*) nests: a novel commensalism. *Oecologia*, **71**, 560–3.

George, R.S. (1959). Fleas from nests of the pied flycatcher and other species in the Forest of Dean. *Bird Study*, **6**, 132–6.

Gergely, Z., Mészáros, L.A., Szabad, J. and Székely, T. (2009). Old nests are cues for suitable breeding sites in the Eurasian penduline tit *Remiz pendulinus*. *Journal of Avian Biology*, **40**, 2–6.

Giansanti, F., Leboffe, L., Pitari, G., Ippoliti, R. and Antonini, G. (2012). Physiological roles of ovotransferrin. *Biochimica et Biophysica Acta*, **1820**, 218–25.

Gibbs, H.M., Chambers, L.E. and Bennett, A.F. (2011). Temporal and spatial variability of breeding in Australian birds and the potential implications of climate change. *Emu*, **111**, 283–91.

Gil, D. (2008). Hormones in avian eggs: Physiology, ecology and behavior. *Advances in the Study of Behavior*, **38**, 337–98.

Gil, D., Biard, C., Lacroix, A., Spottiswoode, C.N., Saino, N., Puerta, M. and Møller, A.P. (2007). Evolution of yolk androgens in birds: Development, coloniality, and sexual dichromatism. *The American Naturalist*, **169**, 802–19.

Gilardi, K.V.K., Gilardi, J.D., Frank, A., Goff, M.L. and Boyce, W.M. (2001). Epidermoptid mange in Laysan albatross fledglings in Hawaii. *Journal of Wildlife Diseases*, **37**, 490–8.

Gilbert, L., Bulmer, E., Arnold, K.E. and Graves, J.A. (2007). Yolk androgens and embryo sex: maternal effects or confounding factors? *Hormones and Behavior*, **51**, 231–8.

Gilchrist, H.G. and Gaston, A.J. (1997). Effects of murre nest site characteristics and wind conditions on predation by glaucous gulls. *Canadian Journal of Zoology*, **75**, 518–24.

Gill, B.J. (2006). A catalogue of moa eggs (Aves: Dinornithiformes). *Records of Auckland Museum*, **43**, 55–80.

Gill, B.J. (2007). Eggshell characteristics of moa eggs (Aves: Dinornithiformes). *Journal of the Royal Society of New Zealand*, **37**, 139–50.

Gill, F. and Wright, M. (2006). *Birds of the world: recommended English names*. A. and C. Black, London.

Gillis, H., Gauffre, B., Huot, R. and Bretagnolle, V. (2012). Vegetation height and egg coloration differentially affect predation rate and overheating risk: an experimental test mimicking a ground-nesting bird. *Canadian Journal of Zoology*, **90**, 694–703.

Giordano, M., Groothuis, T.G.G. and Tschirren, B. (2014). Interactions between prenatal maternal effects and post-hatching conditions in a wild bird population. *Behavioral Ecology*, **25**, 1459–66.

Giorgi, F. and Lionello, P. (2008). Climate change projections for the Mediterranean region. *Global and Planetary Change*, **63**, 90–104.

Giraudeau, M., Czirják, G.Á., Duval, C., Bretagnolle, V., Gutierrez, C. and Heeb, P. (2014). An experimental test in Mallards (*Anas platyrhynchos*) of the effect of incubation and maternal preen oil on eggshell microbial load. *Journal of Ornithology*, **155**, 671–7.

Gjerdrum, C., Elphick, C.S. and Rubega, M. (2005). Nest site selection and nesting success in saltmarsh breeding sparrows: The importance of nest habitat, timing, and study site differences. *The Condor*, **107**, 849–62.

Gjerdrum, C., Sullivan-Wiley, K., King, E., Rubega, M.A. and Elphick, C.S. (2008). Egg and chick fates during tidal flooding of Saltmarsh Sharp-tailed Sparrow nests. *The Condor*, **110**, 579–84.

Glue, D.E. (1977). Breeding biology of Long-eared Owls. *British Birds*, **70**, 318–31.

Glue, D.E. (1990). Breeding biology of the Grasshopper Warbler in Britain. *British Birds*, **83**, 131–45.

Glue, D.E. and Boswell, T. (1994). Comparative nesting ecology of the three British breeding woodpeckers. *British Birds*, **87**, 253–69.

Glue, D.E. and Morgan, R. (1972). Cuckoo hosts in British habitats. *Bird Study*, **19**, 187–92.

Glue, D.E. and Morgan, R. (1974). Breeding statistics and movements of the stone curlew. *Bird Study*, **21**, 21–8.

Glück, E. (1983). Nest-habitat separation of six European finch species in orchards. *Journal of Ornithology*, **124**, 369–92.

Godsave, S.F., Lohmann, R., Vloet, R.P.M. and Gahr, M. (2002). Androgen receptors in the embryonic zebra finch hindbrain suggest a function for maternal androgens in perihatching survival. *Journal of Comparative Neurology*, **453**, 57–70.

Gold, C.S. and Dahlsten, D.L. (1983). Effects of parasitic flies (*Protocalliphora* spp.) on nestlings of mountain and chestnut-backed chickadees. *The Wilson Bulletin*, **95**, 560–72.

Goldsmith, A.R. (1991). Prolactin and avian reproductive strategies. *Acta XX Congressus Internationalis Ornithologici*, **4**, 2063–71.

Goldstein, D.L. (1988). Estimates of daily energy expenditure in birds: the time-energy budget as an integrator of laboratory and field studies. *American Zoologist*, **28**, 829–44.

Gong, D., Wilson, P.W., Bain, M.M., McDade, K., Kalina, J., Hervé-Grépinet, V., Nys, Y. and Dunn, I.C. (2010). Gallin; an antimicrobial peptide member of a new avian defensin family, the ovodefensins, has been subject to recent gene duplication. *BMC Immunology*, **11**, 12.

Gooch, S., Baillie, S.R. and Birkhead, T.R. (1991). Magpie *Pica pica* and songbird populations. Retrospective investigation of trends in population density and breeding success. *Journal of Applied Ecology*, **28**, 1068–86.

Goodale, W., Attix, L. and Evers, D. (2005). Common Loon, *Gavia immer*, nest attendance patterns recorded by remote video camera. *Canadian Field-Naturalist*, **119**, 455–6.

Goodenough, A.E. and Stallwood, B. (2010). Intraspecific variation and interspecific differences in the bacterial and fungal assemblages of blue tit (*Cyanistes caeruleus*) and great tit (*Parus major*) nests. *Microbial Ecology*, **59**, 221–32.

Goodfellow, P. (2011). *Avian architecture: How birds design, engineer, and build*. Princeton University Press, Princeton.

Goodship, N.M. and Buchanan, K.L. (2006). Nestling testosterone is associated with begging behaviour and fledging success in the pied flycatcher, *Ficedula hypoleuca*. *Proceedings of the Royal Society B*, **273**, 71–6.

Gorchein, A., Lim, C.K. and Cassey, P. (2009). Extraction and analysis of colourful eggshell pigments using HPLC and HPLC/electrospray ionization tandem mass spectrometry. *Biomedical Chromatography*, **23**, 602–6.

Gorman, H.E., Orr, K.J., Adam, A. and Nager, R.G. (2005). Effects of incubation conditions and offspring sex on embryonic development and survival in the Zebra Finch (*Taeniopygia guttata*). *The Auk*, **122**, 1239–48.

Gorman, K.B. and Williams, T.D. (2005). Correlated evolution of maternally derived yolk testosterone and early developmental traits in passerine birds. *Biology Letters*, **1**, 461–4.

Gosler, A.G., Barnett, P.R. and Reynolds, S.J. (2000). Inheritance and variation in eggshell patterning in the great tit *Parus major*. *Proceedings of the Royal Society of London Series B: Biological Sciences*, **267**, 2469–73.

Gosler, A.G., Connor, O.R. and Bonser, R.H.C. (2011). Protoporphyrin and eggshell strength: preliminary findings from a passerine bird. *Avian Biology Research*, **4**, 214–23.

Gosler, A.G., Higham, J.P. and Reynolds, S.J. (2005). Why are birds' eggs speckled? *Ecology Letters*, **8**, 1105–13.

Gosnell, H.T. (1947). *The science of birdnesting*. Jones and Co., Liverpool.

Göth, A. and Booth, B.T. (2005). Temperature-dependent sex ratio in a bird. *Biology Letters*, **1**, 31–3.

Götmark, F. (1992). Blue eggs do not reduce nest predation in the song thrush, *Turdus philomelos*. *Behavioral Ecology and Sociobiology*, **30**, 245–52.

Götmark, F. (1993). Conspicuous nests may select for non-cryptic eggs: a comparative study of avian families. *Ornis Fennica*, **70**, 102–5.

Götmark, F., Blomqvist, D., Johansson, O.C. and Bergkvist, J. (1995). Nest site selection: a trade-off between concealment and view of the surroundings? *Journal of Avian Biology*, **26**, 305–12.

Gottlieb, G. and Vandenbergh, J.G. (1968). Ontogeny of vocalisation in duck and chick embryos. *Journal of Experimental Zoology*, **168**, 307–26.

Grãns, A. and Altimiras, J. (2007). Ontogeny of vocalizations and movements in response to cooling in chicken fetuses. *Physiology and Behavior*, **91**, 229–39.

Graul, W.D. (1975). Breeding biology of the Mountain Plover. *The Wilson Bulletin*, **87**, 6–31.

Green, A.K., Barnes, D.M. and Karasov, W.H. (2005). A new method to measure intestinal activity of P-glycoprotein in avian and mammalian species. *Journal of Comparative Physiology*, **175B**, 57–66.

Green, J.A., Boyd, I.L., Woakes, A.J., Warren, N.L. and Butler, P.J. (2009). Evaluating the prudence of parents: daily energy expenditure throughout the annual cycle of a free-ranging bird, the macaroni penguin *Eudyptes chrysolophus*. *Journal of Avian Biology*, **40**, 529–38.

Green, R.E., Collingham, Y.C., Willis, S.G., Gregory, R.D., Smith, K.W. and Huntley, B. (2008). Performance of climate envelope models in retrodicting recent changes in bird population size from observed climatic change. *Biology Letters*, **4**, 599–602.

Greenberg, R., Elphick, C., Nordby, J.C., Gjerdrum, C., Spautz, H., Shriver, G., Schmeling, B., Olsen, B., Marra, P., Nur, N. and Winter, M. (2006). Flooding and predation: trade-offs in the nesting ecology of tidal-marsh sparrows. *Studies in Avian Biology*, **32**, 96–109.

Greenwood, J.J.D. (2007). Citizens, science and bird conservation. *Journal of Ornithology*, **148** (Supplement 1), S77–S124.

Greenwood, J.J.D. (2012). Citizens, science, and environmental policy: a British perspective. In JL Dickinson, R Bonney, eds. *Citizen science: public participation in environmental research*, pp. 150–64. Cornell University Press, Ithaca.

Gregoire, A., Garnier, S., Dréano, N. and Faivre, B. (2003). Nest predation in Blackbirds (*Turdus merula*) and the influence of nest characteristics. *Ornis Fennica*, **80**, 1–10.

Greig-Smith, P.W. (1982). Dispersal between nest-sites by stonechats *Saxicola torquata* in relation to previous breeding success. *Ornis Scandinavica*, **13**, 232–8.

Grellet-Tinner, G., Chiappe, L., Norell, M. and Bottjer, D. (2002). Dinosaur eggs and nesting behaviours: a paleobiological investigation. *Palaeogeography, Palaeoclimatology, Paleoecology*, **232**, 294–321.

Grellet-Tinner, G. and Dyke, G. (2005). The eggshell of the Eocene bird *Lithornis*. *Acta Palaeontologica Polonica*, **50**, 831–5.

Grellet-Tinner, G. and Norell, M. (2002). An avian egg from the Campanian of Bayn Dzak, Mongolia. *Journal of Vertebrate Paleontology*, **22**, 719–21.

Grellet-Tinner, G., Murelaga, X., Larrasoaña, J.C., Silveira, L.F., Olivares, M., Ortega, L.A., Trimby, P.W. and Pascual, A. (2012). The first occurrence in the fossil record of an aquatic avian twig-nest with Phoenicopteriformes eggs: evolutionary implications. *PLoS ONE*, **7**(10), e46972.

Griffith, S.C. and Gilby, A.J. (2013). Egg development time in the Zebra Finch *Taeniopygia guttata* varies with laying order and clutch size. *Ibis*, **155**, 725–33.

Griffiths, D., Dickinson, A. and Clayton, N. (1999). Episodic memory: what can animals remember about their past? *Trends in Cognitive Sciences*, **3**, 74–80.

Griffiths, P. (1993). The question of *Compsognathus* eggs. *Revue du Paleobiologie*, Special Issue **7**, 85–94.

Grindstaff, J.L. (2008). Maternal antibodies reduce costs of an immune response during development. *The Journal of Experimental Biology*, **211**, 654–60.

Grindstaff, J.L., Brodie, E.D. III and Ketterson, E.D. (2003). Immune function across generations: integrating mechanisms and evolutionary process in maternal antibody transmission. *Proceedings of the Royal Society of London Series B: Biological Sciences*, **270**, 2309–19.

Grindstaff, J.L., Hasselquist, D., Nilsson, J.-Å., Sandell, M.I., Smith, H.G. and Stjernman, M. (2006). Transgenerational priming of immunity: maternal exposure to a bacterial antigen enhances offspring humoral immunity. *Proceedings of the Royal Society B*, **273**, 2551–7.

Groothuis, T.G.G. and Carere, C. (2005). Avian personalities: characterization and epigenesis. *Neuroscience and Biobehavioral Reviews*, **29**, 137–50.

Groothuis, T.G.G., de Jong, B. and Müllera, M. (2014). In search for a theory of testosterone in female birds: a comment on Goymann and Wingfield. *Behavioral Ecology*, **25**, 702–4.

Groothuis, T.G.G., Eising, C., Blount, J., Surai, P., Apanius, V., Dijkstra C. and Muller, W. (2006). Multiple pathways of maternal effects in black-headed gull eggs: constraint and adaptive compensatory adjustment. *Journal of Evolutionary Biology*, **19**, 1304–13.

Groothuis, T.G.G., Muller, W., von Engelhardt, N., Carere, C. and Eising, C. (2005). Maternal hormones as a tool to adjust offspring phenotype in avian species. *Neuroscience and Biobehavioral Reviews*, **29**, 329–52.

Groothuis, T.G.G. and Schwabl, H. (2002). Determinants of within- and among-clutch variation in levels of maternal hormones in Black-Headed Gull eggs. *Functional Ecology*, **16**, 81–9.

Groothuis, T.G.G. and Schwabl, H. (2008). Hormone-mediated maternal effects in birds: mechanisms matter but what do we know of them? *Philosophical Transactions of the Royal Society of London B*, **363**, 1647–61.

Gross, A.O. (1912). Observations of the yellow-billed tropicbird (*Phaethon americanus* Grant) at the Bermuda Islands. *The Auk*, **29**, 49–71.

Grubbauer, P. and Hoi, H. (1996). Female penduline tits (*Remiz pendulinus*) choosing high quality nests benefit by decreased incubation effort and increased hatching success. *Ecoscience*, **3**, 274–9.

Guigueno, M.F. and Sealy, S.G. (2012). Nest sanitation in passerine birds: implications for egg rejection in hosts of brood parasites. *Journal of Ornithology*, **153**, 35–52.

Gula, R., Theuerkauf, J., Rouys, S. and Legault, A. (2010). An audio/video surveillance system for wildlife. *European Journal of Wildlife Research*, **56**, 803–7.

Gunnarsson, G., Elmberg, J., Pöysä, H., Nummi, P., Sjöberg, K., Dessborn, L. and Arzel, C. (2013). Density dependence in ducks: a review of the evidence. *European Journal of Wildlife Research*, **59**, 305–21.

Gurr, L. (1955). A pneumatic nest-recording device. *Ibis*, **97**, 584–6.

Guthrie, D.A., Thomas, H.W. and Kennedy, G.L. (2000). A new species of extinct Late Pleistocene puffin (Aves: Alcidae) from the Southern California Channel Islands. In *Proceedings of a Fifth California Islands Symposium*. U.S. Dept. of the Interior Minerals Management Service, Pacific OCS Region (published as CD) MMS 99-0038, 525–30.

Gwinner, H. (1997). The function of green plants in nests of European starlings (*Sturnus vulgaris*). *Behaviour*, **134**, 337–51.

Gwinner, H. and Berger, S. (2005). European starlings: nestling condition, parasites and green nest material during the breeding season. *Journal of Ornithology*, **146**, 365–71.

Gwinner, H. and Berger, S. (2008). Starling males select green nest material by olfaction using experience-independent and experience-dependent cues. *Animal Behaviour*, **75**, 971–6.

Gwinner, H., Oltrogge, M., Trost, L. and Nienaber, U. (2000). Green plants in starling nests: effects on nestlings. *Animal Behaviour*, **59**, 301–9.

Haas, C.A. (1998). Effects of prior nesting success on site fidelity and breeding dispersal: an experimental approach. *The Auk*, **115**, 929–36.

Haemig, P.D. (1999). Predation risk alters interactions among species: competition and facilitation between ants and nesting birds in a boreal forest. *Ecology Letters*, **2**, 178–84.

Haff, T.M. and Magrath, R.D. (2011). Calling at a cost: elevated nestling calling attracts predators to active nests. *Biology Letters*, **7**, 493–5.

Haffer, J. (2007). The development of ornithology in central Europe. *Journal of Ornithology*, **148** (Supplement 1), S125–53.

Haftorn, S. (1978). Egg-laying and regulation of egg temperature during incubation in the goldcrest *Regulus regulus*. *Ornis Scandinavica*, **9**, 2–21.

Haftorn, S. (1979). Incubation and regulation of egg temperature in the willow tit *Parus montanus*. *Ornis Scandinavica*, **10**, 220–34.

Haftorn, S. (1981). Incubation during the egg-laying period in relation to clutch-size and other aspects of reproduction in the Great Tit *Parus major*. *Ornis Scandinavica*, **12**, 169–85.

Haftorn, S. (1983). Egg temperature during incubation in the great tit *Parus major*, in relation to ambient temperature, time of day, and other factors. *Fauna Norvegica Series C Cinclus*, **6**, 22–38.

Haftorn, S. (1988). Incubating female passerines do not let the egg temperature fall below the physiological zero temperature during their absences from the nest. *Ornis Scandinavica*, **19**, 97–110.

Haftorn, S. (1994). The act of tremble-thrusting in tit nests: performance and possible function. *Fauna Norvegica Series C Cinclus*, **17**, 55–74.

Haftorn, S. and Reinertsen, R.E. (1985). The effect of temperature and clutch size on the energetic cost of incubation in a free-living blue tit (*Parus caeruleus*). *The Auk*, **102**, 470–8.

Haftorn, S. and Reinertsen, R.E. (1990). Thermoregulatory and behavioral responses during incubation of free-living female pied flycatchers *Ficedula hypoleuca*. *Ornis Scandinavica*, **21**, 255–65.

Haftorn, S. and Slagsvold, T. (1995). Egg covering in birds: Description of the behaviour in tits (*Parus* spp.) and a test of hypotheses of its function. *Fauna Norvegica Series C Cinclus*, **18**, 85–106.

Hagelin, J.C. and Miller, G.D. (1997). Nest-site selection in South Polar Skuas: Balancing nest safety and access to resources. *The Auk*, **114**, 638–45.

Hall, R. (1986). The freckled duck—an ancient enigma. *Wildfowl World, The Wildfowl Trust Magazine*, **94**, 20–1.

Hall, Z.J., Bertin, M., Bailey, I.E., Meddle, S.L. and Healy, S.D. (2014). Neural correlates of nesting behavior in zebra finches (*Taeniopygia guttata*). *Behavioural Brain Research*, **264**, 26–33.

Hall, Z.J., Street, S. and Healy, S.D. (2013). The evolution of cerebellum structure correlates with nest complexity. *Biology Letters*, **9**, 20130687.

Halliwell, B.H. and Gutterige, J.M.C. (2007). *Free radicals in biology and medicine*, 4th edition. Oxford University Press, Oxford.

Halupka, K. (1994). Incubation feeding in meadow pipit *Anthus pratensis* affects female time budget. *Journal of Avian Biology*, **25**, 251–3.

Halupka, K. (1998). Nest-site selection and nest predation in meadow pipits. *Folia Zoologica*, **47**, 29–37.

Hamdoun, A. and Epel, D. (2007). Embryo stability and vulnerability in an always changing world. *Proceedings of the National Academy of Sciences of the USA*, **104**, 1745–50.

Hamer, K.C. and Furness, R.W. (1991). Age-specific breeding performance and reproductive effort in great skuas *Catharacta skua*. *Journal of Animal Ecology*, **60**, 693–704.

Hamilton, W.D. and Zuk, M. (1982). Heritable true fitness and bright birds: a role for parasites? *Science*, **218**, 384–7.

Hanley, D. and Doucet, S.M. (2012). Does environmental contamination influence egg coloration? A long-term study in herring gulls. *Journal of Applied Ecology*, **49**, 1055–63.

Hanley, D., Doucet, S.M. and Dearborn, D.C. (2010). A blackmail hypothesis for the evolution of conspicuous egg coloration in birds. *The Auk*, **127**, 453–9.

Hanley, D., Heiber, G. and Dearborn, D.C. (2008). Testing an assumption of the sexual-signalling hypothesis: does blue-green egg color reflect maternal antioxidant capacity? *The Condor*, **110**, 767–71.

Hanley, D., Samaš, P., Hauber, M.E. and Grim, T. (2015). Who moved my eggs? An experimental test of the egg arrangement hypothesis for the rejection of brood parasitic eggs. *Animal Cognition*, **18**, 299–305.

Hanley, D., Stoddard, M.C., Cassey, P. and Brennan, P.L.R. (2013). Eggshell conspicuousness in ground nesting birds: do conspicuous eggshells signal nest location to conspecifics? *Avian Biology Research*, **6**, 147–56.

Hannu, P., Eadie, J.M. and Lyon B.E. (2014). Conspecific brood parasitism in waterfowl and cues parasites use. *Wildfowl* [Special Issue], **4**, 192–219.

Hansell, M. and Ruxton, G.D. (2002). An experimental study of the availability of feathers for avian nest building. *Journal of Avian Biology*, **33**, 319–21.

Hansell, M. H. (1984). *Animal architecture and building behaviour*. Longman, London.

Hansell, M.H. (1995). The demand for feathers as building material by woodland nesting birds. *Bird Study*, **42**, 240–5.

Hansell, M.H. (1996). The function of lichen flakes and white spider cocoons on the outer surface of birds' nests. *Journal of Natural History*, **30**, 303–11.

Hansell, M.H. (2000). *Bird nests and construction behaviour*. Cambridge University Press, Cambridge.

Hansell, M.H. (2007). *Built by animals*. Oxford University Press, Oxford.

Hansell, M.H. and Deeming, D.C. (2002). Location, structure and function of incubation sites. In DC Deeming, ed. *Avian incubation: behavior, environment, and evolution*. pp. 8–27. Oxford University Press, Oxford.

Hanssen, S.A. (2006). Costs of an immune challenge and terminal investment in a long-lived bird. *Ecology*, **87**, 2440–6.

Hanssen, S.A., Erikstad, K.E., Johnsen, V. and Bustnes, J.O. (2003). Differential investment and costs during avian incubation determined by individual quality: an experimental study of the Common Eider (*Somateria mollissima*). *Proceedings of the Royal Society of London Series B: Biological Sciences*, **270**, 531–7.

Hanssen, S.A., Hasselquist, D., Folstad, I. and Erikstad, K.E. (2004). Costs of immunity: immune responsiveness reduces survival in a vertebrate. *Proceedings of the Royal Society of London Series B: Biological Sciences*, **271**, 925–30.

Hanssen, S.A., Hasselquist, D., Folstad, I. and Erikstad, K.E. (2005). Cost of reproduction in a long-lived bird: incubation effort reduces immune function and future reproduction. *Proceedings of the Royal Society of London Series B: Biological Sciences*, **272**, 1039–46.

Hargitai, R., Mateo, R. and Török, J. (2011). Shell thickness and pore density in relation to shell colouration, female characteristics, and environmental factors in the Collared Flycatcher *Ficedula albicollis*. *Journal of Ornithology*, **152**, 579–88.

Hargitai, R., Matus, Z., Hegyi, G., Michl, G., Tóth, G. and Török, J. (2006). Antioxidants in the egg yolk of a wild passerine: differences between breeding seasons. *Comparative Biochemistry and Physiology B*, **143**, 145–52.

Harper, D.G.C. (1999). Feather mites, pectoral muscle condition, wing length and plumage coloration of passerines. *Animal Behaviour*, **58**, 553–62.

Harper, G.H., Marchant, A. and Boddington, D.G. (1992). The ecology of the hen flea *Ceratophyllus gallinae* and the moorhen flea *Dasypsyllus gallinulae* in nestboxes. *Journal of Animal Ecology*, **61**, 317–27.

Harris, S.J., Risely, K., Massimino, D., Newson, S.E., Eaton, M.A., Musgrove, A.J., Noble, D.G., Procter, D. and Baillie, S.R. (2014). *The Breeding Bird Survey 2013. BTO Research Report No. 658*. British Trust for Ornithology, Thetford.

Harrison, C. (1978). *A field guide to the nests, eggs, and nestlings of North American birds*. Collins, New York.

Harrison, C.J.O. (1966). Variation in the distribution of pigment within the shell structure of birds' eggs. *Journal of Zoology*, **148**, 526–39.

Harrison, T. (2005). Fossil bird eggs from the Pliocene of Laetoli, Tanzania: their taxonomic and paleoecological relationships. *Journal of African Earth Sciences*, **41**, 289–302.

Hart, B.L. (1988). Biological basis of the behavior of sick animals. *Neuroscience and Biobehavioral Reviews*, **12**, 123–37.

Hart, B.L. (1990). Behavioral adaptations to pathogens and parasites: five strategies. *Neuroscience and Biobehavioral Reviews*, **14**, 273–94.

Hart, B.L. (1997). Behavioural defence. In DH Clayton, J Moore, eds. *Host–parasite evolution: General principles and avian models*, pp. 59–77. Oxford University Press, Oxford.

Hartman, C.A. and Oring L.W. (2003). Orientation and microclimate of Horned Lark nests: the importance of shade. *The Condor*, **105**, 158–63.

Hartman, C.A. and Oring, L.W. (2006). An inexpensive method for remotely monitoring nest activity. *Journal of Field Ornithology*, **77**, 418–24.

Hartshorne, J.M. (1962). Behavior of the Eastern Bluebird at the nest. *Living Bird*, **1**, 131–50.

Harvey, P.H. and Pagel, M.D. (1991). *The comparative method in evolutionary biology*. Oxford University Press, Oxford.

Harvey, P.H., Stenning, M.L. and Campbell, B. (1988). Factors influencing reproductive success in the pied flycatcher. In TH Clutton-Brock, ed. *Reproductive success: Studies of individual variation in contrasting breeding systems*, pp. 189–200. University of Chicago, Chicago and London.

Hasday, J.D. and Singh, I. (2000). Fever and the heat shock response: distinct, partially overlapping processes. *Cell Stress and Chaperones*, **5**, 471–80.

Hasselquist, D. and Nilsson, J.-Å. (2009). Maternal transfer of antibodies in vertebrates: trans-generational effects on offspring immunity. *Philosophical Transactions of the Royal Society of London B*, **364**, 51–60.

Hasson, O. (1994). Cheating signals. *Journal of Theoretical Biology*, **167**, 223–38.

Hatchwell, B.J., Chamberlain, D.E. and Perrins, C.M. (1996). The reproductive success of blackbirds *Turdus merula* in relation to habitat structure and choice of nest site. *Ibis*, **138**, 256–62.

Hatchwell, B.J., Russell, A.F., Fowlie, M.K. and Ross, D.J. (1999). Reproductive success and nest-site selection in a cooperative breeder: Effect of experience and a direct benefit of helping. *The Auk*, **116**, 355–63.

Hauber, M.E. (2000). Nest predation and cowbird parasitism in song sparrows. *Journal of Field Ornithology*, **71**, 389–98.

Hauber, M.E. and Sherman, P.W. (2001). Self-referent phenotype matching: theoretical considerations and empirical evidence. *Trends in Neurosciences*, **24**, 609–16.

Haussmann, M.F., Longenecker, A.S., Marchetto, N.M., Juliano, S.A. and Bowden, R.M. (2012). Embryonic exposure to corticosterone modifies the juvenile stress response, oxidative stress and telomere length. *Proceedings of the Royal Society B*, **279**, 1447–56.

Hawkes, L.A., Broderick, A.C., Godfrey, M.H. and Godley, B.J. (2007). Investigating the potential impact of climate change on a marine turtle population. *Global Change Biology*, **13**, 923–32.

Hawkins, L.L. (1986). Nesting behavior of male and female whistling swans and implications of male incubation. *Wildfowl*, **37**, 5–27.

Hawley, D.M., DuRant, S.E., Wilson, A.F., Adelman, J.S. and Hopkins, W.A. (2012). Additive metabolic costs of thermoregulation and pathogen infection. *Functional Ecology*, **26**, 701–10.

Hayward, L., Richardson, J., Grogan, M. and Wingfield, J. (2006). Sex differences in the organizational effects of corticosterone in the egg yolk of quail. *General and Comparative Endocrinology*, **146**, 144–8.

Haywood, S. (1993). Sensory and hormonal control of clutch size in birds. *Quarterly Review of Biology*, **68**, 33–60.

Haywood, S. (2013). Origin of evolutionary change in avian clutch size. *Biological Reviews of the Cambridge Philosophical Society*, **88**, 895–911.

Heaney, V. and Monaghan, P. (1996). Optimal allocation of effort between reproductive phases: the trade-off between incubation costs and subsequent brood rearing capacity. *Proceedings of the Royal Society of London Series B: Biological Sciences*, **263**, 1719–24.

Heather, B.D. and Robertson, H.A. (1996). *The field guide to the birds of New Zealand*. Viking, Auckland.

Hebda, G., Pochrząst, K., Mitrus, S. and Wesołowski, T. (2013). Disappearance rates of old nest material from tree cavities: An experimental study. *Scandinavian Journal of Forest Research*, **28**, 445–50.

Hector, J. (1867). Notice of an egg of the great moa (*Dinornis gigantea*) containing remains of an embryo, found in the province of Otago, New Zealand. *Proceedings of the Zoological Society of London*, **3**, 991–2.

Heeb, P., Kölliker, M. and Richner, H. (2000). Bird–ectoparasite interactions, nest humidity and ectoparasite community structure. *Ecology*, **81**, 958–68.

Heeb, P., Werner, I., Kölliker, M. and Richner, H. (1998). Benefits of induced host responses against an ectoparasite. *Proceedings of the Royal Society of London Series B: Biological Sciences*, **265**, 51–6.

Heeb, P., Werner, I., Richner, H. and Kölliker, M. (1996). Horizontal transmission and reproductive rates of hen fleas in great tit nests. *Journal of Animal Ecology*, **65**, 474–84.

Heenan, C.B. (2013). An overview of the factors influencing the morphology and thermal properties of avian nests. *Avian Biology Research*, **6**, 104–18.

Heenan, C.B. and Seymour, R.S. (2011). Structural support, not insulation, is the primary driver for avian cup-shaped nest design. *Proceedings of the Royal Society B*, **278**, 2924–9.

Heenan, C.B. and Seymour, R.S. (2012). The effect of wind on the rate of heat loss from avian cup-shaped nests. *PLoS ONE*, **7**(2), e32252.

Hegyi, G., Herényi, M., Szöllősi, E., Rosivall, B., Török, J. and Groothuis, T.G.G. (2011). Yolk androstenedione, but not testosterone, predicts offspring fate and reflects parental quality. *Behavioral Ecology*, **22**, 29–38.

Heidinger, B.J., Blount, J.D., Boner, W., Griffiths, K., Metcalfe, N.B. and Monaghan, P. (2012). Telomere length in early life predicts lifespan. *Proceedings of the National Academy of Sciences of the USA*, **109**, 1743–8.

Heinroth, O. (1922). Die Beziehungen zwischen Vogelgewicht, Eigewicht, Gelegegewicht und Brutdauer. *Journal of Ornithology*, **70**, 172–285.

Heisenberg, W. (1927). Über den anschaulichen Inhalt der quantentheoretischen Kinematik und Mechanik. *Zeitschrift für Physik*, **43**, 172–98.

Helfenstein, F., Berthouly, A., Tanner, M., Karadas, F. and Richner, H. (2008). Nestling begging intensity and parental effort in relation to prelaying carotenoid availability. *Behavioral Ecology*, **19**, 108–15.

Heneberg, P. (2009). Soil penetrability as a key factor affecting the nesting of burrowing birds. *Ecological Research*, **24**, 453–9.

Heneberg, P. (2013). Decision making in burrowing birds: Sediment properties in conflict with biological variables. *Quaternary International*, **296**, 227–30.

Henriksen, R., Rettenbacher, S. and Groothuis, T.G.G. (2011). Prenatal stress in birds: pathways, effects, function and perspectives. *Neuroscience and Biobehavioral Reviews*, **35**, 1484–501.

Henson, P. and Grant, T.A. (1991). The effects of human disturbance on Trumpeter Swan breeding behavior. *Wildlife Society Bulletin*, **19**, 248–57.

Hepp, G.R. and Kennamer, R.A. (1993). Effects of age and experience on reproductive performance of Wood Ducks. *Ecology*, **74**, 2027–36.

Hepp, G.R. and Kennamer, R.A. (2011). Date of nest initiation mediates incubation costs of Wood Ducks (*Aix sponsa*). *The Auk*, **128**, 258–64.

Hepp, G.R. and Kennamer, R.A. (2012). Warm is better: incubation temperature influences apparent survival and recruitment of Wood Ducks (*Aix sponsa*). *PLoS ONE*, **7**(10), e47777.

Hepp, G.R., Kennamer, R.A. and Johnson, M.H. (2006). Maternal effects in Wood Ducks: incubation temperature influences incubation period and neonate phenotype. *Functional Ecology*, **20**, 307–14.

Herlugson, C. (1981). Nest site selection in mountain bluebirds. *The Condor*, **83**, 253–5.

Herranz, J., Traba, J., Morales, M.B. and Suárez, F. (2004). Nest size and structure variation in two ground nesting passerines, the Skylark *Alauda arvensis* and the Short-toed lark *Calandrella brachydactyla*. *Ardea*, **92**, 209–18.

Hervé, V., Meudal, H., Labas, V., Réhault-Godbert, S., Gautron, J., Berges, M., Guyot, N., Delmas, A.F., Nys, Y. and Landon, C. (2014). Three-dimensional NMR structure of Hen Egg Gallin (Chicken Ovodefensin) reveals a new variation of the β-defensin fold. *The Journal of Biological Chemistry*, **289**, 7211–20.

Hervé-Grépinet, V., Réhault-Godbert, S., Labas, V., Magallon, T., Derache, C., Lavergne, M., Gautron, J., Lalmanach, A.C. and Nys, Y. (2010). Purification and characterization of avian beta-defensin 11, an antimicrobial peptide of the hen egg. *Antimicrobial Agents and Chemotherapy*, **54**, 4401–9.

Hickey J.J. and Anderson D.W. (1968). Chlorinated hydrocarbons and eggshell changes in raptorial and fish-eating birds. *Science*, **162**, 271–3.

Higham, J.P. and Gosler, A.G. (2006). Speckled eggs: water-loss and incubation behaviour in the great tit *Parus major*. *Oecologia*, **149**, 561–70.

Hill, G.E. (2002). *A red bird in a brown bag*. Oxford University Press, New York.

Hill, W.L. (1995). Intraspecific variation in egg composition. *The Wilson Bulletin*, **107**, 382–7.

Hilton, G.M., Hansell, M.H., Ruxton, G.D., Reid, J.M. and Monaghan, P. (2004). Using artificial nests to test importance of nesting material and nest shelter for incubation energetics. *The Auk*, **121**, 777–87.

Hincke, M.T. (1995). Ovalbumin is a component of the chicken eggshell matrix. *Connective Tissue Research*, **31**, 227–33.

Hinde, R.A. and Harrison Matthews, L. (1958). The nest-building behaviour of domesticated canaries. *Proceedings of the Zoological Society of London*, **131**, 1–48.

Hinde, R. and Steel, E. (1972). Reinforcing events in the integration of canary nest building. *Animal Behaviour*, **54**, 1383–92.

Hirsch, K.F. (1989). Interpretations of Cretaceous and Pre-Cretaceous eggs and shell fragments. In DD Gillette, MG Lockley, eds. *Dinosaur tracks and traces*, pp. 89–97. Cambridge University Press, Cambridge.

Hirsch, K.F., Kihm, A.J. and Zelenitsky, D.K. (1996). New eggshell of ratite morphotype with predation marks from the Eocene of Colorado. *Journal of Vertebrate Paleontology*, **17**, 360–9.

Hirsch, K.F. and Packard, M.J. (1987). Review of fossil eggs and their shell structure. *Scanning Microscopy*, **1**, 383–400.

Hirsch-Jagobson, R., Cox, W.A., Tewes, E.E., Thompson, F.R. and Faaborg, J. (2012). Parents or predators: examining intraseasonal variation in nest survival for a migratory passerine. *The Condor*, **114**, 358–64.

Ho, D.H., Reed, W.L. and Burggren, W.W. (2011). Egg yolk environment differentially influences physiological and morphological development of broiler and layer chicken embryos. *The Journal of Experimental Biology*, **214**, 619–28.

Hochachka, W.M., Fink, D., Hutchinson, R.A., Sheldon, D., Wong, W. and Kelling, S. (2012). Data-intensive science applied to broad-scale citizen science. *Trends in Ecology and Evolution*, **27**, 130–7.

Hochbaum, H.A. (1955). *Travels and traditions of waterfowl*. University of Minnesota Press, Minneapolis.

Hodgson, Z.G., Meddle, S.L., Roberts, M.L., Buchanan, K.L., Evans, M.R., Metzdorf, R., Gahr, M. and Healy, S.D. (2007). Spatial ability is impaired and hippocampal mineralocorticoid receptor mRNA expression reduced in zebra finches (*Taeniopygia guttata*) selected for acute high corticosterone response to stress. *Proceedings of the Royal Society B*, **274**, 239–45.

Hoekman, S.T., Ball, I.J. and Fondell, T.F. (2002). Grassland birds orient nests relative to nearby vegetation. *The Wilson Bulletin*, **114**, 450–6.

Hogstad, O. (1995). Do avian and mammalian nest predators select for different nest dispersion patterns of fieldfares *Turdus pilaris*? A 15-year study. *Ibis*, **137**, 484–9.

Hoi, H., Krištín, A., Valera, F. and Hoi, C. (2012). Traditional versus non-traditional nest-site choice: alternative decision strategies for nest-site selection. *Oecologia*, **169**, 117–24.

Hoi, H., Kristofik, J., Darolová, A. and Hoi, C. (2010). Are parasite intensity and related costs of the milichiid fly *Carnus hemapterus* related to host sociality? *Journal of Ornithology*, **151**, 907–13.

Hoi, H., Schleicher, B. and Valera, F. (1994). Female mate choice and nest desertion in penduline tits, *Remiz pendulinus*; the importance of nest quality. *Animal Behaviour*, **48**, 743–6.

Holm, T.E. and Laursen, K. (2009). Experimental disturbance by walkers affects behaviour and territory density of nesting Black-tailed Godwit *Limosa limosa*. *Ibis*, **151**, 77–87.

Holveck, M.-J., Doutrelant, C., Guerreiro, R., Perret, P., Gomez, D. and Grégoire, A. (2010). Can eggs in a cavity be a female secondary sexual signal? Male nest visits and modeling of egg visual discrimination in blue tits. *Biology Letters*, **6**, 453–7.

Hone, D.W.E., Dyke, G.J., Haden, M. and Benton, M.J. (2008). Body size evolution in Mesozoic birds. *Journal of Evolutionary Biology*, **21**, 618–24.

Honza, M., Procházka, P., Morongová, K., Čapek, M. and Jelínek, V. (2011). Do nest light conditions affect rejection of parasitic eggs? A test of the light environment hypothesis. *Ethology*, **117**, 539–46.

Hoover, J.P. (2006). Water depth influences nest predation for a wetland-dependent bird in fragmented bottomland forests. *Biological Conservation*, **127**, 37–45.

Hoover, J.P. and Brittingham, M.C. (1993). Regional variation in cowbird parasitism of Wood Thrushes. *The Wilson Bulletin*, **105**, 228–38.

Hopkins, B.C., DuRant, S.E., Hepp, G.R. and Hopkins, W.A. (2011). Incubation temperature influences locomotor performance in young Wood Ducks (*Aix sponsa*). *Journal of Experimental Zoology*, **315**, 274–9.

Hopla, C.E., Durden, L.A. and Keirnans, J.E. (1994). Ectoparasites and classification. *Scientific and Technical Review of the Office International des Épizooties*, **13**, 985–1017.

Horie, S. and Takagi, M. (2012). Nest positioning by male Daito White-eyes *Zosterops japonicus daitoensis* improves with age to reduce nest predation risk. *Ibis*, **154**, 285–95.

Hoskins, J.D. and Cupp, E.W. (1988). Ticks of veterinary importance. Part II. The Argasidae family: identification, behavior and associated diseases. *Compendium on Continuing Education for the Practising Veterinarian*, **10**, 699–709.

Hötker, H., Kölsch, G. and Visser, G.H. (1996). Der Energieumsatz brütender Säbelschnäbler *Recurvirostra avosetta*. *Journal of Ornithology*, **137**, 203–12.

Houston, A.I. and Davies, N.B. (1985). The ecological consequences of adaptive behaviour. In RM Sibley, RM Smith, eds. *Behavioural ecology*, pp. 471–87. Blackwell Scientific, Oxford.

Houston, A.I., Székely, T. and McNamara, J.M. (2005). Conflict between parents over care. *Trends in Ecology and Evolution*, **20**, 33–8.

Howard, D.V. (1967). Variation in the breeding season and clutch-size of the Robin in the northeastern United States and the maritime provinces of Canada. *The Wilson Bulletin*, **79**, 432–40.

Howell, J.C. (1943). Notes on the nesting habits of the American robin (*Turdus migratorius* L.). *The American Midland Naturalist*, **28**, 529–603.

Howell, T.R. and Dawson, W.R. (1954). Nest temperatures and attentiveness in the Anna Hummingbird. *The Condor*, **56**, 93–7.

Howey, P., Board, R.G., Davis, D.H. and Kear, J. (1984). The microclimate of the nests of waterfowl. *Ibis*, **126**, 16–32.

Howlett, J.S. and Stutchbury, B.J. (1996). Nest concealment and predation in hooded warblers: experimental removal of nest cover. *The Auk*, **113**, 1–9.

Howlett, J.S. and Stutchbury, B.J.M. (1997). Within-season dispersal, nest-site modification, and predation in renesting hooded warblers. *The Wilson Bulletin*, **109**, 643–9.

Hoyt, D.F. (1979). Practical methods of estimating volume and fresh weight of bird eggs. *The Auk*, **96**, 73–7.

Hudson, K. (1994). American robins nesting in *Phragmites*. *Bird Observer*, **22**, 153–5.

Hudson, P.J. and Dobson, A.P. (1997). Host-parasite processes and demographic consequences. In DH Clayton, J Moore, eds. *Host–parasite evolution: General principles and avian models*, pp. 128–54. Oxford University Press, Oxford.

Huggins, R.A. (1941). Egg temperature of wild birds under natural conditions. *Ecology*, **22**, 148–57.

Hughes, A.L. and Yeager, M. (1998). Natural selection at major histocompatibility complex loci of vertebrates. *Annual Review of Genetics*, **32**, 415–35.

Hughes, B.J., Martin, G.R., Wearn, C.P. and Reynolds, S.J. (2013). Sublingual fistula in a Masked Booby (*Sula dactylatra*) and possible role of ectoparasites in its etiology. *Journal of Wildlife Diseases*, **49**, 455–7.

Hulbert, A.J., Pamplona, R., Buffenstein, R. and Buttemer, W.A. (2007). Life and death: Metabolic rate, membrane composition, and life span of animals. *Physiological Reviews*, **87**, 1175–213.

Hulet, R., Gladys, G., Hill, D., Meijerhof, R. and El-Shiekh, T. (2007). Influence of egg shell embryonic incubation temperature and broiler breeder flock age on posthatch growth performance and carcass characteristics. *Poultry Science*, **86**, 408–12.

Humphrey, T.J. (1994). Contamination of egg shell and contents with *Salmonella enteritidis*: a review. *International Journal of Food Microbiology*, **21**, 31–40.

Humphreys, S., Elphick, C.S., Gjerdrum, C. and Rubega, M. (2007). Testing the function of the domed nests of Saltmarsh Sharp-tailed Sparrows. *Journal of Field Ornithology*, **78**, 152–8.

Humphries, D.A. (1968). The host-finding behavior of the hen flea *Ceratophyllus gallinae* (Schrank) (Siphonaptera). *Parasitology*, **58**, 403–14.

Hunter, D.B., Rohner, C. and Currie, D.C. (1997). Mortality in fledging great horned owls from black fly hematophaga and Leucocytozoonosis. *Journal of Wildlife Diseases*, **33**, 486–91.

Huntley, B., Green, R.E., Collingham, Y.C. and Willis, S.G. (2007). *A climatic atlas of European breeding birds*. Lynx Edicions, Barcelona.

Hunton, P. (2005). Research on eggshell structure and quality: an historical overview. *Revista Brasileira de Ciência Avícola*, **7**, 67–71.

Hurtrez-Boussès, S. (2000). Effects of ectoparasites of young on parents' behaviour in a Mediterranean population of Blue Tits. *Journal of Avian Biology*, **31**, 266–9.

Hurtrez-Boussès, S., Blondel, J., Perret, P., Fabreguettes, J. and Renaud, F. (1998). Chick parasitism by blowflies affects feeding rates in a Mediterranean population of blue tits. *Ecology*, **1**, 17–20.

Huynen, L., Gill, B.J., Millar, C.D. and Lambert, D.M. (2010). Ancient DNA reveals extreme egg morphology and nesting behavior in New Zealand's extinct moa. *Proceedings of the National Academy of Sciences of the USA*, **107**, 16201–6.

Ibáñez-Álamo, J.D., Ruiz-Raya, F., Roncalli, G. and Soler, M. (2014). Is nest predation an important selective pressure determining fecal sac removal? The effect of olfactory cues. *Journal of Ornithology*, **155**, 491–6.

Ibáñez-Álamo, J.D., Sanllorente, O. and Soler, M. (2012). The impact of researcher disturbance on nest predation rates: a meta-analysis. *Ibis*, **154**, 5–14.

Ibrahim, H.R., Iwamori, E., Sugimoto, Y. and Aoki, T. (1998). Identification of a distinct antibacterial domain within the N-lobe of ovotransferrin. *Biochimica et Biophysica Acta*, **1401**, 289–303.

Ibrahim, H.R., Sugimoto, Y. and Aoki, T. (2000). Ovotransferrin antimicrobial peptide (OTAP-92) kills bacteria through a membrane damage mechanism. *Biochimica et Biophysica Acta*, **1523**, 196–205.

Igic, B., Cassey, P., Samas, P., Grim, T. and Hauber, M.E. (2009). Cigarette butts form a perceptually cryptic component of song thrush (*Turdus philomelos*) nests. *Notornis*, **56**, 134–8.

Igic, B., Fecheyr-Lippens, D., Xiao, M., Chan, A., Hanley, D., Brennan, P.R.L., Grim, T., Waterhouse, G.I.N., Hauber, M.E. and Shawkey, M.D. (2015). A nanostructural basis for gloss of avian eggshells. *Journal of The Royal Society Interface*, **12**, 20141210.

Igic, B., Greenwood, D.R., Palmer, D.J., Cassey, P., Gill, B.J., Grim, T., Brennan, P.L., Bassett, S.M., Battley, P.F. and Hauber, M.E. (2010). Detecting pigments from colourful eggshells of extinct birds. *Chemoecology*, **20**, 43–8.

Ilmonen, P., Taarna, T. and Hasselquist, D. (2000). Experimentally activated immune defence in female pied flycatchers results in reduced breeding success. *Proceedings of the Royal Society of London Series B: Biological Sciences*, **267**, 665–70.

Ingram, D.R. and Homan, K.D. (2008). A study on the relationship between eggshell color and eggshell quality in commercial broiler breeders. *International Journal of Poultry Science*, **7**, 700–3.

Ipek, A., Sahan, U., Baycan, S.C. and Sozcu, A. (2014). The effects of different eggshell temperatures on embryonic development, hatchability, chick quality, and first-week broiler performance. *Poultry Science*, **93**, 464–72.

Ishikawa, S., Suzuki, K., Fukuda, E., Arihara, K., Yamamoto, Y., Mukai, T. and Itoh, M. (2010). Photodynamic antimicrobial activity of avian eggshell pigments. *FEBS Letters*, **584**, 770–4.

Jackson, F.D., Varricchio, D.V. and Corsini, J.A. (2013). Avian eggs from the Eocene Willwood and Chadron formations of Wyoming and Nebraska. *Journal of Vertebrate Paleontology*, **33**, 1190–201.

Jacob, J. and Ziswiler, V. (1982). The uropygial gland. In D Farner, J King, K Parkes, eds. *Avian biology*, pp. 359–62. Academic Press, London.

Jacot, A., Romero-Diaz, C., Tschirren, B., Richner, H. and Fitze, P.S. (2010). Dissecting carotenoid from structural components of carotenoid-based coloration: a field experiment with Great Tits (*Parus major*). *The American Naturalist*, **176**, 55–62.

Jacquin, L., Blottiere, L., Haussy, C., Perret, S. and Gasparin, J. (2012). Prenatal and postnatal parental effects on immunity and growth in 'lactating' pigeons. *Functional Ecology*, **26**, 866–75.

Jagannath, A., Shore, R.F., Walker, L.A., Ferns, P.N. and Gosler, A.G. (2008). Eggshell pigmentation indicates pesticide contamination. *Journal of Applied Ecology*, **45**, 133–40.

James, F.C. and Shugart, H.H. Jr. (1974). The phenology of the nesting season of the American Robin (*Turdus migratorius*) in the United States. *The Condor*, **76**, 159–68.

Jarrett, N. (1992). First captive-bred Freckled Duck reared at Slimbridge. *Wildfowl & Wetlands*, **107** (Autumn–Winter), 10.

Jarvis, M.J.F., Keffen, R.H. and Jarvis, C. (1985). Some physical requirements for ostrich egg incubation. *Ostrich*, **56**, 42–51.

Jelínek, V., Procházka, P., Požgayová, M. and Honza, M. (2014). Common Cuckoos *Cuculus canorus* change their nest-searching strategy according to the number of available host nests. *Ibis*, **156**, 189–97.

Jenner, E. (1788). Observations on the natural history of the Cuckoo. *Philosophical Transactions of the Royal Society of London*, **78**, 219–37.

Jetz, W., Sekercioglu, C.H. and Böhning-Gaese, K. (2008). The worldwide variation in avian clutch size across species and space. *PLoS BIOLOGY*, **6**(12), e303.

Jetz, W., Thomas, G.H., Joy, J.B., Hartmann, K. and Mooers, A.O. (2012). The global diversity of birds in space and time. *Nature*, **491**, 444–8.

Jetz, W., Thomas, G.H., Joy, J.B., Redding, D.W., Hartmann, K. and Mooers, A.O. (2014). Global distribution and conservation of evolutionary distinctness in birds. *Current Biology*, **24**, 1–12.

Jia, C.-X., Sun, Y.-H. and Swenson, J.E. (2010). Unusual incubation behavior and embryonic tolerance of hypothermia by the Blood Pheasant (*Ithaginis cruentus*). *The Auk*, **127**, 926–31.

Jiménez-Franco, M.V., Martínez, J.E. and Calvo, J.F. (2014). Lifespan analyses of forest raptor nests: patterns of creation, persistence and reuse. *PLoS ONE*, **9**(4), e93628.

Johnsen, A., Vesterkjær, K. and Slagsvold, T. (2011). Do male Pied Flycatchers (*Ficedula hypoleuca*) adjust their feeding effort according to egg colour? *Ethology*, **117**, 309–17.

Johnson, A.L. (2000). Reproduction in the female. In GC Whittow, ed. *Sturkie's avian physiology*, 5th edition, pp. 569–96. Academic Press, London.

Johnson, A.R. and Kikkawa, J. (2003). Flamingos. In CM Perrins, ed. *The new encyclopedia of birds*, pp. 123–9. Oxford University Press, Oxford.

Johnson, A.W., Behn, F. and Millie, W.R. (1958). The South American flamingos. *The Condor*, **60**, 289–99.

Johnson, L.S., Napolillo, F.M., Kozlovsky, D.Y., Hebert, R.M. and Allen, A. (2013). Variation in incubation effort during egg laying in the Mountain Bluebird and its association with hatching asynchrony. *Journal of Field Ornithology*, **84**, 242–54.

Johnstone, R.A. and Hinde, C.A. (2006). Negotiation over offspring care—how should parents respond to each other's efforts? *Behavioral Ecology*, **17**, 818–27.

Jones, C. (1985). Heavy hippoboscid infestations on buzzards. *British Birds*, **78**, 592.

Jones, C.G., Heck, W., Lewis, R.E., Mungroo, Y., Slade, G. and Cade, T. (1995). The restoration of the Mauritius kestrel *Falco punctatus* population. *Ibis*, **137** (Supplement 1), 173–80.

Jones, D.N. (1988). Hatching success of the Australian Brush-turkey *Alectura lathami* in South-east Queensland. *Emu*, **88**, 260–2.

Jones, K.C., Roth, K.L., Islam, K., Hamel, P.B. and Smith, C.G. (2007). The incidence of nest material kleptoparasitism involving Cerulean Warblers. *The Wilson Journal of Ornithology*, **119**, 271–5.

Jongejan, F. and Uilenberg, G. (2004). The global importance of ticks. *Parasitology*, **129**, S3–S14.

Jordan, R. (1989). *Parrot incubation procedures*. Mattachione, Silvio, and Company, Pickering.

Joseph, N.S., Lourens, A. and Moran, E.T. Jr. (2006). The effects of suboptimal eggshell temperature during incubation on broiler chick quality, live performance, and further processing yield. *Poultry Science*, 85, 932–8.

Joys, A.C. and Crick, H.Q.P. (2004). *Breeding periods for selected bird species in England. BTO Research Report No. 352*. British Trust for Ornithology, Thetford.

Juul-Madsen, H.R., Viertlboeck, B., Smith, A.L. and Göbel, T.W.F. (2008). Avian innate immune responses. In F Davison, B Kaspers, KA Schat, eds. *Avian immunology*, pp. 129–58. Elsevier, Amsterdam.

Kaiser, G.W. (2007). *The inner bird: anatomy and evolution*. University of British Columbia Press, Vancouver.

Kaliński, A., Wawrzyniak, J., Bańbura, M., Skwarska, J., Zieliński, P., Gladalski, M. and Bańbura, J. (2014). Does the threat of European Pine Marten (*Martes martes*) predation influence the height of nests built by Blue Tits (*Cyanistes caeruleus*) and Great Tits (*Parus major*)? *Avian Biology Research*, 7, 83–90.

Kania, W. (1992). Safety of catching adult European birds at the nest. Ringers' opinions. *The Ring*, 14, 5–50.

Karl, T.R. and Trenberth, K.E. (2003). Modern global climate change. *Science*, 302, 1719–23.

Kasielke, S. (2007). Incubation of eggs. In LJ Gage, RS Duerr, eds. *Hand-rearing birds*, pp. 39–54. Blackwell, Ames.

Kaur, H., Hughes, M.N., Green, C.J., Naughton, P., Foresti, R. and Motterlini, R. (2003). Interaction of bilirubin and biliverdin with reactive nitrogen species. *FEBS Letters*, 543, 113–9.

Kearns, L.J. and Rodewald, A.D. (2013). Within-season use of public and private information on predation risk in nest-site selection. *Journal of Ornithology*, 154, 163–72.

Kelling, S., Hochachka, W.M., Fink, D., Riedewald, M., Caruana, R., Ballard, G. and Hooker, G. (2009). Data-intensive science: a new paradigm for biodiversity studies. *BioScience*, 59, 613–20.

Kemp, A.C. and Woodcock, M. (1995). *The hornbills: Bucerotiformes*. Oxford University Press, Oxford.

Kendeigh, S.C. (1952). Parental care and its evolution in birds. *Illinois Biological Monographs*, 12.

Kendeigh, S.C. (1963). Thermodynamics of incubation in the house wren, *Troglodytes aedon*. *Proceedings of the 13th International Ornithological Congress*, 884–904.

Kendeigh, S.C. and Baldwin, S.P. (1930). The mechanical recording of the nesting activities of birds. *The Auk*, 47, 471–80.

Kennamer, R.A., Harvey, W.F. IV and Hepp, G.R. (1990). Embryonic development and nest attentiveness of wood ducks during egg laying. *The Condor*, 92, 587–92.

Kennedy, G.Y. and Vevers, H.G. (1973). Eggshell pigments of the araucano fowl. *Comparative Biochemistry and Physiology Part B Comparative Biochemistry*, 44, 11–25.

Kennedy, G.Y. and Vevers, H.G. (1976). Survey of avian eggshell pigments. *Comparative Biochemistry and Physiology Part B Biochemistry and Molecular Biology*, 55, 117–23.

Kern, M. (1984). Racial differences in nests of white-crowned sparrows. *The Condor*, 86, 455–66.

Kern, M. and van Riper, C. (1984). Altitudinal variations in nests of the Hawaiian Honeycreeper, *Hemignathus virens virens*. *The Condor*, 86, 443–54.

Kershner, E.L., Walk, J.W. and Warner, R.E. (2004). Breeding-season decisions, renesting, and annual fecundity of female eastern meadowlarks (*Sturnella magna*) in Southeastern Illinois. *The Auk*, 121, 796–805.

Kersten, M. (1996). Time and energy budgets of Oystercatchers *Haematopus ostralegus* occupying territories of different quality. *Ardea*, 84, 291–310.

Kessel, B. (1957). A study of the breeding biology of the European starling (*Sturnus vulgaris*) in North America. *The American Midland Naturalist*, 58, 257–331.

Kessler, F. (1962). Measurement of nest attentiveness in the Ring-necked Pheasant. *The Auk*, 79, 702–5.

Keyel, A.C., Strong, A.M., Perlut, N.G. and Reed, J.M. (2013). Evaluating the roles of visual openness and edge effects on nest-site selection and reproductive success in grassland birds. *The Auk*, 130, 161–70.

Khorrami, S., Tazawa, H. and Burggren, W. (2008). 'Blood-doping' effects on hematocrit regulation and oxygen consumption in late-stage chicken embryos (*Gallus gallus*). *The Journal of Experimental Biology*, 211, 883–9.

Kilner, R. (1999). Family conflicts and the evolution of nestling mouth colour. *Behaviour*, 136, 779–804.

Kilner, R.M. (2006). The evolution of egg colour and patterning in birds. *Biological Reviews of the Cambridge Philosophical Society*, 81, 383–406.

King, J.R. (1973). Energetics of reproduction in birds. In DS Farner, ed. *Breeding biology of birds*, pp. 78–107. National Academy of Sciences, Washington D.C.

King, M.O., Owen, J.P. and Schwabl, H. (2011). Injecting the mite into ecological immunology: Measuring the antibody response of House Sparrows (*Passer domesticus*) challenged with hematophagous mites. *The Auk*, 128, 340–5.

Kinsky, F. (1971). The consistent presence of paired ovaries in the kiwi (*Apteryx*) with some discussion of this condition in other birds. *Journal für Ornithologie*, 112, 334–57.

Klasing, K.C. (1998). *Comparative avian nutrition*. CAB International, Wallingford.

Klasing, K.C. (2004). The costs of immunity. *Acta Zoologica Sinica*, 50, 961–9.

Klonglan, E.D., Coleman, I.A. and Kozicky, E.L. (1956). A pheasant nest activity recording instrument. *Journal of Wildlife Management*, 20, 173–7.

Klopfer, P.H. (1962). The development of foliage preferences in birds. *American Zoologist*, 2, 534.

Klopfer, P.H. (1963). Behavioral aspects of habitat selection: The role of early experience. *The Wilson Bulletin*, 75, 15–22.

Knudsen, E., Lindén, A., Both, C., Jonzén, N., Pulido, F., Saino, N., Sutherland, W.J., Bach, L.A., Coppack, T., Ergon, T., Gienapp, P., Gill, J.A., Gordo, O., Hedenström, A., Lehikoinen, E., Marra, P.P., Møller, A.P., Nilsson, A.L.K., Péron, G., Ranta, E., Rubolini, D., Sparks, T.H., Spina, F.,

Studds, C.E., Sæther, S.A., Tryjanowski, P. and Stenseth, N.C. (2011). Challenging claims in the study of migratory birds and climate change. *Biological Reviews of the Cambridge Philosophical Society*, **86**, 928–46.

Kohring, R. and Hirsch, K.F. (1996). Crocodilian and avian eggshells from the Middle Eocene of the Geiseltal, Eastern Germany. *Journal of Vertebrate Paleontology*, **16**, 67–80.

Kolaczkowska, A., Kolaczkowski, M., Sokolowska, A., Miecznikowska, H., Kubiak, A., Rolka, K. and Polanowski, A. (2010). The antifungal properties of chicken egg cystatin against Candida yeast isolates showing different levels of azole resistance. *Mycoses*, **53**, 314–20.

Komdeur, J. (1996). Influence of age on reproductive performance in the Seychelles warbler. *Behavioral Ecology*, **7**, 417–25.

Komdeur, J., Wiersma, P. and Magrath, M. (2002). Paternal care and male mate-attraction effort in the European starling is adjusted to clutch size. *Proceedings of the Royal Society of London Series B: Biological Sciences*, **269**, 1253–61.

Korant, B.D., Brzin, J. and Turk, V. (1985). Cystatin, a protein inhibitor of cysteine proteases alters viral protein cleavages in infected human cells. *Biochemical and Biophysical Research Communications*, **127**, 1072–6.

Korschgen, C.E. (1997). Breeding stress of female eiders in Maine. *Journal of Wildlife Management*, **41**, 360–73.

Korschgen, C.E., Kenow, K.P., Green, W.L., Johnson, D.H., Samuel, M.D. and Sileo, L. (1996). Survival of radio-marked Canvasback ducklings in northwestern Minnesota. *Journal of Wildlife Management*, **60**, 120–32.

Kose, M. and Møller, A.P. (1999). Sexual selection, feather breakage and parasites: the importance of white spots in the tail of the barn swallow (*Hirundo rustica*). *Behavioral Ecology and Sociobiology*, **45**, 430–6.

Kosztolanyi, A., Cuthill, I.C. and Székely, T. (2009). Negotiation between parents over care: reversible compensation during incubation. *Behavioral Ecology*, **20**, 446–52.

Kovarik, P., Pavel, V. and Chutny, B. (2009). Incubation behaviour of the Meadow Pipit (*Anthus pratensis*) in an alpine ecosystem of Central Europe. *Journal of Ornithology*, **150**, 549–56.

Kowalczyk, K., Daiss, J., Halpern, J. and Roth, T.F. (1985). Quantitation of maternal-fetal IgG transport in the chicken. *Immunology*, **54**, 755–62.

Kozlovic, D.R., Knapton, R.W. and Barlow, J.C. (1996). Unsuitability of the House Finch as a host of the Brown-headed Cowbird. *The Condor*, **98**, 253–8.

Kreisinger, J. and Albrecht, T. (2008). Nest protection in mallards *Anas platyrhynchos*: untangling the role of crypsis and parental behaviour. *Functional Ecology*, **22**, 872–9.

Krijgsveld, K.L., Ricklefs, R.E. and Visser, G.H. (2012). Daily energy expenditure in precocial chicks: smaller species perform at higher levels. *Journal of Ornithology*, **153**, 1203–14.

Krist, M. (2011). Egg size and offspring quality: a meta-analysis in birds. *Biological Reviews of the Cambridge Philosophical Society*, **86**, 692–716.

Krist, M. and Grim, T. (2007). Are blue eggs a sexually selected signal of female collared flycatchers? A cross-fostering experiment. *Behavioral Ecology and Sociobiology*, **61**, 863–76.

Kristan, D.M., Golightly, R.T. Jr and Tomkiewicz, S.M. Jr. (1996). A solar-powered transmitting video camera for monitoring raptor nests. *Wildlife Society Bulletin*, **24**, 284–90.

Krištín, A., Hoi, H., Valera, F. and Hoi, C. (2007). Philopatry, dispersal patterns and nest-site reuse in Lesser Grey Shrikes (*Lanius minor*). *Biodiversity and Conservation*, **16**, 987–95.

Krištofík, J., Darolová, A., Griggio, M., Majtán, J., Okuliarová, M., Zeman, M., Zídková, L. and Hoi, H. (2013). Does egg colouration signal female and egg quality in reed warbler (*Acrocephalus scirpaceus*)? *Ethology, Ecology and Evolution*, **25**, 129–43.

Krystianiak, S., Kozuszek, R., Kontecka, H. and Nowaczewski, S. (2005). Quality and ultrastructure of eggshell and hatchability of eggs in relation to eggshell colour in pheasants. *Animal Science Papers and Reports*, **23**, 5–14.

Ksepka, D. (2011). *March of the Fossil Penguins*. https://fossilpenguins.wordpress.com/tag/eggs/.

Kubelka, V., Sladecek, M. and Salek, M. (2014). Inter-specific nest scrape reuse in waders: Little Ringed Plovers taking over the nest scrapes of Northern Lapwings. *Bird Study*, **61**, 282–6.

Kuby, J. (1998). *Immunology*. WH Freeman and Company, New York.

Kuehler, C. and Good, J. (1990). Artificial incubation of bird eggs at the Zoological Society of San Diego. *International Zoo Yearbook*, **29**, 118–36.

Kuehler, C., Lieberman, A., Oesterle, P., Powers, T., Kuhn, M., Kuhn, J., Nelson, J., Snetsinger, T., Herrmann, C., Harrity, P., Tweed, E., Fancy, S., Woodworth, B. and Telfer, T. (2000). Development of restoration techniques for Hawaiian thrushes: collection of wild eggs, artificial incubation, hand-rearing, captive-breeding, and re-introduction to the wild. *Zoo Biology*, **19**, 263–77.

Kuehler, C.M. and Witman, P.N. (1988). Artificial incubation of California condor *Gymnogyps californianus* eggs removed from the wild. *Zoo Biology*, **7**, 123–32.

Kull, R.C. (1977). Color selection of nesting material by killdeer. *The Auk*, **94**, 602–4.

Kurochkin, E.N. (1996). *A new enantiornithid of the Mongolian Late Cretaceous, and a general appraisal of the infraclass Enantiornithes (Aves)*. Special Issue, Palaeontological Institute, Russian Academy of Sciences, Moscow.

Kurochkin, E.N., Chatterjee, S. and Mikhailov, K.E. (2013). An embryonic enantiornithine bird and associated eggs from the Cretaceous of Mongolia. *Paleontological Journal*, **47**, 1252–69.

Kwieciński, Z., Krawczyk, A. and Ćwiertnia, P. (2009). Effect of ambient temperature on selected aspects of incubation behaviour in the Golden Eagle *Aquila chrysaetos* under aviary conditions. *Notatki Ornitologiczne*, **50**, 1–8.

Kyo, S., Takakura, M., Kanaya, T., Zhuo, W., Fujimoto, K., Nishio, Y., Orimo, A. and Inoue, M. (1999). Estrogen activates telomerase. *Cancer Research*, **59**, 5917–21.

Lack, D. (1946). Clutch and brood size in the Robin. *British Birds*, **39**, 98–109.

Lack, D. (1947). The significance of clutch size Part I, II. *Ibis*, **89**, 302–52.

Lack, D. (1948a). The significance of clutch size. Part III.—Some interspecific comparisons. *Ibis*, **90**, 25–45.

Lack, D. (1948b). Further notes on clutch size in the Robin. *British Birds*, **41**, 98–104.

Lack, D. (1948c). Natural selection and family size in the Starling. *Evolution*, **2**, 95–110.

Lack, D. (1955a). British tits (*Parus* spp.) in nesting boxes. *Ardea*, **43**, 50–84.

Lack, D. (1955b). Summary report on nesting tits. *Bird Study*, **2**, 199–201.

Lack, D. (1958a). The significance of the colour of turdine eggs. *Ibis*, **100**, 145–66.

Lack, D. (1958b). A quantitative breeding study of British tits. *Ardea*, **46**, 92–124.

Lack, D.L. (1968). *Ecological adaptations for breeding in birds*. Methuen, London.

Lafuma, L., Lambrechts, M.M. and Raymond, M. (2001). Aromatic plants in bird nests as a protection against bloodsucking flying insects? *Behavioural Processes*, **56**, 113–20.

Lambrechts, M.M., Blondel, J., Hurtrez-Bousses, S., Maistre, M. and Perret, P. (1997). Adaptive inter-population differences in blue tit life-history traits on Corsica. *Evolutionary Ecology*, **11**, 599–612.

Lamprecht, I. and Schmolz, E. (2004). Thermal investigations of some bird nests. *Thermochimica Acta*, **415**, 141–8.

Lantz, S.J. and Conway, C.J. (2009). Factors affecting daily nest survival of burrowing owls within black-tailed prairie dog colonies. *Journal of Wildlife Management*, **73**, 232–41.

Latif, Q.S., Heath, S.K. and Rotenberry, J.T. (2012). How avian nest site selection responds to predation risk: testing an 'adaptive peak hypothesis'. *Journal of Animal Ecology*, **81**, 127–38.

Lavelin, I., Meiri, N., Einat, M., Genina, O. and Pines, M. (2002). Mechanical strain regulation of the chicken glypican-4 gene expression in the avian eggshell gland. *American Journal of Physiology—Regulatory, Integrative and Comparative Physiology*, **283**, R853–61.

Lavelin, I., Meiri, N., Genina, O., Alexiev, R. and Pines, M. (2001). Na⁺-K⁺-ATPase gene expression in the avian eggshell gland: distinct regulation in different cell types. *American Journal of Physiology—Regulatory, Integrative and Comparative Physiology*, **281**, R1169–76.

Lavelin, I., Meiri, N. and Pines, M. (2000). New insight in eggshell formation. *Poultry Science*, **79**, 1014–7.

Law-Brown, J. (2003). *Enterococcus phoeniculicola* sp. nov., a novel member of the enterococci isolated from the uropygial gland of the Red-billed Woodhoopoe, *Phoeniculus purpureus*. *International Journal of Systematic and Evolutionary Microbiology*, **53**, 683–5.

Lawver, D.R. and Jackson, F.D. (2014). A review of the fossil record of turtle reproduction: eggs, embryos, nests and copulating pairs. *Bulletin of the Peabody Museum of Natural History*, **55**, 215–36.

Lea, R.W. and Klandorf, H. (2002). The brood patch. In DC Deeming, ed. *Avian incubation: behaviour, environment, and evolution*, pp. 100–18. Oxford University Press, Oxford.

Lea, R.W. and Sharp, P.J. (1989). Concentrations of plasma prolactin and luteinizing-hormone following nest deprivation and renesting in ring doves (*Streptopelia risoria*). *Hormones and Behavior*, **23**, 279–89.

Leader, N. and Yom-Tov, Y. (1998). The possible function of stone ramparts at the nest entrance of the blackstart. *Animal Behaviour*, **56**, 207–17.

Leck, C.F. and Cantor, F.L. (1979). Seasonality, clutch size, and hatching success in the Cedar Waxwing. *The Auk*, **96**, 196–8.

Ledón-Rettig, C.C., Richards, C.L. and Martin, L.B. (2013). Epigenetics for behavioral ecologists. *Behavioral Ecology*, **24**, 311–24.

Lee, J.W., Kim, D.W. and Yoo, J.C. (2005). Egg rejection by both male and female vinous-throated parrotbills *Paradoxornis webbianus*. *Integrative Biosciences*, **9**, 211–3.

Lee, P.L.M. and Prys-Jones, R.P. (2008). Extracting DNA from museum bird eggs, and whole genome amplification of archive DNA. *Molecular Ecology Resources*, **8**, 551–60.

Lee, W.S., Kwon, Y.S. and Yoo, J.C. (2010). Egg survival is related to the colour matching of eggs to nest background in Black-tailed Gulls. *Journal of Ornithology*, **151**, 765–70.

Lee, W.Y., Lee, K., Chun, J., Choe, J.C., Jablonski, P.G. and Lee, S. (2013). Comparison of a culture-based and a PCR-based methods for estimating bacterial abundance on eggshells, with comments on statistical analyses. *Journal of Field Ornithology*, **84**, 304–15.

Leech, D.I., Shawyer, C.R., Barimore, C.J. and Crick, H.Q.P. (2009). The barn owl monitoring programme: establishing a protocol to assess temporal and spatial variation in productivity at a national scale. *Ardea*, **97**, 421–8.

Legagneux, P., Emeriau, S., Giraudeau, M., Duval, C. and Caizergues, A. (2011). Combining field and aviary approaches to monitor incubation in ducks: importance of clutch size, body mass and weather. *Bird Study*, **58**, 421–34.

Leggitt, V.L., Buchheim, H.P. and Biaggi, R.E. (1998). *The stratigraphic setting of three* Presbyornis *nesting sites: Eocene fossil lake, Lincoln County, Wyoming*. National Park Service, Geologic Resources Division Technical Report 98/1, 61–8.

Lehman, C.P., Rumble, M.A., Flake, L.D. and Thompson, D.J. (2008). Merriam's turkey nest survival and factors affecting nest predation by mammals. *Journal of Wildlife Management*, **72**, 1765–74.

Lent, R.A. (1992). Variation in Gray Catbird nest morphology. *Journal of Field Ornithology*, **63**, 411–9.

Leonard, M.L. and Picman, J. (1987). The adaptive significance of multiple nest building by male Marsh Wrens. *Animal Behaviour*, **35**, 271–7.

Lessells, C. and McNamara, J.M. (2012). Sexual conflict over parental investment in repeated bouts: negotiation reduces overall care. *Proceedings of the Royal Society B*, **279**, 1506–14.

Levey, D.J., Duncan, R.S. and Levins, C.F. (2004). Use of dung as a tool by burrowing owls. *Nature*, **431**, 39.

Li, J.Q., Lin, S.T., Wang, Y. and Zhang, Z.W. (2009). Nest-dismantling behavior of the hair-crested drongo in Central China: an adaptive behavior for increasing fitness? *The Condor*, **111**, 197–201.

Li, P.J. and Martin, T.E. (1991). Nest-site selection and nesting success of cavity-nesting birds in high elevation forest drainages. *The Auk*, **108**, 405–18.

Li, T., Zhao, B., Zhou, Y.K., Hu, R. and Du, W.G. (2014). Thermoregulatory behavior is widespread in the embryos of reptiles and birds. *The American Naturalist*, **183**, 445–51.

Licht, D.S., McAuley, D.G., Longcore, J.R. and Sepik, G.F. (1989). An improved method to monitor nest attentiveness using radio-telemetry. *Journal of Field Ornithology*, **60**, 251–8.

Lickliter, R., Bahrick, L.E. and Markham, R.G. (2006). Intersensory redundancy educates selective attention in bobwhite quail embryos. *Developmental Science*, **9**, 604–15.

Lickliter, R., Bahrick, L.E. and Honeycutt, H. (2004). Intersensory redundancy enhances memory in bobwhite quail embryos. *Infancy*, **5**, 253–69.

Lickliter, R. and Hellewell, T.B. (1992). Contextual determinants of auditory learning in bobwhite quail embryos and hatchlings. *Developmental Psychobiology*, **25**, 7–31.

Liker, A., Márkus, M., Vozár, A., Zemankovics, E. and Rózsa, L. (2001). Distribution of *Carnus hemapterus* in a starling colony. *Canadian Journal of Zoology*, **79**, 574–80.

Liljeström, M., Schiavini, A. and Reboreda, J.C. (2009). Chilean swallows (*Tachycineta meyeni*) adjust the number of feathers added to the nest with timing of breeding. *Journal of Ornithology*, **121**, 783–8.

Lill, A. (1979). Nest inattentiveness and its influence on development of the young in the Superb Lyrebird. *The Condor*, **81**, 225–31.

Lill, A. and Fell, P.J. (2007). Microclimate of nesting burrows of the Rainbow Bee-eater. *Emu*, **107**, 108–14.

Lima, S.L. (1987). Clutch size in birds: A predation perspective. *Ecology*, **68**, 1062–70.

Lima, S.L. (2009). Predators and the breeding bird: behavioural and reproductive flexibility under the risk of predation. *Biological Reviews of the Cambridge Philosophical Society*, **84**, 485–513.

Lindström, J. (1999). Early development and fitness in birds and mammals. *Trends in Ecology and Evolution*, **14**, 343–8.

Little, S.E. (2008). Myiasis in wild birds. In TC Atkinson, NJ Thomas, DB Hunter, eds. *Parasitic diseases of wild birds*, pp. 546–56. Wiley-Blackwell, Ames.

Lloyd, J. (2002). Louse flies, keds, and related flies (Hippoboscoidea). In GR Mullen, LA Durden, eds. *Medical and veterinary entomology*, pp. 349–62. Elsevier Science, San Diego.

Lloyd, J.D. and Martin, T.E. (2004). Nest-site preference and maternal effects on offspring growth. *Behavioral Ecology*, **15**, 816–23.

Lloyd, P., Plagányi, E., Lepage, D., Little, R.M. and Crowe, T.M. (2000). Nest-site selection, egg pigmentation and clutch predation in the ground-nesting Namaqua Sandgrouse *Pterocles namaqua*. *Ibis*, **142**, 123–31.

Lochmiller, R.L. and Deerenberg, C. (2000). Trade-offs in evolutionary immunology: just what is the cost of immunity? *Oikos*, **88**, 87–98.

Löhrl, H. (1959). Zur frage des zeipunktes einer prägung auf die heimatregion beim halsbandschnäpper (*Ficedula albicollis*). *Journal für Ornithologie*, **100**, 132–40.

Lombardo, M.P., Bosman, R.M., Faro, C.A., Houtteman, S.G. and Kluisza, T.S. (1995). Effect of feathers as nest insulation on incubation behavior and reproductive performance of Tree Swallows (*Tachycineta bicolor*). *The Auk*, **112**, 973–81.

Londono, G.A., Levey, D.J. and Robinson, S.K. (2008). Effects of temperature and food on incubation behaviour of the northern mockingbird, *Mimus polyglottos*. *Animal Behaviour*, **76**, 669–77.

Long, A.M., Jensen, W.E. and With, K.A. (2009). Orientation of Grasshopper Sparrow and Eastern Meadowlark nests in relation to wind direction. *The Condor*, **111**, 395–9.

Longcore, J.R., Samson, F.T. and Whittendale, T.W. Jr. (1971). DDE thins eggshells and lowers reproductive success of captive black ducks. *Bulletin of Environmental Contamination and Toxicology*, **6**, 485–90.

López-Arrabé, J., Cantarero, A., González-Braojos, S., Ruiz-de-Castañeda, R. and Moreno, J. (2012). Only some ectoparasite populations are affected by nest re-use: an experimental study on pied flycatchers. *Ardeola*, **59**, 253–66.

López-Iborra, G.M., Pinheiro, R.T., Sancho, C. and Martínez, A. (2004). Nest size influences nest predation risk in two coexisting *Acrocephalus* warblers. *Ardea*, **92**, 85–91.

López-Rull, I., Gil, M. and Gil, D. (2007). Spots in starling *Sturnus unicolor* eggs are good indicators of ectoparasite load by *Carnus hemapterus* (Diptera: Carnidae). *Ardeola*, **54**, 131–4.

López-Rull, I., Mikšík, I. and Gil, D. (2008). Egg pigmentation reflects female and egg quality in the spotless starling *Sturnus unicolor*. *Behavioral Ecology and Sociobiology*, **62**, 1877–84.

López-Rull, I., Salaberría, C. and Gil, D. (2010). Seasonal decline in egg size and yolk androgen concentration in a double brooded passerine. *Ardeola*, **57**, 321–32.

Lord, A.M., McCleery, R.H. and Cresswell, W. (2011). Incubation prior to clutch completion accelerates embryonic development and so hatch date for eggs laid earlier in a clutch in the great tit *Parus major*. *Journal of Avian Biology*, **42**, 187–91.

Lotem, A., Nakamura, H. and Zahavi, A. (1992). Rejection of cuckoo eggs in relation to host age: a possible evolutionary equilibrium. *Behavioral Ecology*, **3**, 128–32.

Loukola, O.J., Laaksonen, T., Seppanen, J.T. and Forsman, J.T. (2014). Active hiding of information from information-parasites. *BMC Evolutionary Biology*, **14**, 32.

Lourens, A., van den Brand, H., Meijerhof, R. and Kemp, B. (2005). Effect of eggshell temperature during incubation on embryo development, hatchability, and posthatch development. *Poultry Science*, **84**, 914–20.

Louw, G.N. and Seely, M.K. (1982). *Ecology of desert organisms*. Longman Group, New York.

Love, O.P., Gilchrist, G., Bêty, J., Wynne-Edwards, K.E., Berzins, L. and Williams, T.D. (2009). Using life-histories to predict and interpret variability in yolk hormones. *General and Comparative Endocrinology*, **163**, 169–74.

Love, O.P. and Williams, T.D. (2008a). The adaptive value of stress-induced phenotypes: Effects of maternally derived corticosterone on sex-biased investment, cost of reproduction, and maternal fitness. *The American Naturalist*, **172**, E135–49.

Love, O.P. and Williams, T.D. (2008b). Plasticity in the adrenocortical response of a free-living vertebrate: the role of pre- and post-natal developmental stress. *Hormones and Behavior*, **54**, 496–505.

Lovell, P.G., Ruxton, G.D., Langridge, K.V. and Spencer, K.A. (2013). Egg-laying substrate selection for optimal camouflage by Quail. *Current Biology*, **23**, 260–4.

Loye, J.E. and Carroll, S.P. (1991). Nest ectoparasite abundance and cliff swallow colony site selection, nestling development and departure time. In JE Loye, M Zuk, eds. *Bird–parasite interactions: ecology, evolution, and behaviour*, pp. 221–41. Oxford University Press, Oxford.

Loye, J.E. and Zuk, M. (1991). *Bird–parasite interactions: ecology, evolution, and behaviour*. Oxford University Press, Oxford.

Lü, J.C., Unwin, D.M., Deeming, D.C., Jin, X., Liu, Y. and Ji, Q. (2011). An egg–adult association and its implications for pterosaur reproductive biology. *Science*, **331**, 321–4.

Lu, W., Zhang, W., Molloys, S.S., Thomas, G., Ryan, K., Chiang, Y., Anderson, S. and Laskowski, M. (1993). Arg15-Lys17-Arg18 turkey ovomucoid third domain inhibits human furin. *The Journal of Biological Chemistry*, **268**, 14583–5.

Lubbock, M. (1978). Success of collecting waterfowl from the Arctic. In J Barzdo, ed. *Management of polar birds and mammals in captivity*, Proceedings of Symposium 3 of the Association of British Wild Animal Keepers, 3–8.

Lubbock, M. (1980a). Australian Adventure. *Wildfowl World, The Wildfowl Trust Magazine*, **82**, 22–3.

Lubbock, M. (1980b). Wildfowl avicultural report 1979. *Wildfowl*, **31**, 174–5.

Lubbock, M. (1981a). Pink-ears hatch a success story. *Wildfowl World, The Wildfowl Trust Magazine* Summer **83**, 26–7.

Lubbock, M. (1981b). Wildfowl avicultural report 1980. *Wildfowl*, **32**, 174–5.

Ludvig, É., Vanicsek, L., Török, J. and Csörgö, T. (1995). The effect of nest-height on the seasonal pattern of breeding success in blackbirds *Turdus merula*. *Ardea*, **83**, 411–8.

Lundholm, C.E. (1997). DDE-induced eggshell thinning in birds: effects of p,p′-DDE on the calcium and prostaglandin metabolism of the eggshell gland. *Comparative Biochemistry and Physiology Part C Pharmacology, Toxicology and Endocrinology*, **118**, 113–28.

Lundy, H. (1969). A review of the effects of temperature, humidity, turning, and gaseous environment in the incubator on the hatchability of the hen's egg. In TC Carter, BM Freeman, eds. *The fertility and hatchability of the hen's egg*, pp. 143–76. Oliver and Boyd, Edinburgh.

Lung, N., Thompson, J.P., Kollias, G. Jr, Olsen, J., Zdziarski, J. and Klein, P. (1996). Maternal immunoglobulin G antibody transfer and development of immunoglobulin G antibody responses in blue and gold macaw (*Ara ararauna*) chicks. *American Journal of Veterinary Research*, **57**, 1162–7.

Lyon, B.E. (2003). Egg recognition and counting reduce costs of avian conspecific brood parasitism. *Nature*, **422**, 495–9.

Lyon, B.E. and Eadie, J.M. (2008). Conspecific brood parasitism in birds: a life history perspective. *Annual Review of Ecology and Systematics*, **39**, 343–63.

Lyon, B.E. and Montgomerie, R.D. (1985). Incubation feeding in Snow Buntings: female manipulation or indirect male parental care? *Behavioral Ecology and Sociobiology*, **17**, 279–84.

MacDonald, E.C., Camfield, A.F., Jankowski, J.E. and Martin, K. (2013). Extended incubation recesses by alpine-breeding Horned Larks: a strategy for dealing with inclement weather? *Journal of Field Ornithology*, **84**, 58–68.

MacDonald, M.A. and Bolton, M. (2008). Predation on wader nests in Europe. *Ibis*, **150** (Supplement 1), 54–73.

MacLachlan, I., Nimpf, J. and Schneider, W.J. (1994). Avian riboflavin binding protein binds to lipoprotein receptors in association with vitellogenin. *Journal of Biological Chemistry*, **269**, 24127–32.

Madigan, M., Martinko, J., Stahl, D. and Clark, D. (2010). *Brock biology of microorganisms*, 13th edition. Benjamin Cummings, San Francisco.

Magana, M., Alonso, J.C., Martin, C.A., Bautista, L.M. and Martin, B. (2010). Nest-site selection by Great Bustards *Otis tarda* suggests a trade-off between concealment and visibility. *Ibis*, **152**, 77–89.

Magrath, M.J.L., van Overveld, T. and Komdeur, J. (2005). Contrasting effects of reduced incubation cost on clutch attendance by male and female European starlings. *Behaviour*, **142**, 1479–93.

Magrath, R.D. (1990). Hatching asynchrony in altricial birds. *Biological Reviews of the Cambridge Philosophical Society*, **65**, 587–622.

Magrath, R.D., Haff, T.M., Fallow, P.M. and Radford, A.N. (2014). Eavesdropping on heterospecific alarm calls: from mechanisms to consequences. *Biological Reviews of the Cambridge Philosophical Society*. doi:10.1111/brv.12122.

Mahler, B., Confalonieri, V.A., Lovette, I.J. and Reboreda, J.C. (2008). Eggshell spotting in brood parasitic shiny cowbirds (*Molothrus bonariensis*) is not linked to the female sex chromosome. *Behavioral Ecology and Sociobiology*, **62**, 1193–9.

Mainwaring, M.C., Deeming, D.C., Jones, C.I. and Hartley, I.R. (2014b). Adaptive latitudinal variation in common blackbird *Turdus merula* nest characteristics. *Ecology and Evolution*, **4**, 851–61.

Mainwaring, M.C. and Hartley, I.R. (2008). Seasonal adjustments in nest cup lining in Blue Tits *Cyanistes caeruleus*. *Ardea*, **96**, 278–82.

Mainwaring, M.C. and Hartley, I.R. (2009). Experimental evidence for state-dependent nest weight in the blue tit, *Cyanistes caeruleus*. *Behavioural Processes*, **81**, 144–6.

Mainwaring, M.C. and Hartley, I.R. (2013). The energetic costs of nest building birds. *Avian Biology Research*, **6**, 12–7.

Mainwaring, M.C., Hartley, I.R., Bearhop, S., Brulez, K., du Feu, C.R., Murphy, G., Plummer, K., Webber, S.L., Reynolds, S.J. and Deeming, D.C. (2012). Latitudinal variation in blue and great tit nest characteristics indicates environmental adjustment. *Journal of Biogeography*, **39**, 1669–77.

Mainwaring, M.C., Hartley, I.R., Lambrechts, M.M. and Deeming, D.C. (2014a). The design and function of birds' nests. *Ecology and Evolution*, **4**, 3909–28.

Mallord, J.W., Dolman, P.M., Brown, A. and Sutherland, W.J. (2007). Nest-site characteristics of Woodlarks *Lullula arborea* breeding on heathlands in southern England: are there consequences for nest survival and productivity? *Bird Study*, **54**, 307–14.

Mallory, M.L. and Weatherhead, P.J. (1992). A comparison of three techniques for monitoring avian nest attentiveness and weight. *Journal of Field Ornithology*, **63**, 428–35.

Mallory, M.L. and Weatherhead, P.J. (1993). Incubation rhythms and mass loss of Common Goldeneyes. *The Condor*, **95**, 849–59.

Marais, M., Maloney, S.K. and Gray, D.A. (2011). The metabolic cost of fever in Pekin ducks. *Journal of Thermal Biology*, **36**, 116–20.

Marchant, S. (1974). Analysis of nest-records of the Willie Wagtail. *Emu*, **74**, 149–60.

Marchant, S. (1984). Nest-records of the Eastern Yellow Robin *Eopsaltria australis*. *Emu*, **84**, 167–74.

Marchant, S. and Fullagar, P.J. (1983). Nest records of the Welcome Swallow. *Emu*, **83**, 66–74.

Marchi, G.D., Chiozzi, G., Dell'Omo, G. and Fasola, M. (2015). Low incubation investment in the burrow-nesting Crab Plover *Dromas ardeola* permits extended foraging on a tidal food resource. *Ibis*, **157**, 31–43.

Margalida, A., Ecolan, S., Boudet, J., Bertran, J., Martinez, J.-M. and Heredia, R. (2006). A solar-powered transmitting video camera for monitoring cliff-nesting raptors. *Journal of Field Ornithology*, **77**, 7–12.

Markham, R.G., Toth, G. and Lickliter, R. (2006). Prenatally elevated physiological arousal interferes with perceptual learning in bobwhite quail embryos. *Behavioral Neuroscience*, **120**, 1315–25.

Marples, B.J. and Gurr, L. (1943). A mechanism for recording automatically the nesting habits of birds. *Emu*, **43**, 67–71.

Marshall, A.G. (1981). *The ecology of ectoparasitic insects*. Academic Press, New York.

Marsola, J.C.A., Grellet-Tinner, G., Montefeltro, F.C., Sayão, J.M., Hsiou, A.S. and Langer, M.C. (2014). The first fossil avian egg from Brazil. *Alcheringa*, **38**, 563–7.

Martin, J.M., Bailey, R.L., Phillips, T., Cooper, C., Dickinson, J., Lowe, J., Rietsma, R., Gifford, K. and Bonney, R. (2013b). *NestWatch nest monitoring manual*. Cornell Lab of Ornithology, Ithaca.

Martin, L.B., Scheuerlein, A. and Wikelski, M. (2003). Immune activity elevates energy expenditure of House Sparrows: a link between direct and indirect costs? *Proceedings of the Royal Society of London Series B: Biological Sciences*, **270**, 153–8.

Martin, P. and Bateson, P. (1993). *Measuring behaviour*. Cambridge University Press, Cambridge.

Martin, R.D., Genoud, M. and Hemelrijk, C.K. (2005). Problems of allometric scaling analysis: examples from mammalian reproductive biology. *The Journal of Experimental Biology*, **208**, 1731–47.

Martin, T.E. (1988). Processes organizing open-nesting bird assemblages: competition or nest predation. *Evolutionary Ecology*, **2**, 37–50.

Martin, T.E. (1993a). Nest predation and nest sites. New perspectives on old patterns. *BioScience*, **43**, 523–32.

Martin, T.E. (1993b). Nest predation among vegetation layers and habitat types: revising the dogmas. *The American Naturalist*, **141**, 897–913.

Martin, T.E. (1993c). Evolutionary determinants of clutch size in cavity-nesting birds: nest predation or limited breeding opportunities? *The American Naturalist*, **142**, 937–46.

Martin, T.E. (1995). Avian life-history evolution in relation to nest sites, nest predation, and food. *Ecological Monographs*, **65**, 101–27.

Martin, T.E. (1996). Fitness costs of resource overlap among coexisting bird species. *Nature*, **380**, 338–40.

Martin, T.E. (2002). A new view of avian life-history evolution tested on an incubation paradox. *Proceedings of the Royal Society of London Series B: Biological Sciences*, **269**, 309–16.

Martin, T.E. (2008). Egg size variation among tropical and temperate songbirds: An embryonic temperature hypothesis. *Proceedings of the National Academy of Sciences of the USA*, **105**, 9268–71.

Martin, T.E., Arriero, E. and Majewska, A. (2011). A trade-off between embryonic development rate and immune function of avian offspring is revealed by considering embryonic temperature. *Biology Letters*, **7**, 425–8.

Martin, T.E., Auer, S.K., Bassar, R.D., Niklison, A.M. and Lloyd, P. (2007). Geographic variation in avian incubation periods and parental influences on embryonic temperature. *Evolution*, **61**, 2558–69.

Martin, T.E. and Ghalambor, C.K. (1999). Males feeding females during incubation. I. Required by microclimate or constrained by nest predation? *The American Naturalist*, **153**, 131–9.

Martin, T.E. and Li, P. (1992). Life history traits of open- vs. cavity-nesting birds. *Ecology*, **73**, 579–92.

Martin, T.E. and Maron, J.L. (2012). Climate impacts on bird and plant communities from altered animal-plant interactions. *Nature Climate Change*, **2**, 195–200.

Martin, T.E. and Roper, J.J. (1988). Nest predation and nest-site selection of western population of the Hermit Thrush. *The Condor*, **90**, 51–7.

Martin, T.E. and Schwabl, H. (2008). Variation in maternal effects and embryonic development rates among passerine species. *Philosophical Transactions of the Royal Society B*, **363**, 1663–74.

Martin, T.E., Scott, J. and Menge, C. (2000). Nest predation increases with parental activity: separating nest site and parental activity effects. *Proceedings of the Royal Society Series B: Biological Sciences*, **267**, 2287–93.

Martin, T.E., Ton, R. and Niklison, A. (2013a). Intrinsic vs. extrinsic influences on life history expression: metabolism and parentally induced temperature influences on embryo development rate. *Ecology Letters*, **16**, 738–45.

Martínez-de la Puente, J., Merino, S., Lobato, E., Rivero-de Aguilar, J., Del Cerro, S., Ruiz-de-Castañeda, R. and Moreno, J. (2009). Does weather affect biting fly abundance in avian nests? *Journal of Avian Biology*, **40**, 653–7.

Martínez-de la Puente, J., Merino, S., Lobato, E. Rivero-de Aguilar, J., del Cerro, S., Ruiz-de-Castañeda, R. and Moreno, J. (2010). Nest-climatic factors affect the abundance of biting flies and their effects on nestling condition. *Acta Oecologica*, **36**, 543–7.

Martinez-Vargas, M.C. and Erickson, C.J. (1973). Some social and hormonal determinants of nest-building behaviour in the ring dove (*Streptopelia risoria*). *Behaviour*, **45**, 12–37.

Martín-Platero, A.M., Valdivia, E., Ruíz-Rodríguez, M., Soler, J.J., Martín-Vivaldi, M., Maqueda, M. and Martínez-Bueno, M. (2006). Characterization of antimicrobial substances produced by *Enterococcus faecalis* MRR 10-3, isolated from the uropygial gland of the hoopoe (*Upupa epops*). *Applied and Environmental Microbiology*, **72**, 4245–9.

Martín-Vivaldi, M., Peña, A., Peralta-Sánchez, J.M., Sánchez, L., Ananou, S., Ruiz-Rodríguez, M. and Soler, J.J. (2010). Antimicrobial chemicals in hoopoe preen secretions are produced by symbiotic bacteria. *Proceedings of the Royal Society B*, **277**, 123–30.

Martín-Vivaldi, M., Soler, J.J., Peralta-Sánchez, J.M., Arco, L., Martín-Platero, A.M., Martínez-Bueno, M., Ruiz-Rodríguez, M. and Valdivia, E. (2014). Special structures of hoopoe eggshells enhance the adhesion of symbiont-carrying uropygial secretion that increase hatching success. *Journal of Animal Ecology*, **83**, 1289–301.

Martyka, R., Rutkowska, J. and Cichoń, M. (2011). Sex-specific effects of maternal immunization on yolk antibody transfer and offspring performance in zebra finches. *Biology Letters*, **7**, 50–3.

Marvel, K. and Bonfils, C. (2013). Identifying external influences on global precipitation. *Proceedings of the National Academy of Sciences of the USA*, **110**, 19301–6.

Marzluff, J.M. (1988). Do pinyon jays alter nest placement based on prior experience. *Animal Behaviour*, **36**, 1–10.

Marzluff, J.M., Bowman, R. and Donnelly, R.E. (2001). *Avian conservation and ecology in an urbanizing world*. Kluwer Academic, Norwell.

Masman, D., Daan, S. and Beldhuis, H.J.A. (1988). Ecological energetics of the Kestrel: daily energy expenditure throughout the year based on time-energy budget, food intake and doubly labeled water methods. *Ardea*, **76**, 64–81.

Master, T.L. and Allen, M.C. (2012). Acadian Flycatcher (*Empidonax virescens*) nest tail structure and function in temperate forests. *The American Midland Naturalist*, **167**, 136–49.

Matthysen, E., Adriaensen, F. and Dhondt, A.A. (2011). Multiple responses to increasing spring temperatures in the breeding cycle of blue and great tits (*Cyanistes caeruleus, Parus major*). *Global Change Biology*, **17**, 1–16.

Matysiokova, B., Cockburn, A. and Remeš, V. (2011). Male incubation feeding in songbirds responds differently to nest predation risk across hemispheres. *Animal Behaviour*, **82**, 1347–56.

Matysiokova, B. and Remeš, V. (2014). The importance of having a partner: male help releases females from time limitation during incubation in birds. *Frontiers in Zoology*, **11**, 24.

Mauer, B.A. (1998). The evolution of body size in birds. I. Evidence for non-random diversification. *Evolutionary Ecology*, **12**, 925–34.

Maurer, B. (1996). Energetics of avian foraging. In C Carey, ed. *Avian energetics and nutritional ecology*, pp. 250–79. Chapman and Hall, New York.

Maurer, G., Portugal, S.J. and Cassey, P. (2011b). An embryo's eye view of avian eggshell pigmentation. *Journal of Avian Biology*, **42**, 494–504.

Maurer, G., Portugal, S.J. and Cassey, P. (2012). A comparison of indices and measured values of eggshell thickness of different shell regions using museum eggs of 230 European bird species. *Ibis*, **154**, 714–24.

Maurer, G., Portugal, S.J., Hauber, M.E., Mikšík, I., Russell, D.G.D. and Cassey, P. (2015). First light for avian embryos: eggshell thickness and pigmentation mediate variation in development and UV exposure in wild bird eggs. *Functional Ecology*, **29**, 209–18.

Maurer, G., Portugal, S.J., Mikšík, I. and Cassey, P. (2011a). Speckles of cryptic black-headed gull eggs show no mechanical or conductance structural function. *Journal of Zoology*, **285**, 194–204.

Maxim Integrated Products (2014). http://www.maximintegrated.com/en/products/comms/ibutton.html. [Accessed: 27 August 2014]

Mayer, P.M., Smith, L.M., Ford, R.G., Watterson, D.C., McCutchen, M.D. and Ryan, M.R. (2009). Nest construction by a ground-nesting bird represents a potential trade-off between egg crypticity and thermoregulation. *Oecologia*, **159**, 893–901.

Mayer-Gross, H. (1964). Late nesting in Britain in 1960. *British Birds*, **57**, 102–18.

Mayer-Gross, H., Crick, H.Q.P. and Greenwood, J.J.D. (1997). The effect of observers visiting the nests of passerines: an experimental study. *Bird Study*, **44**, 53–65.

Mayfield, H.F. (1961). Nesting success calculated from exposure. *The Wilson Bulletin*, **73**, 255–61.

Mayfield, H.F. (1975). Suggestions for calculating nest success. *The Wilson Bulletin*, **87**, 456–66.

Maynard Smith, J. and Harper, D.G.C. (1995). Animal signals: models and terminology. *Journal of Theoretical Biology*, **177**, 305–11.

Mayr, G. and Manegold, A. (2013). Can ovarian follicles fossilize? *Nature*, **499**, E1. (doi:10.1038/nature12367)

Mazgajski, T.D. (2007). Effect of old nest material on nest site selection and breeding parameters in secondary hole nesters: a review. *Acta Ornithologica*, **42**, 1–14.

Mazgajski, T.D. (2009). Breeding success of hole-nesting birds—effects of nest sites characteristics and predators avoidance strategies. *Wiadomosci Ekologiczne*, **55**, 159–83.

Mazgajski, T.D. (2013). Nest site preparation and reproductive output of the European Starling (*Sturnus vulgaris*). *Avian Biology Research*, **6**, 119–26.

Maziarz, M. and Wesołowski, T. (2013). Microclimate of tree cavities used by Great Tits (*Parus major*) in a primeval forest. *Avian Biology Research*, **6**, 47–56.

Maziarz, M. and Wesołowski, T. (2014). Does darkness limit the use of tree cavities for nesting by birds? *Journal of Ornithology*, **155**, 793–9.

McAldowie, A.M. (1886). Observations on the development and the decay of the pigment layer on birds' eggs. *Journal of Anatomy and Physiology*, **20**, 225–37.

McClintock, M.E., Hepp, G.R. and Kennamer, R.A. (2014). Plasticity of incubation behaviors helps Wood Ducks (*Aix sponsa*) maintain an optimal thermal environment for developing embryos. *The Auk: Ornithological Advances*, **131**, 672–80.

McCracken, K.G., Afton, A.D. and Alisauskas, R.T. (1997). Nest morphology and body size of Ross' geese and lesser snow geese. *The Auk*, **114**, 610–8.

McDougall, P. and Milne, H. (1978). The anti-predator function of defecation on their own eggs by female Eiders. *Wildfowl*, **29**, 55–9.

McGowan, A., Sharp, S.P. and Hatchwell, B.J. (2004). The structure and function of nests of Long-tailed Tits *Aegithalos caudatus*. *Functional Ecology*, **18**, 578–83.

McGowan, C.P. and Simons, T.R. (2006). Effects of human recreation on the incubation behavior of American Oystercatchers. *The Wilson Journal of Ornithology*, **118**, 485–93.

McGuire, A. and Kleindorfer, S. (2007). Nesting success and apparent nest-adornment in Diamond Firetails (*Stagonopleura guttata*). *Emu*, **107**, 44–51.

McKitrick, M.C. (1992). Phylogenetic analysis of avian parental care. *The Auk*, **109**, 828–46.

McNab, B.K. (2002). *The physiological ecology of vertebrates—a view from energetics*. Cornell University Press, Ithaca.

McNab, B.K. (2009). Ecological factors affect the level and scaling of avian BMR. *Comparative Biochemistry and Physiology*, **152A**, 22–45.

McNair, D.B. (1984a). Reuse of other species' nests by Lark Sparrows. *Southwestern Naturalist*, **29**, 506–9.

McNair, D.B. (1984b). Clutch-size and nest placement in the Brown-headed Nuthatch. *The Wilson Bulletin*, **96**, 296–301.

McNair, D.B. (1985). A comparison of oology and nest record card data in evaluating the reproductive biology of Lark Sparrows, *Chondestes grammacus*. *Southwestern Naturalist*, **30**, 213–24.

McNair, D.B. (1987). Egg data slips—are they useful for information on egg-laying dates and clutch size? *The Condor*, **89**, 369–76.

McNair, D.B. (2000). Summary of historical breeding and breeding-season records of the Lark Sparrow in Tennessee. *The Migrant*, **71**, 73–8.

McNamara, J.M., Houston, A.I., Barta, Z. and Osorno, J.L. (2003). Should young ever be better off with one parent than with two? *Behavioral Ecology*, **14**, 301–10.

Mead, P.S., Morton, M.L. and Fish, B.E. (1987). Sexual dimorphism in egg size and implications regarding facultative manipulation of sex in mountain white-crowned sparrows. *The Condor*, **89**, 798–803.

Medlin, E.C. and Risch, T.S. (2006). An experimental test of snake skin use to deter nest predation. *The Condor*, **108**, 963–5.

Meijerhof, R. and van Beek, G. (1993). Mathematical modelling of temperature and moisture loss of hatching eggs. *Journal of Theoretical Biology*, **165**, 27–41.

Mellor, P.S., Boorman, J. and Baylis, M. (2000). *Culicoides* biting midges: their role as arbovirus vectors. *Annual Review of Entomology*, **45**, 307–40.

Mennerat, A., Mirleau, P., Blondel, J., Perret, P., Lambrechts, M.M. and Heeb, P. (2009c). Aromatic plants in nests of the blue tit *Cyanistes caeruleus* protect chicks from bacteria. *Oecologia*, **161**, 849–55.

Mennerat, A., Perret, P., Bourgault, P., Blondel, J., Gimenez, O., Thomas, D.W., Heeb, P. and Lambrechts, M.M. (2009b). Aromatic plants in nests of blue tits: positive effects on nestlings. *Animal Behaviour*, **77**, 569–74.

Mennerat, A., Perret, P., Caro, P.S., Heeb, P. and Lambrechts, M.M. (2008). Aromatic plants in blue tit *Cyanistes caeruleus* nests: no negative effect on blood-sucking *Protocalliphora* blow fly larvae. *Journal of Avian Biology*, **39**, 127–32.

Mennerat, A., Perret, P. and Lambrechts, M.M. (2009a). Local individual preferences for nest materials in a passerine bird. *PLoS ONE*, **4**(4), e5104.

Merino, S., Moreno, J., Potti, J., De León, A. and Rodríguez, R. (1998). Nest ectoparasites and maternal effort in pied flycatchers. *Biologia e Conservazione del la Fauna*, **102**, 200–5.

Merino, S., Moreno, J., Tomás, G., Martínez, J., Morales, J., Martínez-de la Puente, J. and Osorno, J.L. (2006). Effects of parental effort on blood stress protein HSP60 and immunoglobulins in female blue tit: a brood size manipulation experiment. *Journal of Animal Ecology*, **75**, 1147–53.

Merino, S. and Møller, A.P. (2010). Host-parasite interactions and climate change. In AP Møller, W Fiedler, P Berthold, eds. *Effects of climate change on birds*, pp. 213–26. Oxford University Press, Oxford.

Merino, S. and Potti, J. (1995). Mites and blowflies decrease growth and survival in nestling pied flycatchers. *Oikos*, **73**, 95–103.

Merino, S. and Potti, J. (1995). Pied flycatchers prefer to nest in clean nest boxes in an area with detrimental nest ectoparasites. *The Condor*, **97**, 828–31.

Mersten-Katz, C., Barnea, A., Yom-Tov, Y. and Ar, A. (2013). The woodpecker's cavity microenvironment: Advantageous or restricting? *Avian Biology Research*, **5**, 227–37.

Mertens, J.A.L. (1977). Thermal conditions for successful breeding in Great Tits (*Parus major* L.). II. Thermal properties of nests and nestboxes and their implications for the range of temperature tolerance of Great Tit broods. *Oecologia*, **28**, 31–56.

Mertens, K., Vaesen, I., Loffel, J., Kemps, B., Kamers, B., Perianu, C., Zoons, J., Darius, P., Decuypere, E. and De Baerdemaeker, J. (2010). The transmission color value: A novel egg quality measure for recording shell color used for monitoring the stress and health status of a brown layer flock. *Poultry Science*, **89**, 609–17.

Metcalfe, J., Stock, M.K. and Ingermann, R.L. (1987). Development of the avian embryo. *Journal of Experimental Zoology*, Supplement **1**.

Metcalfe, N.B. and Monaghan, P. (2001). Compensation for a bad start: grow now, pay later? *Trends in Ecology and Evolution*, **16**, 254–60.

Metcalfe, N.B. and Monaghan, P. (2003). Growth versus lifespan: perspectives from evolutionary ecology. *Experimental Gerontology*, **38**, 935–40.

Metcalfe, N.B. and Monaghan, P. (2013). Does reproduction cause oxidative stress? An open question. *Trends in Ecology and Evolution*, **28**, 347–50.

Metz, K.J. (1991). The enigma of multiple nest building by male marsh wrens. *The Auk*, **108**, 170–3.

Mezquida, E.T. (2004). Nest site selection and nesting success of five species of passerines in a South American open *Prosopis* woodland. *Journal of Ornithology*, **145**, 16–22.

Middleton, A.L.A. and Prigoda, E. (2001). What does 'fledging' mean? *Ibis*, **143**, 296–8.

Midtgård, U., Sejrsen, P. and Johansen, K. (1985). Blood-flow in the brood patch of bantam hens—evidence of cold vasodilatation. *Journal of Comparative Physiology B Biochemical Systemic and Environmental Physiology*, **155**, 703–9.

Mikhailov, K.E. (1992). The microstructure of avian and dinosaurian eggshell: phylogenetic implications. In KE Campbell, ed. *Papers in avian paleontology: honoring Pierce Brodkorb*, pp. 361–73. Science Series No. 36, Natural History Museum of LA County, Los Angeles.

Mikhailov, K.E. (1997a). *Avian eggshells: an atlas of scanning electron micrographs*. British Ornithologists' Club Occasional Publications, No. 3, 87 pp.

Mikhailov, K.E. (1997b). Fossil and recent eggshell in amniotic vertebrates: fine structure, comparative morphology and classification. *Special Papers in Palaeontology*, **56**, 1–80.

Mikhailov, K.E., Bray, E.S. and Hirsch, K.F. (1996). Parataxonomy of fossil egg remains (Vertebrata): principles and applications. *Journal of Vertebrate Paleontology*, **16**, 763–9.

Miller, G.H., Magee, J.W., Johnson, B.J., Fogel, M.L., Spooner, N.A., McCulloch, M.T. and Aycliffe, L.K. (1999). Pleistocene extinction of *Genyornis newtoni*: human impact on Australian megafauna. *Science*, **283**, 181–205.

Mills, A.M. (1986). The influence of moonlight on the behavior of goatsuckers (Caprimulgidae). *The Auk*, **103**, 370–8.

Milne, K.A. and Hejl, S.J. (1989). Nest-site characteristics of white-headed woodpeckers. *Journal of Wildlife Management*, **53**, 50–5.

Miskelly, C.M. (1989). Flexible incubation system and prolonged incubation in New Zealand Snipe. *The Wilson Bulletin*, **101**, 127–32.

Mitchell, T.S., Warner, D.A. and Janzen, F.J. (2013). Phenotypic and fitness consequences of maternal nest-site choice across multiple early life stages. *Ecology*, **94**, 336–45.

Mock, D.W. and Parker, G.A. (1997). *The evolution of sibling rivalry*. Oxford University Press, Oxford.

Moczydłowski, E. (1986). Microclimate of the nest-sites of pygoscelid penguins (Admiralty Bay, South Shetland Islands). *Polish Polar Research*, **7**, 377–94.

Moczydłowski, E. (1989). Protection of eggs and chicks against flooding as a part of nesting strategy of pygoscelid penguins at King George Island, South Shetlands. *Polish Polar Research*, **10**, 163–81.

Mohabey, D.M., Udhoji, S.G. and Verma, K.K. (1993). Palaeontological and sedimentological observations on non-marine Lameta Formation (Upper Cretaceous) of Maharashtra, India: their palaeoecological and paleaoenvironmental significance. *Palaeogeography, Palaeoclimatology, Palaeoecology*, **105**, 83–94.

Moisiadis, V.G. and Matthews, S.G. (2014). Glucocorticoids and fetal programming part 2: mechanisms. *Nature Reviews Endocrinology*, **10**, 403–11.

Møller, A.P. (1984). On the use of feathers in birds' nests: predictions and tests. *Ornis Scandinavica*, **15**, 38–42.

Møller, A.P. (1987). Egg predation as a selective factor for nest design: an experiment. *Oikos*, **50**, 91–4.

Møller, A.P. (1988). Nest predation and nest site choice in passerine birds in habitat patches of different size: a study of magpies and blackbirds. *Oikos*, **53**, 215–21.

Møller, A.P. (1990a). Nest predation selects for small nest size in the blackbird. *Oikos*, **57**, 237–40.

Møller, A.P. (1990b). Effects of parasitism by a hematophagous mite on reproduction in the barn swallow. *Ecology*, **71**, 2345–57.

Møller, A.P. (2006). Rapid change in nest size of a bird related to change in a secondary sexual character. *Behavioral Ecology*, **17**, 108–16.

Møller, A.P. (2012). Reproductive behaviour. In U Candolin, BBM Wong, eds. *Behavioural responses to a changing world*, pp. 106–18. Oxford University Press, Oxford.

Møller, A.P., Allander, K. and Dufva, R. (1990). Fitness effects of parasites on passerine birds: A review. *Population Biology of Passerine Birds*, **24**, 269–80.

Møller, A.P., de Lope, F. and Saino, N. (2004). Parasitism, immunity, and arrival date in a migratory bird, the barn swallow. *Ecology*, **85**, 206–19.

Møller, A.P., Flensted-Jensen, E., Klarborg, K., Mardal, W. and Nielsen, J.T. (2010). Climate change affects the duration of the reproductive season in birds. *Journal of Animal Ecology*, **79**, 777–84.

Møller, A.P., Flensted-Jensen, E., Mardal, W. and Soler, J.J. (2013). Host-parasite relationship between colonial terns and bacteria is modified by a mutualism with a plant with antibacterial defenses. *Oecologia*, **173**, 169–78.

Møller, A.P. and Saino, N. (2004). Immune response and survival. *Oikos*, **104**, 299–304.

Monaghan, P. (2008). Early growth conditions, phenotypic development and environmental change. *Philosophical Transactions of the Royal Society B*, **363**, 1635–45.

Monaghan, P., Bolton, M. and Houston, D.C. (1995). Egg production constraints and the evolution of avian clutch size. *Proceedings of the Royal Society of London Series B: Biological Sciences*, **259**, 189–91.

Monaghan, P., Metcalfe, N.B. and Torres, R. (2009). Oxidative stress as a mediator of life history trade-offs: mechanisms, measurements and interpretation. *Ecology Letters*, **12**, 75–92.

Monaghan, P. and Nager, R.G. (1997). Why don't birds lay more eggs? *Trends in Ecology and Evolution*, **12**, 270–4.

Monaghan, P., Nager, R.G. and Houston, D.C. (1998). The price of eggs: increased investment in egg production reduces the offspring rearing capacity of parents. *Proceedings of the Royal Society of London Series B: Biological Sciences*, **265**, 1731–5.

Mönkkönen, M., Forsman, J.T., Kananoja, T. and Ylönen, H. (2009). Indirect cues of nest predation risk and avian reproductive decisions. *Biology Letters*, **5**, 176–8.

Montevecchi, W.A. (1976). Field experiments on the adaptive significance of avian eggshell pigmentation. *Behaviour*, **58**, 26–39.

Moore, J. (2002). *Parasites and the behaviour of animals*. Oxford University Press, Oxford.

Moore, M.C. and Johnston, G.I.H. (2008). Toward a dynamic model of deposition and utilization of yolk steroids. *Integrative and Comparative Biology*, **48**, 411–8.

Moore, N. (1966). A pesticide monitoring system with special reference to the selection of indicator species. *Journal of Applied Ecology*, **3**, 261–9.

Morales, J., Sanz, J.J. and Moreno, J. (2006). Egg colour reflects the amount of yolk maternal antibodies and fledging success in a songbird. *Biology Letters*, **2**, 334–6.

Morales, J., Velando, A. and Moreno, J. (2008). Pigment allocation to eggs decreases plasma antioxidants in a songbird. *Behavioral Ecology and Sociobiology*, **63**, 227–33.

Moreno, J. (1989). Energetic constraints on uniparental incubation in the wheatear *Oenanthe oenanthe* (L.). *Ardea*, **77**, 107–15.

Moreno, J. (2012). Avian nests and nest-building as signals. *Avian Biology Research*, **5**, 238–51.

Moreno, J. and Carlson, A. (1989). Clutch size and the costs of incubation in the pied flycatcher *Ficedula hypoleuca*. *Ornis Scandinavica*, **20**, 123–8.

Moreno, J., Gustafsson, L., Carlson, A. and Pärt, T. (1991). The cost of incubation in relation to clutch size in the collared flycatcher *Ficedula albicollis*. *Ibis*, **133**, 186–93.

Moreno, J., Lobato, E., Merino, S. and Martínez-de la Puente, J. (2008b). Blue–green eggs in pied flycatchers: an experimental demonstration that a supernormal stimulus elicits improved nestling condition. *Ethology*, **114**, 1078–83.

Moreno, J., Lobato, E., Morales, J., Merino, S., Tomás, G., Martinez-de la Puente, J., Sanz, J.J., Mateo, R. and Soler, J.J. (2006a). Experimental evidence that egg color indicates female condition at laying in a songbird. *Behavioral Ecology*, **17**, 651–5.

Moreno, J., Martínez, J., Corral, C., Lobato, E., Merino, S., Morales, J., Martínez-de la Puente, J. and Tomás, G. (2008a). Nest construction rate and stress in female Pied Flycatchers *Ficedula hypoleuca*. *Acta Ornithologica*, **43**, 57–64.

Moreno, J., Merino, S., Lobato, E., Ruiz-de Castañeda, R., Martínez-de la Puente, J., del Cerro, S. and Rivero-de Aguilar, J. (2009). Nest-dwelling ectoparasites of two sympatric hole-nesting passerines in relation to nest composition: an experimental study. *Ecoscience*, **16**, 418–27.

Moreno, J., Morales, J., Lobato, E., Merino, S., Tomás, G. and Martinez-de la Puente, J. (2005). Evidence for the signaling function of egg color in the pied flycatcher *Ficedula hypoleuca*. *Behavioral Ecology*, **16**, 931–7.

Moreno, J., Morales, J., Lobato, E., Merino, S., Tomás, G. and Martínez-de la Puente, J. (2006b). More colourful eggs induce a higher relative paternal investment in the pied flycatcher *Ficedula hypoleuca*: a cross-fostering experiment. *Journal of Avian Biology*, **37**, 555–60.

Moreno, J. and Osorno, J.L. (2003). Avian egg colour and sexual selection: does eggshell pigmentation reflect female condition and genetic quality? *Ecology Letters*, **6**, 803–6.

Moreno, J., Osorno, J.L., Morales, J., Merino, S. and Tomás, G. (2004). Egg colouration and male parental effort in the pied flycatcher *Ficedula hypoleuca*. *Journal of Avian Biology*, **35**, 300–4.

Moreno, J. and Sanz, J.J. (1994). The relationship between the energy expenditure during incubation and clutch size in the pied flycatcher *Ficedula hypoleuca*. *Journal of Avian Biology*, **25**, 125–30.

Moreno-Rueda, G. (2014). Body-mass-dependent trade-off between immune response and uropygial gland size in house sparrows *Passer domesticus*. *Journal of Avian Biology*, **45**, 1–6.

Moreno-Rueda, G. and Hoi, H. (2012). Female house sparrows prefer big males with a large white wing bar and fewer feather holes caused by chewing lice. *Behavioral Ecology*, **23**, 271–7.

Morgan, R. (1982a). The breeding biology of the Nightingale *Luscinia megarhynchos* in Britain. *Bird Study*, **29**, 67–72.

Morgan, R. and Shorten, M. (1974). Breeding of the woodcock in Britain. *Bird Study*, **21**, 193–9.

Morgan, R.A. (1982b). *Breeding seasons of some British waders*. BTO Research Report No 7. British Trust for Ornithology, Tring.

Morgan, R.A. and Glue, D.E. (1977). Breeding, mortality and movements of Kingfishers. *Bird Study*, **24**, 15–24

Morgan, R.A. and Glue, D.E. (1981). Breeding survey of Black Redstarts in Britain 1977. *Bird Study*, **28**, 163–8.

Morosinotto, C., Thomson, R.L., Hänninen, M. and Korpimäki, E. (2012). Higher nest predation risk in association with a top predator: mesopredator attraction? *Oecologia*, **170**, 507–15.

Morton, M.L. and Pereyra, M.E. (1985). The regulation of egg temperatures and attentiveness patterns in the Dusky Flycatcher (*Empidonax oberholseri*). *The Auk*, **102**, 25–37.

Morton, M.L., Sockman, K.W. and Peterson, L.E. (1993). Nest predation in the Mountain White-Crowned Sparrow. *The Condor*, **95**, 72–82.

Moskát, C., Székely, T., Kisbenedek, T., Karcza, Z. and Bártol, I. (2003). The importance of nest cleaning in egg rejection behaviour of Great Reed Warblers *Acrocephalus arundinaceus*. *Journal of Avian Biology*, **34**, 16–9.

Moss, D., Joys, A.C., Clark, J.A., Kirby, A., Smith, A., Baines, D. and Crick, H.Q.P. (2005). *Timing of breeding of moorland birds*. BTO Research Report No. 362, British Trust for Ornithology, Thetford.

Mougeot, F., Benítez-López, A., Casas, F., Garcia, J.T. and Viñuela, J. (2014). A temperature-based monitoring of nest attendance patterns and disturbance effects during incubation by ground-nesting sandgrouse. *Journal of Arid Environments*, **102**, 89–97.

Mouritsen, K.N. and Madsen, J. (1994). Toxic birds: Defence against parasites? *Oikos*, **69**, 357–8.

Mousseau, T.A. and Fox, C.W. (1998a). *Maternal effects as adaptations*. Oxford University Press, Oxford.

Mousseau, T.A. and Fox, C.W. (1998b). The adaptive significance of maternal effects. *Trends in Ecology and Evolution*, **13**, 403–7.

Moyer, B.R. and Wagenbach, G.E. (1995). Sunning by Black Noddies (*Anous minutus*) may kill chewing lice (*Quadraceps hopkinsi*). *The Auk*, **112**, 1073–7.

Moyer, R.B., Drown, D.M. and Clayton D.H. (2002). Low humidity reduces ectoparasite pressure: implications for host life history evolution. *Oikos*, **97**, 223–8.

Mullen, G.R. (2002). Biting midges (Ceratopogonidae). In GR Mullen, LA Durden, eds. *Medical and veterinary entomology*, pp. 163–83. Elsevier Science, San Diego.

Müller, M. and Groothuis, T.G.G. (2013). Within-clutch variation in yolk testosterone as an adaptive maternal effect to modulate avian sibling competition: Evidence from a comparative study. *The American Naturalist*, **181**, 125–36.

Müller, W., Deptuch, K., López-Rull, I. and Gil, D. (2007b). Elevated yolk androgen levels benefit offspring development in a between-clutch context. *Behavioral Ecology*, **18**, 929–36.

Müller, W. and Eens, M. (2009). Elevated yolk androgen levels and the expression of multiple sexually selected male characters. *Hormones and Behavior*, **55**, 175–81.

Müller, W., Goerlich, V.C., Vergauwen, J., Groothuis, T.G.G. and Eens, M. (2012). Sources of variation in yolk hormone deposition: Consistency, inheritance and developmental effects. *General and Comparative Endocrinology*, **175**, 337–43.

Müller, W., Groothuis, T.G.G., Dijkstra, C., Siitari, H. and Alatalo, R.V. (2004). Maternal antibody transmission and breeding densities in the black-headed gull (*Larus ridibundus*). *Functional Ecology*, **18**, 719–24.

Müller, W., Groothuis, T.G.G., Kasprzik, A., Dijkstra, C., Alatalo, L.V. and Siitari, H. (2005). Prenatal androgen exposure modulates cellular and humoral immune function of black-headed gull chicks. *Proceedings of the Royal Society B*, **272**, 1971–7.

Müller, W., Lessells, C.M., Korsten, P. and von Engelhardt, N. (2007a). Manipulative signals in family conflict? On the function of maternal yolk hormones in birds. *The American Naturalist*, **169**, E84–96.

Murai, A. (2013). Maternal transfer of immunoglobulins into egg yolks of birds. *Journal of Poultry Science*, **50**, 185–93.

Murphy, M.P., Holmgren, A., Larsson, N.-G., Halliwell, B., Chang, C.J., Kalyanaraman, B., Rhee, S.G., Thornalley, P.J., Partridge, L., Gems, D., Nyström, T., Belousov, V., Schumacker, P.T and Winterbourn, C.C. (2011). Unraveling the biological roles of reactive oxygen species. *Cell Metabolism*, **13**, 361–6.

Murray, M.D. (1990). Influence of host behaviour on some ectoparasites of birds and mammals. In CJ Barnard, JM Behnke, eds. *Parasitism and host behaviour*, pp. 290–315. Taylor and Francis, New York.

Murray, P.F. and Vickers-Rich, P. (2004). *Magnificent mihirungs: the colossal flightless birds of the Australian dreamtime*. Indiana University Press, Bloomington and Indianapolis.

Muth, F. and Healy, S.D. (2011). The role of adult experience in nest building in the zebra finch, *Taeniopygia guttata*. *Animal Behaviour*, **82**, 185–9.

Muth, F. and Healy, S.D. (2013). Zebra Finches build nests that do not resemble their natal nest. *Avian Biology Research*, **5**, 218–26.

Muth, F. and Healy, S.D. (2014). Zebra finches select nest material appropriate for a building task. *Animal Behaviour*, **90**, 237–44.

Muth, F., Steele, M. and Healy, S.D. (2013). Colour preferences in nest-building zebra finches. *Behavioural Processes*, **99**, 106–11.

Nagata, K. and Yoshida, N. (1984). Interaction between trypsin-like enzyme from *Streptomyces erythraeus* and chicken ovomucoid. *Journal of Biochemistry*, **96**, 1041–9.

Nager, R.G., Monaghan, P. and Houston, D.C. (2000). Within-clutch trade-offs between the number and quality of eggs: experimental manipulations in gulls. *Ecology*, **81**, 1339–50.

Naish, D. (2014). The fossil record of bird behaviour. *Journal of Zoology*, **292**, 268–80.

Naugle, D.E., Aldridge, C.L., Walker, B.L., Cornish, T.E., Moynahan, B.J., Holloran, M.J., Brown, K., Johnson, G.D., Schmidtmann, E.T., Mayer, R.T., Kato, C.Y., Matchett, M.R., Christiansen, T.J., Cook, W.E., Creekmore, T., Falise, R.D., Rinkes, E.T. and Boyce, M.S. (2004). West Nile virus: pending crisis for greater sage-grouse. *Ecology Letters*, **7**, 704–13.

Navara, K.J. (2013). The role of steroid hormones in the adjustment of primary sex ratio in birds: Compiling the pieces of the puzzle. *Integrative and Comparative Biology*, **53**, 923–37.

Navara, K.J., Hill, G.E. and Mendonca, M.T. (2006). Yolk testosterone stimulates growth and immunity in house finch chicks. *Physiological and Biochemical Zoology*, **79**, 550–5.

Nelson, K.J. and Martin, K. (1999). Thermal aspects of nest-site location for Vesper Sparrows and Horned Larks in British Columbia. *Studies in Avian Biology*, **19**, 137–43.

Nettleship, D.N. (1972). Breeding success of the common puffin (*Fratercula arctica* L.) on different habitats at Great Island, Newfoundland. *Ecological Monographs*, **42**, 239–68.

Neudorf, D.L. and Sealy, S.G. (1994). Sunrise nest attentiveness in cowbird hosts. *The Condor*, **96**, 162–9.

Newell, F.L. and Rodewald, A.D. (2011). Role of topography, canopy structure, and floristics in nest-site selection and nesting success of canopy songbirds. *Forest Ecology and Management*, **262**, 739–49.

Newson, S., Crick, H.Q.P., Clark, J.A. and Moss, D. (2007). *Timing of Breeding of Moorland Birds in Wales. BTO Research Report No. 468*, British Trust for Ornithology, Thetford.

Newson, S.E., Leech, D.I., Hewson, C.M., Crick, H.Q.P. and Grice, P.V. (2010). Potential impact of grey squirrels *Sciurus carolinensis* on woodland bird populations in England. *Journal of Ornithology*, **151**, 211–8.

Newson, S.E., Woodburn, R.J.W., Noble, D.G., Baillie, S.R. and Gregory, R.D. (2005). Evaluating the Breeding Bird Survey for producing national population size and density estimates. *Bird Study*, **52**, 42–54.

Newton, A.V. (1896). *A dictionary of birds*. A. and C. Black, London.

Newton, I. (1973). Egg breakage and breeding failure in British merlins. *Bird Study*, **20**, 241–4.

Newton, I. (1989). *Lifetime reproduction in birds*. Academic Press, London.

Newton, I. (2013). *Bird populations*. Harper Collins, London.

Newton, I. and Bogan, J. (1974). Organochlorine residues, eggshell thinning, and hatching success in British sparrowhawks. *Nature*, **249**, 582–3.

Ng, H. and Garibaldi, J. (1975). Death of *Staphylococcus aureus* in liquid whole egg near pH 8. *Applied Microbiology*, **29**, 782–6.

Nickell, W.P. (1956). Variations in engineering features of the nests of several species of birds in relation to nest sites and nesting materials. *Butler University Botanical Study*, **13**, Article 14.

Niizuma, Y., Takagi, M., Senda, M., Chochi, M. and Watanuki, Y. (2005). Incubation capacity limits maximum clutch size in black-tailed gulls *Larus crassirostris. Journal of Avian Biology*, **36**, 421–7.

Nilsson, J.-Å. and Smith, H.G. (1988). Incubation feeding as a male tactic for early hatching. *Animal Behaviour*, **36**, 641–7.

Nilsson, J.F., Stjernman, M. and Nilsson, J.-Å. (2008). Experimental reduction of incubation temperature affects both nestling and adult blue tits *Cyanistes caeruleus. Journal of Avian Biology*, **39**, 553–9.

Nord, A. and Nilsson, J.-Å. (2011). Incubation temperature affects growth and energy metabolism in Blue Tit nestlings. *The American Naturalist*, **178**, 639–51.

Nord, A. and Nilsson, J.-Å. (2012). Context-dependent costs of incubation in the Pied Flycatcher. *Animal Behaviour*, **84**, 427–36.

Nord, A., Sandell, M.L. and Nilsson, J.-Å. (2010). Female Zebra Finches compromise clutch temperature in energetically demanding incubation conditions. *Functional Ecology*, **24**, 1031–6.

Nord, A., Sköld-Chiriac, S., Hasselquist, D. and Nilsson, J.-Å. (2014). A tradeoff between perceived predation risk and energy conservation revealed by an immune challenge experiment. *Oikos*, **123**, 1091–100.

Norell, M.A., Clark, J.M., Dashzeveg, D., Barsbold, R., Chiappe, L.M., Davidson, A.R., McKenna, M.C., Perle, A. and Novacek M.J. (1994). A theropod dinosaur embryo and the affinities of the Flaming Cliffs dinosaur eggs. *Science*, **266**, 779–82.

Nores, A.I. and Nores, M. (1994). Nest building and nesting behavior of the brown cacholote. *The Wilson Bulletin*, **106**, 106–20.

Norris, K. and Evans, M.R. (2000). Ecological immunology: life history trade-offs and immune defenses in birds. *Behavioral Ecology*, **11**, 19–26.

Noske, R.A., Mulyani, Y.A. and Lloyd, P. (2013). Nesting beside old nests, but not over water, increases current nest survival in a tropical mangrove-dwelling warbler. *Journal of Ornithology*, **154**, 517–23.

Nowicki, S., Searcy, W.A. and Peters, S. (2002). Brain development, song learning and mate choice in birds: a review and experimental test of the 'nutritional stress hypothesis'. *Journal of Comparative Physiology A*, **188**, 1003–14.

O'Brien, V.A., Moore, A.T., Young, G.R., Komar, N., Reisen, W.K. and Brown, C.R. (2011). An enzootic vector-borne virus is amplified at epizootic levels by an invasive avian host. *Proceedings of the Royal Society B*, **278**, 239–46.

Ockendon, N., Leech, D. and Pearce-Higgins, J.W. (2013). Climatic effects on breeding grounds are more important drivers of breeding phenology in migrant birds than carry-over effects from wintering grounds. *Biology Letters*, **9**, 20130669.

O'Connor, J., Zheng, X., Wang, X., Wang, Y. and Zhou, Z. (2013a). Ovarian follicles shed new light on dinosaur reproduction during the transition towards birds. *National Science Review*, **1**, 15–17.

O'Connor, J., Zheng, X. and Zhou, Z. (2013b). Can ovarian follicles fossilize? *Nature*, **499**, E1–2. (doi:10.1038/nature12368).

O'Connor, R.J. and Morgan, R.A. (1982). Some effects of weather conditions on the breeding of the Spotted Flycatcher *Muscicapa striata* in Britain. *Bird Study*, **29**, 41–8.

O'Connor, R.S. and Ritchison, G. (2013). Notes on the incubation, brooding, and provisioning behaviour of Chuckwill's-widows. *The Wilson Journal of Ornithology*, **125**, 815–9.

Ohto-Fujita, E., Konno, T., Shimizu, M., Ishihara, K., Sugitate, T., Miyake, J., Yoshimura, K., Taniwaki, K., Sakurai, T., Hasebe, Y. and Atomi, Y. (2011). Hydrolyzed eggshell membrane immobilized on phosphorylcholine polymer supplies extracellular matrix environment for human dermal fibroblasts. *Cell and Tissue Research*, **345**, 177–90.

Okuliarova, M., Groothuis, T.G.G., Škrobánek, P. and Zeman, M. (2011). Experimental evidence for genetic heritability of maternal hormone transfer to offspring. *The American Naturalist*, **177**, 824–34.

Ollason, J.C. and Dunnet, G.M. (1988). *Variation in breeding success in fulmars*. In TH Clutton-Brock, ed. *Reproductive success: Studies in individual variation in contrasting breeding systems*, pp. 263–78. University of Chicago Press, Chicago.

Olson, C.R., Vleck, C.M. and Vleck, D. (2006). Periodic cooling of bird eggs reduces embryonic growth efficiency. *Physiological and Biochemical Zoology*, **79**, 927–36.

Olson, S.L. and Hearty, P.J. (2003). Probable extirpation of a breeding colony of Short-tailed Albatross (*Phoebastria albatrus*) on Bermuda by Pleistocene sea-level rise. *Proceedings of the National Academy of Sciences of the USA*, **100**, 12825–9.

Olson, S.L. and Hearty, P.J. (2013). Fossilized egg indicates probable breeding of Brown Pelican (*Pelecanus occidentalis*) on Bermuda in the Middle Pleistocene. *Proceedings of the Biological Society of Washington*, **126**, 169–77.

Olsson, K. and Allander, K. (1995). Do fleas and/or old nest material, influence nest site preference in hole-nesting passerines? *Ethology*, **101**, 160–70.

Ontiveros, D., Caro, J. and Pleguezuelos, M.J. (2008). Green plant material versus ectoparasites in nests of Bonelli's eagle. *Journal of Zoology*, **274**, 99–104.

Oppenheim, R.W. (1973). Prehatching and hatching behavior: a comparative and physiological consideration. In G Gottlieb, ed. *Behavioral embryology*, pp. 163–244. Academic Press, New York.

Oppliger, A., Richner, H. and Christie, P. (1994). Effect of an ectoparasite on lay date, nest-site choice, desertion, and hatching success in the great tit (*Parus major*). *Behavioral Ecology*, **5**, 130–4.

Ormerod, S.J. and Tyler, S.J. (1990). Environmental pollutants in the eggs of Welsh Dipper *Cinclus cinclus*: a potential monitor of organochlorine and mercury contamination in upland rivers. *Bird Study*, **37**, 171–6.

Orr, Y. (1970). Temperature measurements at the nest of the desert lark (*Ammomanes deserti deserti*). *The Condor*, **72**, 476–8.

Oschadleus, D.H. (2012). Trapped! Weaver nests as death traps. *Ornithological Observations*, **3**, 38–43.

Ospina, E., Cooper, C.B., Liljesthrom, M., Ardia, D. and Winkler, W. (2015). Biparental nest attendance in Chilean Swallows (*Tachycineta meyeni*) breeding in Ushuaia, Argentina. *Emu*, **115**, 76–9.

Ottinger, M.A., Dean, K., McKernan, M. and Quinn, M.J. (2011). Endocrine disruption of reproduction in birds. In DO Norris, KH Lopez, eds. *Hormones and reproduction of vertebrates—Birds*, pp. 239–60. Elsevier, Amsterdam.

Ounsted, M. (1983). Wildfowl avicultural report 1982. *Wildfowl*, **34**, 174–5.

Ounsted, M. (1988). Attempts by the Wildfowl Trust to reestablish the white-winged wood duck and white-headed duck *Cairina scutulata* and *Oxyura leucocephala*. *International Zoo Yearbook*, **27**, 216–22.

Owen, J. (1991). Principles and problems of incubator design. In SG Tullett, ed. *Avian incubation*, pp. 205–24. Butterworth-Heinnemann, London.

Owen, J.P., Nelson, A.C. and Clayton, D.H. (2010). Ecological immunology of bird-ectoparasite systems. *Trends in Parasitology*, **26**, 530–9.

Pacejka, A.J., Santana, E., Harper, R.G. and Thompson, C.F. (1996). House wrens *Troglodytes aedon* and nest-dwelling ectoparasites: mite population growth and feeding patterns. *Journal of Avian Biology*, **27**, 273–8.

Packard, M.J. and DeMarco, V.G. (1991). Eggshell structure and formation in eggs of oviparous reptiles. In DC Deeming, MWJ Ferguson, eds. *Egg incubation: its effects on embryonic development in birds and reptiles*, pp. 53–69. Cambridge University Press, Cambridge.

Paclík, M., Misík, J. and Weidinger, K. (2009). Nest predation and nest defence in European and North American woodpeckers: a review. *Annales Zoologici Fennici*, **46**, 361–79.

Paclík, M. and Weidinger, K. (2007). Microclimate of tree cavities during winter nights—implications for roost site selection in birds. *International Journal of Biometeorology*, **51**, 287–93.

Paganelli, C.V. (1980). The physics of gas exchange across the avian eggshell. *American Zoologist*, **20**, 329–38.

Paganelli, C.V. (1991). The avian eggshell as a mediating barrier: respiratory gas fluxes and pressures during development. In DC Deeming, MWJ Ferguson, eds. *Egg incubation: its effects on embryonic development in birds and reptiles*, pp. 261–75. Cambridge University Press, Cambridge.

Pahl, R., Winkler, D.W., Graveland, J. and Batterman, B.W. (1997). Songbirds do not create long-term stores of calcium in their legs prior to laying: results from high-resolution radiography. *Proceedings of the Royal Society of London Series B: Biological Sciences*, **264**, 239–44.

Paitz, R.T. and Bowden, R.M. (2013). Sulfonation of maternal steroids is a conserved metabolic pathway in vertebrates. *Integrative and Comparative Biology*, **53**, 895–901.

Paitz, R.T. and Casto, J.M. (2012). The decline in yolk progesterone concentrations during incubation is dependent on embryonic development in the European starling. *General and Comparative Endocrinology*, **176**, 415–9.

Paitz, R.T., Bowden, R.M. and Casto, J.M. (2011). Embryonic modulation of maternal steroids in European starlings (*Sturnus vulgaris*). *Proceedings of the Royal Society B*, **278**, 99–106.

Palmgren, M. and Palmgren, P. (1939). Über die wärmeisolierungskapazität verschiedener kleinvogelnester. *Ornis Fennica*, **16**, 1–6.

Palomino, J.J., Martín-Vivaldi, M., Soler, M. and Soler, J.J. (1998). Functional significance of nest size variation in the Rufous Bush Robin *Cercotrichas galactotes*. *Ardea*, **86**, 178–85.

Paradis, E., Baillie, S.R., Sutherland, W.J., Dudley, C., Crick, H.Q.P. and Gregory, R.D. (2000). Large-scale spatial variation in the breeding performance of song thrushes *Turdus philomelos* and blackbirds *T. merula* in Britain. *Journal of Applied Ecology*, **37**, 73–87.

Paradis, E., Baillie, S.R., Sutherland, W.J. and Gregory, R.D. (2002). Exploring density-dependent relationships in demographic parameters in populations of birds at a large spatial scale. *Oikos*, **97**, 293–307.

Parejo, D., Amo, L., Rodríguez, J. and Avilés, J. (2012). Rollers smell the fear of nestlings. *Biology Letters*, **8**, 502–4.

Partecke, J. and Schwabl, H. (2008). Organizational effects of maternal testosterone on reproductive behavior of adult house sparrows. *Developmental Neurobiology*, **68**, 1538–48.

Patnaik, R., Sahni, A., Cameron, D., Pillans, B., Chatrath, P., Simons, E., Williams, M. and Bibi, F. (2009). Ostrich-like eggshells from a 10.1 million-yr-old Miocene ape locality, Haritalyangar, Himachal Pradesh, India. *Current Science*, **96**, 1485–95.

Peach, W.J., Buckland, S.T. and Baillie, S.R. (1996). The use of constant effort mist-netting to measure between-year changes in the abundance and productivity of common passerines. *Bird Study*, **43**, 142–56.

Peach, W.J., Crick, H.Q.P and Marchant, J.H. (1995). The demography of the decline in the British willow warbler population. *Journal of Applied Statistics*, **22**, 905–22.

Peach, W.J., Siriwardena, G.M. and Gregory, R.D. (1999). Long-term changes in over-winter survival rates explain the decline of reed buntings *Emberiza schoeniclus* in Britain. *Journal of Applied Ecology*, **36**, 798–811.

Peakall, D.B. (1970). The Eastern Bluebird: its breeding season, clutch size, and nesting success. *Living Bird*, **9**, 239–55.

Pechacek, P. (2005). Use of non-stop video surveillance to monitor breeding activity of primary cavity nesters in remote areas. *Acta Ethologica*, **8**, 1–4.

Pelayo, J.T. and Clark, R.G. (2003). Consequences of egg size for offspring survival: a cross fostering experiment in Ruddy Ducks (*Oxyura jamaicensis*). *The Auk*, **120**, 384–93.

Peluc, S.I., Sillett, T.S., Rotenberry, J.T. and Ghalambor, C.K. (2008). Adaptive phenotypic plasticity in an island songbird exposed to a novel predation risk. *Behavioral Ecology*, **19**, 830–5.

Pen, I., Uller, T., Feldmeyer, B., Harts, A., While, G.M. and Wapstra, E. (2010). Climate-driven population divergence in sex-determining systems. *Nature*, **468**, 436–9.

Pendlebury, C.J. and Bryant, D.M. (2005). Night-time behaviour of egg-laying tits. *Ibis*, **147**, 342–5.

Penloup, A., Martin, J.L., Gory, G., Brunstein, D. and Bretagnolle, V. (1997). Distribution and breeding success of pallid swifts, *Apus pallidus*, on Mediterranean islands: nest predation by the roof rat, *Rattus rattus*, and nest site quality. *Oikos*, **80**, 78–88.

Peralta-Sánchez, J.M., Martín-Vivaldi, M., Martín-Platero, A.M., Martínez-Bueno, M., Oñate, M., Ruiz-Rodríguez, M. and Soler, J.J. (2012). Avian life history traits influence eggshell bacterial loads: a comparative analysis. *Ibis*, **154**, 725–37.

Peralta-Sánchez, J.M., Møller, A.P., Martin-Platero, A.M. and Soler, J.J. (2010). Number and colour composition of nest lining feathers predict eggshell bacterial community in barn swallow nests: an experimental study. *Functional Ecology*, **24**, 426–33.

Peralta-Sanchez, J.M., Moller, A.P. and Soler, J.J. (2011). Colour composition of nest lining feathers affects hatching success of barn swallows, *Hirundo rustica*. *Biological Journal of the Linnean Society*, **102**, 67–74.

Peralta-Sánchez, J.M., Soler, J.J., Martín-Platero, A.M., Knight, R., Martínez-Bueno, M. and Møller, A.P. (2014). Eggshell bacterial load is related to antimicrobial properties of feathers lining barn swallow nests. *Microbial Ecology*, **67**, 480–7.

Pérez, J.H., Ardia, D.R., Chad, E.K. and Clotfelter, E.D. (2008). Experimental heating reveals nest temperature affects nestling condition in tree swallows (*Tachycineta bicolor*). *Biology Letters*, **4**, 468–71.

Pérez-Tris, J., Carbonell, R. and Tellería, J.L. (2002). Parasites and the blackcap's tail: implications for the evolution of feather ornaments. *Biological Journal of the Linnean Society*, **76**, 481–92.

Perrins, C.M. and Birkhead, T.R. (1983). *Avian ecology*. Blackie, Glasgow.

Perrins, C.M. and Crick, H.Q.P. (1996). Influence of lunar cycle on laying dates of European Nightjars *Caprimulgus europaeus*. *The Auk*, **113**, 705–8.

Petit, C., Hossaert-McKey, M., Perret, P., Blondel, J. and Lambrechts, M.M. (2002). Blue tits use selected plants and olfaction to maintain an aromatic environment for nestlings. *Ecology Letters*, **5**, 585–9.

Pfannkuche, K.A., Gahr, M., Weites, I.M., Riedstra, B., Wolf, C. and Groothuis, T.G.G. (2011). Examining a pathway for hormone mediated maternal effects—Yolk testosterone affects androgen receptor expression and endogenous testosterone production in young chicks (*Gallus gallus domesticus*). *General and Comparative Endocrinology*, **172**, 487–93.

Phillips, T. and Dickinson, J.L. (2009). Tracking the nesting success of North America's breeding birds through public participation in NestWatch. *Proceedings of the 4th International Partners in Flight Conference: Tundra to Tropics*, 633–40.

Pierce, D.J. and Simons, T.R. (1986). The influence of human disturbance on Tufted Puffin breeding success. *The Auk*, **103**, 214–6.

Piersma, T., Lindström, Å., Drent, R.H., Tulp, I., Jukema, J., Morrison, R.I.G., Reneerkens, J., Schekkerman, H. and Visser, G.H. (2003). High daily energy expenditure of incubating shorebirds on the High Arctic tundra: a circumpolar study. *Functional Ecology*, **17**, 356–62.

Pietz, P.J., Granfors, D.A. and Ribic, C.A. (2012). Knowledge gained from video-monitoring grassland passerine nests. In CA Ribic, FR Thompson III, PJ Pietz, eds. *Video surveillance of nesting birds*. Studies in Avian Biology, Vol. 43, pp. 3–22. University of California Press, Berkeley.

Pihlaja, M., Siitari, H. and Alatalo, R.V. (2006). Maternal antibodies in a wild altricial bird: effects on offspring immunity, growth and survival. *Journal of Animal Ecology*, **75**, 1154–64.

Pike, T.W. (2011). Egg recognition in Japanese quail. *Avian Biology Research*, **4**, 231–6.

Pilz, K.M., Quiroga, M., Schwabl, H. and Adkins-Regan, E. (2004). European starling chicks benefit from high yolk testosterone levels during a drought year. *Hormones and Behavior*, **46**, 179–92.

Pinowski, J., Haman, A., Jerzak, L., Pinowski, B., Barkowska, B., Grodzki, A. and Haman, K. (2006). The thermal properties of some nests of the Eurasian Tree Sparrow *Passer montanus*. *Journal of Thermal Biology*, **31**, 573–81.

Pires, A.B., Belo, F.A. and Rabaça, E.J. (2012). Aromatic plants in Eurasian Blue Tit nests: the 'nest protection hypothesis' revisited. *The Wilson Journal of Ornithology*, **124**, 162–5.

Pitala, N., Ruuskanen, S., Laaksonen, T., Doligez, B., Tschirren, B. and Gustafsson, L. (2009). The effects of experimentally manipulated yolk androgens on growth and immune function of male and female nestling collared flycatchers *Ficedula albicollis*. *Journal of Avian Biology*, **40**, 225–30.

Pitts, T.D. (1988). Effects of nest box size on Eastern Bluebird nests. *Journal of Field Ornithology*, **59**, 309–13.

Poinar, G. Jr, Voisin, C. and Voisin, J.-F. (2007). Bird eggshell in Dominican amber. *Palaeontology*, **50**, 1381–3.

Poisbleau, M., Demongin, L., Angelier, F., Dano, S., Lacroix, A. and Quillfeldt, P. (2013). Hatching vocalisations in free-living Rockhopper Penguins. *Ardea*, **101**, 39–44.

Polačiková, L., Takasu, F., Stokke, B., Moksnes, A., Røskaft, E., Cassey, P., Hauber, M. and Grim, T. (2013). Egg arrangement in avian clutches covaries with the rejection of foreign eggs. *Animal Cognition*, **16**, 819–28.

Poole, A. (2005). The Birds of North America Online: http://bna.birds.cornell.edu/BNA/. Cornell Lab of Ornithology, Ithaca. [Accessed: 24 November 2014]

Poole, H.K. (1965). Spectrophotometric identification of eggshell pigments and timing of superficial pigment deposition in the Japanese quail. *Proceedings of the Society for Experimental Biology and Medicine. Society for Experimental Biology and Medicine (New York, N.Y.)*, **119**, 547–51.

Pope, T.L., Conkling, T.J., Smith, K.N., Colon, M.R., Morrison, M.L. and Wilkins, R.N. (2013). Effects of adult behavior and nest-site characteristics on black-capped vireo nest survival. *The Condor*, **115**, 155–62.

Porter, R.D. and Wiemeyer, S.N. (1969). Dieldrin and DDT: effects on sparrow hawk eggshells and reproduction. *Science*, **165**, 199–200.

Portner, H.O. and Farrell, A.P. (2008). Physiology and climate change. *Science*, **322**, 690–2.

Portugal, S.J., Hauber, M.E., Maurer, G., Stokke, B.G., Grim, T. and Cassey, P. (2014b). Rapid development of brood-parasitic embryos cannot be explained by increased gas exchange through the eggshell. *Journal of Zoology*, **293**, 219–26.

Portugal, S.J., Maurer, G., Thomas, G.H., Hauber, M.E., Grim, T. and Cassey, P. (2014a). Nesting behaviour influences species-specific gas exchange across avian eggshells. *The Journal of Experimental Biology*, **217**, 3326–32.

Postma, E., Siitari, H., Schwabl, H., Richner, H. and Tschirren, B. (2014). The multivariate egg: quantifying within- and among-clutch correlations between maternally derived yolk immunoglobulins and yolk androgens using multivariate mixed models. *Oecologia*, **174**, 631–8.

Potter, B.A., Carlson, B.M., Adams, A.E. and Voss, M.A. (2013). An assessment of the microbial diversity present on the surface of naturally incubated house wren eggs. *The Open Ornithological Journal*, **6**, 32–9.

Potti, J. and Merino, S. (1995). Louse loads of Pied Flycatchers: effects of host's sex, age, condition and relatedness. *Journal of Avian Biology*, **26**, 203–8.

Poulton, E.B. (1890). *The colours of animals*. Trübner and Co. Ltd, London.

Powell, L.A. and Rangen, K.L. (2000). Variation in wood thrush nest dimensions and construction. *North American Band Birder*, **25**, 89–96.

Powlesland, R.G. (1977). Effects of the haematophagous mite *Ornithonyssus bursa* on nestling starlings in New Zealand. *New Zealand Journal of Zoology*, **4**, 85–94.

Prinzinger, R., Pressmar, A. and Schleucher, E. (1991). Body temperature in birds. *Comparative Biochemistry and Physiology A*, **99**, 499–506.

Price, P. (1980). *Evolutionary biology of parasites*. Princeton University Press, Princeton.

Proctor, H. and Owens, I. (2000). Mites and birds: diversity, parasitism and coevolution. *Trends in Ecology and Evolution*, **15**, 358–64.

Proctor, H.C. (2003). Feather mites (Acari: Astigmata): ecology, behavior, and evolution. *Annual Reviews of Entomology*, **48**, 185–209.

Proffitt, F.M., Newton, I., Wilson, J.D. and Siriwardena, G.M. (2004). Bullfinch *Pyrrhula pyrrhula* breeding ecology in lowland farmland and woodland: comparisons across time and habitat. *Ibis*, **146**, 78–86.

Prokop, P. (2004). The effect of nest usurpation on breeding success of the black-billed magpie *Pica pica*. *Biologia, Bratislava*, **59**, 213–7.

Prokop, P. and Trnka, A. (2011). Why do grebes cover their nests? Laboratory and field tests of two alternative hypotheses. *Journal of Ethology*, **29**, 17–22.

Prum, R.O., Andersson, S. and Torres, R.H. (2003). Coherent scattering of ultraviolet light by avian feather barbs. *The Auk*, **120**, 163–70.

Prum, R.O. and Torres, R.H. (2003). A Fourier tool for the analysis of coherent light scattering by bio-optical nanostructures. *Integrative and Comparative Biology*, **43**, 591–602.

Prum, R.O., Torres, R.H., Williamson, S. and Dyck, J. (1998). Coherent light scattering by blue feather barbs. *Nature*, **396**, 28–9.

Puchala, P. (2004). Detrimental effects of larval blow flies (*Protocalliphora azurea*) on nestlings and breeding success of tree sparrows (*Passer montanus*). *Canadian Journal of Zoology*, **82**, 1285–90.

Purvis, A. and Rambaut, A. (1995). Comparative analysis by independent contrasts (CAIC): An Apple Macintosh application for analysing comparative data. *Computer Applications in the Biosciences*, **11**, 247–51.

Pytte, C.L., Ficken, M.S. and Moiseff, A. (2004). Ultrasonic singing by the blue-throated hummingbird: a comparison between production and perception. *Journal of Comparative Physiology A*, **190**, 665–73.

Quader, S. (2006). What makes a good nest? Benefits of nest choice to female Baya Weavers (*Ploceus philippinus*). *The Auk*, **123**, 475–86.

Quader, S., Isvaran, K., Hale, R.E., Miner, B.G. and Seavy, N.E. (2004). Nonlinear relationships and phylogenetically independent contrasts. *Journal of Evolutionary Biology*, **17**, 709–15.

Quesada, J. (2007). The different roles of the roof density and nest size in the Iberian magpie nest. *Acta Ethologica*, **10**, 41–5.

Quinn, J.L., Reynolds, S.J. and Bradbury, R.B. (2008). Birds as predators and as prey. *Ibis*, **150** (Supplement 1), 1–8.

Quinn, J.L. and Ueta, M. (2008). Protective nesting associations in birds. *Ibis*, **150**, 146–67.

Quintana, R.D., Cirelli, V. and Benitez, O. (2001). Nest materials of skuas (*Catharacta* spp.) and kelp gulls (*Larus dominicanus*) at Cierva Point, Antarctic Peninsula. *Notornis*, **48**, 235–41.

Quiroga, M.A., Reboreda, J.C. and Beltzer, A.H. (2012). Host use by *Philornis* sp. in a passerine community in central Argentina. *Revista Mexicana de Biodiversidad*, **83**, 110–6.

R Core Development Team (2012). *R: A language and environment for statistical computing*. R Foundation for Statistical Computing, Vienna.

Råberg, L., Nilsson, J.-Å., Ilmonen, P., Stjernman, M. and Hasselquist, D. (2000). The cost of an immune response: vaccination reduces parental effort. *Ecology Letters*, **3**, 382–6.

Radford, D.J. (1980). Pair of song thrushes with four broods in a season. *Bird Study*, **27**, 121.

Rae, S. and Rae, D. (2013). Orientation of tawny frogmouth (*Podargus strigoides*) nests and their position on branches optimises thermoregulation and cryptic concealment. *Australian Journal of Zoology*, **61**, 469–74.

Rahn, H. (1991). Why birds lay eggs. In DC Deeming, MWJ Ferguson, eds. *Egg incubation: its effects on embryonic development in birds and reptiles*, pp. 345–60. Cambridge University Press, Cambridge.

Rahn, H., Ackerman, R. and Paganelli, C. (1977). Humidity in the avian nest and egg water loss during incubation. *Physiological Zoology*, **50**, 269–83.

Rahn, H. and Ar, A. (1974). The avian egg: incubation time and water loss. *The Condor*, **76**, 147–52.

Rahn, H., Ar, A. and Paganelli, C.V. (1979). How bird eggs breathe. *Scientific American*, **240**, 46–55.

Rahn, H. and Paganelli, C.V. (1981). *Gas exchange of avian eggs*. State University of New York at Buffalo, Buffalo.

Rahn, H. and Paganelli, C.V. (1990). Gas fluxes in avian eggs: Driving forces and the pathway for exchange. *Comparative Biochemistry and Physiology*, **95A**, 1–15.

Rahn, H., Paganelli, C.V. and Ar, A. (1975). Relation of avian egg weight to body weight. *The Auk*, **92**, 759–65.

Rahn, H., Whittow, G.C. and Paganelli, C.V. (1985). *Gas exchange of avian eggs. Volume 2*. State University of New York at Buffalo, Buffalo.

Rastogi, A.D., Zanette, L. and Clinchy, M. (2006). Food availability affects diurnal nest predation and adult antipredator behaviour in song sparrows, *Melospiza melodia*. *Animal Behaviour*, **72**, 933–40.

Ratcliffe, D.A. (1967). Decrease in eggshell weight in certain birds of prey. *Nature*, **215**, 208–10.

Ratcliffe, D.A. (1970). Changes attributed to pesticides in egg breakage frequency and eggshell thickness in some British birds. *Journal of Applied Ecology*, **7**, 67–115.

Rauter, C.M., Reyer, H.-U. and Bollmann, K. (2002). Selection through predation, snowfall and microclimate on nest-site preferences in the Water Pipit *Anthus spinoletta*. *Ibis*, **144**, 433–44.

Réale, D., Garant, D., Humphries, M.M., Bergeron, P., Careau, V. and Montiglio, P.O. (2010). Personality and the emergence of the pace-of-life syndrome concept at the population level. *Philosophical Transactions of the Royal Society B*, **365**, 4051–63.

Rebstock, G.A. and Boersma, P.D. (2011). Parental behavior controls incubation period and asynchrony of hatching in Magellanic Penguins. *The Condor*, **112**, 316–25.

Redmond, L.J., Murphy, M.T. and Dolan, A.C. (2007). Nest reuse by Eastern Kingbirds: Adaptive behavior or ecological constraint? *The Condor*, **109**, 463–8.

Redpath, S., Madders, M., Donnelly, E., Anderson, B., Thirgood, S., Martin, A. and McLeod, D. (1998). Nest site selection by hen harriers in Scotland. *Bird Study*, **45**, 51–61.

Reed, W.L. and Clark, M.E. (2011). Beyond maternal effects in birds: responses of the embryo to the environment. *Integrative and Comparative Biology*, **51**, 73–80.

Reed, W.L., Clark, M.E., Parker, P.G., Raouf, S.A., Arguedas, N., Monk, D.S., Snadjr, E., Nolan, V.J. and Ketterson, E.D. (2006). Physiological effects on demography: A long-term experimental study of testosterone's effects on fitness. *The American Naturalist*, **167**, 667–83.

Reed, T.E., Grøtan, V., Jenouvrier, S., Sæther, B.E. and Visser, M.E. (2013). Population growth in a wild bird is buffered against phenological mismatch. *Science*, **340**, 488–91.

Refsnider, J.M., Bodensteiner, J.L., Reneker, J.J. and Janzen, F.J. (2013). Nest depth may not compensate for sex ratio skews caused by climate change in turtles. *Animal Conservation*, **16**, 481–90.

Réhault-Godbert, S., Labas, V., Helloin, E., Hervé-Grépinet, V., Slugocki, C., Berges, M., Bourin, M.-C., Brionne, A., Poirier, J.-C., Gautron, J., Coste, F. and Nys, Y. (2013). Ovalbumin-related protein X is a heparin-binding ov-serpin exhibiting antimicrobial activities. *Journal of Biological Chemistry*, **288**, 17285–95.

Reid, J.M., Cresswell, W., Holt, S., Mellanby, R.J., Whitfield, D.P. and Ruxton, G.D. (2002a). Nest scrape design and clutch heat loss in Pectoral Sandpipers (*Calidris melanotos*). *Functional Ecology*, **16**, 305–12.

Reid, J.M., Monaghan, P. and Nager, R.G. (2002b). Incubation and the costs of reproduction. In DC Deeming, ed. *Avian incubation: behaviour, environment, and evolution*, pp. 314–25. Oxford University Press, Oxford.

Reid, J.M., Monaghan, P. and Ruxton, G.D. (2000a). Resource allocation between reproductive phases: the importance of thermal conditions in determining the cost of incubation. *Proceedings of the Royal Society of London Series B: Biological Sciences*, **267**, 37–41.

Reid, J.M., Monaghan, P. and Ruxton, G.D. (2000b). The consequences of clutch size for incubation conditions and hatching success in starlings. *Functional Ecology*, **14**, 560–5.

Reidy, J.L., Stake, M.M. and Thompson, F.R. (2009). Nocturnal predation of females on nests: An important source of mortality for Golden-cheeked Warblers? *The Wilson Journal of Ornithology*, **121**, 416–21.

Reinhardt, K. and Siva-Jothy, M.T. (2007). Biology of bed bugs (Cimicidae). *Annual Review of Entomology*, **52**, 351–74.

Reiss, M.J. (1989). *The allometry of growth and reproduction*. Cambridge University Press, Cambridge.

Reitsma, L.R. and Whelan, C.J. (2000). Does vertical partitioning of nest sites decrease nest predation? *The Auk*, **117**, 409–15.

Rékási, J., Rózsa, L., and Kiss, B.J. (1997). Patterns in the distribution of avian lice (Phthiraptera: Amblycera, Ischnocera). *Journal of Avian Biology*, **28**, 150–6.

Remeš, V. (2005a). Birds and rodents destroy different nests: a study of Blackcap *Sylvia atricapilla* using the removal of nest concealment. *Ibis*, **147**, 213–6.

Remeš, V. (2005b). Nest concealment and parental behaviour interact in affecting nest survival in the blackcap (*Sylvia atricapilla*): an experimental evaluation of the parental compensation hypothesis. *Behavioral Ecology and Sociobiology*, **58**, 326–32.

Remeš, V. and Martin, T.E. (2002). Environmental influences on the evolution of growth and developmental rates in passerines. *Evolution*, **56**, 2505–18.

Rendell, W.B. and Robertson, R.J. (1994). Cavity-entrance orientation and nest-site use by secondary hole-nesting birds. *Journal of Field Ornithology*, **65**, 27–35.

Réné de Roland, L.ŔA., Sam, T.S., Rakotondratsima, M.P.H. and Thorstrom, R. (2007). Rediscovery of the Madagascar Pochard *Aythya innotata* in northern Madagascar. *Bulletin of the African Bird Club*, **14**, 171–4.

Rensel, M.A., Wilcoxen, T.E. and Schoech, S.J. (2010). The influence of nest attendance and provisioning on nestling stress physiology in the Florida Scrub-Jay. *Hormones and Behavior*, **57**, 162–8.

Reynolds, P.S. and Lee, R.M. (1996). Phylogenetic analysis of avian energetics: passerines and nonpasserines do not differ. *The American Naturalist*, **147**, 735–59.

Reynolds, S.J. (1997). Uptake of ingested calcium during egg production in the Zebra Finch (*Taeniopygia guttata*). *The Auk*, **114**, 562–9.

Reynolds, S.J. (2001). The effects of low dietary calcium during egg-laying on eggshell formation and skeletal calcium reserves in the Zebra Finch *Taeniopygia guttata*. *Ibis*, **143**, 205–15.

Reynolds, S.J. (2003). Mineral retention, medullary bone formation, and reproduction in the White-tailed Ptarmigan (*Lagopus leucurus*): A critique of Larison *et al.* (2001). *The Auk*, **120**, 224–8.

Reynolds, S.J., Martin, G.R. and Cassey, P. (2009). Is sexual selection blurring the functional significance of eggshell coloration hypotheses? *Animal Behaviour*, **78**, 209–15.

Reynolds, S.J. and Perrins, C.M. (2010). Dietary calcium availability and reproduction in birds. *Current Ornithology*, **17**, 31–74.

Reynolds, S.J. and Schoech, S.J. (2012). A known unknown: elaboration of the 'observer effect' on nest success? *Ibis*, **154**, 1–4.

Reynolds, S.J. and Waldron, S. (1999). Body water dynamics at the onset of egg-laying in the Zebra Finch *Taeniopygia guttata*. *Journal of Avian Biology*, **30**, 1–6.

Rhymer, J.M. (1988). The effect of egg size variability on thermoregulation of Mallard (*Anas platyrhynchos*) offspring and its implications for survival. *Oecologia*, **75**, 20–4.

Ribic, C.A., Thompson, F.R. III and Pietz, P.J. (2012). *Video surveillance of nesting birds*. Studies in Avian Biology, Vol. 43. University of California Press, Berkeley.

Richards, P.D.G. and Deeming, D.C. (2001). Correlation between shell colour and ultrastructure in pheasant eggs. *British Poultry Science*, **42**, 338–43.

Richardson, A. (1997). Aviculture at WWT over half a century. *Wildfowl*, **47**, 264–72.

Richner, H., Oppliger, A. and Christe, P. (1993). Effect of an ectoparasite on reproduction in great tits. *Journal of Animal Ecology*, **62**, 703–10.

Ricklefs, R.E. (1969). An analysis of nesting mortality in birds. *Smithsonian Contributions to Zoology*, **9**, 1–48.

Ricklefs, R.E. (1977). Composition of eggs of several bird species. *The Auk*, **94**, 350–6.

Ricklefs, R.E. (1980). Geographical variation in clutch size among passerine birds: Ashmole's hypothesis. *The Auk*, **97**, 38–49.

Ricklefs, R.E. (2000). Density dependence, evolutionary optimization, and the diversification of avian life histories. *The Condor*, **102**, 9–22.

Ricklefs, R.E. and Starck, J.M. (1998a). Embryonic growth and development. In JM Starck, RE Ricklefs, eds. *Avian growth and development: evolution within the altricial-precocial spectrum*, pp. 31–58. Oxford University Press, Oxford.

Ricklefs, R.E. and Starck, J.M. (1998b). The evolution of the developmental mode in birds. In JM Starck, RE Ricklefs, eds. *Avian growth and development: evolution within the altricial-precocial spectrum*, pp. 366–80. Oxford University Press, Oxford.

Ricklefs, R.E. and Williams, J.B. (1984). Daily energy expenditure and water-turnover rate of adult European starlings (*Sturnus vulgaris*) during the nesting cycle. *The Auk*, **101**, 707–16.

Riehl, C. (2011). Paternal investment and the 'Sexually Selected Hypothesis' for the evolution of eggshell coloration: revisiting the assumptions. *The Auk*, **128**, 175–9.

Ringelman, J.K., Longcore, J.R. and Owen, R.B. Jr. (1982). Nest and brood attentiveness in female black ducks. *The Condor*, **84**, 110–6.

Ringelman, K.M., Eadie, J.M. and Ackerman, J.T. (2014). Adaptive nest clustering and density-dependent nest survival in dabbling ducks. *Oikos*, **123**, 239–47.

Ringhofer, M. and Hasegawa, T. (2014). Social cues are preferred over resource cues for breeding-site selection in Barn Swallows. *Journal of Ornithology*, **155**, 531–8.

Robb, G.N., McDonald, R.A., Chamberlain, D.E. and Bearhop, S. (2008). Food for thought: supplementary feeding as a driver of ecological change in avian populations. *Frontiers in Ecology and the Environment*, **6**, 476–84.

Roberts, J. and Brackpool, C. (1994). The ultrastructure of avian egg shells. *Poultry Science Reviews*, **5**, 245–72.

Robertson, B.A. (2009). Nest-site selection in a postfire landscape: do parents make tradeoffs between microclimate and predation risk? *The Auk*, **126**, 500–10.

Robertson, R.J. and Norman, R.F. (1976). Behavioral defenses to brood parasitism by potential hosts of the brown-headed cowbird. *The Condor*, **78**, 166–73.

Robinson, R.A., Baillie, S.R. and King, R. (2012). Population processes in European blackbirds *Turdus merula*: a state–space approach. *Journal of Ornithology*, **152**, 419–33.

Robinson, R.A. and Clark, J.A. (2014). The Online Ringing Report: Bird ringing in Britain & Ireland in 2013 British Trust for Ornithology, Thetford (http://www.bto.org/ringing-report, created on 30-May-2014).

Robinson, R.A., Freeman, S.N., Balmer, D.E. and Grantham, M.J. (2007). Cetti's Warbler *Cettia cetti*: analysis of an expanding population. *Bird Study*, **54**, 230–5.

Robinson, R.A., Morrison, C.A. and Baillie, S.R. (2014a). Integrating demographic data: towards a framework for monitoring wildlife populations at large spatial scales. *Methods in Ecology and Evolution*, **5**, 1361–72.

Robinson, S.K., Thompson, F.R. III, Donovan, T.M., Whitehead, D.R. and Faaborg, J. (1995). Regional forest fragmentation and the nesting success of migratory birds. *Science*, **267**, 1987–90.

Robinson, T.J., Siefferman, L., Bentz, A.B. and Risch, T.S. (2014c). The separate effects of egg size and parental quality on the development of ornamental plumage coloration. *Ibis*, **156**, 415–23.

Robinson, W.D., Styrsky, J.D., Payne, B.J., Harper, R.G. and Thompson, C.F. (2008). Why are incubation periods longer in the Tropics? A common-garden experiment with House Wrens reveals it is all in the egg. *The American Naturalist*, **171**, 532–5.

Robinson, W.D., Austin, S.H., Robinson, T.R. and Ricklefs, R.E. (2014b). Incubation temperature does not explain variation in the embryo development periods in a sample of Neotropical passerine birds. *Journal of Ornithology*, **155**, 45–51.

Rochester, J., Heiblum, R., Rozenboim, I. and Millam, J. (2008). Post-hatch oral estrogen exposure reduces oviduct and egg mass and alters nest-building behavior in adult zebra finches (*Taeniopygia guttata*). *Physiology & Behavior*, **95**, 370–80.

Rockweit, J.T., Franklin, A.B., Bakken, G.S. and Gutiérrez, R.J. (2012). Potential influences of climate and nest structure on spotted owl reproductive success: a biophysical approach. *PLoS ONE*, **7**(7), e41498.

Rodgers, J.A. Jr, Wenner, A.S. and Schwikert, S.T. (1988). The use and function of green nest material by wood storks. *The Wilson Bulletin*, **103**, 411–23.

Rodrigues, M. and Crick, H.Q.P. (1997). The breeding biology of the Chiffchaff *Phylloscopus collybita* in Britain: a comparison of an intensive study with records of the BTO Nest Record Scheme. *Bird Study*, **44**, 374–83.

Rohwer, V.G. and Law, J.S.Y. (2010). Geographic variation in nests of Yellow Warblers breeding in Churchill, Manitoba, and Elgin, Ontario. *The Condor*, **112**, 596–604.

Rollack, C.E., Wiebe, K. and Stoffel, M.J. (2013). Turkey Vulture breeding behavior studied with trail cameras. *Journal of Raptor Biology*, **47**, 153–60.

Romanoff, A.L. (1960). *The avian embryo*. Macmillan, New York.

Romanoff, A.L. (1967). *Biochemistry of the avian embryo*. John Wiley and Sons, New York.

Romanoff, A.L. and Romanoff, A.J. (1949). *The avian egg*. John Wiley and Sons, New York.

Romanoff, A.L. and Romanoff, A.J. (1972). *Pathogenesis of the avian egg*. John Wiley and Sons, New York.

Roos, S. and Pärt, T. (2004). Nest predators affect spatial dynamics of breeding red-backed shrikes (*Lanius collurio*). *Journal of Animal Ecology*, **73**, 117–27.

Roselaar, K. (2003). An inventory of major European bird collections. *Bulletin of the British Ornithologists' Club Supplement*, **123A**, 253–337.

Rose-Martel, M., Du, J. and Hincke, M.T. (2012). Proteomic analysis provides new insight into the chicken eggshell cuticle. *Journal of Proteomics*, **75**, 2697–706.

Rosenweig, C., Karoly, D., Vicarelli, M., Neofotis, P., Wu, Q., Casassa, G., Menzel, A., Root, T.L., Estrella, N., Seguin, B., Tryjanowski, P., Liu, C., Rawlins, S. and Imeson, A. (2008). Attributing physical and biological impacts to anthropogenic climate change. *Nature*, **453**, 353–7.

Rothstein, S.I. (1974). Mechanisms of avian recognition: possible learned and innate factors. *The Auk*, **91**, 796–807.

Rothstein, S.I. (1975). Mechanisms of avian egg-recognition: Do birds know their own eggs? *Animal Behaviour*, **23**, 268–78.

Rothstein, S.I. (1978). Mechanisms of avian egg-recognition: additional evidence for learned components. *Animal Behaviour*, **26**, 671–7.

Rothstein, S.I. (1982). Mechanisms of avian egg recognition: which egg parameters elicit responses by rejecter species? *Behavioral Ecology and Sociobiology*, **11**, 229–39.

Rothstein, S.I. (1990a). A model system for coevolution: avian brood parasitism. *Annual Review of Ecology and Systematics*, **21**, 481–508.

Rothstein, S.I. (1990b). Brood parasitism and clutch-size determination in birds. *Trends in Ecology and Evolution*, **5**, 101–2.

Rowley, I. (1970). The use of mud in nest-building—a review of the incidence and taxonomic importance. *Ostrich*, Supplement **8**, 141–8.

Royama, T. (1996). A fundamental problem in key factor analysis. *Ecology*, **77**, 87–93.

Royle, N.J., Surai, P.F. and Hartley, I.R. (2001). Maternally derived androgens and antioxidants in bird eggs: complementary but opposing effects. *Behavioral Ecology*, **12**, 381–5.

Roznik, E.A. and Alford, R.A. (2012). Does waterproofing Thermochron iButton dataloggers influence temperature readings? *Journal of Thermal Biology*, **37**, 260–4.

Rózsa, L., Rékási, J. and Reiczigel, J. (1996). Relationship of host coloniality to the population ecology of avian lice (Insecta: Phthiraptera). *Journal of Animal Ecology*, **65**, 242–8.

Rubolini, D., Martinelli, R., von Engelhardt, N., Romano, M., Groothuis, T.G.G., Fasola, M. and Saino, N. (2007). Consequences of prenatal androgen exposure for the reproductive performance of female pheasants (*Phasianus colchicus*). *Proceedings of the Royal Society of London Series B: Biological Sciences*, **274**, 137–42.

Rubolini, D., Romano, M., Martinelli, R., Leoni, B. and Saino, N. (2006). Effects of prenatal yolk androgens on armaments and ornaments of the ring-necked pheasant. *Behavioral Ecology and Sociobiology*, **59**, 549–60.

Rubolini, D., Romano, M., Navara, K., Karadas, F., Ambrosini, R., Caprioli, M. and Saino, N. (2011). Maternal effects mediated by egg quality in the Yellow-legged Gull *Larus michahellis* in relation to laying order and embryo sex. *Frontiers in Zoology*, **8**, 24.

Rueda, L.M., Patel, K.J., Axtell, R.C. and Stinner, R.E. (1990). Temperature-dependent development and survival rates of *Culex quinquefasciatus* and *Aedes aegypti* (Diptera: Culicidae). *Journal of Medical Entomology*, **27**, 892–8.

Ruiz-de-Castañeda, R., Vela, A.I., González-Braojos, S., Briones, V. and Moreno, J. (2011a). Drying eggs to inhibit bacteria: Incubation during laying in a cavity nesting passerine. *Behavioural Processes*, **88**, 142–8.

Ruiz-de-Castañeda, R., Vela, A.I., Lobato, E., Briones, V. and Moreno, J. (2011b). Bacterial loads on eggshells of the pied flycatcher: environmental and maternal factors. *The Condor*, **113**, 200–8.

Rumpf, M. and Nichelmann, M. (1993). Development of prenatal acoustic interaction in the muscovy duck (*Cairina moschata*). *British Poultry Science*, **34**, 287–96.

Rumpf, M. and Tzschentke, B. (2010). Perinatal acoustic communication in birds: why do birds vocalize in the egg? *The Open Ornithology Journal*, **3**, 141–9.

Russell, D.G.D., Hansell, M. and Reilly, M. (2010). Constructive behaviour: developing a UK nest collection resource. *Proceedings of the 5th International Meeting of European Bird Curators*, 77–96. Natural History Museum, Vienna.

Russell, D.G.D., Hansell, M. and Reilly, M. (2013). Bird nests in museum collections: a rich resource for research. *Avian Biology Research*, **6**, 178–82.

Rutkowska, J., Dubiec, A. and Nakagawa, S. (2014). All eggs are made equal: meta-analysis of egg sexual size dimorphism in birds. *Journal of Evolutionary Biology*, **27**, 153–60.

Rutkowska, J., Martyka, R., Aneta, A. and Cichoń, M. (2012). Offspring survival is negatively related to maternal response to sheep red blood cells in zebra finches. *Oecologia*, **168**, 355–9.

Ruuskanen, S., Doligez, B., Gustafsson, L. and Laaksonen, T. (2012a). Long-term effects of yolk androgens on phenotype and parental feeding behavior in a wild passerine. *Behavioral Ecology and Sociobiology*, **66**, 1201–11.

Ruuskanen, S., Doligez, B., Pitala, N., Gustafsson, L. and Laaksonen, T. (2012b). Long-term fitness consequences of high yolk androgen levels: sons pay the costs. *Functional Ecology*, **26**, 884–94.

Ruuskanen, S. and Laaksonen, T. (2010). Yolk hormones have sex-specific long-term effects on behavior in the pied flycatcher (*Ficedula hypoleuca*). *Hormones and Behavior*, **57**, 119–27.

Ruuskanen, S. and Laaksonen, T. (2013). Sex-specific effects of yolk androgens on begging behavior and digestion in pied flycatchers. *Journal of Avian Biology*, **44**, 331–8.

Ruuskanen, S., Lehikoinen, E., Nikinmaa, M., Siitari, H., Waser, W. and Laaksonen, T. (2013). Long-lasting effects of yolk androgens on phenotype in the pied flycatcher (*Ficedula hypoleuca*). *Behavioral Ecology and Sociobiology*, **67**, 361–72.

Ruuskanen, S., Siitari, H., Eeva, T., Belskii, E., Järvinen, A., Kerimov, A., Krams, I., Moreno, J., Morosinotto, C., Mänd, R., Möstl, E., Orell, M., Qvarnström, A., Salminen, J.-P., Slater, F., Tilgar, V., Visser, M.E., Winkel, W., Zang, H. and Laaksonen, T. (2011). Geographical variation in egg mass and egg content in a passerine bird. *PLoS ONE*, **6**, e25360.

Ryder, T.B., Reitsma, R., Evans, B. and Marra, P.P. (2010). Quantifying avian nest survival along an urban gradient using citizen- and scientists-generated data. *Ecological Applications*, **20**, 419–26.

Saalfeld, S.T., Conway, W.C., Haukos, D.A. and Johnson, W.P. (2012). Snowy plover nest site selection, spatial patterning, and temperatures in the Southern High Plains of Texas. *Journal of Wildlife Management*, **76**, 1703–11.

Sabath, K. (1991). Upper Cretaceous amniotic eggs from the Gobi Desert. *Acta Palaeontologica Polonica*, **36**, 151–92.

Safran, R.J. (2004). Adaptive site selection rules and variation in group size of barn swallows: Individual decisions predict population patterns. *The American Naturalist*, **164**, 121–31.

Safran, R.J., Pilz, K.M., McGraw, K.J., Correa, S.M. and Schwabl, H. (2008). Are yolk androgens and carotenoids in barn swallow eggs related to parental quality? *Behavioral Ecology and Sociobiology*, **62**, 427–38.

Sage, R.B., Hoodless, A.N., Hewson, C.M., Wilson, S., Le Clare, C., Marchant, J.H., Draycott, R.A.H. and Fuller, R.J. (2011). Assessing breeding success in common woodland birds using a novel method. *Bird Study*, **58**, 409–20.

Saino, N., Ara, P.D., Martinelli, R. and Møller, A.P. (2002). Early maternal effects and antibacterial immune factors in the eggs, nestlings and adults of the barn swallow. *Journal of Evolutionary Biology*, **15**, 735–43.

Saino, N., Dall'ara, P. and Møller, A.P. (2001). Immunoglobulin plasma concentration in relation to egg laying and mate ornamentation of female barn swallows (*Hirundo rustica*). *Journal of Evolutionary Biology*, **14**, 95–109.

Saino, N. and Fasola, M. (1996). The function of embryonic vocalization in the little tern (*Sterna albifrons*). *Ethology*, **102**, 265–71.

Saino, N., Romano, M., Ambrosini, R., Rubolini, D., Boncoraglio, G., Caprioli, M. and Romano, A. (2012). Longevity and lifetime reproductive success of Barn Swallow offspring are predicted by their hatching date and phenotypic quality. *Journal of Animal Ecology*, **81**, 1004–12.

Saino, N., Romano, M., Ferrari, R.P., Martinelli, R. and Møller, A.P. (2003). Maternal antibodies but not carotenoids in barn swallow eggs covary with embryo sex. *Journal of Evolutionary Biology*, **16**, 516–22.

Saint Jalme, M. (1999). Endangered avian species captive propagation: an overview of functions and techniques. *Avian and Poultry Biology Reviews*, **13**, 187–202.

Saint Jalme, M. and van Heezik, Y. (1996). *Propagation of the houbara bustard*. Kegan Paul International, London.

Salaberria, C., Celis, P., López-Rull, I. and Gill, D. (2014). Effects of temperature and nest heat exposure on nestling growth, dehydration and survival in a Mediterranean hole-nesting passerine. *Ibis*, **156**, 265–75.

Salido, L., Purse, B.V., Marrs, R., Chamberlain, D.E. and Shultz, S. (2012). Flexibility in phenology and habitat use act as buffers to long-term population declines in UK passerines. *Ecography*, **35**, 604–13.

Salifou, S., Offoumon, O.T.L.F., Gouissi, F.M. and Pangui, L.J. (2013). Endogenous recipes for controlling arthropod ectoparasites of domestic poultry. *Revista Brasileira de Parasitologia Veterinária*, **22**, 119–23.

Salzer, D.W. and Larkin, G.J. (1990). Impact of courtship feeding on clutch and third-egg size in glaucous-winged gulls. *Animal Behaviour*, **39**, 1149–62.

Samaranayake, Y.H., Samaranayake, L.P., Pow, E.H.N., Beena, V.T. and Yeung, K.W.S. (2001). Antifungal effects of lysozyme and lactoferrin against genetically similar, sequential *Candida albicans* isolates from a human immunodeficiency virus-infected southern Chinese cohort. *Journal of Clinical Microbiology*, **39**, 3296–302.

Samaš, P., Grim, T., Hauber, M.E., Cassey, P., Weidinger, K. and Evans, K.L. (2013). Ecological predictors of reduced avian reproductive investment in the southern hemisphere. *Ecography*, **36**, 809–18.

Samiullah, S. and Roberts, J. (2013). The location of protoporphyrin in the eggshell of brown-shelled eggs. *Poultry Science*, **92**, 2783–8.

Sánchez, J.M., Corbacho, C., Muñoz del Viejo, A. and Parejo, D. (2004). Colony-site tenacity and egg color crypsis in the Gull-billed Tern. *Waterbirds*, **27**, 21–30.

Santisteban, L., Sieving, K.E. and Avery, M.L. (2002). Use of sensory cues by fish crows *Corvus ossifragus* preying on artificial bird nests. *Journal of Avian Biology*, **33**, 245–52.

Sanz, J.J. (1996). Effect of food availability on incubation period in the pied flycatcher (*Ficedula hypoleuca*). *The Auk*, **113**, 249–53.

Sanz, J.J. and García-Navas, V. (2009). Eggshell pigmentation pattern in relation to breeding performance of blue tits *Cyanistes caeruleus*. *Journal of Animal Ecology*, **78**, 31–41.

Sanz, J.J., Kranenbarg, S. and Tinbergen, J.M. (2000). Differential response by males and females to manipulation of partner contribution in the great tit (*Parus major*). *Journal of Animal Ecology*, **69**, 74–84.

Sanz-Aguilar, A., Tavecchia, G., Pradel, R., Minguez, E. and Oro, D. (2008). The cost of reproduction and experience-dependent vital rates in a small petrel. *Ecology*, **89**, 3195–203.

Sargent, T.D. (1965). Role of experience in nest building of zebra finch. *The Auk*, **82**, 48–61.

Sato, T., Cheng, Y., Wu, X., Zelenitsky, D.K. and Hsiao, Y. (2005). A pair of shelled eggs inside a female dinosaur. *Science*, **308**, 375.

Saunders, S.P., Roche, E.A., Arnold, T.W. and Cuthbert, F.J. (2012). Female site familiarity increases fledging success in Piping Plovers (*Charadrius melodus*). *The Auk*, **129**, 329–37.

Savalli, U.M. (1994). Mate choice in the Yellow-shouldered Widowbird: correlates of male attractiveness. *Behavioral Ecology and Sociobiology*, **35**, 227–34.

Savill, P., Perrins, C., Kirby, K. and Fisher, N. (2011). *Wytham Woods: Oxford's ecological laboratory*. Oxford University Press, Oxford.

Sayler, R.D. (1992). Ecology and evolution of brood parasitism in waterfowl. In BDJ Batt, AD Afton, MG Anderson, CD Ankney, DH Johnson, JA Kadlec, GL Krapu, eds. *Ecology and management of breeding waterfowl*, pp. 290–322. University of Minnesota Press, Minneapolis.

Schaefer, V.H. (1980). Geographic variation in the insulative qualities of nests of the Northern Oriole. *The Wilson Bulletin*, **92**, 466–74.

Schaffner, F.C. (1990). Egg recognition by elegant terns (*Sterna elegans*). *Colonial Waterbirds*, **13**, 25–30.

Scharlemann, J.P.W. (2001). Museum egg collections as stores of long-term phenological data. *International Journal of Biometeorology*, **45**, 208–11.

Schiermann, G. (1939). 'Stammesgenossenschaften' bei Vogeln. *Ornithologische Monatsberichte*, **47**, 1–3.

Schlaepfer, M.A., Runge, M.C. and Sherman, P.W. (2002). Ecological and evolutionary traps. *Trends in Ecology and Evolution*, **17**, 474–80.

Schleich, H.H. and Kästle, W. (1988). *Reptile egg-shells SEM atlas*. Gustav Fischer, Stuttgart.

Schmidt, K.A. (1999). Foraging theory as a conceptual framework for studying nest predation. *Oikos*, **85**, 151–60.

Schmidt, K.A., Nelis, L.C., Briggs, N. and Ostfeld, R.S. (2005). Invasive shrubs and songbird nesting success: Effects of climate variability and predator abundance. *Ecological Applications*, **15**, 258–65.

Schmidt, K.A. and Whelan, C.J. (1999). Nest predation on woodland songbirds: when is nest predation density dependent? *Oikos*, **87**, 65–74.

Schneider, E.G. and McWilliams, S.R. (2007). Using nest temperatures to estimate nest attendance of Piping Plovers. *Journal of Wildlife Management*, **71**, 1998–2006.

Schoech, S.J., Bridge, E.S., Boughton, R.K., Reynolds, S.J., Atwell, J.W. and Bowman, R. (2008). Food supplementation: a tool to increase reproductive output? A case study in the threatened Florida Scrub-Jay. *Biological Conservation*, **141**, 162–73.

Schoech, S.J., Rensel, M.A., Bridge, E.S., Boughton, R.K. and Wilcoxen, T.E. (2009). Environment, glucocorticoids, and the timing of reproduction. *General and Comparative Endocrinology*, **163**, 201–7.

Schoech, S.J., Rensel, M.A. and Wilcoxen, T.E. (2012). Here today, not gone tomorrow: long-term effects of corticosterone. *Journal of Ornithology*, **153** (Supplement 1), S217–26.

Schönwetter, M. (1960–71). *Handbuch der oologie*, Lief. 1–19. Akademie-Verlag, Berlin.

Schramm, M. (1983). Automatic recording of nest visits by burrow-nesting birds. *Journal of Field Ornithology*, **54**, 192–4.

Schuetz, J.G. (2005). Common waxbills use carnivore scat to reduce the risk of nest predation. *Behavioral Ecology*, **16**, 133–7.

Schwabl, H. (1993). Yolk is a source of maternal testosterone for developing birds. *Proceedings of the National Academy of Sciences of the USA*, **90**, 11446–50.

Schwabl, H. (1996). Maternal testosterone in the avian egg enhances postnatal growth. *Comparative Biochemistry and Physiology A*, **114**, 271–6.

Schwabl, H., Holmes, D., Strasser, R. and Scheuerlein, A. (2012). Embryonic exposure to maternal testosterone influences age-specific mortality patterns in a captive passerine bird. *Age*, **34**, 87–94.

Schwabl, H., Palacios, M.G. and Martin, T.E. (2007). Selection for rapid embryo development correlates with embryo exposure to maternal androgens among passerine birds. *The American Naturalist*, **170**, 196–206.

Schwagmeyer, P.L., Mock, D.W., Lamey, T.C., Lamey, C.S. and Beecher, M.D. (1991). Effects of sibling contact on hatch timing in an asynchronously hatching bird. *Animal Behaviour*, **41**, 887–94.

Schwarzbach, S.E., Albertson, J.D. and Thomas, C.M. (2006). Effects of predation, flooding, and contamination on reproductive success of California Clapper Rails (*Rallus longirostris obsoletus*) in San Francisco Bay. *The Auk*, **12**, 45–60.

Schweitzer, M.H., Jackson, F.D., Chiappe, L.M., Schmitt, J.G., Calvo, J.O. and Rubilar, D.E. (2002). Late Cretaceous avian eggs with embryos from Argentina. *Journal of Vertebrate Paleontology*, **22**, 191–5.

Schweitzer, M.H., Wittmeyer, J.L. and Horner, J.R. (2005). Gender-specific reproductive tissue in ratites and *Tyrannosaurus rex*. *Science*, **308**, 1456–60.

Scott, W.E.D. (1902). Instinct in song birds. Methods of breeding in hand-reared Robins (*Merula migratoria*). *Science*, **16**, 70–1.

Scott, W.E.D. (1904). The inheritance of song in passerine birds. Further observations on the development of song and nest-building in hand-reared Rose-breasted Grosbeaks, *Zamelodia ludoviciana* (Linnaeus). *Science*, **20**, 282–3.

Sealy, S.G. and Lorenzana, J.C. (1997). Yellow Warblers (*Dendroica petechia*) do not recognize their own eggs. *Bird Behavior*, **12**, 57–66.

Sealy, S.G. and Neudorf, D.L. (1995). Male northern orioles eject cowbird eggs: implications for the evolution of rejection behavior. *The Condor*, **97**, 369–75.

Sedinger, J.S. and Flint, P.L. (1991). Growth rate is negatively correlated with hatch date in Black Brant. *Ecology*, **72**, 496–502.

Sedinger, J.S., Flint, P.L. and Lindberg, M.S. (1995). Environmental influence on life-history traits: growth, survival, and fecundity in Black Brant (*Branta bernicla*). *Ecology*, **76**, 2404–14.

Sedinger, J.S., Herzog, M.P. and Ward, D.H. (2004). Early environment and recruitment of Black Brant (*Branta bernicla nigricans*) into the breeding population. *The Auk*, **121**, 68–73.

Selman, C., Blount, J.D., Nussey, D.H. and Speakman, J.R. (2012). Oxidative damage, ageing, and life-history evolution: where now? *Trends in Ecology and Evolution*, **27**, 570–7.

Seltmann, M.W., Jaatinen, K., Steele, B.B. and Öst, M. (2014). Boldness and stress responsiveness as drivers of nest-site selection in a ground-nesting bird. *Ethology*, **120**, 77–89.

Seppänen, J.-T. and Forsman J.T. (2007). Interspecific social learning: novel preference can be acquired from a competing species. *Current Biology*, **17**, 1248–52.

Seppänen, J.-T., Forsman, J.T., Monkkonen, M., Krams, I. and Salmi, T. (2011). New behavioural trait adopted or rejected by observing heterospecific tutor fitness. *Proceedings of the Royal Society B*, **278**, 1736–41.

Sergio, F., Blas, J., Blanco, G., Tanferna, A., López, L., Lemus, J.A. and Hiraldo, F. (2011). Raptor nest decorations are a reliable threat against conspecifics. *Science*, **331**, 327–30.

Seviour, E., Sykes, F. and Board, R. (1972). A microbiological survey of the incubated eggs of chickens and waterfowl. *British Poultry Science*, **13**, 549–56.

Seymour, R.S. (1984). *Respiration and metabolism of embryonic vertebrates*. Dr. W. Junk, Dordrecht.

Seymour, R.S., Vleck, D. and Vleck, C.M. (1986). Gas exchange in the incubation mounds of megapode birds. *Journal of Comparative Physiology B*, **156**, 773–82.

Shafey, T.M., Ghannam, M.M., Al-Batshan, H.A. and Al-Ayed, M.S. (2004). Effect of pigment intensity and region of eggshell on the spectral transmission of light that passes the eggshell of chickens. *International Journal of Poultry Science*, **3**, 228–33.

Shaffer, S.A., Clatterbuck, C.A., Kelsey, E.C., Naiman, A.D., Young, L.C., Vanderwerf, E.A., Warzybok, P., Bradley, R., Jahncke, J. and Bower, G.C. (2014). As the egg turns: monitoring egg attendance behavior in wild birds using novel data logging technology. *PLoS ONE*, **9**(6), e97898.

Shaffer, S.A., Costa, D.P. and Weimerskirch, H. (2004). Field metabolic rates of black-browed albatrosses *Thalassarche melanophrys* during the incubation stage. *Journal of Avian Biology*, **35**, 551–8.

Sharon, N. (1967). The chemical structure of lysozyme substrates and their cleavage by the enzyme. *Proceedings of the Royal Society of London Series B: Biological Sciences*, **167**, 402–15.

Sharp, P.J., Dawson, A. and Lea, R.W. (1998). Control of luteinizing hormone and prolactin secretion in birds. *Comparative Biochemistry and Physiology C Pharmacology, Toxicology and Endocrinology*, **119**, 275–82.

Shaw, L.M., Chamberlain, D. and Evans, M. (2008). The house sparrow *Passer domesticus* in urban areas: reviewing a possible link between post-decline distribution and human socioeconomic status. *Journal of Ornithology*, **149**, 293–9.

Shawkey, M.D., Estes, A.M., Siefferman, L.M. and Hill, G.E. (2003b). Nanostructure predicts intraspecific variation in ultraviolet–blue plumage colour. *Proceedings of the Royal Society of London Series B: Biological Sciences*, **270**, 1455–60.

Shawkey, M.D., Firestone, M.K., Brodie, E.L. and Beissinger, S.R. (2009). Avian incubation inhibits growth and diversification of bacterial assemblages on eggs. *PLoS ONE*, **4**(2), e4522.

Shawkey, M.D., Kosciuch, K.L., Liu, M., Rohwer, F.C., Loos, E.R., Wang, J.M. and Beissinger, S.R. (2008). Do birds differentially distribute antimicrobial proteins within clutches of eggs? *Behavioral Ecology*, **19**, 920–7.

Shawkey, M.D., Pillai, S.R. and Hill, G.E. (2003a). Chemical warfare? Effects of uropygial oil on feather-degrading bacteria. *Journal of Avian Biology*, **34**, 345–9.

Sheldon, B.C. and Verhulst, S. (1996). Ecological immunology: costly parasite defenses and trade-offs in evolutionary ecology. *Trends in Ecology and Evolution*, **11**, 317–21.

Shields, W.M. (1984). Factors affecting nest and site fidelity in Adirondack barn swallows (*Hirundo rustica*). *The Auk*, **101**, 780–9.

Shrubb, M. (1990). Effects of agricultural change on nesting lapwings *Vanellus vanellus* in England and Wales. *Bird Study*, **37**, 115–27.

Shutler, D. and Campbell, A.A. (2007). Experimental addition of greenery reduces flea loads in nests of a non-greenery using species, the tree swallow *Tachycineta bicolor*. *Journal of Avian Biology*, **38**, 7–12.

Sibly, R.M., Witt, C.C., Wright, N.A., Venditti, C., Jetz, W. and Brown, J.H. (2012). Energetics, lifestyle, and reproduction in birds. *Proceedings of the National Academy of Sciences of the USA*, **109**, 10937–41.

Sidis, Y., Zilberman, R. and Ar, A. (1994). Thermal aspects of nest placement in the Orange-tufted Sunbird (*Nectarinia osea*). *The Auk*, **111**, 1001–5.

Sih, A., Bell, A.M., Johnson, J.C. and Ziemba, R.E. (2004). Behavioral syndromes: an integrative overview. *Quarterly Review of Biology*, **79**, 241–77.

Siikamäki, P. (1995). Are large clutches costly to incubate—the case of the pied flycatcher. *Journal of Avian Biology*, **26**, 76–80.

Siikamäki, P., Rätti, O., Hovi, M. and Bennett, G.F. (1997). Association between haematozoan infections and reproduction in the Pied Flycatcher. *Functional Ecology*, **11**, 176–83.

Silva, B., Correia, J., Nunes, F., Tavares, P., Varum, H. and Pinto, J. (2010). Bird nest construction—lessons for building with earth. *WSEAS Transactions on Environment and Development*, **6**, 83–92.

Silva, E.T. (1949). Nest records of the song thrush. *British Birds*, **42**, 97–111.

Simon, A., Thomas, D., Blondel, J., Perret, P. and Lambrechts, M.M. (2004). Physiological ecology of Mediterranean blue tits (*Parus caeruleus*): Effects of ectoparasites (*Protocalliphora* spp.) and food abundance on metabolic

capacity of nestlings. *Physiological and Biochemical Zoology*, **77**, 492–501.

Simons, P.C.M. and Wiertz, G. (1966). The ultra-structure of the surface of the cuticle of the hen's egg in relation to egg-cleaning. *Poultry Science*, **45**, 1153–62.

Sinclair, B.J. and Chown, S.L. (2006). Caterpillars benefit from thermal ecosystem engineering by Wandering Albatrosses on sub-Antarctic Marion Island. *Biology Letters*, **2**, 51–4.

Sinclair, J.R., Tuke, L. and Opiang, M. (2010). What the locals know: comparing traditional and scientific knowledge of megapodes in Melanesia. In S Tidemann, A Gosler, eds. *Ethno-ornithology: birds, indigenous peoples, culture and society*, pp. 115–37. Earthscan, London.

Siriwardena, G.M., Crick, H.Q.P., Baillie, S.R. and Wilson, J.D. (2000a). Agricultural habitat-type and the breeding performance of granivorous farmland birds in Britain. *Bird Study*, **47**, 66–81.

Siriwardena, G.M., Crick, H.Q.P., Baillie, S.R. and Wilson, J.D. (2000b). The importance of variation in the breeding performance of seed-eating birds in determining their population trends on farmland. *Journal of Applied Ecology*, **37**, 128–48.

Siriwardena, G.M., Crick, H.Q.P., Baillie, S.R. and Wilson, J.D. (2001). Changes in agricultural land-use and breeding performance of some granivorous farmland passerines in Britain. *Agriculture, Ecosystems & Environment*, **84**, 191–206.

Skowron, C. and Kern, M. (1980). The insulation in nests of selected North American songbirds. *The Auk*, **97**, 816–24.

Skutch, A.F. (1949). Do tropical birds rear as many young as they can nourish? *Ibis*, **91**, 430–58.

Skutch, A.F. (1957). The incubation patterns of birds. *Ibis*, **99**, 69–93.

Skutch, A.F. (1962). The constancy of incubation. *The Wilson Bulletin*, **71**, 115–52.

Skutch, A.F. (1976). *Parent birds and their young*. University of Texas Press, Austin and London.

Slager, D.L., McDermott, M.E. and Rodewald, A.D. (2012). Kleptoparasitism of nesting material from a Red-faced Spinetail (*Cranioleuca erythrops*) nest site. *The Wilson Journal of Ornithology*, **124**, 812–5.

Slagsvold, T. (1982). Clutch size variation in passerine birds: the nest predation hypothesis. *Oecologia*, **54**, 159–69.

Slagsvold, T. (1989a). On the evolution of clutch size and nest size in passerine birds. *Oecologia*, **79**, 300–5.

Slagsvold, T. (1989b). Experiments on clutch size and nest size in passerine birds. *Oecologia*, **80**, 297–302.

Slagsvold, T., Wigdahl Kleiven, K., Eriksen, A. and Johannessen, L.E. (2013). Vertical and horizontal transmission of nest site preferences in titmice. *Animal Behaviour*, **85**, 323–8.

Slater, P. (2001). Breeding ecology of a suburban population of Woodpigeons *Columba palumbus* in northwest England. *Bird Study*, **48**, 361–6.

Sleigh, M.J. and Casey, M.B. (2014). Prenatal sensory experience affects hatching behavior in domestic chicks (*Gallus*

gallus) and Japanese quail chicks (*Coturnix coturnix japonica*). *Developmental Psychobiology*, **56**, 935–42.

Smiseth, P.T., Scott, M.P. and Andrews, C. (2011). Hormonal regulation of offspring begging and mediation of parent-offspring conflict. *Animal Behaviour*, **81**, 507–17.

Smith, H.G. (1989). Larger clutches take longer to incubate. *Ornis Scandinavica*, **20**, 156–8.

Smith, H.G. and Montgomerie, R. (1992). Male incubation in Barn Swallows: the influence of nest temperature and sexual selection. *The Condor*, **94**, 750–9.

Smith, J.A. (2011). *From nest building to life-history patterns: Does food supplementation influence reproductive behaviour of birds?* PhD thesis, University of Birmingham, Birmingham.

Smith, J.A., Harrison, T.J.E., Martin, G.R. and Reynolds, S.J. (2013). Feathering the nest: food supplementation influences nest construction by Blue (*Cyanistes caeruleus*) and Great Tits (*Parus major*). *Avian Biology Research*, **6**, 18–25.

Smith, M.D. and Conway, C. (2011). Collection of mammal manure and other debris by nesting burrowing owls. *Journal of Raptor Research*, **45**, 220–8.

Smith, N.G. (1968). The advantage of being parasitized. *Nature*, **219**, 690–4.

Smith, P.A., Tulp, I., Schekkerman, H., Gilchrist, H. and Forbes, M.R. (2012). Shorebird incubation behaviour and its influence on the risk of nest predation. *Animal Behaviour*, **84**, 835–42.

Smith, W.K., Roberts, S.W. and Miller, P.C. (1974). Calculating the nocturnal energy expenditure of an incubating Anna's Hummingbird. *The Condor*, **76**, 176–83.

Snow, D.W. (1955a). The breeding of Blackbird, Song Thrush and Mistle Thrush in Great Britain. Part II. Clutch size. *Bird Study*, **2**, 72–84.

Snow, D.W. (1955b). The breeding of Blackbird, Song Thrush and Mistle Thrush in Great Britain. Part III. Nesting success. *Bird Study*, **2**, 169–78.

Snow, D.W. (1958). *A study of blackbirds*. Allen and Unwin, London.

Snow, D.W. (1966). Population dynamics of the blackbird. *Nature*, **211**, 1231–3.

Snow, D.W. (1969a). Moult of British thrushes and chats. *Bird Study*, **16**, 115–29.

Snow, D.W. (1969b). An analysis of breeding success in the blackbird, *Turdus merula*. *Ardea*, **57**, 163–71.

Snow, D.W. (1969c). Some vital statistics of British mistle thrushes. *Bird Study*, **16**, 34–44

Snyder, N. and Snyder, H. (2000). *The Californian condor: a saga of natural history and conservation*. Academic Press, San Diego.

Sockman, K.W., Sharp, P.J. and Schwabl, H. (2006). Orchestration of avian reproductive effort: an integration of the ultimate and proximate bases for flexibility in clutch size, incubation behaviour, and yolk androgen deposition. *Biological Reviews of the Cambridge Philosophical Society*, **81**, 629–66.

Sockman, K.W., Weiss, J., Webster, M.S., Talbott, V. and Schwabl, H. (2008). Sex-specific effects of yolk-androgens on growth of nestling American kestrels. *Behavioral Ecology and Sociobiology*, **62**, 617–25.

Söderström, B., Pärt, T. and Rydén, J. (1998). Different nest predator faunas and nest predation risk on ground and shrub nests at forest ecotones: an experiment and a review. *Oecologia*, **117**, 108–18.

Sol, D., Jovani, R. and Torres, J. (2003). Parasite mediated mortality and host immune response explain age-related differences in blood parasitism in birds. *Oecologia*, **135**, 542–7.

Soler, J.J., Cuervo, J.J., Møller, A.P. and de Lope, F. (1998a). Nest building is a sexually selected behaviour in the barn swallow. *Animal Behaviour*, **56**, 1435–42.

Soler, J.J., de Neve, L., Martínez, J.G. and Soler, M. (2001). Nest size affects clutch size and the start of incubation in magpies: an experimental study. *Behavioral Ecology*, **12**, 301–7.

Soler, J.J., Martín-Vivaldi, M., Peralta-Sánchez, J.M., Arco, L. and Juárez-García-Pelayo, N. (2014). Hoopoes color their eggs with antimicrobial uropygial secretions. *Naturwissenschaften*, **101**, 697–705.

Soler, J.J., Martín-Vivaldi, M., Ruiz-Rodríguez, M., Valdivia, E., Martín-Platero, A.M., Martínez-Bueno, M., Peralta-Sánchez, J.M. and Méndez, M. (2008a). Symbiotic association between hoopoes and antibiotic-producing bacteria that live in their uropygial gland. *Functional Ecology*, **22**, 864–71.

Soler, J.J., Møller, A.P. and Soler, M. (1998b). Nest building, sexual selection and parental investment. *Evolutionary Ecology*, **12**, 427–41.

Soler, J.J., Moreno, J., Avilés, J.M., Møller, A.P. and Sorenson, M. (2005). Blue and green egg-color intensity is associated with parental effort and mating system in passerines: support for the sexual selection hypothesis. *Evolution*, **59**, 636–44.

Soler, J.J., Navarro, C., Contreras, T.P., Avilés, J.M. and Cuervo, J.J. (2008b). Sexually selected egg coloration in spotless starlings. *The American Natutalist*, **171**, 183–94.

Soler, J.J., Peralta-Sánchez, J.M., Flensted-Jensen, E., Martín-Platero, A.M. and Møller, A.P. (2011a). Innate humoural immunity is related to eggshell bacterial load of European birds: a comparative analysis. *Naturwissenschaften*, **98**, 807–13.

Soler, J.J., Peralta-Sánchez, J.M., Martínez-Bueno, M., Martín-Vivaldi, M., Martín-Gálvez, D., Vela, A.I., Briones, V. and Pérez-Contreras, T. (2011b). Brood parasitism is associated with increased bacterial contamination of host eggs: bacterial loads of host and parasitic eggs. *Biological Journal of the Linnean Society*, **103**, 836–48.

Soler, J.J., Peralta-Sánchez, J.M., Martín-Platero, A.M., Martín-Vivaldi, M., Martínez-Bueno, M. and Møller, A.P. (2012). The evolution of size of the uropygial gland: mutualistic feather mites and uropygial secretion reduce bacterial loads of eggshells and hatching failures of European birds. *Journal of Evolutionary Biology*, **25**, 1779–91.

Solís, J.C. and de Lope, F. (1995). Nest and egg crypsis in the ground-nesting stone curlew *Burhinus oedicnemus*. *Journal of Avian Biology*, **26**, 135–8.

Solomon, S.E. (1987). Egg shell pigmentation. In RG Wells, CG Belyarin, eds. *Egg quality—Current problems and recent advances*, pp. 147–57. Butterworths and Co. Ltd, London.

Solomon, S.E. (1991). *Egg and eggshell quality*. Wolfe Publishing Ltd, London.

Sonerud, G.A. (1985). Nest hole shift in Tengmalm's owl *Aegolius funereus* as defense against nest predation involving long-term-memory in the predator. *Journal of Animal Ecology*, **54**, 179–92.

Sotherland, P.R. and Rahn, H. (1987). On the composition of bird eggs. *The Condor*, **89**, 48–65.

Sparks, N.H.C. (2011). Eggshell pigments from formation to deposition. *Avian Biology Research*, **4**, 162–7.

Sparks, N.H.C. and Board, R.G. (1984). Cuticle, shell porosity and water uptake through hens' eggshells. *British Poultry Science*, **25**, 267–76.

Sparks, T., Huber, K. and Tryjanowski, P. (2008). Something for the weekend? Examining the bias in avian phenological recording. *International Journal of Biometerology*, **52**, 505–10.

Spee, M., Beaulieu, M., Dervaux, A., Chastel, O., Le Maho, Y. and Raclot, T. (2010). Should I stay or should I go? Hormonal control of nest abandonment in a long-lived bird, the Adelie penguin. *Hormones and Behavior*, **58**, 762–8.

Spee, M., Marchal, L., Lazin, D., Le Maho, Y., Chastel, O., Beaulieu, M. and Raclot, T. (2011). Exogenous corticosterone and nest abandonment: A study in a long-lived bird, the Adelie penguin. *Hormones and Behavior*, **60**, 362–70.

Spencer, K.A., Heidinger, B.J., D'Alba, L.B., Evans, N.P. and Monaghan, P. (2010). Then versus now: effect of developmental and current environmental conditions on incubation effort in birds. *Behavioral Ecology*, **21**, 999–1004.

Sperry, J.H., Barron, D.G. and Weatherhead, P.J. (2012). Snake behavior and seasonal variation in nest survival of northern cardinals *Cardinalis cardinalis*. *Journal of Avian Biology*, **43**, 496–502.

Spillner, E., Braren, I., Greunke, K., Seismann, H., Blank, S. and du Plessis, D. (2012). Avian IgY antibodies and their recombinant equivalents in research, diagnostics and therapy. *Biologicals*, **40**, 313–22.

Spotswood, E.N., Goodman, K.R., Carlisle, J., Cormier, R.L., Humple, D.L., Rousseau, J., Guers, S.L. and Barton, G.G. (2012). How safe is mist netting? Evaluating the risk of injury and mortality to birds. *Methods in Ecology and Evolution*, **3**, 29–38.

Spottiswoode, C.N. (2009). Fine-scale life-history variation in sociable weavers in relation to colony size. *Journal of Animal Ecology*, **78**, 504–12.

Spottiswoode, C.N. and Koorevaar, J. (2012). A stab in the dark: chick killing by brood parasitic honeyguides. *Biology Letters*, **8**, 241–4.

Spurr, E.B. (1999). Developing a long-life toxic bait and lures for mustelids. *Science for Conservation*, **127A**, 1–24.

Stanback, M.T. and Dervan, A.A. (2001). Within-season nest-site fidelity in Eastern Bluebirds: Disentangling effects of nest success and parasite avoidance. *The Auk*, **118**, 743–5.

Stanback, M.T., Mercadante, A.N., Cline, E.L., Burke, H. and Roth, J.E. (2013). Cavity depth, not experience, determines

nest height in eastern bluebirds. *The Wilson Journal of Ornithology*, **125**, 301–6.

Stanevičius, V. and Balevičius, A. (2005). Factors influencing nest material selection in Marsh Harrier (*Circus aeruginosus*). *Acta Zoologica Lituanica*, **15**, 21–9.

Starck, J.M. and Ricklefs, R.E. (1998). *Avian growth and development: Evolution within the altricial-precocial spectrum*. Oxford University Press, New York.

Staszewski, V., Gasparini, J., McCoy, K.D., Tveraa, T. and Boulinier, T. (2007a). Evidence of an interannual effect of maternal immunization on the immune response of juveniles in a long-lived colonial bird. *Journal of Animal Ecology*, **76**, 1215–23.

Staszewski, V., McCoy, K.D., Tveraa, T. and Boulinier, T. (2007b). Interannual dynamics of antibody levels in naturally infected long-lived colonial birds. *Ecology*, **88**, 3183–91.

Stauffer, G.E., Diefenbach, D.R., Marshall, M.R. and Brauning, D.W. (2011). Nest success of grassland sparrows on reclaimed surface mines. *Journal of Wildlife Management*, **75**, 548–57.

Steel, E. and Hinde, R.A. (1972). Influence of photoperiod on oestrogenic induction of nest-building in canaries. *Journal of Endocrinology*, **55**, 265–78.

Stenseth, N.C., Mysterud, A., Ottersen, G., Hurrell, J.W., Chan, K.-S. and Lima, M. (1999). Ecological effects of climate fluctuations. *Science*, **285**, 1071–3.

Stoddard, M.C., Fayet, A.L., Kilner, R.M. and Hinde, C.A. (2012). Egg speckling patterns do not advertise offspring quality or influence male provisioning in great tits. *PLoS ONE*, **7**(7), e40211.

Stoddard, M.C., Kilner, R.M. and Town, C. (2014). Pattern recognition algorithm reveals how birds evolve individual egg pattern signatures. *Nature Communications*. doi:10.1038/ncomms5117.

Stoddard, M.C., Marshall, K.L.A. and Kilner, R.M. (2011). Imperfectly camouflaged avian eggs: artefact or adaptation? *Avian Biology Research*, **4**, 196–213.

Stoddard, M.C. and Stevens, M. (2011). Avian vision and the evolution of egg color mimicry in the common cuckoo. *Evolution*, **65**, 2004–13.

Stokes, D.L. and Boersma, P.D. (1998). Nest-site characteristics and reproductive success in Magellanic penguins (*Spheniscus magellanicus*). *The Auk*, **115**, 34–49.

Stoleson, S.H. and Beissinger, S.R. (1995). Hatching synchrony and the onset of incubation in birds, revisited. *Current Ornithology*, **12**, 191–270.

Stoleson, S.H. and Beissinger, S.R. (1999). Egg viability as a constraint on hatching synchrony at high ambient temperatures. *Journal of Animal Ecology*, **68**, 951–62.

Storer, N.P. and Hansell, M.H. (1992). Specialisation in the choice and use of spider silk in the nest of the chaffinch (*Fringella coelebs*) (Aves, Fringillidae). *Journal of Natural History*, **26**, 1421–30.

Strasser, R. and Schwabl, H. (2004). Yolk testosterone organizes behavior and male plumage coloration in house sparrows (*Passer domesticus*). *Behavioral Ecology and Sociobiology*, **56**, 491–7.

Streby, H.M., Refsnider, J.M. and Andersen, D.E. (2014). Redefining reproductive success in songbirds: Moving beyond the nest success paradigm. *The Auk*, **131**, 718–26.

Stucker, J.H., Buhl, D.A. and Sherfy, M.H. (2013). Consequences of least tern (*Sternula antillarum*) microhabitat nest-site selection on natural and mechanically constructed sandbars in the Missouri river. *The Auk*, **130**, 753–63.

Styrsky, J.N. (2005). Influence of predation on nest-site reuse by an open-cup nesting Neotropical passerine. *The Condor*, **107**, 133–7.

Suarez, A.V. and Tsutsui, N.D. (2004). The value of museum collections for research and society. *BioScience*, **54**, 66–74.

Suárez, F., Morales, M.B., Mínguez, I. and Herranz, J. (2005). Seasonal variation in nest mass and dimensions in an open-cup ground-nesting shrub-steppe passerine: The Tawny Pipit *Anthus campestris*. *Ardeola*, **52**, 43–51.

Suárez-Rodríguez, M., López-Rull, I. and Macías Garcia, C. (2013). Incorporation of cigarette butts into nests reduces nest ectoparasite load in urban birds: new ingredients for an old recipe? *Biology Letters*, **9**, 20120931.

Suárez-Rodríguez, M. and Macías Garcia, C. (2014). There is no such thing as a free cigarette; lining nests with discarded butts brings short-term benefits, but causes toxic damage. *Journal of Evolutionary Biology*, **27**, 2719–26.

Sugiarto, H. and Yu, P.-L. (2004). Avian antimicrobial peptides: the defense role of beta-defensins. *Biochemical and Biophysical Research Communications*, **323**, 721–7.

Sullivan, B.L., Aycrigg, J.L., Barry, J.H., Bonney, R.E., Bruns, N., Cooper, C.B., Damoulas, T., Dhondt, A.A., Dietterich, T., Farnsworth, A., Fink, D., Fitzpatrick, J.W., Fredericks, T., Gerbracht, J., Gomes, C., Hochachka, W.M., Iliff, M.J., Lagoze, C., La Sorte, F.A., Merrifield, M., Morris, W., Phillips, T.B., Reynolds, M., Rodewald, A.D., Rosenberg, K.V., Trautmann, N.M., Wiggins, A., Winkler, D.W., Wong, W. and Wood, C.L. (2014). The eBird enterprise: An integrated approach to development and application of citizen science. *Biological Conservation*, **169**, 31–40.

Summers, R.W., Humphreys, E., Newell, M. and Donald, C. (2002). Nest-site selection by crossbills *Loxia* spp. in ancient native pinewoods at Abernethy Forest, Strathspey, Highland. *Bird Study*, **49**, 258–62.

Surai, P.F. (2002). *Natural antioxidants in avian nutrition and reproduction*. Nottingham University Press, Nottingham.

Surgey, J., du Feu, C.R. and Deeming D.C. (2012). Opportunistic use of a wool-like artificial material as lining of tit (Paridae) nests. *The Condor*, **114**, 1–9.

Sutter, G.C. (1997). Nest-site selection and nest-entrance orientation in Sprague's Pipit. *The Wilson Bulletin*, **109**, 462–9.

Svensson, E., Råberg, L., Koch, C. and Hasselquist, D. (1998). Energetic stress, immunosuppression and the costs of an antibody response. *Functional Ecology*, **12**, 912–9.

Swaileh, K.M. and Sansur, R. (2005). Monitoring urban heavy metal pollution using the House Sparrow (*Passer domesticus*). *Journal of Environmental Monitoring*, **8**, 209–13.

Sykes, A.H. (1991). An introduction to the history of incubators. In SG Tullett, ed. *Avian incubation*, pp. 297–303. Butterworth-Heinnemann, London.

Szentirmai, I. and Székely, T. (2002). Do Kentish plovers regulate the amount of their nest material? An experimental test. *Behaviour*, **139**, 847–59.

Szentirmai, I. and Székely, T. (2004). Diurnal variation in nest material use by the Kentish Plover *Charadrius alexandrinus*. *Ibis*, **146**, 535–7.

Szentirmai, I., Székely, T. and Liker, A. (2005). The influence of nest size on heat loss of penduline tit eggs. *Acta Zoologica Academiae Scientarum Hungaricae*, **51**, 59–66.

Talent, L.G., Jarvis, R.L. and Krapu, G.L. (1983). Survival of Mallard broods in south-central North Dakota. *The Condor*, **85**, 74–8.

Tamarit, R., Viñals, A., Gómez, J. and Gil-Delgado, J.A. (2012). Use of blackbird nests as a resource by garden dormice (*Eliomys quercinus*). *Peckiana*, **8**, 135–40.

Tanaka, K. and Zelenitsky, D.K. (2014). Comparisons between experimental and morphometric water vapour conductance in the eggs of extant birds and crocodiles: implications for predicting nest type in dinosaurs. *Canadian Journal of Zoology*, **92**, 1049–58.

Tatner, P. and Bryant, D.M. (1993). Interspecific variation in daily energy expenditure during avian incubation. *Journal of Zoology*, **231**, 215–32.

Tauber, M.J., Tauber, C.A. and Masaki, S. (1986). *Seasonal adaptations of insects*. Oxford University Press, Oxford.

Taylor, G. and Thomas, A. (2014). *Evolutionary biomechanics: selection, phylogeny, and constraint*. Oxford University Press, Oxford.

Taylor, L.W., Kreutzigel, G.O. and Abercrombie, G.L. (1971). The gaseous environment of the chick embryo in relation to its development and hatchability. 5. Effect of carbon dioxide and oxygen levels during the terminal days of incubation. *Poultry Science*, **50**, 66–78.

Taylor, L.W., Sjodin, R.A. and Gunns, C.A. (1956). The gaseous environment of the chick embryo in relation to its development and hatchability. 1. Effect of carbon dioxide and oxygen levels during the first four days of incubation upon hatchability. *Poultry Science*, **35**, 1206–15.

Temple, S.A. (1972). A portable time-lapse camera for recording wildlife activity. *Journal of Wildlife Management*, **36**, 944–7.

Thackeray, S.J., Sparks, T.H., Frederiksen, M., Burthe, S., Bacon, P.J., Bell, J.R., Botham, M.S., Brereton, T.M., Bright, P.W., Carvalho, L., Clutton-Brock, T., Dawson, A., Edwards, M., Elliott, J.M., Harrington, R., Johns, D., Jones, I.D., Jones, J.T., Leech, D.I., Roy, D.B., Scott, W.A., Smith, M., Smithers, R.J., Winfield, I.J. and Wanless, S. (2010). Trophic level asynchrony in rates of phenological change for marine, freshwater and terrestrial environments. *Global Change Biology*, **16**, 3304–13.

Thaler, E. (1976). Observations on the nesting behaviour of goldcrest and firecrest *Regulus regulus and Regulus ignicapillus*. *Journal für Ornithologie*, **117**, 121–44.

Thibault, J.C. (1994). Nest-site tenacity and mate fidelity in relation to breeding success in Cory's Shearwater *Calonectris diomedea*. *Bird Study*, **41**, 25–8.

Thierry, A.M., Brajon, S., Massemin, S., Handrich, Y., Chastel, O. and Raclot, T. (2013a). Decreased prolactin levels reduce parental commitment, egg temperatures, and breeding success of incubating male Adelie penguins. *Hormones and Behavior*, **64**, 737–47.

Thierry, A.M., Massemin, S., Handrich, Y. and Raclot, T. (2013b). Elevated corticosterone levels and severe weather conditions decrease parental investment of incubating Adelie penguins. *Hormones and Behavior*, **63**, 475–83.

Thomas, B.T. (1983). The plain-fronted thornbird: nest construction, material choice, and nest defense behavior. *The Wilson Bulletin*, **95**, 106–17.

Thompson, C.W., Hillgarth, N., Leu, M. and McClure, H.E. (1997). High parasite load in house finches (*Carpodacus mexicanus*) is correlated with reduced expression of a sexually selected trait. *The American Naturalist*, **149**, 270–94.

Thompson, K.R. and Furness, R.W. (1991). The influence of rainfall and nest-site quality on the population dynamics of the Manx Shearwater *Puffinus puffinus* on Rhum. *Journal of Zoology*, **225**, 427–37.

Thomson, D.L., Monaghan, P. and Furness, R.W. (1998). The demands of incubation and avian clutch size. *Biological Reviews of the Cambridge Philosophical Society*, **73**, 293–304.

Thomson, R.L., Sirkia, P.M., Villers, A. and Laaksonen, T. (2013). Temporal peaks in social information: prospectors investigate conspecific nests after a simulated predator visit. *Behavioral Ecology and Sociobiology*, **67**, 905–11.

Thorpe, W.H. (1956). *Learning and instinct in animals.* Methuen, London.

Tiainen, J., Hanski, I.K. and Mehtälä, J. (1983). Insulation of nests and the northern limits of three *Phylloscopus* warblers in Finland. *Ornis Scandinavica*, **14**, 149–53.

Tidemann, S. and Gosler, A. (2010). *Ethno-ornithology: birds, indigenous peoples, culture and society.* Earthscan, London.

Tieleman, B.I., van Noordwijk, H.J. and Williams, J.B. (2008). Nest site selection in a hot desert: trade-off between microclimate and predation risk. *The Condor*, **110**, 116–24.

Tieleman, B.I. and Williams, J.B. (2005). To breed or not to breed: that is the question. Decisions facing hoopoe larks in the Arabian desert. In R Drent, JP Bakker, T Piersma, JM Tinbergen, eds. *Seeking nature's limits: Ecologists in the field*, pp. 154–61. KNNV Publishing, Utrecht.

Tieleman, B.I., Williams, J.B. and Buschur, M.E. (2002). Physiological adjustments to arid and mesic environments in larks (Alaudidae). *Physiological and Biochemical Zoology*, **75**, 305–13.

Tinbergen, J.M. and Dietz, M.W. (1994). Parental energy expenditure during brood rearing in the great tit (*Parus major*) in relation to body mass, temperature, food availability and clutch size. *Functional Ecology*, **8**, 563–72.

Tinbergen, J.M., and Williams, J.B. (2002). Energetics of incubation. In DC Deeming, ed. *Avian incubation: behaviour,*

environment, and evolution, pp. 299–313. Oxford University Press, Oxford.

Tinbergen, N., Broekhuysen, G.J., Feekes, F., Houghton, J.C.W., Kruuk, H. and Szulc, E. (1962). Egg shell removal by the black-headed gull, *Larus ridibundus* L.; a behaviour component of camouflage. *Behaviour*, **19**, 74–116.

Tissier, M.L., Williams, T.D. and Criscuolo, F. (2014). Maternal effects underlie ageing costs of growth in the Zebra Finch (*Taeniopygia guttata*). *PLoS ONE*, **9**(5), e97705.

Tobler, M., Nilsson, J.Å. and Nilsson, J.F. (2007). Costly steroids: egg testosterone modulates nestling metabolic rate in the zebra finch. *Biology Letters*, **3**, 408–10.

Tobler, M. and Sandell, M.I. (2007). Yolk testosterone modulates persistence of neophobic responses in adult zebra finches, *Taeniopygia guttata*. *Hormones and Behavior*, **52**, 640–5.

Tøien, Ø., Aulie, A. and Steen, J.B. (1986). Thermoregulatory responses to egg cooling in incubating bantam hens. *Journal of Comparative Physiology B*, **156**, 303–7.

Tomás, G., Merino, S., Martínez-de la Puente, J., Moreno, J., Morales, J. and Lobato, E. (2008). Determinants of abundance and effects of blood-sucking flying insects in the nest of a hole-nesting bird. *Oecologia*, **156**, 305–12.

Tomás, G., Merino, S., Martínez-de la Puente, J., Moreno, J., Morales, J., Lobato, E., Rivero-de Aguilar, J. and del Cerro, S. (2012). Interacting effects of aromatic plants and female age on nest-dwelling ectoparasites and blood-sucking flies in avian nests. *Behavioural Processes*, **90**, 246–53.

Tomás, G., Merino, S., Martinez-de la Puente, J., Moreno, J., Morales, J. and Rivero-de Aguilar, J. (2013). Nest size and aromatic plants in the nest as sexually selected female traits in blue tits. *Behavioral Ecology*, **24**, 926–34.

Tomás, G., Merino, S., Sanz, J.J., Morales, J. and García-Fraile, S. (2006). Nest weight and female health in the blue tit (*Cyanistes caeruleus*). *The Auk*, **123**, 1013–21.

Tombre, I.M. and Erikstad, K.E. (1996). An experimental study of incubation effort in high Arctic Barnacle Geese. *Journal of Animal Ecology*, **65**, 325–31.

Tombre, I.M., Erikstad, K.E. and Bunes, V. (2012). State-dependent incubation behavior in high arctic Barnacle Geese. *Polar Biology*, **35**, 985–92.

Tomialojc, L. (1992). Colonization of dry habitats by the Song Thrush *Turdus philomelos*: is the type of nest material an important constraint? *Bulletin of the British Ornithologists' Club*, **112**, 27–34.

Toms, M.P. (2011). Legislation and good practice. In J Ferguson-Lees, R Castell, D Leech, eds. *A field guide to monitoring nests*, pp. 17–20. British Trust for Ornithology, Thetford.

Töpfer, T. and Gedeon, K. (2012). The construction and thermal insulation of Ethiopian Bush-crow (*Zavattariornis stresemanni*) nests: a preliminary study. *Avian Biology Research*, **5**, 198–202.

Torti, V.M. and Dunn, P.O. (2005). Variable effects of climate change on six species of North American birds. *Oecologia*, **145**, 486–95.

Tranter, H.S. and Board, R.G. (1984). The influence of incubation temperature and pH on the antimicrobial properties of hen egg albumen. *Journal of Applied Bacteriology*, **56**, 53–61.

Traylor, J.J. and Alisauskas, R.T. (2006). Effects of intrinsic and extrinsic factors on survival of White-winged Scoter (*Melanitta fusca deglandi*) ducklings. *The Auk*, **123**, 67–81.

Tripet, F., Glaser, M. and Richner, H. (2002). Behavioural responses to ectoparasites: time-budget adjustments and what matters to Blue Tits *Parus caeruleus* infested by fleas. *Ibis*, **144**, 461–9.

Tripet, F. and Richner, H. (1997). Host responses to ectoparasites: food compensation by parent blue tits. *Oikos*, **78**, 557–61.

Tripet, F. and Richner, H. (1999). Dynamics of hen flea *Ceratophyllus gallinae* subpopulations in blue tit nests. *Journal of Insect Behavior*, **115**, 159–74.

Trivers, R.L. (1972). Parental investment and sexual selection. In B Campbell, ed. *Sexual selection and the descent of man*, pp. 136–79. Aldine, Chicago.

Trnka, A. and Prokop, P. (2011). The use and function of snake skins in the nests of Great Reed Warblers *Acrocephalus arundinaceus*. *Ibis*, **153**, 627–30.

Tschirren, B., Fitze, P.S. and Richner, H. (2007). Maternal modulation of natal dispersal in a passerine bird: An adaptive strategy to cope with parasitism? *The American Naturalist*, **169**, 87–93.

Tuba, Z., Slack, N.G. and Stark, L.R. (2012). *Bryophyte ecology and climate change*. Cambridge University Press, Cambridge.

Tullett, S.G. (1984). The porosity of avian eggshells. *Comparative Biochemistry and Physiology A*, **78**, 5–13.

Tullett, S.G. (1991). *Avian incubation*. Butterworths-Heinnemann, London.

Tullett, S.G. and Deeming, D.C. (1982). The relationship between eggshell porosity and oxygen consumption of the embryo in the domestic fowl. *Comparative Biochemistry and Physiology A*, **72**, 529–33.

Tullett, S.G. and Deeming, D.C. (1987). Failure to turn eggs during incubation: effects on embryo weight, development of the chorioallantois and absorption of albumen. *British Poultry Science*, **28**, 239–49.

Tulp, I., Schekkerman, H., Bruinzeel, L.W., Jukema, J., Visser, G. and Piersma, T. (2009). Energetic demands during incubation and chick rearing in a uniparental and a biparental shorebird breeding in the High Arctic. *The Auk*, **126**, 155–64.

Tulp, I., Schekkerman, H., Chylarecki, P., Tomkovich, P., Soloviev, M., Bruinzeel, L., van Dijk, K., Hildén, O., Hötker, H., Kania, W., Van Roomen, M., Sikora, A. and Summers, R. (2002). Body mass patterns of Little Stints at different latitudes during incubation and chick-rearing. *Ibis*, **144**, 122–34.

Tulp, I., Schekkerman, H. and de Leeuw, J. (2012). Eggs in the freezer: energetic consequences of nest site and nest design in Arctic breeding shorebirds. *PLoS ONE*, **7**(6), e38041.

Tun-Lin, W., Burkot, T.R. and Kay, B.H. (2000). Effects of temperature and larval diet on development rates and survival of the dengue vector *Aedes aegypti* in north Queensland, Australia. *Medical and Veterinary Entomology*, **14**, 31–7.

Turner, A.K. (2006). *The barn swallow*. T. and A.D. Poyser, London.

Turner, J.S. (1987). Blood circulation and the flows of heat in an incubated egg. *Journal of Experimental Zoology*, Supplement **1**, 99–104.

Turner, J.S. (1994). Thermal impedance of a contact-incubated bird's egg. *Journal of Thermal Biology*, **19**, 237–43.

Turner, J.S. (1991). The thermal energetics of incubated bird eggs. In DC Deeming, MWJ Ferguson, eds. *Egg incubation: its effects on embryonic development in birds and reptiles*, pp. 117–46. Cambridge University Press, Cambridge.

Turner, J.S. (2002). Maintenance of egg temperature. In DC Deeming, ed. *Avian incubation: behaviour, environment, and evolution*, pp. 119–42. Oxford University Press, Oxford.

Ueta, M. (1998). Azure-winged magpies avoid nest predation by nesting near a Japanese lesser sparrowhawk's nest. *The Condor*, **100**, 400–2.

Uller, T. (2008). Developmental plasticity and the evolution of parental effects. *Trends in Ecology and Evolution*, **23**, 432–8.

Underwood, T.J. and Sealy, S.G. (2002). Adaptive significance of egg colouration. In DC Deeming, ed. *Avian incubation. behavior, environment, and evolution*, pp. 280–98. Oxford University Press, Oxford.

Underwood, T.J. and Sealy, S.G. (2006). Grasp-ejection in two small ejectors of cowbird eggs: a test of bill-size constraints and the evolutionary equilibrium hypothesis. *Animal Behaviour*, **71**, 409–16.

United Nations (2014). World Urbanization Prospects, The 2014Revision:Highlights.http://esa.un.org/unpd/wup/Highlights/WUP2014-Highlights.pdf. [Accessed: 14 December 2014]

Utter, J. (1971). *Daily energy expenditure of free-living purple martins* (Progne subis*) and mockingbirds* (Mimus polyglottos*) with a comparison of two northern populations of mockingbirds*. Unpublished PhD thesis, Rutgers University, New Brunswick.

Václav, R., Calero-Torralbo, M.A. and Valera, F. (2008). Ectoparasite load is linked to ontogeny and cell-mediated immunity in an avian host system with pronounced hatching asynchrony. *Biological Journal of the Linnean Society*, **94**, 463–73.

Vadehra, D.V., Baker, R.C. and Naylor, H.B. (1972). Distribution of lysozyme activity in the exteriors of eggs from *Gallus gallus*. *Comparative Biochemistry and Physiology*, **43B**, 503–8.

Valenti, P., Visca, P., Antonini, G. and Orsi, N. (1985). Antifungal activity of ovotransferrin towards genus *Candida*. *Mycopathologia*, **89**, 169–75.

Valera, F., Casas Crivillé, A. and Calero-Torralbo, M.A. (2006). Prolonged diapause in the ectoparasite *Carnus hemapterus* (Diptera: Cyclorrhapha, Acalyptratae)—how frequent is it in parasites? *Parasitology*, **133**, 179–86.

Välikangas, J. (1933). Finnische zugvogel aus englischen vo-geleiern. *Vogelzug*, **4**, 159–66.

Valkiínas, G., Liutkevicius, G. and Iezhova, T.A. (2002). Complete development of three species of *Haemoproteus* (Haemosporida, Haemoproteidae) in the biting midge *Culicoides impunctatus* (Diptera, Ceratopogonidae). *Journal of Parasitology*, **88**, 864–8.

van de Pol, M., Ens, B.J., Heg, D.H., Brouwer, L., Krol, J., Maier, M., Exo, K.-M., Oosterbeek, K., Lok, T., Eising, C.M. and Koffijberg, K. (2010). Do changes in the frequency, magnitude and timing of extreme climatic events threaten the population viability of coastal birds? *Journal of Applied Ecology*, **47**, 720–30.

van den Steen, E., Dauwe, T., Covaci, A., Jaspers, V.L., Pinxten, R. and Eens, M. (2006). Within- and among-clutch variation of organohalogenated contaminants in eggs of great tits (*Parus major*). *Environmental Pollution*, **144**, 355–9.

van den Steen, E., Jaspers, V.L.B., Covaci, A., Neels, H., Eens, M. and Pinxten, R. (2009). Maternal transfer of organochlorines and brominated flame retardants in blue tits (*Cyanistes caeruleus*). *Environment International*, **35**, 69–75.

van den Steen, E., Pinxten, R., Covaci, A., Carere, C., Eeva, T., Heeb, P., Kempenaers, B., Lifjeld, J.T., Massa, B., Norte, A.C., Orell, M., Sanz, J.J., Senar, J.C., Sorace, A. and Eens, M. (2010). The use of blue tit eggs as a biomonitoring tool for organohalogenated pollutants in the European environment. *Science of The Total Environment*, **408**, 1451–7.

Vanderwerf, E.A. (2012). Evolution of nesting height in an endangered Hawaiian forest bird in response to a non-native predator. *Conservation Biology*, **26**, 905–11.

van Dijk, R.E., Kaden, J.C., Argüelles-Ticó, A., Beltran, L.M., Paquet, M., Covas, R., Doutrelant, C. and Hatchwell, B.J. (2013). The thermoregulatory benefits of the communal nest of social weavers *Philetairus socius* are spatially structured within nests. *Journal of Avian Biology*, **44**, 102–10.

van Noordwijk, A.J., McCleery, R.H. and Perrins, C.M. (1995). Selection for the timing of great tit breeding in relation to caterpillar growth and temperature. *Journal of Animal Ecology*, **64**, 451–8.

van Rhijn, J.G. and Groothuis, T. (1987). On the mechanism of mate selection in black-headed gulls. *Behaviour*, **100**, 134–69.

van Riper III, C. (1977). The use of sheep wool in nest construction by Hawaiian birds. *The Auk*, **94**, 646–51.

Van Thyne, J. and Berger, A.J. (1959). *Fundamentals of ornithology*. John Wiley and Sons, New York.

Van Valen, L. (1976). Energy and evolution. *Evolutionary Theory*, **1**, 179–229.

Varney, J.R. and Ellis, D.H. (1974). Telemetering egg for use in incubation and nesting studies. *Journal of Wildlife Management*, **38**, 142–8.

Varpe, Ø. and Tveraa, T. (2005). Chick survival in relation to nest site: is the Antarctic petrel hiding from its predator? *Polar Biology*, **28**, 388–94.

Varricchio, D.J. (2011). A distinct dinosaur life history? *Historical Biology*, **23**, 91–107.

Varricchio, D.J. and Barta, D.E. (2015). Revisiting Sabath's 'larger avian eggs' from the Gobi Cretaceous. *Acta Palaeontolgica Polonica*, **60**, 11–25.

Varricchio, D.J., Jackson, F. and Trueman, C.N. (1997). A nesting trace with eggs for the Cretaceous theropod dinosaur *Troodon formosus*. *Journal of Vertebrate Paleontology*, **19**, 91–100.

Varricchio, D.J., Moore, J.R., Erickson, G.M., Norell, M.A., Jackson, F.D. and Borkowski, J.J. (2008). Avian paternal care had dinosaur origin. *Science*, **322**, 1826–8.

Vautard, R., Cattiaux, J., Yiou, P., Thépaut, J.-N. and Ciais, P. (2010). Northern Hemisphere atmospheric stilling partly attributed to an increase in surface roughness. *Nature Geoscience*, **3**, 756–61.

Vedder, O. (2012). Individual birds advance offspring hatching in response to increased temperature after the start of laying. *Oecologia*, **170**, 619–28.

Veiga, J.P. and Polo, V. (2005). Feathers at nests are potential female signals in the spotless starling. *Biology Letters*, **1**, 334–7.

Verboven, N., Ens, B.J. and Dechesne, S. (2001). Effect of investigator disturbance on nest attendance and egg predation in Eurasian Oystercatchers. *The Auk*, **118**, 503–8.

Verboven, N., Verreault, J., Letcher, R.J., Gabrielsen, G.W. and Evans, N.P. (2009). Nest temperature and parental behaviour of Arctic-breeding glaucous gulls exposed to persistent organic pollutants. *Animal Behaviour*, **77**, 411–8.

Verhulst, S., Oosterbeek, K. and Ens, B.J. (2001). Experimental evidence for effects of human disturbance on foraging and parental care in oystercatchers. *Biological Conservation*, **101**, 375–80.

Verhulst, S. and Tinbergen, J.M. (1997). Clutch size and parental effort in the Great Tit *Parus major*. *Ardea*, **85**, 111–26.

Vevers, G. (1982). *The colours of animals*. Edward Arnold, London.

Victoria, J.K. (1972). Clutch characteristics and egg discriminative ability of the African village weaverbird *Ploceus cucullatus*. *Ibis*, **114**, 367–76.

Vince, M.A. (1966). Potential stimulation produced by avian embryos. *Animal Behaviour*, **14**, 34–40.

Vince M.A. (1969). Embryonic communication, respiration and synchronisation. In RA Hinde, ed. *Bird vocalization*, pp. 233–60. Cambridge University Press, Cambridge.

Vince, M.A. (1972). Communication between quail embryos and the synchronisation of hatching. *Proceedings of the International Ornithological Congress*, **15**, 357–62.

Viñuela, J. and Sunyer, C. (1992). Nest orientation and hatching success of Black Kites *Milvus migrans* in Spain. *Ibis*, **134**, 340–5.

Visser, G.H. and Ricklefs, R.E. (1995). Relationship between body composition and homeothermy in neonates of precocial and semiprecocial birds. *The Auk*, **112**, 192–200.

Visser, M.E. and Lessells, C.M. (2001). The costs of egg production and incubation in Great Tits (*Parus major*). *Proceedings of the Royal Society of London Series B: Biological Sciences*, **268**, 1271–7.

Visser, M.E., van Noordwijk, A.J., Tinbergen, J.M. and Lessells, C.M. (1998). Warmer springs lead to mis-timed reproduction in Great Tits (*Parus major*). *Proceedings of the Royal Society of London Series B: Biological Sciences*, **265**, 1867–70.

Vleck, C.M. (1981). Energetic cost of incubation in the Zebra Finch. *The Condor*, **83**, 229–37.

Vleck, C.M. (1998). Hormonal control of incubation/brooding behavior: lessons from wild birds. *Proceedings of the WPSA 10th European Poultry Conference, Israel, 1998*, pp. 163–9. World Poultry Science Association, Beekbergen.

Vleck, C.M. (2002). Hormonal control of incubation behaviour. In DC Deeming, ed. *Avian incubation. behaviour, environment, and evolution*, pp. 54–62. Oxford University Press, Oxford.

Vleck, C.M. and Bucher, T.L. (1998). Energy metabolism, gas exchange and ventilation. In JM Stark, RE Ricklefs, eds. *Avian growth and development: evolution within the precocial-altricial spectrum*, pp. 89–116. Oxford University Press, New York.

Vleck, C.M., Ross, L.L., Vleck, D. and Bucher, T.L. (2000). Prolactin and parental behavior in Adelie penguins: Effects of absence from nest, incubation length, and nest failure. *Hormones and Behavior*, **38**, 149–58.

Vleck, C.M. and Vleck, D. (1987). Metabolism and energetics of avian embryos. *Journal of Experimental Zoology*, Supplement **1**, 111–25.

von Engelhardt, N., Carere, C., Dijkstra, C. and Groothuis, T.G.G. (2006). Sex-specific effects of yolk testosterone on survival, begging and growth of zebra finches. *Proceedings of the Royal Society B*, **273**, 65–70.

von Engelhardt, N., Dijkstra, C., Daan, S. and Groothuis, T.G.G. (2004). Effects of 17-beta estradiol treatment of female zebra finches on offspring sex ratio and survival. *Hormones and Behavior*, **45**, 306–13.

von Engelhardt, N. and Groothuis, T.G.G. (2011). Maternal hormones in avian eggs. In DO Norris, KH Lopez, eds. *Hormones and reproduction of vertebrates—Birds*, pp. 91–127. Elsevier, Amsterdam.

von Engelhardt, N., Henriksen, R. and Groothuis, T.G.G. (2009). Steroids in chicken egg yolk: Metabolism and uptake during early embryonic development. *General and Comparative Endocrinology*, **163**, 175–83.

Voss, M.A., Hainsworth, F.R. and Ellis-Felege, S.N. (2006). Use of a new model to quantify compromises between embryo development and parental self-maintenance in three species of intermittently incubating passerines. *Journal of Thermal Biology*, **31**, 453–60.

Votier, S.C., Archibald, K., Morgan, G. and Morgan, L. (2011). The use of plastic debris as nesting material by a colonial seabird and associated entanglement mortality. *Marine Pollution Bulletin*, **62**, 168–72.

Votýpka, J., Oborník, M., Volf, P., Svobodová, M. and Luken, J. (2002). *Trypanosoma avium* of raptors (Falconiformes): phylogeny and identification of vectors. *Parasitology*, **125**, 253–63.

Wada, T. (1994). Effects of height of neighboring nests on nest predation in the rufous turtle-dove (*Streptopelia orientalis*). *The Condor*, **96**, 812–6.

Wagner, K. and Seymour, R.S. (2001). Nesting climate and behaviour of Cape Barren Geese (*Cereopsis novaehollandiae* Latham). *Australian Journal of Zoology*, **49**, 155–70.

Waite, J.L., Henry, A.R. and Clayton, D.H. (2012). How effective is preening against mobile ectoparasites? An experimental test with pigeons and hippoboscid flies. *International Journal for Parasitology*, **42**, 463–7.

Walk, J.W., Wentworth, K., Kershner, E.L., Bollinger, E.K. and Warner, R.E. (2004). Renesting decisions and annual fecundity of female Dickcissels (*Spiza americana*) in Illinois. *The Auk*, **121**, 1250–61.

Wallace, A.R. (1867). The philosophy of birds' nests. *Intellectual Observer*, **11**, 413–20.

Wallace, A.R. (1889). *Darwinism: An exposition of the theory of natural selection with some of its applications*. Macmillan, London.

Walls, J.G., Hepp, G.R. and Eckhardt, L.G. (2011). Effects of incubation delay on viability and microbial growth of wood duck (*Aix sponsa*) eggs. *The Auk*, **128**, 663–70.

Walls, J.G., Hepp, G.R. and Eckhardt, L.G. (2012). Effects of nest reuse and onset of incubation on microbial growth and viability of wood duck eggs. *The Condor*, **114**, 720–5.

Walsberg, G.E. and King, J.R. (1978a). The energetic consequences of incubation for two passerine species. *The Auk*, **95**, 644–55.

Walsberg, G.E. and King, J.R. (1978b). The heat budget of incubating Mountain White-crowned Sparrows (*Zonotrichia leucophrys oriantha*) in Oregon. *Physiological Zoology*, **51**, 92–103.

Walsberg, G.E. and Voss-Roberts, K.A. (1983). Incubation in desert-nesting doves: mechanisms for egg cooling. *Physiological Zoology*, **56**, 88–93.

Walsh, P.T., Evans, D.M., Hansell, M. and Ruxton, G.D. (2007). Factors influencing nocturnal egg-turning frequency in meadow pipits *Anthus pratensis*. *Bird Study*, **54**, 133–6.

Walsh, P.T., Hansell, M., Borello, W.D. and Healy, S.D. (2010). Repeatability of nest morphology in African weaver birds. *Biology Letters*, **6**, 149–51.

Walsh, P.T., Hansell, M., Borello, W.D. and Healy, S.D. (2011). Individuality in nest building: Do southern masked weaver (*Ploceus velatus*) males vary in their nest-building behaviour? *Behavioural Processes*, **88**, 1–6.

Walsh, P.T., Hansell, M., Borello, W.D. and Healy, S.D. (2013). Are elaborate bird nests built using simple rules? *Avian Biology Research*, **6**, 157–62.

Walters, L.A. and Getty, T. (2010). Are brighter eggs better? Egg color and parental investment by House Wrens. *Journal of Field Ornithology*, **81**, 155–66.

Walther, G.-R., Post, E., Convey, P., Menzel, A., Parmesan, C., Beebee, T.J., Fromentin, J.-M., Hoegh-Guldberg, O. and Bairlein, F. (2002). Ecological responses to recent climate change. *Nature*, **416**, 389–95.

Wang, J., Liang, Y., Omana, D.A., Kav, N.N. and Wu, J. (2012). Proteomics analysis of egg white proteins from different egg varieties. *Journal of Agricultural and Food Chemistry*, **60**, 272–82.

Wang, J.M. and Beissinger, S.R. (2009). Variation in the onset of incubation and its influence on avian hatching success and asynchrony. *Animal Behaviour*, **78**, 601–13.

Wang, J.M. and Beissinger, S.R. (2011). Partial incubation: Its occurrence, distribution, and quantification. *The Auk*, **128**, 454–66.

Wang, J.M., Firestone, M.K. and Beissinger, S.R. (2011). Microbial and environmental effects on avian egg viability: do tropical mechanisms act in a temperate environment? *Ecology*, **92**, 1137–45.

Wang, J.M. and Weathers, W.W. (2009). Egg laying, egg temperature, attentiveness, and incubation in the Western Bluebird. *The Wilson Journal of Ornithology*, **121**, 512–20.

Wang, X.T., Zhao, C.J., Li, J.Y., Xu, G.Y., Lian, L.S., Wu, C.X. and Deng, X.M. (2009b). Comparison of the total amount of eggshell pigments in Dongxiang brown-shelled eggs and Dongxiang blue-shelled eggs. *Poultry Science*, **88**, 1735–9.

Wang, Y., Chen, S., Blair, R.B., Jiang, P. and Ding, P. (2009a). Nest composition adjustments by Chinese bulbuls *Pycnonotus sinensis* in an urbanized landscape of Hangzhou (E China). *Acta Ornithologica*, **44**, 185–92.

Wang, Z., Liu, R. and Wang, A. (2013). Comparison of HMOX1 expression and enzyme activity in blue-shelled chickens and brown-shelled chickens. *Genetics and Molecular Biology*, **36**, 282–6.

Wang, Z., Liu, R., Wang, A., Li, J. and Deng, X. (2011c). Expression and activity analysis reveal that heme oxygenase (decycling) 1 is associated with blue egg formation. *Poultry Science*, **90**, 836–41.

Wangensteen, O.D. and Rahn, H. (1971). Respiratory gas exchange by the avian embryo. *Respiration Physiology*, **11**, 31–4.

Ward, S. (1996). Energy expenditure of female barn swallows *Hirundo rustica* during egg formation. *Physiological Zoology*, **69**, 930–51.

Warham, J. (1990). *The petrels: their ecology and breeding systems*. Academic Press, London.

Warner, D.A. and Shine, R. (2008). The adaptive significance of temperature dependent sex determination in a reptile. *Nature*, **451**, 566–8.

Warner, R.E. (1968). The role of introduced diseases in the extinction of the endemic Hawaiian avifauna. *The Condor*, **70**, 101–20.

Warren, R.P. and Hinde, R.A. (1959). The effect of oestrogen and progesterone on the nest-building of domesticated canaries. *Animal Behaviour*, **7**, 209–13.

Watts, B.D. (1987). Old nest accumulation as a possible protective mechanism against search-strategy predators. *Animal Behaviour*, **35**, 1566–8.

Weatherhead, P.J. and Forbes, M.R.L. (1994). Natal philopatry in passerine birds: genetic or ecological influences? *Behavioral Ecology*, **5**, 426–33.

Weathers, W.W. (1985). Energy cost of incubation in the canary. *Comparative Biochemistry and Physiology A*, **81**, 411–3.

Weathers, W.W., Davidson, C.L., Olson, C.R., Morton, M.L., Nur, N. and Famula, T.R. (2002). Altitudinal variation in parental energy expenditure by white-crowned sparrows. *The Journal of Experimental Biology*, **205**, 2915–24.

Weathers, W.W. and Sullivan, K.A. (1989a). Nest attentiveness and egg temperature in the Yellow-eyed Junco. *The Condor*, **91**, 629–33.

Weathers, W.W. and Sullivan, K.A. (1989b). Juvenile foraging proficiency, parental effort, and avian reproductive success. *Ecological Monographs*, **59**, 223–46.

Webb, D.R. (1987). Thermal tolerance of avian embryos: a review. *The Condor*, **89**, 874–98.

Webb, S.L., Olson, C.V., Dzialak, M.R., Harju, S.M., Winstead, J.B. and Lockman, D. (2012). Landscape features and weather influence nest survival of a ground-nesting bird of conservation concern, the greater sage-grouse, in human-altered environments. *Ecological Processes*, **1**, 1–15.

Webber, S.L. (2012). *The role of food availability in determining the energetic and life history costs of reproduction in short-lived birds*. PhD thesis, University of Birmingham, Birmingham.

Webster, B., Hayes, W. and Pike, T.W. (2015). Avian egg odour encodes information on embryo sex, fertility and development. *PLoS ONE*, **10**(1), e0116345.

Węgrzyn, E., Leniowski, K., Rykowska, I. and Wasiak, W. (2011). Is UV and blue-green egg colouration a signal in cavity-nesting birds? *Ethology, Ecology & Evolution*, **23**, 121–39.

Weidinger, K. (1998). Effect of predation by skuas on breeding success of the Cape Petrel *Daption capense* at Nelson Island, Antarctica. *Polar Biology*, **20**, 170–7.

Weidinger, K. (2001). Laying dates and clutch size of open-nesting passerines in the Czech Republic: a comparison of systematically and incidentally collected data. *Bird Study*, **48**, 38–47.

Weidinger, K. (2002). Interactive effects of concealment, parental behaviour and predators on the survival of open passerine nests. *Journal of Animal Ecology*, **71**, 424–37.

Weidinger, K. (2004). Relative effects of nest size and site on the risk of predation in open nesting passerines. *Journal of Avian Biology*, **35**, 515–23.

Weidinger, K. and Kočvara, R. (2010). Repeatability of nest predation in passerines depends on predator species and time scale. *Oikos*, **119**, 138–46.

Weidinger, K. and Pavel, V. (2013). Predator-prey interactions between the South Polar skua *Catharacta maccormicki* and Antarctic tern *Sterna vittata*. *Journal of Avian Biology*, **44**, 89–95.

Weimerskirch, H. (1990). Weight-loss of Antarctic fulmars *Fulmarus glacialoides* during incubation and chick brooding. *Ibis*, **132**, 68–77.

Weiner, S. and Addadi, L. (1991). Acidic macromolecules of mineralized tissues: The controllers of crystal formation. *Trends in Biochemical Sciences*, **16**, 252–6.

Weldon, P.J. (2000). Avian chemical defense: Toxic birds not of a feather. *Proceedings of the National Academy of Sciences of the USA*, **97**, 2948–9.

Weldon, P.J., Carroll, J.F., Kramer, M., Bedoukian, R.H., Coleman, R.E. and Bernier, U.R. (2011). Anointing chemicals and hematophagous arthropods: responses by ticks and mosquitoes to *Citrus* (Rutaceae) peel exudates and monoterpene components. *Journal of Chemical Ecology*, **37**, 348–59.

Weller, M.W. and Derksen, D.V. (1972). Use of time-lapse photography to study nesting activities of birds. *The Auk*, **89**, 196–200.

Wellman-Labadie, O., Lakshminarayanan, R. and Hincke, M.T. (2008a). Antimicrobial properties of avian eggshell-specific C-type lectin-like proteins. *FEBS Letters*, **582**, 699–704.

Wellman-Labadie, O., Picman, J. and Hincke, M.T. (2008b). Antimicrobial activity of the Anseriform outer eggshell and cuticle. *Comparative Biochemistry and Physiology*, **149B**, 640–9.

Wellman-Labadie, O., Picman, J. and Hincke, M.T. (2007). Avian antimicrobial proteins: structure, distribution and activity. *World's Poultry Science Journal*, **63**, 421–38.

Wellnhofer, P. (2009). *Archaeopteryx: the icon of evolution*. Verlag Dr. Friedrich Pfeil, München.

Wernham-Calladine, C.V. (1995). Guillemot preening activity in relation to tick infestation. *The Bulletin of the British Ecological Society*, **26**, 187–95.

Wesołowski, T. (2002). Anti-predator adaptations in nesting Marsh Tits *Parus palustris*: the role of nest-site security. *Ibis*, **144**, 593–601.

Wesołowski, T. (2006). Nest-site re-use: Marsh Tit *Poecile palustris* decisions in a primeval forest. *Bird Study*, **53**, 199–204.

Wesołowski, T. (2012). 'Lifespan' of non-excavated holes in a primeval temperate forest: A 30 year study. *Biological Conservation*, **153**, 118–26.

Wesołowski, T. (2013). Timing and stages of nest-building by Marsh Tits (*Poecile palustris*) in a primeval forest. *Avian Biology Research*, **6**, 31–8.

Wesołowski, T. and Maziarz, M. (2012). Dark tree cavities—a challenge for hole nesting birds? *Journal of Avian Biology*, **43**, 454–60.

West, A.P., Herr, A.B. and Bjorkman, P.J. (2004). The chicken yolk sac IgY receptor, a functional equivalent of the mammalian MHC-related Fc receptor, is a phospholipase A2 receptor homolog. *Immunity*, **20**, 601–10.

Westerterp, K. and Bryant, D. (1984). Energetics of free existence in swallows and martins (Hirundinidae) during breeding: a comparative study using doubly labeled water. *Oecologia*, **62**, 376–81.

Westerterp, K. and Drent, R.H. (1985). Flight energetics of the starling *Sturnus vulgaris* during the parental period. *Proceedings of the 18th International Ornithological Congress*, 392–8.

Westmoreland, D. (2008). Evidence of selection for egg crypsis in conspicuous nests. *Journal of Field Ornithology*, **79**, 263–8.

Westmoreland, D. and Best, L.B. (1985). The effect of disturbance on mourning dove nesting success. *The Auk*, **102**, 774–80.

Westmoreland, D. and Kiltie, R.A. (2007). Egg coloration and selection for crypsis in open-nesting blackbirds. *Journal of Avian Biology*, **38**, 682–9.

Westmoreland, D., Schmitz, M. and Burns, K.E. (2007). Egg color as an adaptation for thermoregulation. *Journal of Field Ornithology*, **78**, 176–83.

Wheelwright, N.T. and Beagley, J.C. (2005). Proficient incubation by inexperienced Savannah Sparrows *Passerculus sandwichensis*. *Ibis*, **147**, 67–76.

Wheelwright, N.T., Lawler, J.J. and Weinstein, J.H. (1997). Nest-site selection in Savannah sparrows: Using gulls as scarecrows? *Animal Behaviour*, **53**, 197–208.

White, D.W. and Kennedy, E.D. (1997). Effect of egg covering and habitat on nest destruction by House Wrens. *The Condor*, **99**, 873–9.

White, F.N. and Kinney, J.L. (1974). Avian incubation. *Science*, **186**, 107–15.

Whitney, B.M., Pacheco, J.F., da Fonseca, P.S.M. and Barth, R.H. (1996). The nest and nesting ecology of *Acrobatornis fonsecai* (Furnariidae), with implications for intrafamilial relationships. *The Wilson Bulletin*, **108**, 434–48.

Whittingham, L.A., Dunn, P.O. and Lifjeld, J.T. (2007). Egg mass influences nestling quality in tree swallows, but there is no differential allocation in relation to laying order or sex. *The Condor*, **109**, 585–94.

Whittingham, M.J., Percival, S.M. and Brown, A.F. (2002). Nest-site selection by golden plover: why do shorebirds avoid nesting on slopes? *Journal of Avian Biology*, **33**, 184–90.

Whittow, G.C. (1980). Physiological and ecological correlates of prolonged incubation in sea birds. *American Zoologist*, **20**, 427–36.

Whittow, G.C. and Berger, A.J. (1977). Heat loss from the nest of the Hawaiian honeycreeper, 'Amakihi'. *The Wilson Bulletin*, **89**, 480–3.

Whittow, G.C. and Rahn, H. (1984). *Seabird energetics*. Plenum Press, New York.

Whitworth, T.L. (2003a). A key to the puparia of 27 species of North American *Protocalliphora* Hough (Diptera: Calliphoridae) from bird nests and two new puparial descriptions. *Proceedings of the Entomological Society of Washington*, **105**, 995–1033.

Whitworth, T.L. (2003b). A new species of North American *Protocalliphora* Hough (Diptera: Calliphoridae) from bird nests. *Proceedings of the Entomological Society of Washington*, **105**, 664–73.

Whitworth, T.L. and Bennett, G.F. (1992). Pathogenicity of larval *Protocalliphora* (Diptera: Calliphoridae) parasitizing nestling birds. *Canadian Journal of Zoology*, **70**, 2184–91.

Wiebe, K.L. (2009). Nest excavation does not reduce harmful effects of ectoparasitism: an experiment with a woodpecker, the northern flicker *Colaptes auratus*. *Journal of Avian Biology*, **40**, 166–72.

Wiebe, K.L., Koenig, W.D. and Martin, K. (2007). Costs and benefits of nest reuse versus excavation in cavity-nesting birds. *Annales Zoologici Fennici*, **44**, 209–17.

Wiebe, K.L. and Martin, K. (1998). Costs and benefits of nest cover for ptarmigan: changes within and between years. *Animal Behaviour*, **56**, 1137–44.

Wiersma, P., Selman, C., Speakman, J.R. and Verhulst, S. (2004). Birds sacrifice oxidative protection for reproduction.

Proceedings of the Royal Society of London Series B: Biological Sciences, **271**, S360–3.

Wigglesworth, V.B. (1944). Action of inert dusts on insects. *Nature*, **153**, 493–4.

Williams, J.B. (1987). Field metabolism and food consumption of savannah sparrows during the breeding season. *The Auk*, **104**, 277–89.

Williams, J.B. (1988). Field metabolism of Tree Swallows during the breeding season. *The Auk*, **105**, 706–14.

Williams, J.B. (1993a). Energetics of incubation in free-living Orange-breasted Sunbirds in South Africa. *The Condor*, **95**, 115–26.

Williams, J.B. (1993b). On the importance of energy considerations to small birds with gynelateral intermittent incubation. In BD Bell, RO Cossee, JEC Flux, BD Heather, RA Hitchmough, CJR Robertson, MJ Williams, eds., *Acta XX Congressus Internationalis Ornithologici*, Christchurch, 2 – 9 December 1990, pp. 1964–75. New Zealand Ornithological Trust Board, Wellington.

Williams, J.B. (1996). Energetics of avian incubation. In C Carey, ed. *Avian energetic and nutritional ecology*, pp. 375–416. Chapman and Hall, New York.

Williams, J.B. (2001). Energy expenditure and water flux of free-living Dune Larks in the Namib: a test of the reallocation hypothesis on a desert bird. *Functional Ecology*, **15**, 175–85.

Williams, J.B. and Dwinnel, B. (1990). Field metabolism of free-living female Savannah Sparrows during incubation—a study using doubly labeled water. *Physiological Zoology*, **63**, 353–72.

Williams, J.B. and Nagy, K.A. (1985). Water flux and energetics of nestling Savannah Sparrows in the field. *Physiological Zoology*, **58**, 515–25.

Williams, J.B. and Tieleman, B.I. (2008). Body temperature of free-living Hoopoe Larks in Saudi Arabia. *Integrative and Comparative Biology*, **47** (Supplement 1), e146.

Williams, T.D. (1994). Intraspecific variation in egg-size and egg composition in birds: effects on offspring fitness. *Biological Reviews of the Cambridge Philosophical Society*, **68**, 35–59.

Williams, T.D. (1995). *The penguins*. Oxford University Press, Oxford.

Williams, T.D. (1999). Parental and first generation effects of exogenous 17beta-estradiol on reproductive performance of female zebra finches (*Taeniopygia guttata*). *Hormones and Behavior*, **35**, 135–43.

Williams, T.D. (2012). *Physiological adaptations for breeding in birds*. Princeton University Press, Princeton.

Williams, T.D., Ames, C.E., Kiparissis, Y. and Wynne-Edwards, K.E. (2005). Laying sequence-specific variation in yolk estrogen levels, and relationship with plasma estrogen in female zebra finches (*Taeniopygia guttata*). *Proceedings of the Royal Society B*, **272**, 173–7.

Willis, C.K.R., Jameson, J.W., Faure, P.A., Boyles, J.G., Brack, V. Jr and Cervone, T.H. (2009). Thermocron iButton and iBBat temperature dataloggers emit ultrasound. *Journal of Comparative Physiology B*, **179**, 867–74.

Wilson, H.R., Neuman, S.L., Eldred, A.R. and Mather, F.B. (2003). Embryonic malpositions in broiler chickens and bobwhite quail. *Journal of Applied Poultry Research*, **12**, 14–23.

Wimberger, H.P. (1984). The use of green plant material in bird nests to avoid ectoparasites. *The Auk*, **101**, 615–8.

Windsor, R.L., Fegely, J.L. and Ardia, D.R. (2013). The effects of nest size and insulation on thermal properties of Tree Swallow nests. *Journal of Avian Biology*, **44**, 1–6.

Wingfield, J.C., Maney, D.L., Breuner, C.W., Jacobs, J.D., Lynn, S., Ramenofsky, M. and Richardson, R.D. (1998). Ecological bases of hormone-behavior interactions: The 'emergency life history stage'. *American Zoologist*, **38**, 191–206.

Winkler, D.W. (1993). Use and importance of feather as nest lining in tree swallows (*Tachycineta bicolor*). *The Auk*, **110**, 29–36.

Winkler, D.W., Dunn, P.O. and McCulloch, C.E. (2002). Predicting the effects of climate change on avian life-history traits. *Proceedings of the National Academy of Sciences of the USA*, **99**, 13595–9.

Wolaver, B.D. and Sharp, J.M. Jr. (2007). Thermochron iButton: limitation of this inexpensive and small-diameter temperature logger. *Ground Water Monitoring and Remediation*, **27**, 127–8.

Wolf, M., van Doorn, G., Leimar, O. and Weissing, F.J. (2007). Life-history trade-offs favour the evolution of animal personalities. *Nature*, **447**, 581–4.

Wood, J.R. (2008). Moa (Aves: Dinornithiformes) nesting material from rockshelters in the semi-arid interior of South Island, New Zealand. *Journal of the Royal Society of New Zealand*, **38**, 115–29.

Woods, C.P. (1993). Variation in loggerhead shrike nest composition between two shrub species in southwest Idaho. *Journal of Field Ornithology*, **64**, 352–7.

Woolf, N.K., Bixby, J.L. and Capranica, R.R. (1976). Prenatal experience and avian development: Brief auditory stimulation accelerates the hatching of Japanese quail. *Science*, **194**, 959–60.

Wootton, J.T. (1986). Clutch-size differences in western and introduced eastern populations of House Finches: patterns and hypotheses. *The Wilson Bulletin*, **98**, 459–62.

Wright, J. and Cuthill, I. (1990). Biparental care: short-term manipulation of partner contribution and brood size in the starling, *Sturnus vulgaris*. *Behavioral Ecology*, **1**, 116–24.

Wu, D., Landsberger, S. and Larson S.M. (1997). Determination of the elemental distribution in cigarette components and smoke by instrumental neutron activation analysis. *Journal of Radioanalytical and Nuclear Chemistry*, **217**, 77–82.

Wu, J. and Acero-Lopez, A. (2012). Ovotransferrin: Structure, bioactivities, and preparation. *Food Research International*, **46**, 480–7.

Xing, J., Wellman-Labadie, O., Gautron, J. and Hincke, M.T. (2007). Recombinant eggshell ovocalyxin-32: expression, purification and biological activity of the glutathione S-transferase fusion protein. *Comparative Biochemistry and Physiology*, **147B**, 172–7.

Xu, Y.-C., Yang, D.-B., Speakman, J.R. and Wang, D.-H. (2014). Oxidative stress in response to natural and experimentally elevated reproductive effort is tissue dependent. *Functional Ecology*, **28**, 402–10.

Yahner, R.H. (1993). Old nests as cues for nest-site selection by birds—an experimental test in small even-aged forest plots. *The Condor*, **95**, 239–41.

Yanes, M., Herranz, J. and Suárez, F. (1996). Nest microhabitat selection in larks from a European semi-arid shrub-steppe: The role of sunlight and predation. *Journal of Arid Environments*, **32**, 469–78.

Yang, C., Wang, L., Cheng, S.J., Hsu, Y.C., Liang, W. and Møller, A.P. (2014). Nest defenses and egg recognition of yellow-bellied prinia against cuckoo parasitism. *Naturwissenschaften*, **101**, 727–34.

Yang, H., Wang, Z. and Lu, J. (2009). Study on the relationship between eggshell colors and egg quality as well as shell ultrastructure in Yangzhou chicken. *African Journal of Biotechnology*, **8**, 2898–902.

Yang, T.-R., Cheng, Y.-N. and Yang, K.-M. (2012). Do brooding and polygamy behaviors exist on Cretaceous oviraptoroid dinosaurs of China: a paleobiological perspective. *Geophysical Research Abstracts*, **14**, EGU2012-9887-1, EGU General Assembly 2012.

Yeh, P.J., Hauber, M.E. and Price, T.D. (2007). Alternative nesting behaviours following colonisation of a novel environment by a passerine bird. *Oikos*, **116**, 1473–80.

Yerkes, T. (1998). The influence of female age, body mass, and ambient conditions on redhead incubation constancy. *The Condor*, **100**, 62–8.

Yom-Tov, Y. (1980). Intraspecific nest parasitism in birds. *Biological Reviews of the Cambridge Philosophical Society*, **55**, 93–108.

Yom-Tov, Y. (2001). An updated list and some comments on the occurrence of intraspecific nest parasitism in birds. *Ibis*, **143**, 133–43.

Young, B.E. (1994b). Geographic and seasonal patterns of clutch-size variation in House Wrens. *The Auk*, **111**, 545–55.

Young, E. (1994a). *Skua and penguin: predator and prey*. Cambridge University Press, Cambridge.

Young, H.G. and Kear, J. (2006). The rise and fall of wildfowl of the western Indian Ocean and Australasia. *Bulletin of the British Ornithologists' Club*, **126A**, 25–39.

Young, I.R., Zieger, S. and Babanin, A.V. (2011). Global trends in wind speed and wave height. *Science*, **332**, 451–5.

Youngren, O.M., Elhalawani, M.E., Silsby, J.L. and Phillips, R.E. (1991). Intracranial prolactin perfusion induces incubation behavior in turkey hens. *Biology of Reproduction*, **44**, 425–31.

Yunick. R.P. (1990). An unusual nestbox nesting of the Winter Wren. *The Kingbird*, **40**, 67–74.

Yuri, T., Kimball, R.T., Harshman, J., Bowie, R.C.K., Braun, M.J., Chojnowski, J.L., Han, K.-L., Hackett, S.J., Huddleston, C.J., Moore, W.S., Reddy, S., Sheldon, F.H., Steadman, D.W., Witt, C.C. and Braun, E.L. (2013). Parsimony and model-based analyses of indels in avian nuclear genes reveal congruent and incongruent phylogenetic signals. *Biology*, **2**, 419–44.

Zahavi, A. (1975). Mate selection—a selection for a handicap. *Journal of Theoretical Biology*, **53**, 205–14.

Zanette, L.Y., White, A.F., Allen, M.C. and Clinchy, M. (2011). Perceived predation risk reduces the number of offspring songbirds produce per year. *Science*, **334**, 1398–401.

Zangmeister, J.L., Haussmann, M.F., Cerchiara, J. and Mauck, R.A. (2009). Incubation failure and nest abandonment by Leach's Storm-Petrels detected using PIT tags and temperature loggers. *Journal of Field Ornithology*, **80**, 373–9.

Zann, R. and Rossetto, M. (1991). Zebra finch incubation: brood patch, egg temperature and thermal properties of the nest. *Emu*, **91**, 107–20.

Zhao, R., Xu, G.-Y., Liu, Z.-Z., Li, J.-Y. and Yang, N. (2006). A study on eggshell pigmentation: biliverdin in blue-shelled chickens. *Poultry Science*, **85**, 546–9.

Zheng, X., O'Connor, J., Huchzermeyer, F., Wang, X., Wang, Y., Wang, M. and Zhou, Z. (2013). Preservation of ovarian follicles reveals early evolution of avian reproductive behaviour. *Nature*, **495**, 507–11.

Zhou, Z. and Zhang, F. (2004). A precocial avian embryo from the Lower Cretaceous of China. *Science*, **306**, 653.

Zieliński, J. (2012). Nest reuse by Eurasian blackcap *Sylvia atricapilla*. *Ardea*, **100**, 98–100.

Zimmerling, J.R. and Ankney, C.D. (2005). Variation in incubation patterns of red-winged blackbirds nesting at lagoons and ponds in eastern Ontario. *The Wilson Bulletin*, **117**, 280–90.

Zölei, A., Hauber, M.E., Geltsch, N. and Moskát, C. (2012). Asymmetrical signal content of egg shape as predictor of egg rejection by great reed warblers, hosts of the common cuckoo. *Behaviour*, **149**, 391.

Zwartjes, P.W. and Nordell, S.E. (1998). Patterns of cavity-entrance orientation by Gilded Flickers (*Colaptes chrysoides*) in cardon cactus. *The Auk*, **115**, 119–26.

Index

Note: Locators in bold indicate illustration or inclusion in table.

A

Acanthiza chrysorrhoa (yellow-rumped thornbill) **62**, 62
Accipiter
　gularis (Japanese sparrowhawk) 60
　nisus (Eurasian sparrowhawk) 133, 209
Acrocephalus
　arundinaceus (great reed warbler) 38
　australis (Australian reed warbler) 39, 58
　schoenobaenus (sedge warbler) 214
　scirpaceus (Eurasian reed warbler) 50, 133, 212
　sechellensis (Seychelles warbler) 145
Actitis hypoleucos (common sandpiper) **155**
Adult
　size 45, 47, **48**, 48
Aegithalos caudatus (long-tailed bushtit) 24, 26, 43, **44**, **45**
Agelaius phoeniceus (red-winged blackbird) 58, 136, **188**, 214
Agriculture 216
Aix
　galericulata (mandarin duck) 77
　sponsa (wood duck) 76, 164, 172, 173, **174**, 174, **175**, 175, **176**, 176, **180**, 181
Alaemon alaudipes (greater hoopoe-lark) 71, 167–168, **169**, 169
Alauda arvensis (Eurasian skylark) **128**, 217
Albumen 10, 98, 110, 114, 117, 118, 119, 151
　egg turning 202
　mass 102
Alcedo atthis (common kingfisher) 53
Alectoris
　rufa (red-legged partridge) 197, **198**
　spp. (partridges) 196
Alectura lathami (Australian brush-turkey) 66, 78, 81, 173

Allometry 9, 30, 97, 98, 100–103, **101**, 102, 103, 222
　comparative analysis using independent contrasts 100
　phylogenetically constrained linear modelling **101**, 101, **102**, 102, 222
Altricial 47, 84, 91, 92, 98, 103, **104**, 104, **105**, 106, 107, **108**, 108, **109**, 109, 111, 146
　egg size 111
　order 103–110
Altricial-precocial spectrum 91, 92, 96, 98–99, 222
　evolution 98, 111–112
Ammodramus
　caudacutus (saltmarsh sparrow) 51, 62, 66, **188**, 190
　maritimus (seaside sparrow) **188**, 191
Ammomanes deserti (desert lark) 71
Anaemia 82, 88, 89, 90
Analysis of covariance 100–103, **102**, 102, 103, **105**, 106
Anas 19
　clypeata (northern shoveler) 143
　platyrhynchos (mallard) 57, 76, 85, 144, 185
　platyrhynchos var. *domestica* (Pekin duck) 150, 196
　rubripes (American black duck) 183
Animal material
　in nest **34–35**, 38, 43
Anseranas semipalmata (magpie goose) 17
Anthobaphes violacea (orange-breasted sunbird) **157**
Anthus
　petrosus (Eurasian rock pipit) **128**
　pratensis (meadow pipit) 56, **128**, 143, 212
　spinoletta (water pipit) 51
　trivialis (tree pipit) 172
Antibodies 113, 114, 117, 120, 123, 125, 136, 163, **176**, 176
　immunoglobulin G (IgG) 115, 118, 122
　immunoglobulin Y (IgY) 118

Antigen 176
Anti-microbials 75, 77, 78, **79**, 80, 118
Antioxidants 113, 115, 117, 125, 129, 133, 147, 162
　carotenoid 115, 116, 118, 136
　vitamin A 115
　vitamin E 114
Aphelocoma coerulescens (Florida scrub-jay) 3, 4, 5
Aptenodytes forsteri (emperor penguin) 78
Apteryx spp. (kiwis) 9, 97
Apus
　apus (common swift) 85
　pallidus (pallid swift) 54
Aquila
　chrysaetos (golden eagle) 185
　spp. (eagles) 85
Arachnid 82
　acari 84, 86
Archeopteryx 8
Arenaria interpres (ruddy turnstone) **155**
Asio otus (long-eared owl) 214
Athene cunicularia (burrowing owl) 54, 59, 60
Audio playback 53
Aviculture 197, 199, 201, 202–203, 206
Aythya
　americana (redhead) 18
　ferina (common pochard) 54, **55**, **188**
　innotata (Madagascar pochard) 204–206, **205**, 207
　spp. (pochards) 204

B

Bacteria 69, 75, 78, 89, 133, 201
Bacillus 76
　cereus 78
　licheniformis 77
Enterococci 77
Escherichia coli 78
　gram-negative 75, 76, **79**, 80
　gram-positive 75, 78, **79**, 80
　growth 80
　infection 82
　motile 76

Bacteria (*continued*)
  non-pathogenic 76
  pathogenic 75, 76
  Pseudomonas 76
  Salmonella enteritidis 76, 78
  Staphylococcus 77
    aureus 78, 80
  taxa 76
Behaviour 50, 70, 208, 215
  adult **146**
  anti-parasite **90**
  anti-predator 59
  broody 202
  climate change 73
  cognition 28, 221
  compensation 57
  courtship 136
  deprivation experiments 16–17
  embryo 149–151
  foraging 154
  imprinting 20, 21, 149
  incubation 9, 39, 40, 42, 73, 82, 87,
    135, 142, 154, 163, 185, 223
  innate 18, 21
  kleptoparasitism 30
  learning 17–18, 22
  location 52–53
  mate feeding 143, 144, 171
  mate-guarding 136
  nest construction 16, 26, 29
  neural basis 27
  noctural 182, 183, 193
  oxidative stress 147, 151
  personality 145
  preening 90, 93–94, 143
  thermoregulation 167, 172, 173, 174,
    175, 177
  tool use 28
  tremble-thrusting 45–46
  vigilance 56–57, 133
Biliverdin 78, 129, 130, 132, 134, 210
  synthesis 129
Bird as incubator 30
Bird, fossil (Cenozoic) **13**, 15
  Aepyornis
    maxima 12, 111
    sp. (elephant bird) 10, 12
  Badistornis aramus 12
  Dinornithiformes 12, 14
    Dinornis
      gigantea 12
      robustus 111
    Megalapteryx didinus (upland
      moa) 130
  Dromaius sp. (emu) 12
  Dromornis? sp. (mihirung) 12
  Eudyptula minor 12
  Fractula dowi 12
  Francolinus sp. (francolin) 12
  Genyornis newtoni 12

Incognitoolithus ramotubulus 12
Lithornis sp. (Lithornithidae) 12
Medioolithus sp. (Medioolithidae) 12
Metoolithus nebraskensis 12
Microolithus wilsoni 10
Numidia sp. (guineafowl) 12
Palaeoflamingo 13
Paleolodus 13
Pelecanus occidentialis 12
Phoebastria albatrus 12
Presbyornis 12, 14
  ratite 14
  size 111
Struthiolithus 12
Trochilidae 12
Bird, fossil (Mesozoic) 9, 10, **13**, 15
  Confuciusornis sanctus 9, 11
  Enantiornithine 10, 11, 14
  Gobipipus reshetovi 10
  Gobipteryx minuta 10, 11, 14
  Jeholornis prima 9, 11
  Laevisoolithus sochavai 11
  Neuquenornis volnans 11
  Ornitholithus 14, 15
  Ornithothoraces 11, 14
  Styloolithus bathi 10, 11, 14
Birdlife Australia 210
  Australian Nest Record Scheme 210,
    211, 212
Bird-nest incubation unit 30, 40, 222
Bird Studies Canada 210
  Nest Record Scheme 210, 212, 214
Birds New Zealand 210
  New Zealand Nest Record
    Scheme 210
Body
  condition 39–40, 65, 88, 92, 129, 133,
    147, 171, 172
  mass 87, 98, 177
    female 103, **108**, **109**
  temperature 167
Bombycilla cedrorum (cedar waxwing) 215
Bone
  fossil 10, 12, 14
  medullary 9, 129
Brain **27**, 27, 121
Branta bernicla (brant goose) 177
Breeding
  attempt 217, **218**, 218, **219**, 219
  performance 145, 216
  timing 222
British Trust for Ornithology (BTO) 1,
    4, 210
  BTO NRS Code of Conduct 219–220
  Nest Record Scheme 210, 211, 212,
    213, 213, 214, 215, 216, 217, 218,
    219
Brood
  multiple 71, 92
  size 162

Brood patch 40, 84, 100, 144, 164
  temperature 102, 199, **200**
  vasodilatation 145
Bucephala clangula (common
    goldeneye) 57, 181
Burhinus oedicnemus (stone curlew) 53
Buteo spp. (buzzards) 85

C

Cacicus cela (yellow-rumped
    cacique) 95
Cairina moschata (Muscovy duck) 149
Calcarius ornatus (chestnut-collared
    longspur) 51
Calidris
  alba (sanderling) **155**
  alpina (dunlin) **155**, 159
  canutus (red knot) 154, **155**
  fuscicollis (white-rumped
    sandpiper) **155**
  maritima (purple sandpiper) **155**
  melanotos (pectoral sandpiper) **46**,
    46, 50
  minuta (little stint) **155**, 159
  pusilla (semipalmated
    sandpiper) 148
  pygmaea (spoon-billed
    sandpiper) 203
Calypte anna (Anna's
    hummingbird) 185
Camouflage 23, 24, 52, 53, 54, **131**,
    131
Caprimulgus
  carolinensis (Chuck-will's-widow)
    188
  europaeus (European nightjar) 211
  vociferus (whip-poor-will) 211
Carbon dioxide 47, 201
Carduelis
  cannabina (common linnet) 216
  carduelis (European goldfinch) **128**
  chloris (European greenfinch) 187
  flammea (common redpoll) **32–35**
  tristis (American goldfinch) **32–35**
Carpodacus mexicanus (house finch) 39,
    62, 95, **188**, 193, 211, 215
Cathartes aura (turkey vulture) 182
Cenozoic 9, 10, 12
Cercomela melanura (blackstart) 38
Cercotrichas galactotes (rufous-tailed
    scrub robin) 42
Cereopsis novaehollandiae (Cape Barren
    goose) 47
Certhilauda erythrochlamys (dune
    lark) **156**, 167
Cettia cetti (Cetti's warbler) **128**
Chalazae 10
Charadrius 60
  alexandrinus (Kentish plover) 53,
    148, 154, **155**

dubius (little ringed plover) 23
hiaticula (ringed plover) 154, **155**
melodus (piping plover) 46, **188**, 192–193
montanus (mountain plover) 51, 52, 61
morinellus (Eurasian dotterel) 50
vociferus (killdeer) 21
Chasiempis sandwichensis (elepaio) 37, **60**, 60
Chelonia 8, **13**
Chen
caerulescens (snow goose) 40, 177
rossii (Ross's goose) 40
Chick rearing 162
energy expenditure 154, 162
field metabolic rate (FMR) 154, **158**, 158, 159
locomotion 164, 174
Chlamydotis
macqueenii (Macqueen's bustard) 201
undulata (houbara bustard) 197, 199
Chondestes grammacus (lark sparrow) 211
Chrysococcyx basalis (Horsfield's bronze cuckoo) 150
Cinclus cinclus (white-throated dipper) 85, **128**, **156**, 161, 210
Cinnyris osea (Palestine sunbird) 40
Circus
aeruginosus (western marsh harrier) 38
cyaneus (northern harrier) 51
Cistothorus palustris (marsh wren) 39
Citizen science 4, 208, 215, 224
bias 213, 216, 217, 218, 219
difficult-to-detect phenomena 213–214
protocol 217
standardisation 213, 215, 217, 218
volunteers 209, 210, 211, **213**, 213, 214, 215, 217–218
website 220
Clamator glandarius (great spotted cuckoo) 61, 76
Climate change 55, 65, 66, 208, 210, 212, 222, 224
adaptations 71–72
Clutch
completion 137
energy 103, **105**, 106–110, 161
evolution 159, 163
initiation date 42, 43, 212, 215, 217
latitude 214, 215
manipulation 152, **160**, 161, 163, 164, **165**, 165, 185, 190
mass 98, 103, **105**, 107–110, 111, 159

optimal size 161
replacement 203, 205
size 5, 80, 97, 98, 107–110, 111, 112, 136, 140, 144, **146**, **160**, 160, 172, 187, 212, 214–215, **223**
thermal consequences 161
Coccothraustes coccothraustes (hawfinch) **128**
Coccyzus erythropthalmus (black-billed cuckoo) 132
Coenocorypha aucklandica huegeli (New Zealand [Snares] snipe) 181
Colinus virginianus (northern bobwhite) 151
Columba
livia (rock dove) 19, 93, 147, 196
palumbus (woodpigeon) 59, 130, 210
Columbiformes 18
Conservation 179, 196, 199, 202, 223
bird translocation 196, 204
breeding programme 196
egg translocation 196, 203, 204
Ex situ 207
extinction 196, 202
In situ 196
Coracias garrulus (European roller) 61, **89**
Cornell Lab of Ornithology 4, 210
nest records 211, 212, 214, 218
nestwatch 210, 218
Nestwatch Nest Monitoring Manual 220
Corvus
corone (carrion crow) 61, **128**
coronix (hooded crow) 85
coronoides (Australian raven) 50
frugilegus (rook) 85
monedula (western jackdaw) **128**, **165**
Coturnix
coturnix (common quail) 41, 150
japonica (Japanese quail) 23, 118, 123, **131**, 131, 133, 134, 138
Cretaceous 9, 10, **13**
Crocodilian 8, **13**
Crotophaga ani (smooth-billed ani) 78
Crypsis 54, 130, **131**, 131–132
Cuculus
canorus (common cuckoo) 5, 56, 132, 212
spp. (cuckoos) 142
Cyanistes caeruleus (blue tit) 19, 20, 25, 29, 31, **32–35**, **36**, 39, 41, 42–43, **45**, 45, 47, 69, 76, 77, 80, **89**, 94, 132, 134, 140, 144, **156**, **160**, 163, 164, **166**, 172, 174, 175, 184, 187, **188**, **189**, **190**, **191**
Cyanopica cyanus (azure-winged magpie) 60, 62

D
Daption capense (cape petrel) 54
Delichon urbicum (house martin) 23, 85, 92, **128**, **156**
Dendrocopos
sp. (woodpecker) 53
syriacus (Syrian woodpecker) 47
Dendroica
cerulea (cerulean warbler) 30
petechia (American yellow warbler) **32–35**, 43, 172, 214
Developmental maturity (mode) 98, 103–112, **105**, 111–112
Dicrurus hottentottus (hair-crested drongo) 58
Dinosaur 8, 9, **13**, 15, 98
Diomedea exulans (wandering albatross) 189
Dolichonyx oryzivorus (bobolink) 212
Dumetella carolinensis (grey catbird) 31, 50

E
Ecological 123, 126
energetics 97
factors **223**
trap 66
Ecology 152, 183, 208
nesting 211, 222
Ectoparasite 39, 58, 82, 222
defense 82
host dwelling 88–90
nest dwelling 85–88
Egg 29
air space 201
albumen 76, 78, **79**, 80
proteins 78, 79, 80
artificial 46, **63**, 196
auditory stimulus 137–138
bird age 145
buried 15
candling 206
chemical stimulus 138
collection 203
colour 23, **52**, 52–53
composition 10, 97, 113, 114
compression 136
conspicuousness 133
contamination 75, **116**, 116
cooling 167, 174, 176
cooling rate 41, **46**, 46, 72, 145, 161
counting 132
covering 26, 61
crypsis 54, 130, **131**, 131–132
dimensions 8–9, 12
disinfectant 204
dummy 40, 182, 185, 190, 198
energy 98, **104**, 104, **105**, 107, **110**, 110, 111, 113
formation 113, 116, 117, 118, 136

Egg (*continued*)
  fossil 8, 10, **11**, 11, 12
  hatching **72**
  heating 130
  infection 75, 76
  laying 72, 115, 210, 212, 213, 215
  manipulation 124, 126
  mass 97, 98, **101**, 101, **105**, 107, **110**,
    110, 144
    loss 47, 201
  metabolism 100
  microbiology 75, 222
  microclimate 53, 131, 192, 222
  morphology 8
  museum collection 208, 209
  neglect 166–167
  pathogen 203
  position 140
  prey 56, 57
  production 199, 202
  pulling 197, 203
  quality 98, 113, 114, 134, 142
  recognition 132
  reflectance 53
  'runt' 214
  shading 70
  shape 12, **13**, 136, 137
  signalling 127, 134, 136, 137, 140, 222
  size 97, 100, 114, 136, **146**
  strength 136
  telemetric egg 185, **186**, 186
  temperature **73**, 144, 150, 161, 164,
    166, 166, **168**, 168, 178, 185, 187,
    199, 203
  turning 10, 71, 143, **146**, 150, 182, 185,
    186, 186, 187, 193, 197, **202**, 202
  unattended 168
  uniformity 132
  viability 71
  visual stimulus 133
  volatile chemical 138
  warming 164
  weighing 201
  weight loss 131, 201
  yolk 78, **79**, 98, 110, 114, 117, 119,
    121, 150, 151
Eggshell **14**, **79**, 80, 98, 127
  accessory material 129
  atypical 134
  Bacillus licheniformis 77
  Background colour 129
  Bacteria 75, 76, 77
  coloration 127, **128**, 130–133, 134,
    135, 137, 139, 140, 141
    artificial 130, 131
    blue-green 133, 134, 139, 141
    change 134
    egg quality 134
  conductance 8, 14, 15

cuticle 77, 78, **79**, 80, 201
dessication 75
filter 134–135
formation 117, 127, 129
fossil 9, 11, 12, 13, **14**, 14, 15
gas exchange 201
immaculate 130, 131
maculation 129, 130, 132, 134, 137
mass 102, 110
membrane 127
morphology 8, 78
patterning 132, 134
pigment 78, 127, 129, 130, 131, 133,
  135, 135, 141
pore 8, **14**, 14, 15, 76, 131, 134, 201
porosity 138
protection 127
recognition 139–140
sexual-selection hypothesis 130,
  132–133, 135
signal 127, 135
strength 133, 134
structure 13, 129
temperature **198**, 198
thermoregulation 130
thickness 12, 14, 15, 97, 102, 131,
  135, 141, 209
UV reflectance 47, 132, 134, 137, 139
vaterite 78
water vapour conductance 47, 100,
  101, 101, **102**, 102, 130, 134, 201
Emberiza
  calandra (corn bunting) **128**
  citrinella (yellowhammer) **128**, 217
  schoeniclus (common reed
    bunting) 85, 216
Embryo
  behaviour **146**, 149–151
  circulation 119
  communication 137, 138
  development 113, 114, 133, 134, **135**,
    173, 197, 203, 222
  energy expenditure 167, **175**, 175
  fossil 9, 10, 11, 12
  learning 150
  metabolism 198, **223**
  mortality 201
  movement 149
  sex 138
  sounds 149
  survival 75, 123
  temperature 71, 73, 130, 138, 168,
    171, 198, 199
  thermal tolerance 198–199
  vocalisation 137
Empidonax virescens (Acadian
  flycatcher) 62
Endoparasite 82
Energetics 97, 98

Energy
  allocation 153
  constraints 171
Environment
  long-term effect 176
  nest construction 22, 26
  nest site 19
  temperature 30
Eopsaltria australis (eastern yellow
  robin) 211
Eremophila alpestris (horned lark) 166,
  173
Erithacus rubecula (European
  robin) **156**, 211
Estrilda astrild (common waxbill) 24,
  38, 61
Ethno-ornithology 6
Eudromia elegans (elegant crested-
  tinamou) 137
Eudyptes
  chrysocome (rockhopper penguin) 93
  chrysolophus (macaroni penguin) 94,
    159
Eudyptula minor (little penguin) 188
Euphagus cyanocephalus (Brewer's
  blackbird) 55
Evolution 8, 97
  eggshell coloration 130
Evolutionary biology 97, 126

F
Falco
  peregrinus (peregrine falcon) 209
  punctatus (Mauritius kestrel) 196
  sparverius (American kestrel) 125
  subbuteo (Eurasian hobby) 59
  tinnunculus (Eurasian kestrel) 134,
    155
Feather 9, 24, 25, **34**, 42, 43, **44**, 45–46,
  61, 77, 82, 137
  bacteria 76, 77
  barrier 91
  colour 25, 46, 137
  fumigation 94
  geographic range 45
  lice 88–89, 93, 94
  mite 77, 90
  pigmentation 77, 91
  predation 45, 61
  quill 90
  UV reflectance 38–39, 80, 125, 132, 137
Ficedula
  albicollis (collared flycatcher) 19, 115,
    124, 134, 139, **157**, 161
  hypoleuca (European pied
    flycatcher) 19, 26, 29, 30, **32–35**,
    43, **45**, 59, 60, 69, **72**, 72, 83, **89**,
    93, 115, **115**, 116, 142, **156**, 161,
    163, **165**, 165, **166**, 187

Fitness 47, 98, 113, 123, 124, 136, 142, 144, 148, 153, 163, 176–177, 178, 223
Fledging 6, 7, 31, 47
    mass 46
    success 124, 210
Food restriction 173, **174**, 174
Food supplementation 5, 30, 144
Foraging 71
    caterpillars 71, 72
Forpus passerinus (green-rumped parrotlet) 70
Fossil record 221
Fratercula arctica (Atlantic puffin) 51
Fringilla coelebs (common chaffinch) 49, **128**, 187
Fulica americana (American coot) 132
Fulmarus
    glacialis (northern fulmar) 85
    glacialoides (southern fulmar) 145
Fungi 76, 78, **79**, 80
    Marasmius spp. (horse-hair fungi) 209
    Pythium ultimum 77
Furnarius rufus (rufous hornero) 85

G

Gallus gallus (domestic fowl) 3, 41, 77, 78, **89**, 95, 118, 127, 134, 138, 149, 164, 173, 185, 196, 198, 201, 202
Garrulus glandarius (Eurasian jay) **128**, 140
Gavia immer (great northern loon) 88, 93, 183
Gene 16, **223**
    c-fos 27
    expression 121
Gerygone magnirostris (large-billed gerygone) 55
Gram-staining 75
Guira guira (guira cuckoo) 137
Gymnogyps californianus (California condor) 5, 197, 199, 203
Gymnorhinus cyanocephalus (Pinyon jay) 19, 56
Gypaetus barbatus (bearded vulture) 183

H

Habitat
    nesting 19, 31
Haematopus ostralegus (Eurasian oystercatcher) 66, **155**
Halobaena caerulea (blue petrel) 147
Hatching 149, 174
    artificial 197
    asynchronous 136
    success 36, **70**, 70, 75, 134, 166, 173, 198, **200**, 201, 203, 206, 210, 224
    synchronous 137, 149

Hatchling
    hyperthermia 177
    mass 100
    phenotype 173, 223
Hawaiian honeycreeper 84
Hemignathus virens (Hawaii amakihi) 37, 40–41
Hirundo
    neoxena (welcome swallow) 211
    rustica (barn swallow) 23, 36, 67, 77, 80, 86, 115, **128**, 132, **156**, 177
    tahitica (Pacific swallow) **156**
Hormone 92, 113, 117, 123
    androgen 115, 117, 118, 119, 121, 122, 124, 151
    androstendione 122, 124
    antagonist 123
    calcitonin 129
    corticosterone 119, 121, 124, 125, 147, 151
    glucocorticoids 147
    growth-promoting effect 144
    manipulation 121
    maternal 120–121, 173
    metabolism 120
    nest-building 19
    oestradiol 121, 125
    oestrogen 19, 119, 121
    oestrone 120
    parathyroid hormone 129
    progesterone 19, 119
    prolactin 146–147
    sex 129
    status 146
    testosterone 19, 114, 115, 119, 120, 121, 124, 151
    thyroid hormone 114
Hydrobates pelagicus (European storm petrel) 85
Hylocichla mustelina (wood thrush) 31, 212
Hypothalamic-pituitary-gonadal axis 121, 124
Hypoxia 201

I

Icterus galbula (Baltimore oriole) 43, 140
Ictinia mississippiensis (Mississippi kite) **180**
Ifrita kowaldi (ifrit) 91
Immunity **90**, 91, 92, 95, 123, 129, 133, 134, 163, 164, 170, **176**, 176
Immunoglobulin 114
    immunoglobulin G (IgG) 115, 118, 122
    immunoglobulin Y (IgY) 118
Incubation
    androparental 142

artificial 5, 69–71, 159, 197–202, **198**, 199, 203, 224
bi-parental 135, 142, 144, 147, 148, **165**, 183
constancy 172
contact 98
energetics 154, 172
    seabird 159
energy expenditure 152, 154, 159
evolution 163
female-only 143, 152, **153**
gyneparental 142, 144, **153**, 159, **165**
humidity 75–76, 134
intermittent 162, 199
male 136, 147
male-only 142
mate feeding 143, 144, **153**, 159
odours 138
pattern 146
period 98, 100, 103, **105**, **110**, 110, 111, 112, **119**, 119, 144, 166, 167, 171, 172–173
recess 143, 144, 161, 166, 167, 171
rhythm 180, 185
session 144, 161
shared 135, 142, 144, 147, 148
temperature 69, 70, 75–76, 130, **153**, 164, **165**, 171, 172, 173, 174, 176, 177, 185, **198**, 198–201, 206
water economy 167–168, **169**, 169
Incubator 197
    contact 197–198, **198**
    egg turning 197, **202**, 202
    fan-assisted 197
    force-draught 198
    gaseous environment 201
    heater 197
    humidity 197, 201
    multi-stage 198
    portable 203, 204, **205**, 205, 206
    still-air 197, 204, **205**
    temperature **198**, 198, 199, **200**
Indicator spp. (honeyguides) 5
Insect 82, 84, 222
    ant 94
    caterpillars 71, 72
    diptera 84
        calliphoridae (blowfly) 87
        carnid 83, 86–87
        Carnus hemapterus (carnid fly) **83**, 86–87, **89**
        Ceratopogonidae (biting midges) 88
        Crataerina melbae (louse fly) **83**
        Culex 82
            quinquefasciatus (mosquito) **83**, 84
        Culicidae 88, **89**
        Culicoides sp. (biting midge) **83**
        Hippoboscidae (louse fly) 89–90

Insect (*continued*)
life history 87
mosquito **83**, **84**, 85, 88, 91, 95
Muscidae (muscid fly) 87
Philornis spp. (Muscidae) 87, 95
Protocalliphora spp. (blowflies) 77, 87, **89**
azurea **83**
Sarcophagidae (flesh fly) 87
Simuliidae (black fly) 87–88
Simulium
annulus 88
aureum (blackfly) **83**
Flea **83**, **84**, 84, 85, 86, 93, 122
Flying **84**
Haematophagous 83, 84, 85, 86, 87, 88, 90, 91
Hemiptera
Acanthocrios furnarii 85
Cimicidae 85
Infestation 85
Oeciacus
hirundinis (house martin bug) **83**, 85
vicarius (swallow bug) 85
life-history 84, 85
non-flying **84**
phthiraptera 88–89
Laemobothrion tinnunculi (chewing louse) **83**
Piagetiella peralis 88
Siphonoptera 85–86
Ceratophyllus columbae (flea) **83**
Insemination, artificial 199

J

Junco phaeonotus (yellow-eyed junco) 69, **157**
Jurassic 10

L

Lagopus
leucura (white-tailed ptarmigan) 192
muta (rock ptarmigan) 57
Lanius
collurio (red-backed shrike) **128**
excubitor (great grey shrike) 58, 62
ludovicianus (loggerhead shrike) 48
minor (lesser grey shrike) 58
Larus 51, 60
argentatus (herring gull) 66, 141, 185, 209
audouinii (Audouin's gull) 147
cachinnans (Caspian gull) 80
crassirostris (black-tailed gull) 23, **165**
dominicanus (kelp gull) 37, 131
fuscus 115, 118, 142
glaucescens (glacous-winged gull) 136

hyperboreus (glacous gull) **165**
michahellis (yellow-legged gull) 115, 151
occidentalis (western-legged gull) 186
ridibundus (common black-headed gull) 66, 114, 115, 125, 134, 148
Leipoa ocellata (malleefowl) 174
Lichen 24, **33**, 36, 37, 43, 49
Lichenostomus chrysops (yellow-faced honeyeater) 53
Life history 7, 111, 143, 145, 152, 177, 191, 221
Light **135**, 139
infrared (IR) 130, 134, **135**, 182, 183
ultraviolet (UV) **135**, 192
Lizard 169
Locustella naevia (common grasshopper warbler) **128**
Loxia curvirostra (red crossbill) **128**
Lullula arborea (woodlark) **128**
Luscinia megarhynchos (common nightingale) **128**

M

Macronutrients 113, 114, 171
Malacorhynchus membranaceus (pink-eared duck) 196
Malaria 84, 88, 94
Malurus cyaneus (superb fairy wren) 150, **188**, 190
Mammal 121
Cervus canadensis (elk) 67
Cynomys ludovicianus (black-tailed prairie dog) 59
Galictis cuja (little grison) 131
Homo sapiens (human) 131
Martes martes (pine marten) 57
Mustela erminea (stoat) 138
Neovison vison (American mink) 19
Procyon lotor (raccoon) 55
Rattus rattus (black rat) 205
rodent 61
scat 24, 38, 60, 61
Sciurus carolinensis (grey squirrel) 216
Spermophilus citellus (European ground squirrel) 43
Tamiasciurus hudsonicus (American red squirrel) 162
Margarops fuscatus (pearly-eyed thrasher) **70**, 70, 87
Megapode 15, 66, 84, 97, 142, 173, 174
Megascops asio (eastern screech owl) 95
Meleagris gallopavo (wild turkey) 147, 196
Melopsittacus undulatus (budgerigar) 150
Melospiza melodia (song sparrow) 53, 144, 212

Merops
apiaster (European bee-eater) 53
ornatus (rainbow bee-eater) 47
viridis (blue-throated bee-eater) **155–156**
Mesozoic 8, 9, 10, 11
Meta-analysis 3, 49, 136
Metabolism 100, 119, 152
basal metabolic rate 152, **153**, 154, **155–157**, 158, 158, 159, 170, 174
body mass 154
field metabolic rate **153**, 154, **155–157**, 158, 158, 159, **165**, 165, 170
incubation metabolic rate **153**, **158**, 158, 159, **160**, 160, 161, **165**, 165, 170
microbial 75
thermoregulation 174–176
waste product 116
Micronutrients 113
Microorganism
growth 75
Milvus migrans (black kite) 25, 38, 61–62
Mimus polyglottos (northern mockingbird) 215
Mite 77, **84**
Acariformes 86
Dermanyssus sp. (haematophagous mite) **83**
feather 86
life history 86
Parasitiformes 86, 90
skin 86
Molecular methods
DNA extraction 209, 210
genome amplification 210
PCR 76
Molothrus
ater (brown-headed cowbird) 211, 212
bonariensis (shiny cowbird) 132, 140
oryzivorus (giant cowbird) 95
spp. (cowbirds) 142
Montifringilla nivalis (white-winged snowfinch) 17
Moss 23, 29, **32–35**, 37, 38, 42, 43, 67
Motacilla
alba (white wagtail) **128**
flava (western yellow wagtail) 18, **128**
Mud 23, 24, **32**, 37, 67
Muscicapa striata (spotted flycatcher) **128**, 212
Museum collections 1, 5
Myadestes obscurus (omao) 199
Mycteria americana (wood stork) 40

N

Nematode 88, 89
Nest **205**, 221
  abandonment 56–57, 147, 166, 181,
    182, 191
  accessibility 54–55
  aromatic plant 25, 61
  artificial materials **35**, 38, 39, 51, **55**,
    56, 61, 62, 95
  attentiveness 4, 143, 144, 148, 171,
    181, 182, 185
  availability 66
  burrow 53, 54, 180
  cavity 46–47, 52, 53, 54, 58, 139, 180,
    201–202
  colour **24**, 24, 46, 52
  composition 21, **32–35**, 172
  compression 36, 45
  concealment 56–57, 58
  construction 4, 16, 17, 18, 21, 25, 26,
    29, 48, 76, 209, 221, 222, **223**
    behaviour 43, 45, 49, 134
    energetics 30
    kleptoparasitism 30
    period 29–30
  cooling 164, 174
  crypsis 46
  cup 30, 36, 69, 184
  decoration 24, 25
  defence 4
  density 56, 57, 59
  desertion 147
  design 60–62
  destruction 58
  dimensions 31, 36
  disturbance 3, 59
  domed 22, 41, 62
  empty 57–58
  energetics 42
  environment 8, 13, 14, 42, 139,
    223
  false **62**, 62
  fossil 12, 13, 14
  fumigation 95
  function 29, 222
  heating **164**, 172, 174
  humidity 47, 61, 93, 201
  hygiene 21
  insulation 21, 26, 29, 41, 42, **45**, 45,
    46, 49
  light **139**, 139, 140
  lining 21, 24, 25
  location 18, 31, 40, 48, 50, 63, 69
    altitude 31, 158
    arctic 154, **155**, **158**, 158
    cliff ledge 54
    desert 167
    direction 51
    flooding 54, 55, 62, 66, 143

    ground 52–53
    height above ground 31, 51,
      55–56, 57, **60**, 60
    latitude 42, 43
    non-arctic 154, 155, **158**, 158
    switching 71
    traditional 58
    tree 53
  maintenance 94–95
  manipulation 148
  mass 30, **31**, 31, **32**, 32, **36**, 36, 42, 43,
    44, **48**, 48, 206
    annual change **36**, 43
  materials 17, **32–35**, 36, 47, 187, **189**,
    191, 210
    choice 20, **21**, 76
    colour 18, **22**, 22, 38, 39, 61, 69
    feather 24, 25, **44**, 46, 77, 172
    removal 93
    structural properties 22, 48
    type 23, 93
    wool 24
  measurements 206
  microclimate 31, 39, 40, 43, 45, 47,
    51, 56, 61, 172
  microflora 76, 222
  monitoring 208–209, 210, 218
  morphology 21
  multiple 39, 40, 58
  museum collection 208, 209
  number 39
  old 58–59, 93
  open 8, 140, 180
  orientation 51–52, 172–173
  outer shell 23
  plasticity 26, 29, 36, 38, 49, 59
  platform 23
  reuse 58, 214
  role 29
  sanitation 94–95, 136
  scrape 47, **189**
  sexual selection 36
  signal 25, 36, 38
  site fidelity 58
  size 30, 62–63, 64, 172
  structure 24, 25, 29, 47–49, 172
  survival **55**
  temperature 69–70, 93, 168
  thermal properties 40, 47–48, 172, 223
    conductance 40–41, 47–48, **48**
    measurement 40–41
  timing 172
  trapped air 43
  type 5–6, 39
  vegetation 53–54, 63
  wall 30, 40, 43, 69
  warming 164
Nestbox 19, 20, 25, 29, 31, 36, 39, 41,
  58, 139, 214

  dimensions 31, 36
  heated 39
  predation 25
  technology **180**, 182
Nestling 71, 84
  behaviour 121, 139
  brooding 146
  composition 173
  detectability 140
  energy 212
  growth 51, 123, 173
  health 133
  mass 65, 124, 173, **174**
  mortality 70, 86, 93, 125, 176
  prey 56, 57
  quality 134–135, 136, 163, 171
Nest Record Scheme 1, 210–219
Nitric oxide (NO) 138
Nothoprocta perdicaria (Chilean
  tinamou) 137
Nothura maculosa (spotted
  nothura) 137
Notiomystis cincta (stitchbird) **188**,
  190
Numenius americanus (long-billed
  curlew) **188**, 188

O

Oceanites oceanicus (Wilson's storm
  petrel) 85
Oceanodroma leucorhoa (Leach's storm
  petrel) 85
Offspring 123, 136
  survival 148, 217
Olfaction 57, 59, 61, 76–77
Oology 209
Oriolus oriolus (Eurasian golden
  oriole) **128**
Ornitholithus sp. (Elongatoolithidae) 12
Otis tarda (great bustard) 51, 57, 203
Ovary 118, 127
  follicles 9–10
Oviduct 118, 127–128
  shell gland 118, 127, 129, 133
Oxidative stress 121, 147, 151, 162, 170
  reproductive costs 162
Oxygen 47, 103, 162, 192, 201
  consumption 100, **101**, 101, **102**, 102,
    119, 174, **175**, 175, 201

P

Pandion haliaetus (osprey) 85, 183, **188**
Paradoxornis webbianus (vinous-throated
  parrotbill) 140
Parasitism 69
  brood 3, 5, 56, 61, 76, 127, 130, 132,
    136, 140, 142, 185, 187, 191,
    211–212
    egg recognition 132, 140

Parasitism (*continued*)
  defence **90**, 90–95
    behavioural 92–95
    external barrier 91
    physiological 91–92
    poison 91
  ectoparasite 58, 61, 82, 122, 124,
    183, 222
  endoparasite 82
  host-parasite interaction 69
  inter-specific **189**
  intra-specific **189**
  kleptoparasitism 30
  nest 25, **84**, 84–85, 192, 210
Parental care 9, 98, 132, 133, 136, 138
Parus major (great tit) 19, 20, 29, 30, **31**,
    31, **32–35**, **36**, 41, 42, **42**, 43, **45**,
    45, 47, 60, 69, 76, 93, 115, **128**, 132,
    133, 134, 139, 145, **156**, 159–160,
    160, 161, 172, 173, 187, **188**
Passer
  domesticus (house sparrow) 39, 78,
    85, **128**, 209, 216
  montanus (tree sparrow) **32–35**, 41
Passerculus sandwichensis (savannah
    sparrow) **32–35**, 60, 61, 145,
    157, **188**
Pelecanus
  erythrorhynchos (American white
    pelican) 88, 149
  occidentalis (brown pelican) 84, 86
Pelvis 8
Perdix perdix (grey partridge) 181, 196
Periparus ater (coal tit) 19
Perisoreus infaustus (Siberian jay) 19
Petrochelidon pyrrhonota (American cliff
    swallow) 85, 93
Phacellodomus rufifrons (rufous-fronted
    thornbird) 17, 36
Phalacrocorax
  auritus (double-crested cormorant) 78
  bougainvillii (Guanay
    cormorant) 83–84, 86
Phalaropus tricolor (Wilson's
    pharalope) 147
Phasianus colchicus (common
    pheasant) 25, 181, 196
Phenology 71, 212, 215
  mismatch 71, 72, 212, 213
Phenotype 113, 116, 173, 178, 223
Pheucticus ludovicianus (rose-breasted
    grosbeak) 17
Philetairus socius (sociable weaver) **188**
Phoebastria immutabilis (Laysan
    albatross) 90, **186**, 186
Phoenicoparrus jamesi (James's
    flamingo) 23
Phoenicopterus roseus (greater
    flamingo) 196

Phoenicurus
  ochruros (black redstart) 19, 59
  phoenicurus (common redstart)
    32–35, 43, **45**
Photorefractoriness 146
Photostimulation 146
Phylloscopus
  collybita (common chiffchaff) 42,
    128, 213
  sibilatrix (wood warbler) 42, **45**
  trochilus (willow warbler) **32–35**,
    37, 42
Phylogenetic
  analysis 4, **27**, 27, 97, 98, **104**, 104
  variation 103
Phylogeny **99**, 101, 222
  allometry 100–103, 222
  family 103
  order 102, 103–110
Physiology **90**, 90, 170, **223**
  body temperature 72
  embryonic 3
  incubation 153
  regulation 153
Pica pica (Eurasian magpie) 30, 62,
    76, **128**
Picoides
  albolarvatus (white-headed
    woodpecker) 211
  borealis (red-cockaded
    woodpecker) 54
  tridactylus (Eurasian three-toed
    woodpecker) 183
Pitohui spp. (pitohuis) 91
Plant materials
  anti-microbial 69, 76–77
  in nest **33**, 43
  twigs 48–49
  volatile compounds 76–77, 94, 95
Plastics 32, 38, 62
Platalea leucorodia (Eurasian
    spoonbill) 66
Ploceus
  capensis (cape weaver) **20**
  cucullatus (village weaver) 17, 23, 26
  philippinus (baya weaver) 53
  velatus (southern masked weaver) 26
Plumage 121, 124, 125, 137
Pluvialis apricaria (European golden
    plover) 51
Podargus strigoides (tawny
    frogmouth) 54
Podiceps nigricollis (black-necked
    grebe) 150
Poecile
  atricapillus (black-capped
    chickadee) **188**
  palustris (marsh tit) 29, 47, 53, 58,
    139

Pollutant 141, 185
  bio-monitoring tool 209
  DDE 133, 141
  DDT 133, 209
  heavy metals 141, 209
Pooecetes gramineus (vesper sparrow) 173
Population
  modelling 216
  size 66, 84, 216, 217
Porphyrin 129, 133
  eggshell strength 133
Porphyrio porphyrio (purple
    swamphen) 142, 182
Post-hatching
  care 98
  phenotype 113
Poultry 3, 41, 71, 77, 94, 129, 173, 196,
    199, 224
Precocial 98, 103, **104**, 104, **105**, 106,
    107, **108**, 108, **109**, 109, 111, 114
  egg size 111
  order 103–110
Predation 39, 127, 131, 133, 138, 140
  avian 55, 59–60
  escape 57
  feathers 45
  foraging behaviour 57–58
  grassland 55
  ground 55
  human 59, 131
  mammalian 55, 59, 131
  nest 19, 24, 25, 45, 46, 50, 62–63, 144,
    146, 171, 222
  risk 3, 19, 25, 183
  snake 60
  woodland 55
Progne subis (purple martin) 181
Protonotaria citrea (prothonotary
    warbler) 31, 37, 54–55
Protoporphyrin 78, 129, 130, 133, 134,
    141, 210
  synthesis 129
Protozoa 82, 86, 88, 90
Prunella modularis (dunnock) **32–35**,
    128, 212
Psaltriparus minimus (American
    bushtit) 142
Psaroclius
  decumanus (crested oropendola) 17,
    95
  montezuma (Montezuma
    oropendola) 95
Pseudoseisura lophotes (brown
    cacholote) 17
Pterocles orientalis (black-bellied
    sandgrouse) **188**, 203
Pterosaur 9, **13**
Ptychoramphus aleuticus (Cassin's
    auklet) 186

Puffinus puffinus (Manx shearwater) 54, 85
Pycnonotus sinensis (light-vented bulbul) 38
Pygoscelis adeliae (Adelie penguin) 23, 50, 52, 59, 147, 148, 185, 199
Pyrrhocorax pyrrhocorax (red-billed chough) **89**
Pyrrhula pyrrhula (Eurasian bullfinch) **128**, 216

Q

Quelea quelea (red-billed quelea) 19
Quiscalus quiscula (common grackle) 212

R

Rallus longirostris (California clapper rail) 66
Ratite 8, 23
Recurvirostra avosetta (pied avocet) 66, **155**
Regulus
    ignicapilla (firecrest) **17**
    regulus (goldcrest) **17**, **128**
Remiz pendulinus (Eurasian penduline tit) 40
Reproduction
    investment 110
    timing 71, 92
Reptile 173
    Chelonia 8, **13**
    crocodilian 8, **13**
    lizard 169
    snake 38, 60, 95
Rhipidura leucophrys (willie wagtail) 211
Rickettsiae 86
Ringing 3, 181, 208, 217
    legislation 209
    licence 184
Riparia riparia (sand martin) 53, 54, **156**, 210
Rissa tridactyla (black-legged kittiwake) 19–20, 58, 93, 137

S

Saxicola rubetra (whinchat) **128**
Scolopax
    minor (American woodcock) 183
    rusticola (Eurasian woodcock) 85
Scopus umbretta (hamerkop) 78
Selection 55
    sexual 58
Semi-precocial 103, **104**, 104, **105**, 106, 107, **108**, 108, **109**, 109, 111
    order 103–110
Senses 137
    auditory 138
    olfaction 57, 59, 61, 76–77

Serinus canarius (Atlantic canary) 17, 18, 123, **160**
Sex determination 66, 80, 121
Sexual
    attraction 89
    dimorphism 9, 12, 69, 92, 103, 136
    display 124
Sialia
    currucoides (mountain bluebird) 19, **188**
    mexicana (western bluebird) 183, **188**
    sialis (eastern bluebird) 31, 76, 136, 137, **188**, 191, 214, 215
Sitta pusilla (brown-headed nuthatch) 215
Snake
    Leptotyphlops dulcis 95
    skin 38, 60
Somateria mollissima (common eider) 46, 163, **165**, 172, 182, 204
Spheniscus
    humboldti (Humboldt penguin) 199, 202
    magellanicus (Magellanic penguin) 54, **188**, 190
Steatornis caripensis (oilbird) 50
Stercorarius 37
    longicaudus (long-tailed jaeger) **155**
    maccormicki (South Polar skua) 59, 185
    skua (great skua) 145
    spp. (skuas) 52
Sterna 93
    albifrons (little tern) 149
    antillarum (least tern) 23, **188**
    fuscata (sooty tern) 83, 86
    hirundinacea (South American tern) 52, 131
    hirundo (common tern) 66
    sandvicensis (sandwich tern) 77
Stictonetta naevosa (freckled duck) 196
Stones 37, 38, 46, 51, 52, 63
Streptopelia
    decaocto (Eurasian collared dove) 85, 216
    orientalis (Oriental turtle dove) 57
    risoria (ring dove) 19
Stress 3, 145
Strigops habroptilus (kakapo) 5, 203
Strix
    aluco (tawny owl) 214, 215
    occidentalis (spotted owl) 41
    uralensis (Ural owl) 59
Struthio camelus (common ostrich) 23, 111, 196, 198, 201
Sturnus **89**
    unicolor (spotless starling) 38, 71

vulgaris (European starling) 25, 76, 77, 86, 93, 120, 125, 140, 148, **157**, 191, 209, 211
Sula
    dactylatra (masked booby) 88
    variegata (Peruvian booby) 84, 86
Sylvia
    atricapilla (Eurasian blackcap) 58, **128**
    curruca (lesser whitethroat) **128**
    spp. (Old World warblers) 39, 64

T

Tachybaptus ruficollis (little grebe) 150
Tachycineta
    bicolor (tree swallow) 25, 39, 42, 69, 70, 136, **156**, 163, **165**, 172, 173, 174, **180**, **188**, 191, 212
    meyeni (Chilean swallow) 46, **188**, 193, 214
    thalassina (violet-green swallow) 94
Tachymarptis melba (alpine swift) 90, 147
Taeniopygia guttata (zebra finch) 18, 19, **21**, **22**, **24**, 24, 27, 37, 38, 41, **116**, 116, **119**, 119, 120, 121, 123, 124, 145, 148, **160**, 160, 162, 163, **165**, 173–174
Technology 179, 223
    copper tambour 181
    data logger 181, 185, 186, 187, **188**, **189**, 192
    egg 182
    global positioning system 193–194
    light sensitive 185
    load cell **180**, 181, 183
    microswitch 182
    NestCam 215
    nest switch 181
    passive integrated transponder (PIT) tags 183–184, **184**, 193, 194, 217
    perch-operated switch 179–180, **180**
    photoelectric cell 181
    photography 182, 184
    position sensor 185
    press-sensitive wireless device 181–182
    radio telemetry 183
    'Raven' software 193
    'Rhythm' software 193
    tag 182, 184
    telemetric egg 185, **186**, 186
    temperature logger, *i*Button® 41, 45, 179, 184, 186–193, **187**, **188**, **189**, **190**, **191**, **194**, 223
    temperature probe 184, 185
    thermal imaging 41, **42**, 45, 194
    thermistor 184, 185
    thermocouple 4, 184–185, **189**
    video 4, 64, 181, 182, 183
    X-ray computed tomography 209

Temperature
ambient 159–160, **160**, 161, 165, **166**, **189**
body 169
egg **73**, 144, 150, 161, 164, **166**, 166, **168**, 168, 178, 185, 187, 199, 203
lethal 168
optimal 199, 201
physiological zero 167
sub-optimal 203
use *ex situ* 41, 45, 187
thermal neutral 160, **161**, 161, 162
Thalassarche melanophrys (black-browed albatross) 159
Thalassoica antarctica (Antarctic petrel) 54
Thermoregulation 42, 88, 98, 130, 173, 174–176, 177
Theropod 8, 9, **13**, 13
Citipati osmolskae 10
Compsognathus longipes 10
egg 8, 10
embryo 8
nest 8
nesting biology 8
oviraptorosaurian 10
parental care 9, 98
Sinosauropteryx prima 10
Tick **84**
hard **83**
infestation 93
Ixodes
ricinus **83**
sp. (hard-bodied tick) 82
life history 86
Parasitiformes 86
Tinamus major (great tinamou) 137
Toxicology 116
Toxin 116
DDE 133
DDT 133
Toxostoma bendirei (Bendire's thrasher) 215
Transmitters
radio 169
Tringa totanus (common redshank) 66, **188**, 188
Trochilidae 12, 23, 43, 49
Troglodytes
aedon (house wren) 25, 93, 124, 134, 152, 180, 185, **188**, 214
troglodytes (winter wren) 39, 58, 76, 214

Turdidae 39
Turdus 23, 211
iliacus (redwing) **128**
merula (common blackbird) 23, 24, **32–35**, 37, 41, 43, **45**, 48, 56, 61, **63**, 63, 136, 181, 187, **188**, 216
migratorius (American robin) 17, **32–35**, 181, 212
philomelos (song thrush) 24, **67**, 67, **68**, 140, 210, 211, 216
viscivorus (mistle thrush) **128**
Tyto alba (barn owl) 94, 215, 216

U

Upupa epops (European hoopoe) 77, 134
Uria
aalge (common murre) 54
lomvia (thick-billed murre) 54
Uropygial secretions 77–78, 90, 93, 134

V

Vanellus vanellus (northern lapwing) 133, 136, **188**, 188, 216
Vermivora
celata (orange-crowned warbler) 55, 67, 143, 144
chrysoptera (golden-winged warbler) 211
pinus (blue-winged warbler) 211
Video 4, 64, 182
digital 182
nestbox 183
surveillance 181, 182
Virus 85, 86, 87, 88, 89, 90

W

Water loss 167–168
evaporative cooling 168, 169
incubation 169
Water vapour conductance 47, 100, **101**, 101, **102**, 102, 130, 134, 201
fossil eggshell 8, 14, 15
Weather 26, 61, 65, **146**, 183
cliff nesting 54
drought 67
ectoparsites 87

extreme 66
flooding 54, 55, 62, 65, 66, 143
insolation 51, 64, 69, 70, 167, 173
latitude 51, 52, 63
light 139
precipitation 43, 51, 61, 72, 143, 166, 172
shading 40
snow cover 56
storm 66, 166
sunlight 130, 167, 181, 192
technology 180–183
temperature 42, 43, 45, 51, 66, 69, **70**, 70, 71, 72, 143, 145, 167, 172, 212
wind direction 40, 46, 51, 52
wind speed 40, 46, 56, 72, 93
Weaverbirds 17, 18, 23, 25, 26, 30, 39
Web
spider's 23
silk **34**, 49
Welfare 219–220

X

X-ray computerised tomography 10

Y

Yolk 78, **79**, 98, 110, 114, 117, 119, 121, 150, 151
antibodies 115
carotenoid 116
energy 119
formation 117
hormones 115, 117, 119, 121–122, 123, 124, 125, 151
lipid 173
mass 102

Z

Zarhynchus wagleri (chestnut-headed oropendola) 95
Zavattariornis stresemanni (Stresemann's bush crow) 41
Zenaida macroura (mourning dove) 72, **73**
Zonotrichia
albicollis (white-throated sparrow) **188**
leucophrys (white-crowned sparrow) 37, 41, 56, 154, **157**, 158

Printed and bound by CPI Group (UK) Ltd, Croydon, CR0 4YY